Quasi-exactly solvable models in quantum mechanics

Quasi-exactly solvable models in quantum mechanics

Alexander G Ushveridze

Arnold Sommerfeld Institute of Mathematical Physics, Clausthal, Germany and Institute of Physics, Tbilisi, Georgia

CRC Press
Taylor & Francis Group
Boca Raton   London   New York

CRC Press is an imprint of the
Taylor & Francis Group, an **informa** business

CRC Press
Taylor & Francis Group
6000 Broken Sound Parkway NW, Suite 300
Boca Raton, FL 33487-2742

First issued in paperback 2019

ISBN-13: 978-0-7503-0266-1 (hbk)
ISBN-13: 978-0-367-40216-7 (pbk)
Library of Congress catalog number: 93-5655

Library of Congress Cataloging-in-Publication Data

Catalog record is available from the Library of Congress

**Visit the Taylor & Francis Web site at
http://www.taylorandfrancis.com**

**and the CRC Press Web site at
http://www.crcpress.com**

To George and Eugenia

Contents

Appendices

Preface

It is known that the number and diversity of exactly solvable models (i.e., models with explicitly diagonalizable hamiltonians) is quite small and by no means meets the requirements of modern quantum physics. The methods of their construction currently known are also almost exhausted. There is no doubt about the need to look for new methods and new ideas, and in this connection it seems to me that there is a very promising direction which came to light several years ago and is associated with the discovery and study of models that realize a fundamentally new type of exact solvability in quantum theory. These models, which have been dubbed "quasi-exactly solvable", are characterized by the fact that their spectral problems can be solved exactly only for certain limited parts of the spectrum, but not for the whole spectrum.

There are three reasons for which the quasi-exactly solvable models are interesting to us.

First of all, they possess all the advantages of ordinary exactly solvable models, i.e., they enable one to model real physical situations and observe non-pertubative phenomena; they can be used as "reference points" in the realization of various approximate methods; and they reflect deep symmetry properties of spectral equations. At the same time, their number is appreciably greater than that of exactly solvable models and this circumstance makes them especially important from the practical point of view.

The second reason is the intimate connection between the quasi-exactly solvable models of quantum mechanics and other rather distant and seemingly unrelated branches of quantum and classical physics. For example, these models are equivalent to quantum tops based on finite-dimensional representations of Lie algebras; they are in one-to-one correspondence with completely integrable models of magnetic chains associated with solutions of Yang–Baxter equations; and they are connected with systems of classical Coulomb particles moving in an external electrostatic field. Note also that quasi-exactly solvable models naturally arise not only in the ordinary one- and multi-dimensional quantum mechanics, but also in quantum mechanics on non-trivial curved manifolds.

The third reason is the elegance and simplicity of theories explaining the phenomenon of quasi-exact solvability and making it possible to construct large classes of quasi-exactly solvable models in non-relativistic quantum mechanics. I hope that a critical analysis of these theories may stimulate the creation of new more general schemes including the field theoretical case.

Although the first examples of quasi-exactly solvable models were described only a few years ago, the results in this field are rather numerous and their flow grows from day to day. In this book I have made an attempt to collect most of these results together and expound them in a unified and accessible form. The material is discussed in sufficient detail to enable the reader to follow every step and is supplemented by an exhaustive list of references.

The book has the following structure.

Chapter 1 introduces the concept of quasi-exact solvability and discusses an illustrative example (the one-dimensional sextic anharmonic oscillator) that can be analysed by comparatively simple methods but, at the same time, manifests many characteristic features of more complex quasi-exactly solvable systems. This example refutes the common opinion that all quantal models with polynomial anharmonicity are exactly non-solvable. The phenomenon of quasi-exact solvability of the sextic oscillator appears for certain discrete (quantized) values of parameters characterizing the potential, and the reason for their quantization is explained in the chapter from different points of view. Each of these explanations enables one to formulate a certain method for constructing large classes of quasi-exactly solvable systems, so that the reader can find here the basic ideas of almost all the existing approaches to the problem. The long list of key words which could characterize the subject of chapter 1 (hidden symmetries, finite- and infinite-dimensional representations of Lie algebras, complete integrability, Bethe *ansatz*, Gaudin models, classical multi-particle Coulomb problem, Gelfand–Levitan equation, Witten's supersymmetric quantum mechanics, strong-coupling problem, convergent perturbation theory, Bender and Wu singularities, and so on) indicates both the riches of the phenomenon of quasi-exactly solvability and the diversity of its possible physical and mathematical applications.

Chapter 2 presents some new examples of quasi-exactly solvable models of one- and multi-dimensional quantum mechanics and discusses their properties and methods of construction. Here, by means of a very elementary analytic technique, we step by step go over from the one-dimensional case to the multi-dimensional one. The transition to the infinite-dimensional (field theoretical) case is also considered.

Chapter 3 is one of the most important from the methodological point of view and, although it is not directly devoted to the phenomenon of quasi-exact solvability, its results are essential for the subsequent chapters. It introduces a new type of object of study, the multi-parameter spectral equations, and shows that any such equation, after applying to it the so-called inverse procedure of separation of variables, can be reduced to a certain completely integrable quantum system. In this chapter the main attention is devoted to the methods for constructing exactly (algebraically) solvable multi-parameter spectral equations and to the classification of cases in which the corresponding completely integrable (and, simultaneously, exactly solvable) systems become physically meaningful.

Chapter 4 expounds the method for constructing quasi-exactly solvable models with separable variables. The idea of this method immediately follows from the results of chapter 3 and is based on the observation that if the multi-parameter spectral equation admits a finite number of exact (algebraic) solutions only, then the corresponding completely integrable model becomes quasi-exactly solvable. A simple analytic procedure for building partially solvable multi-parameter spectral equations is proposed. The chapter starts with the detailed analysis of the simplest one-dimensional case, presents a reference list of the most interesting one-dimensional quasi-exactly solvable models, and ends with the discussion of multi-dimensional models defined on, in general, curved manifolds.

Chapter 5 starts with noting a deep relationship between the quasi-exactly solvable models with separable variables discussed in preceding chapters and the well known completely integrable Gaudin models associated with algebra $sl(2)$. It is noted that the former can be obtained from the latter by means of a special reduction procedure (called in chapter 1 the projection method). It turns out (and this is the main subject of the chapter) that the same reduction procedure being applied to other Gaudin models (based on other simple Lie algebras) leads to new wide classes of quasi-exactly solvable second-order differential equations with non-separable variables. The classification problem for these equations is solved and their group-theoretical properties are discussed in detail. It is remarkable that, because the Gaudin models are solvable by means of the Bethe *ansatz* method, all solutions of the corresponding quasi-exactly solvable equations can be represented in the closed Bethe form. This fact especially simplifies their analysis and classification and makes it possible to establish an exact correspondence between these equations and those of (2+1)-dimensional classical electrodynamics with a magnetic monopole. The reduction of the obtained quasi-exactly solvable equations to the Schrödinger form is also discussed.

The book ends with four appendices containing some results lying outside its general line but devoted to no less important and interesting methods for constructing exactly and quasi-exactly solvable models.

Appendix A discusses the problem of constructing Schrödinger equations having several *a priori* given solutions.

Appendix B is devoted to the problem of constructing exactly solvable models on non-trivial curved manifolds.

Appendix C presents a simple method for building second-order differential operators having infinite number of polynomial solutions.

Appendix D deserves a special comment. When the work on the book had already been completed I learnt of a series of remarkable results of A González-López, N Kamran and P J Olver, concerning the classification of quasi-exactly solvable models whose hamiltonians are treated as elements of universal enveloping algebras associated with various Lie algebras of vector fields. I enquired of these authors regarding the possibility of contributing a summary of their results to my book and they kindly agreed to write an article concerning the study of two-dimensional problems. I take this opportunity to thank them for their kindness and hope that the reader will be impressed by this very interesting article which is reproduced in appendix D without editing.

I would like to stress that this book is not an exhaustive review and can be considered only as a more or less detailed introduction to the theory of quasi-exact solvability. At present, there are several new intriguing results and observations manifesting that the theme is far from being exhausted, namely such new aspects of the problem as the remarkable parallels between quasi-exactly solvable models of quantum mechanics and two-dimensional conformal field theories, the multi-channel quasi-exactly solvable problems associated with graded Lie algebras, problems based on quantum groups, and some others. Many of these aspects have been discussed in very recent publications and it was practically impossible to cover all them in this book.

Nevertheless, I hope that the book (despite its incompleteness) will be helpful for physicists and mathematicians interested in the problem of quasi-exact solvability in quantum mechanics and I will be happy if it stimulates the reader to make his own contribution to this very interesting field of mathematical physics.

Alexander Ushveridze
7 November 1992

Acknowledgments

Let me first express my gratitude to the Alexander von Humboldt Foundation for supporting my research in Germany during the period in which this book was being written.

I wish to thank all those physicists and mathematicians (I Batalin, I Gelfand, G Goldin, A Gonzalez-Lopez, V Fainberg, E Fradkin, V Kadyshevsky, N Kamran, G Kharadze, P Leach, L Lipatov, A Leznov, A Nersesyan, T Maglaperidze, J Mandzhavidze, V Manko, M Marinov, S Matinyan, M Olshanetsky, P Olver, V Ogievetski, I and N Paziashvili, M Savelyev, W Scherer, A Scotti, M Shifman, L Takhtadjan, K Ter-Martirosyan, A Turbiner, G Vachnadze, A Vainstein, P Wiegmann, B Zakhariev, A Zamolodchikov and others) who have commented, verbally or in writing, on the various aspects of the programme presented in this book.

I am especially grateful to Professor H D Doebner for useful discussions and valuable remarks. This book could not have been completed without his encouragement and helpful advice, which I especially needed in the final stages of the work.

Also special thanks go to the staff of the Institute of Physics Publishing (especially to Ms Emily Wolfe, Mr Neal Marriott and Mr Jim Revill) for their encouragement for me to turn my manuscript into book form, and for their patience during the process of its revision.

And finally, I take this opportunity to thank my mother, Elisabeth Ushveridze, my father, George Ushveridze, for their unfailing support and influence in my life, and, of course, my wife, Alina Zamikhovskaya, whose love, patience and understanding were essential to it all!

Alexander Ushveridze
12 September 1993

Chapter 1

Quasi-exact solvability. What does that mean?

1.1 Introduction

We have become accustomed to calling a quantum model exactly solvable if for all its energy levels and corresponding wavefunctions convenient (explicit) expressions can be obtained, or, more precisely, if the spectral problem for this model can be reduced to a problem of algebra (in the case of quantum mechanics) or classical analysis (in the field theoretical case).

It is well known that exactly solvable models play an extremely important role in many fields of quantum physics.

First of all, they may be interesting in themselves as models of actual physical situations. For example, the behaviour of many quantum systems near their equilibrium can be described by the harmonic oscillator, the spectrum of the hydrogen atom can be found from the Coulomb problem, which is also exactly solvable, and one-dimensional completely integrable non-relativistic field theoretical models (see e.g. Thacker 1981, Wiegmann 1981, Tsvelick and Wiegmann 1983) have proved to be good approximations for the so-called quasi-one-dimensional systems (Bulayevsky 1975, Toombs 1978, Jerome and Schulz 1982).

Second, exactly solvable models can be successfully used as a training ground for elaborating various approximate and qualitative methods of studying exactly non-solvable systems, and for testing theoretical hypotheses of a general nature (see e.g. Solyom 1979). Sometimes the analysis of exact solutions allows one to reveal certain properties of the system which do not change even after it has undergone considerable deformation (universality). The study of such properties is very helpful for a better understanding of the general structure of quantum models (Luther 1976, 1977, Tsvelick and Wiegmann 1983, Japaridze *et al*

1984). By analogy with the cases of the simple harmonic oscillator and the Coulomb problem, the study of which led to comprehension of many fundamental principles of quantum mechanics, the analysis of such completely integrable systems as magnetic chains made a valuable contribution to the understanding of the physical nature of excitations in systems with many degrees of freedom (Bethe 1931, Bonner and Fischer 1964, Sutherland 1970, Baxter 1972, Takhtajan and Faddeev 1981, Gaudin 1983).

Third, exactly solvable models can be used as zeroth-order approximations by constructing various perturbative schemes. For example, the rapid progress of quantum field theory in the fifties was attained in the framework of perturbation theory: the role of the unperturbed problem was played in this case by the free field or, in other words, by the infinite-dimensional harmonic oscillator. At present, there are some preconditions for constructing perturbation theory near the completely integrable one-dimensional field theoretical models and magnetic chains (see e.g. Gaudin *et al* 1981, Korepin 1982, 1984).

Finally, exactly solvable models are also interesting from a purely mathematical point of view, since the phenomenon of exact solvability can often be explained as being a consequence of certain hidden symmetry present in the model under consideration. It is remarkable that the group describing this symmetry appears not only as a necessary attribute of exact solvability, but as a unique language in which this phenomenon has simple and transparent mathematical sense. As an example, it is sufficient to recall the simple harmonic oscillator and the Heisenberg algebra associated with it (Bargmann and Moshinsky 1961, Moshinsky 1962). In the same sense, the various completely integrable systems obtained by means of the inverse scattering method (Takhtajan and Faddeev 1979) turn out to be connected with various finite- and infinite-dimensional Lie algebras (see e.g. Zamolodchikov and Zamolodchikov 1978, Kulish *et al* 1981, Bazhanov 1985, Reshetikhin 1985, Gaudin 1983, Ogievetski *et al* 1987).

The last fifteen years have been marked by the discovery of a number of remarkable methods of building and solving exactly solvable models. They are: the quantum inverse scattering method (Takhtajan and Faddeev 1979), various modifications of the Bethe *ansatz* (Gaudin 1983), the Leznov–Savelyev approach (Leznov and Savelyev 1985), the projection method (Olshanetsky and Perelomov 1983), the method of Gelfand–Levitan–Marchenko equations (Zakhariev and Suzko 1985), the methods based on the use of Witten's supersymmetric quantum mechanics and the Darboux theorem (Infeld and Hull 1951, Gendenshtein 1983, Andrianov *et al* 1984, Gendenshtein and Krive 1985), and so on (Bargmann 1949, Plekhanov *et al* 1982, Leznov 1984, Rudyak and Zakhariev 1987).

Unfortunately, in spite of the numerous merits of the methods listed here, the number and the diversity of models which can be constructed by means of them are relatively small, from the point of view of the requirements of modern quantum physics. Brute-force attempts to find new exactly solvable models and methods of their construction encounter serious difficulties, connected probably with the fact that the usual requirement of exact solvability, understood as the possibility of writing down the *entire* spectrum of the hamiltonian in more or less closed form, is too strict.

A possible way out of this impasse might involve only an essentially new approach to the problem, based on some constructive expansion of the concept of exact solvability in quantum physics. One such approach was proposed several years ago. It was based on the idea of relaxing the standard requirements of exact solvability and seeking models for which the spectral problem could be solved exactly only for certain limited parts of the spectrum but not for the whole spectrum. This idea has gradually crystallized in the last decade. The critical impetus has been given in the works of Zaslavsky and Ulyanov (1984), Bagrov and Vshivtsev (1986), Turbiner and Ushveridze (1987) and, especially, Turbiner (1988a) and Ushveridze (1988c,d). Following the terminology introduced by Turbiner and Ushveridze, we shall call these models "quasi-exactly solvable" referring to their "order" as the number of states for which exact results can be obtained.

The essence of the phenomenon of quasi-exact solvability can be explained as follows. It is known that any hamiltonian H can be represented as an infinite-dimensional hermitian matrix

$$H = \begin{bmatrix} H_{00} & H_{01} & \cdots & H_{0M} & \cdots \\ H_{10} & H_{11} & \cdots & H_{1M} & \cdots \\ \hdotsfor{5} \\ H_{M0} & H_{M1} & \cdots & H_{MM} & \cdots \\ \hdotsfor{5} \end{bmatrix} \qquad (1.1.1)$$

whose elements $H_{nm} = \langle \psi_n | H | \psi_m \rangle$ depend on the concrete choice of orthogonal functions ψ_n forming a basis in the Hilbert space.

In this language, the solution of the spectral problem for H is reduced simply to a diagonalization of the matrix (1.1.1). Unfortunately, unlike what is found in the cases of finite matrices, there are no general algebraic rules which would allow one to diagonalize the infinite-dimensional matrix (1.1.1) in a finite number of steps. In general case the spectral problem for such matrices is non-algebraic. This is a typical situation for the so-called exactly non-solvable models.

The famous exactly solvable problems are distinguished by the fact that in these problems there is known a "natural" basis in the Hilbert space in which the matrix (1.1.1) is very specific and can be reduced explicitly to the diagonal form

$$
H = \begin{bmatrix}
H_{00} & 0 & 0 & 0 & \cdots \\
0 & H_{11} & 0 & 0 & \cdots \\
0 & 0 & H_{22} & 0 & \cdots \\
0 & 0 & 0 & H_{33} & \cdots \\
\hdotsfor{5}
\end{bmatrix}
\tag{1.1.2}
$$

with the aid of an algebraic procedure[1]. There are only few examples of the models for which such a diagonalization is possible. In the one-dimensional case they are: the simple harmonic oscillator, the harmonic oscillator with centrifugal barrier, the Morse potential, the hyperbolic and trigonometric Pöschel-Teller potential wells, the Coulomb and Kratzer potentials, the Natanzon potentials and some others (see e.g. Landau and Lifshitz 1977, Flügge 1971, Natanson 1971, 1978).

Summarizing, the standard theory of the Schrödinger equation includes two contrasting cases: the complete diagonalizability of the hamiltonian by algebraic methods as in equation (1.1.2) (a very rare situation), and the typical case presented in equation (1.1.1), with non-vanishing off-diagonal elements. As a rule the diagonalization of such matrices cannot be carried out algebraically.

A new direction in the spectral theory, an intermediate link between (1.1.1) and (1.1.2), has become apparent recently. Assume that the hamiltonian matrix has a block structure

$$
H = \left[
\begin{array}{cccc|ccccc}
H_{00} & H_{01} & \cdots & H_{0M} & 0 & 0 & \cdots & 0 & \cdots \\
H_{10} & H_{11} & \cdots & H_{1M} & 0 & 0 & \cdots & 0 & \cdots \\
\multicolumn{9}{c}{\cdots\cdots\cdots\cdots\cdots} \\
H_{M0} & H_{M1} & \cdots & H_{MM} & 0 & 0 & \cdots & 0 & \cdots \\
\hline
0 & 0 & \cdots & 0 & & & & & \\
0 & 0 & \cdots & 0 & & \text{Non-vanishing} & & & \\
\multicolumn{4}{c|}{\cdots\cdots\cdots\cdots} & & \text{elements} & & & \\
0 & 0 & \cdots & 0 & & & & & \\
\multicolumn{4}{c|}{\cdots\cdots\cdots\cdots} & & & & & \\
\end{array}
\right],
\tag{1.1.3}
$$

where M is some fixed integer. The block in the upper left corner is an $M+1$ by $M+1$ matrix, while the second non-vanishing block (in the lower right corner) is an infinite-dimensional matrix. Then, quite obviously, one can immediately diagonalize the finite block without touching the infinite one. The operation is performed in exactly the same way as for any finite matrix

[1] As a rule, the existence of such a basis is a consequence of a certain hidden symmetry present in the model under consideration.

and is purely algebraic. This means that in the case of matrices (1.1.3) one determines explicitly only a part of the spectrum, $M + 1$ eigenvalues and the corresponding eigenfunctions of the hamiltonian H.

In other words we are led to models which occupy an intermediate position between exactly solvable models and exactly non-solvable ones. We call such models "quasi-exactly solvable". We see that the term "quasi-exact solvability" implies the situation where the infinite-dimensional hamiltonian matrix can be reduced explicitly to block-diagonal form with one of the blocks being finite. In this case the infinite-dimensional matrix version of the Schrödinger problem breaks up into two completely independent spectral problems, one of which is finite dimensional and can be solved algebraically, while the other is infinite dimensional and nothing about its solutions is known (partial algebraization).

It is absolutely obvious that the block diagonalization of a given random infinite-dimensional matrix is, in principle, a much more simple procedure than its total diagonalization. This gives us reason to assert that the number of quasi-exactly solvable models must exceed the number of ordinary exactly solvable ones (Ushveridze 1989c).

In order to make sure that making this assumption is justified, consider the Schrödinger equation

$$\{-\Delta + V(\vec{x})\}\psi(\vec{x}) = E\psi(\vec{x}), \tag{1.1.4}$$

complemented by the condition of normalizability of the wavefunction $\psi(\vec{x})$

$$\int \psi^2(\vec{x})\mathrm{d}^{\mathcal{D}}\vec{x} = 1. \tag{1.1.5}$$

Assume that $\rho(\vec{x})$ is an arbitrary regular sign-definite function, satisfying the analogous normalization condition

$$\int \rho^2(\vec{x})\mathrm{d}^{\mathcal{D}}\vec{x} = 1. \tag{1.1.6}$$

It is not difficult to see that the Schrödinger equation (1.1.4) with the potential

$$V(\vec{x}) = \frac{\Delta\rho(\vec{x})}{\rho(\vec{x})} \tag{1.1.7}$$

has at least one explicit normalizable solution of the form:

$$\psi(\vec{x}) = \rho(\vec{x}), \quad E = 0. \tag{1.1.8}$$

From the condition of sign-definiteness of the function $\rho(\vec{x})$, it follows that the wavefunction $\psi(\vec{x})$ has no nodes and therefore, according to the oscillator theorem, describes the ground state. In general, the other solutions of this equation cannot be found exactly. Thus, we deal with the set of quasi-exactly solvable models of order one.

Obviously, this set is functionally large. At the same time we know of only a few examples of exactly solvable multi-dimensional Schrödinger equations, and this circumstance proves our assertion.

Note that quasi-exactly solvable models of the type (1.1.7) are trivial to construct not only for the ground state. Of course, when choosing the function $\rho(\vec{x})$ we must take care that the potential obtained is physically sensible. However, all difficulties connected with construction of such potentials are easily overcome (at least for the one-dimensional case). This problem was solved by Gershenson and Turbiner (1982) (for the ground state) and by Rampal and Datta (1984) (for excited states) in models with polynomial potentials.

Much more non-trivial is the procedure for constructing quasi-exactly solvable models of higher orders. Nevertheless, the first examples of these models clearly demonstrated that their potentials are not necessarily exotic "monsters", but can be quite simple and ordinary looking. Many of these models have been known for a long time as exactly non-solvable and the proof of their quasi-exact solvability can be seen as a very intriguing result.

Note that the simplest quasi-exactly solvable models are described by polynomial potentials. Their construction and analysis (especially in the one-dimensional case) does not require any technical effort. At the same time, they have rather interesting physical and mathematical properties and this allows them to be used as very convenient objects for a preliminary acquaintance with the phenomenon of quasi-exact solvability.

From the pedagogical point of view it is reasonable to start our discussion with a brief review of the properties of models with the simplest polynomial potentials. This will be done in sections 1.2 and 1.3 where we consider the simple harmonic oscillator (which is exactly solvable) and quartic anharmonic oscillator (the exact solutions of which are not known). This consideration is helpful for achieving a better understanding of the properties of quasi-exactly solvable models which will be discussed in detail in the remaining sections of this chapter.

1.2 Completely algebraizable spectral problems. The simple harmonic oscillator

The simplest quantum model with a polynomial potential — the harmonic oscillator — is described by the hamiltonian

$$H = -\frac{\partial^2}{\partial x^2} + \alpha x^2, \tag{1.2.1}$$

in which α is a positive parameter. This model is exactly solvable. The simplest way to demonstrate this fact is to use the well known Heisenberg method which is based on the introduction of operators $a^+ = \sqrt{\alpha}x - \frac{\partial}{\partial x}$ and $a^- = \sqrt{\alpha}x + \frac{\partial}{\partial x}$, forming the Heisenberg algebra and generating a basis in which the hamiltonian (1.2.1) takes an explicit diagonal form. Unfortunately, the Heisenberg method is specific only for the harmonic oscillator and cannot be easily extended to other exactly solvable models.

In this section we discuss a more general algebraic method which is free from this demerit. Its main idea is to use more non-trivial differential realizations of the generators of the Heisenberg algebra.

Following this method, note that the hamiltonian (1.2.1) can be rewritten in the form

$$H = \sqrt{\alpha}(2a^+a^- + a^0) - (a^-)^2, \tag{1.2.2}$$

where $a^\pm$ and a^0 are the operators defined as

$$a^+ = x, \quad a^0 = 1, \quad a^- = \sqrt{\alpha}x + \frac{\partial}{\partial x} \tag{1.2.3}$$

and satisfying the following commutation relations

$$[a^-, a^+] = a^0, \quad [a^\pm, a^0] = 0. \tag{1.2.4}$$

We see that these operators form again the Heisenberg algebra. Therefore, it is reasonable to refer to a^+ and a^- as the raising and lowering operators.

We know that the Heisenberg algebra allows representations with lowest weight defined by the formulas

$$a^-|0\rangle = 0, \quad a^0|0\rangle = |0\rangle, \tag{1.2.5}$$

where $|0\rangle$ is the lowest (vacuum) vector. The corresponding representation space Φ is formed by the linear combinations of the vectors

$$|n\rangle = (a^+)^n|0\rangle, \tag{1.2.6}$$

where n takes the values $0,1,2,\ldots$. This space is infinite dimensional and therefore the spectral problem for H in Φ

$$H\varphi = E\varphi, \quad \varphi \in \Phi \tag{1.2.7}$$

is also infinite dimensional.

Note, however, that from the formulas

$$a^+|n\rangle = |n+1\rangle, \quad a^0|n\rangle = |n\rangle, \quad a^-|n\rangle = n|n-1\rangle \tag{1.2.8}$$

and (1.2.2) it follows that any finite-dimensional subspace Φ_n of the space Φ formed by the linear combinations of the first $n + 1$ basis vectors $|0\rangle, |1\rangle, \ldots, |n\rangle$ is an invariant subspace for H:

$$H\Phi_n \subset \Phi_n. \tag{1.2.9}$$

This means that the infinite-dimensional spectral problem (1.2.7) is equivalent to an infinite set of finite-dimensional spectral problems

$$H\varphi = E\varphi, \quad \varphi \in \Phi_n, \quad n = 0, 1, 2, \ldots \tag{1.2.10}$$

each of which, evidently, can be solved algebraically. Therefore, according to the definition given in the previous section, the model (1.2.1) is exactly solvable.

In order to obtain explicit solutions of this problem, note that

$$H|n\rangle = \sqrt{\alpha}(2n + 1)|n\rangle - n(n - 1)|n - 2\rangle. \tag{1.2.11}$$

From (1.2.11) it follows that the linear span of the vectors $|n\rangle$ of a given parity is invariant under the action of the operator H. This gives us reason to assert that the parity of the state is a conserved quantum number. Therefore, instead of the subspaces Φ_n introduced above, we can consider two different sets of invariant subspaces

$$\Phi_M^p = \text{linear span of}\{|p\rangle, |2 + p\rangle, \ldots, |2M + p\rangle\}, \tag{1.2.12}$$

characterized by the parities $p = 0, 1$.

Choosing the solution of the problem (1.2.7) in the form

$$\Phi = \eta_0|2M + p\rangle + \eta_1|2(M - 1) + p\rangle + \ldots + \eta_M|p\rangle \tag{1.2.13}$$

and using the relation (1.2.11) we obtain the explicit expression for the energy

$$E = \sqrt{\alpha}(4M + 2p + 1) \tag{1.2.14}$$

and, simultaneously, the recurrence relations for the coefficients η_m:

$$-4\sqrt{\alpha}\eta_{m+1} = [2(M - m) + p][2(M - m) + p + 1]\eta_m. \tag{1.2.15}$$

From (1.2.15) we find

$$\eta_m = \left(-\frac{1}{4\sqrt{\alpha}}\right)^m \frac{[2M + p]!}{m![2(M - m) + p]!}, \tag{1.2.16}$$

and, thus, the solution of (1.2.7) corresponding to the eigenvalue (1.2.14) is

$$\varphi \sim \sum_{m=0}^{M} \frac{(-4\sqrt{\alpha})^m}{(M-m)!} \frac{(2M+p)!}{(2m+p)!} |2m+p\rangle. \qquad (1.2.17)$$

The last step is to reduce the obtained solution to a coordinate form. To this end we must solve the equation (1.2.5) for $|0\rangle$. Using (1.2.3) we obtain

$$|0\rangle = \exp\left(-\frac{\sqrt{\alpha}x^2}{2}\right) \qquad (1.2.18)$$

and, consequently,

$$|n\rangle = x^n \exp\left(-\frac{\sqrt{\alpha}x^2}{2}\right). \qquad (1.2.19)$$

The substitution of (1.2.19) into (1.2.17) gives us the final result for φ:

$$\varphi \sim \sum_{m=0}^{M} \frac{(2M+p)!(-4\sqrt{\alpha}x^2)^m}{(M-m)!(2m+p)!} x^p \exp\left(-\frac{\sqrt{\alpha}x^2}{2}\right), \qquad (1.2.20)$$

which completes the diagonalization of the hamiltonian H of the simple harmonic oscillator.

Let us now look at our derivation from more general point of view. We denote by $\mathcal{H}$ the Heisenberg algebra and by $\mathcal{U}(\mathcal{H})$ its universal enveloping algebra. Remember that $\mathcal{U}(\mathcal{H})$ consists of all linear combinations of monomials $(a^+)^n(a^-)^m$ in which n and m are arbitrary non-negative integers. Let $\mathcal{U}_0(\mathcal{H})$ be a subalgebra of the algebra $\mathcal{U}(\mathcal{H})$ defined as a linear span of monomials $(a^+)^n(a^-)^m$ satisfying the constraints $n \le m$. It is clear that the elements of this subalgebra act as non-raising operators and therefore, for any $u_0 \in \mathcal{U}_0(\mathcal{H})$ and n we have $u_0\Phi_n \subset \Phi_n$. In this language, the fact of the exact solvability of the model (1.2.1) becomes especially obvious. Indeed, this model is exactly solvable since its hamiltonian H is an element of the subalgebra $\mathcal{U}_0(\mathcal{H})$. Our reasonings result also in a more general assertion: any element of the subalgebra $\mathcal{U}_0(\mathcal{H})$ can be viewed as the "hamiltonian" of a certain exactly solvable model.

Thus, we have essentially formulated a method of constructing exactly solvable models. Of course, this formulation is still rather abstract and requires some comments concerning the problem of making the resulting quantum "hamiltonians" physically meaningful. First of all, these "hamiltonians" must be differential operators, and for this the generators of the Heisenberg algebra also must have a differential form.

The simplest differential realization of these generators is

$$a^0 = 1, \quad a^+ = t, \quad a^- = \frac{\partial}{\partial t}. \qquad (1.2.21)$$

It differs from the standard (Heisenberg) one by a canonical transformation. Transforming homogeneously the operators (1.2.21) and changing the variable t, we obtain a class of new more complex realizations

$$a^0 = 1, \quad a^+ = A(x), \quad a^- = \frac{1}{A'(x)}\left(\frac{\partial}{\partial x} - B(x)\right), \qquad (1.2.22)$$

depending on two arbitrary functions $A(x)$ and $B(x)$.

Note that the form of these realizations makes them very convenient for constructing differential operators of *a priori* given order. In particular, for the "hamiltonians" belonging to the algebra $\mathcal{U}_0(\mathcal{H})$ to have the form of second-order differential operators, it is sufficient to take various linear combinations of monomials $(a^+)^n(a^-)^m$ satisfying the constraint $n \leq m \leq 2$. The most general expression for such operators is

$$H = A_1(a^+)^2(a^-)^2 + A_2 a^+(a^-)^2 + A_3(a^-)^2 + A_4 a^+ a^- + A_5 a^- + A_6. \qquad (1.2.23)$$

Substitution of (1.2.22) into (1.2.23) reduces H to an explicit differential form:

$$H = P(x)\frac{\partial^2}{\partial x^2} + Q(x)\frac{\partial}{\partial x} + R(x). \qquad (1.2.24)$$

Here $P(x)$, $Q(x)$ and $R(x)$ are functions depending on $A(x)$ and $B(x)$.

For spectral equations for H to be Schrödinger type equations, we must require

$$P(x) = -1, \quad Q(x) = 0, \qquad (1.2.25)$$

which gives us two equations for two unknown functions $A(x)$ and $B(x)$. Solving these equations we can easily recover the form of the free term $R(x)$, which, obviously, plays the role of the potential. As a result, we obtain a six-parameter family of one-dimensional quantum mechanical models admitting a complete algebraization of the spectral problem.

It is not difficult to verify that many famous exactly solvable models such as the Morse potential, the Pöschel–Teller potential well, the harmonic oscillator with centrifugal barrier (and some others), can be constructed in the framework of this rather general method. It is remarkable that for all these potentials, the corresponding hamiltonians can be written as special combinations of generators of the Heisenberg algebra (for more details see chapter 3).

The method described above will be referred to below as the method of raising and lowering operators. Its generalization to the multi-dimensional case will be discussed in appendix C.

1.3 The quartic oscillator. The absence of exact solutions

In the twenties it became evident that, unlike the classical harmonic oscillator, the quantum one is far from being always a good model for describing the behaviour of various systems near their equilibrium. Recall that in classical mechanics low-energy motion in any potential $V(x)$ having the minimum at the origin can be approximated by the motion in the harmonic potential αx^2 with $\alpha = \frac{1}{2}V''(0)$. In quantum mechanics the situation is more complex. The reason is that, due to the vacuum oscillations, the wavefunction is always diffused, even for the ground state, and this blocks the possibility of the quantum particle moving always in the vicinity of the minimum. In order to describe the quantum motion in this case, it is necessary to take into account the anharmonic corrections to the harmonic term, which leads naturally to the problem of the anharmonic oscillator. This is one of the ancient problems in quantum mechanics.

The most simple and, therefore, most popular model of this sort is the quartic anharmonic oscillator with the hamiltonian

$$H = -\frac{\partial^2}{\partial x^2} + \alpha x^2 + \beta x^4. \qquad (1.3.1)$$

Its many facets have been discussed in countless publications (see e.g. Bender and Wu 1969, Simon 1970, Hioe *et al* 1978, Bogomolny *et al* 1980, Koudinov and Smondyrev 1983, Shaverdyan and Ushveridze 1983, Shanley 1986, Turbiner and Ushveridze 1988a, Ushveridze 1989b) and there is no reason for their flow to stop because on the one hand, the model has numerous important applications in many fields and on the other hand, it is a perfect testing ground for any novel approximative methods.

In order to estimate the measure of proximity of the model (1.3.1) to the model of the harmonic oscillator, note that the lowest energy level in (1.2.1) is described by the wavefunction

$$\psi(x) = \exp\left\{-\frac{\sqrt{\alpha}x^2}{2}\right\}. \qquad (1.3.2)$$

This means that the probability of finding the particle with coordinate x

essentially differs from zero only in the domain

$$x^2 \lesssim \frac{1}{\sqrt{\alpha}}. \tag{1.3.3}$$

Comparing the harmonic term, αx^2, with the anharmonic one, βx^4, in this domain we come to two essentially different cases

$$\frac{\beta}{\alpha^{\frac{3}{2}}} \ll 1 \quad \text{and} \quad \frac{\beta}{\alpha^{\frac{3}{2}}} \gtrsim 1 \tag{1.3.4}$$

which lead to two drastically different physical pictures.

In the first case the anharmonic term is small in comparison with the harmonic one and, therefore, can be considered as a small correction. In this case the model (1.3.1) is, generally, similar to the model (1.2.1) (at least for low excitations), and the slight quantitative difference can be taken into account by means of perturbation theory.

In the second case the anharmonic term is at least of the same order as the harmonic one and we come to systems with strong anharmonicity. The physics of this system strongly differs from the physics of the harmonic oscillator and therefore it cannot be studied in the framework of the perturbative approach.

Using field theoretical terminology we shall call the first and second cases the weak- and strong-coupling regimes, respectively.

First of all, let us consider the model (1.3.1) in the weak-coupling regime (when the effective coupling constant $\beta/\alpha^{\frac{3}{2}}$ is small).

Since the creation of quantum mechanics, hundreds of physicists have studied the model (1.3.1) in this limit, but only in the middle of the fifties did it become evident that perturbation theory, having been for a long time a basic tool in this study, is not free from essential deficiencies. The substance of these deficiencies is the existence of effects which, at first sight, belong to the range of applicability of perturbation theory but, nevertheless, cannot be seen in any finite order of it (so called non-perturbative effects).

The origin of this phenomenon was studied in detail by Dyson (1952), who argued that the perturbation series are asymptotic and diverge for any values of the coupling parameter $\beta/\alpha^{\frac{3}{2}}$ irrespective of how small it is.

This observation, which is now known under the generic name of "Dyson's instability argument", can be clarified as follows. The perturbation potential βx^4 for large x becomes larger than the unperturbed one, αx^2. Therefore, the change of sign of β transforms the stable system (1.3.1) into an unstable one, and energy levels in it obtain an imaginary part (see figure 1.1). Such a picture takes place for any arbitrarily small

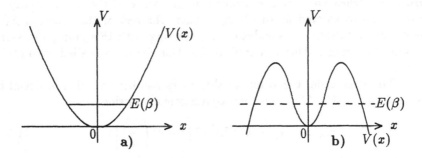

Figure 1.1. The form of the potential (1.3.1) for a) $\beta > 0$ and b) $\beta < 0$.

β, and this means that the point $\beta = 0$ is singular and this singularity is of the branch type. The complex β-plane with this singularity and the corresponding cut is depicted in figure 1.2. The discontinuity on the cut

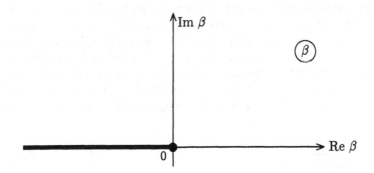

Figure 1.2. Complex β plane for the function $E(\beta)$ for $-\pi < \arg \beta < \pi$.

(the measure of a non-analyticity) is equal to a probability of penetration through the barrier and is exponentially small if β is small:

$$\text{Dis } E(\beta) \sim \left(\beta \alpha^{-\frac{3}{2}}\right)^{-\frac{1}{2}} \exp\left\{-S\left(\beta \alpha^{-\frac{3}{2}}\right)^{-1}\right\}. \qquad (1.3.5)$$

(Here S is a classical action for a subbarrier trajectory.) Thus, in the vicinity of the origin the function $E(\beta)$ is "almost analytic" and, therefore, its expansion in powers of β

$$E(\beta) = E_0 + \beta E_1 + \beta^2 E_2 + \dots \qquad (1.3.6)$$

behaves as a convergent one. This is so until the values of the terms $E_n \beta^n$ in this expression exceed considerably the value of the discontinuity.

However, when these terms become of the order of the discontinuity, it cannot be considered as small any longer; the rest of the function $E(\beta)$ becomes essentially non-analytic and, starting with this point, the series begins to diverge. This is a typical situation for the so-called asymptotic series.

The large-order behaviour of this series has been studied in detail by Vainshtein (1964). The resulting asymptotic expression

$$E_n = (-1)^{n+1} n! A^n n^B C \left[1 + O\left(\frac{1}{n}\right)\right], \qquad (1.3.7)$$

in which A, B and C are certain computable parameters, shows that the terms of the weak-coupling expansion grow factorially, in full accordance with the fact that the series has zero radius of convergence.

The next step in understanding the structure of the model (1.3.1) was done in the important work of Bender and Wu (1969), who showed that the analytic properties of the function $E(\beta)$ in the vicinity of the origin are significantly more complex than might be seen from figure 1.2. Bender and Wu demonstrated that the global Riemann surface for the function $E(\beta)$ (i.e. the Riemann surface at large β) consists of three sheets. On the first (physical) sheet, depicted in figure 1.2, the unique singularity is the branch point $\beta = 0$. But it is not isolated: on the second and third sheets there is an infinite number of branch-point singularities of the square root type which accumulate at zero forming the so-called "horn structure" (see figure 1.3).   Later, the Bender and Wu singularities were discovered also in

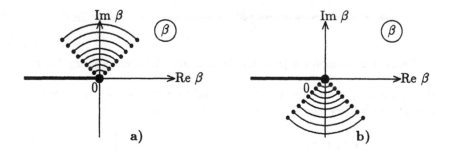

Figure 1.3. Complex β-plane for the function $E(\beta)$. The second and third sheets of the Riemann surface. a) $\pi < \arg \beta < 3\pi$, b) $3\pi < \arg \beta < 5\pi$.

other models (see e.g. Blanch and Glemm 1969, Bender *et al* 1974, Avron and Simon 1978, Hunter and Guerrieri 1981, 1982, Ushveridze 1988a).

The physical meaning of these singularities is that they are the double-crossing (plaiting) points for energy levels as analytic functions of β. Bender

and Wu proved that all the energy levels with equal parities in the model
(1.3.1) are plaited, forming a common Riemann surface with an infinite
number of sheets. This means, for example, that starting with the ground
state $E_0(\beta)$ (with β belonging to the positive half axis of the physical sheet)
and continuing it analytically along the closed contour which passes round
the Bender and Wu singularity and goes back to the initial point, we can
obtain the second energy level $E_2(\beta)$, the fourth level $E_4(\beta)$ and so on.

Note that these results were obtained by means of very refined methods
in the framework of semi-classical approximation. Therefore, the locations
of the plaiting points found by Bender and Wu are asymptotically true only
in the limit when β tends to zero.

The analytic properties of the energy levels $E(\beta)$ listed above relate
to the typical non-perturbative effects which cannot be seen in any finite
order of perturbation theory. In fact, any partial sum of the perturbation
series is a polynomial in β and is therefore regular everywhere. If the
effective coupling $\beta/\alpha^{\frac{3}{2}}$ is small (in other words, if we deal with the
weak-coupling regime) these effects are also exponentially small and, from
the quantitative point of view, are non-essential for us. But if $\beta/\alpha^{\frac{3}{2}}$
increases and becomes of the order of unity or larger, non-perturbative
effects also become appreciable and may, in principle, change considerably
the physics of the system. The problem of describing the behaviour
of quantum models in the strong-coupling regime (the so-called strong-
coupling problem) is an important one in both quantum mechanics and
quantum field theory, as almost all interesting phenomena belong to the
strong-coupling regime, beyond the applicability of perturbation theory
and semi-classical approximation. This problem is so involved that, until
recently, there have been very few attempts to formulate its general
solution. For example, only five years ago the constructive non-semiclassical
methods allowing the study of analytic properties of energy levels in
the quartic anharmonic oscillator (1.3.1) were proposed. The above is
practically all that we know about the analytic structure of the quartic
anharmonic oscillator (1.3.1).

Of course, during the long history of this problem there were
many attempts to find its exact solutions. However, all attempts were
unsuccessful. This led gradually to the opinion that the model (1.3.1) is
exactly non-solvable. This became especially clear after the work of Bender
and Wu who demonstrated an extraordinary complexity of the analytic
function describing the energy levels in the model. At present, it seems
absolutely unrealistic to find an exact solution for (1.3.1) ensuring, for
example, the "horn structure" shown in figure 1.3.

1.4 The sextic oscillator. Exact solvability for low excitations

Apparently, all the efforts of physicists were spent in countless attempts to understand the structure of the quartic anharmonic oscillator. Other polynomial models such as, for example, the sextic anharmonic oscillator with the hamiltonian

$$H = -\frac{\partial^2}{\partial x^2} + \alpha x^2 + \beta x^4 + \gamma x^6 \qquad (1.4.1)$$

were almost forgotten. A few publications on this topic (see e.g. Hioe *et al* 1976) were devoted more to numerical computation of energy levels in (1.4.1) than to study of their analytic properties as functions of the parameters α, β, γ. The position of physicists was clear: if the model (1.3.1) has such terrible analytic properties, what can one expect from more complex models with sextic anharmonicity? The conviction that all models of such a sort are exactly non-solvable was unshakable.

Surprisingly enough, this assertion turned out to be wrong! This was demonstrated in the paper of Turbiner and Ushveridze (1987), where quasi-exact solvability of the model (1.4.1) was proved. The result of this paper can be formulated as follows: there are special cases when the model (1.4.1) allows very simple analytic solutions. These cases are realized when the parameters α, β and γ entering into the potential satisfy the condition

$$\frac{1}{\sqrt{\gamma}}\left(\frac{\beta^2}{4\gamma} - \alpha\right) = 3 + 2n, \qquad (1.4.2)$$

where n is an arbitrary non-negative integer. If n is fixed, the model (1.4.1) has $[\frac{n}{2}]+1$ exact solutions which can be found by means of a simple algebraic procedure. At the same time, other solutions of the model remain unknown to us, so that we deal with the infinite series of quasi-exactly solvable models of the orders $[\frac{n}{2}] + 1$, $n = 0, 1, 2, \ldots$.

This fact can be proved in the same way as the fact of exact solvability of the simple harmonic oscillator (1.2.1). We stress, however, that the derivation given below differs from the one used in the original paper of Turbiner and Ushveridze (1987).

This derivation is purely algebraic. We start with the observation that the hamiltonian (1.4.1) can be rewritten in the form

$$\begin{aligned}
H &= (a^+)^2\left\{2\sqrt{\gamma}a^+a^- + \left(\alpha - \frac{\beta^2}{4\gamma} + 3\sqrt{\gamma}\right)a^0\right\} \\
&\quad + \frac{\beta}{2\sqrt{\gamma}}(2a^+a^- + a^0) - (a^-)^2,
\end{aligned} \qquad (1.4.3)$$

where $a^{\pm}$ and a^0 are the operators defined as

$$a^+ = x, \quad a^0 = 1, \quad a^- = \sqrt{\gamma}x^3 + \frac{\beta}{2\sqrt{\gamma}}x + \frac{\partial}{\partial x}. \tag{1.4.4}$$

It is evident, that, as in the case of the harmonic oscillator, they form the Heisenberg algebra with commutation relations (1.2.4). Infinite-dimensional representations of this algebra can be defined by formula (1.2.5). Conserving all notations of section 1.2 we can write the relation

$$H|n\rangle = \left\{\alpha - \frac{\beta^2}{4\gamma} + \sqrt{\gamma}(3 + 2n)\right\}|n + 2\rangle$$

$$+ \frac{\beta}{2\sqrt{\gamma}}(2n + 1)|n\rangle + n(n - 1)|n - 2\rangle \tag{1.4.5}$$

which, in some sense, is an analogue of the relation (1.2.11). We see that the operator H transforms, as before, even vectors into even ones and odd vectors into odd ones. Therefore, the parity of the vector is a conserved quantum number. This gives us reason to introduce, instead of the vectors $|n\rangle, n = 0, 1, 2, \ldots$, two separate sequences of the vectors:

$$|m, p\rangle = |2m + p\rangle, \quad m = 0, 1, 2, \ldots \tag{1.4.6}$$

where $p = 0, 1$ is the parity. In terms of the new vectors, formula (1.4.5) takes the form:

$$H|m, p\rangle = \left\{\alpha - \frac{\beta^2}{4\gamma} + \sqrt{\gamma}(4m + 2p + 3)\right\}|m + 1, p\rangle$$

$$+ \frac{\beta}{2\sqrt{\gamma}}(4m + 2p + 1)|m, p\rangle$$

$$+ (2m + p)(2m + p - 1)|m - 1, p\rangle. \tag{1.4.7}$$

By analogy with the case of the harmonic oscillator, let us consider the set of $(m + 1)$-dimensional subspaces Φ_m^p of the representation space Φ formed by linear combinations of vectors $|0, p\rangle, \ldots, |m, p\rangle$. It is easy to understand that, unlike the harmonic case, the spaces Φ_m^p are not invariant under the action of the operator H. Indeed, from (1.4.7) we have

$$H\Phi_m^p \subset \Phi_{m+1}^p, \tag{1.4.8}$$

and, therefore, the reduction of the infinite-dimensional spectral problem for H to a series of finite-dimensional problems is, generally speaking, impossible.

What does this mean? At first sight, the situation is unfavourable since our hopes of obtaining the exact solutions for the model (1.4.1) in the same way as in section 1.2 were not realized. However, more careful analysis shows that this first impression is not quite right.

Let us assume that the parameters α, β and γ determining the form of the hamiltonian (1.4.1) satisfy the condition

$$\alpha - \frac{\beta^2}{4\gamma} + \sqrt{\gamma}(4M + 2p + 3) = 0 \qquad (1.4.9)$$

with certain fixed M and p. Then for $m = M$ the coefficient of the leading vector $|m + 1, p\rangle$ in the right hand side of (1.4.7) vanishes and formula (1.4.8) takes the form:

$$H\Phi_M^p \subset \Phi_M^p. \qquad (1.4.10)$$

We see that in this case the space Φ_M^p becomes invariant for the operator H, and this makes the spectral problem for H in Φ_M^p become algebraically solvable. Since the space Φ_M^p is $(M + 1)$ dimensional, we can obtain $M + 1$ exact solutions of the Schrödinger equation

$$H\psi = E\psi, \quad \psi \in \Phi. \qquad (1.4.11)$$

At the same time, other solutions of this equation (lying outside the space Φ_M^p) remain unknown. This is a trivial consequence of the fact that the invariance of the space Φ_m^p for H takes place only for $m = M$ but not for $m > M$. This observation completes the proof of the assertion that the model (1.4.1) is a quasi-exactly solvable model of order $M + 1$ if the parameters α, β, γ satisfy the constraint (1.4.9).

Note that the chosen parametrization of the potential (1.4.1) is far from being most convenient for us. It seems more natural to introduce, instead of the parameters α, β, γ, the new parameters a, b and M by the formulas:

$$\gamma = a^2, \quad \beta = 2ab, \quad \alpha = b^2 - a(4M + 2p + 3). \qquad (1.4.12)$$

Then the hamiltonian (1.4.1) takes the form

$$H = -\frac{\partial^2}{\partial x^2} + [b^2 - a(4M + 2p + 3)]x^2 + 2abx^4 + a^2x^6 \qquad (1.4.13)$$

which is more suitable from the practical point of view, since the non-negative integer M, showing how many solutions of the corresponding Schrödinger equation can be found exactly, enters into the potential (1.4.13) explicitly.

Note now that there are two possible ways to construct these solutions. On the one hand, we could try to complete the algebraic manipulations with the hamiltonian (1.4.3), obtain the needed solutions in algebraic form and then write down the coordinate representation for them. On the other hand, we can construct the appropriate *ansatz* for the wavefunctions immediately in the coordinate form and, substituting it into the Schrödinger equation (1.4.13), obtain the needed solutions analytically. We choose the second way which is more helpful from the pedagogical point of view than the first one.

In order to construct the correct *ansatz* for the wavefunctions we need the coordinate representation for the vectors $|n\rangle$. The lowest vector $|0\rangle$ can be found from the condition that it is annihilated by the operator a^-, the differential form of which is given in (1.4.4). Solving the corresponding first-order differential equation we obtain

$$|0\rangle = \exp\left\{-\frac{ax^4}{4} - \frac{bx^2}{2}\right\}. \tag{1.4.14}$$

Then, using the coordinate representation (1.4.4) of the operator a^+ and definitions (1.4.6), (1.2.6) of the vectors $|m, p\rangle$, we find

$$|m, p\rangle = x^p x^{2m} \exp\left\{-\frac{ax^4}{4} - \frac{bx^2}{2}\right\}. \tag{1.4.15}$$

Finally, recalling the definition of the spaces Φ_M^p, we come to the following most general form of their elements:

$$\psi(x) = x^p P_M(x^2) \exp\left\{-\frac{ax^4}{4} - \frac{bx^2}{2}\right\}. \tag{1.4.16}$$

Here, by $P_M(t)$ we have denoted the polynomials of degree M.

Formula (1.4.16) determines a correct *ansatz* for wavefunctions in the model (1.4.13). Substitution of (1.4.16) into the Schrödinger equation for (1.4.13) reduces it to the form

$$\left\{-\left[\frac{\partial^2}{\partial x^2} + \frac{2p}{x}\frac{\partial}{\partial x}\right] + \left[2bx\frac{\partial}{\partial x} + b(2p+1) - E\right]\right.$$
$$\left. + \left[2ax^3\frac{\partial}{\partial x} - 4Max^2\right]\right\} P_M(x^2) = 0. \tag{1.4.17}$$

Now we are ready to consider some simplest cases of this equation.

1. One explicit solution
Let $M = 0$. In this case the polynomial $P_0(x^2)$ is a constant. Taking

$$P_0(x^2) = 1 \tag{1.4.18}$$

and substituting (1.4.18) into (1.4.17) we obtain the equation for E:

$$(2p+1)b - E = 0. \qquad (1.4.19)$$

Therefore, the Schrödinger equation for

$$H = -\frac{\partial^2}{\partial x^2} + [b^2 - a(2p+3)]x^2 + 2abx^4 + a^2x^6 \qquad (1.4.20)$$

has the following explicit solution:

$$E = (2p+1)b, \qquad (1.4.21a)$$

$$\psi(x) = x^p \exp\left\{-\frac{ax^4}{4} - \frac{bx^2}{2}\right\}. \qquad (1.4.21b)$$

We see that the obtained wavefunction has no nodes if $p = 0$, and has one node at $x = 0$ if $p = 1$. According to the oscillator theorem, this means that for $p = 0$ we deal with the ground state and for $p = 1$ with the first excited state.

2. Two explicit solutions
Now let $M = 1$. In this case the polynomial $P_1(x^2)$ can be written in the form

$$P_1(x^2) = x^2 + q, \qquad (1.4.22)$$

where q is a certain unknown number. Substituting (1.4.22) into (1.4.17) and equating the terms proportional to x^2 and 1 we obtain two equations for q and E:

$$
\begin{aligned}
(2p+5)b - E &= 4aq, \\
(2p+1)b - E &= \frac{2+4p}{q},
\end{aligned}
\qquad (1.4.23)
$$

from which we find the single quadratic equation for E:

$$[E - (2p+1)b][E - (2p+5)b] = 8a(2p+1). \qquad (1.4.24)$$

This is simply the ordinary secular equation. It is of second degree since the invariant subspace Φ_1^p of the Hilbert space Φ is two dimensional. In other words, we diagonalize a 2×2 matrix in the block decomposition (1.1.3).

Solving equation (1.4.24) and finding the corresponding values of q, we come to the assertion that the Schrödinger equation for the hamiltonian

$$H = -\frac{\partial^2}{\partial x^2} + [b^2 - a(2p+7)]x^2 + 2abx^4 + a^2x^6 \qquad (1.4.25)$$

has two explicit solutions:

$$E_\pm = (2p+3)b \pm 2\sqrt{b^2 + 2a(2p+1)}, \qquad (1.4.26a)$$

$$\psi_\pm(x) = [(b \pm \sqrt{b^2 + 2a(2p+1)})x^2 - (2p+1)]$$

$$\times x^p \exp\left\{-\frac{ax^4}{4} - \frac{bx^2}{2}\right\}. \qquad (1.4.26b)$$

Using the oscillator theorem, it is very easy to see that for $p = 0$ formulas (1.4.26) describe the ground state and the second excitation, while for $p = 1$ they describe the first and third exitations.

Similarly, explicit solutions involving radicals can also be written for $M = 2$ and $M = 3$. Note, however, that for larger values of M this is possible only after solving a certain algebraic equation for E of order $M+1$. The reason is that for given M, the solutions belong to the $(M+1)$-dimensional invariant subspace Φ_M^p, and this implies that the corresponding secular equation is an algebraic equation of order $M + 1$.

It is not difficult•to understand that, in general, the procedure described above gives us the energy levels with numbers $0, 2, \ldots, 2M$ for $p = 0$ and $1, 3, \ldots, 2M+1$ for $p = 1$.

In order to prove this important assertion, consider the *ansatz* (1.4.16) for given M and p. We see that the wavefunctions are proportional to polynomials of order $2M + p$ and, therefore, they have exactly $2M + p$ zeros in the complex x-plane. Due to the evenness of these polynomials, the number of real zeros (playing the role of the wavefunction nodes) is $2K+p$, where K depends on the sort of solution and satisfies the inequality $0 \le K \le M$. We know that the different wavefunctions must have different number of nodes. In our case the number of different wavefunctions is $M + 1$ since all $M + 1$ exact solutions belonging to the class (1.4.16) are normalizable and satisfy the needed boundary conditions. But this means that the number K must take $M+1$ different values, which is possible only if $K = 0, 1, \ldots, M$. Using the oscillator theorem, we can conclude that the numbers of states are described by the formula $2K+p$ with $K = 0, 1, \ldots, M$, and this proves the assertion.

Note that the algebraic equations, from which the exactly calculable energy levels and corresponding wavefunctions can be obtained, take an especially simple and compact form if we rewrite them in terms of wavefunction zeros. Let $\pm\sqrt{\xi_1}, \ldots, \pm\sqrt{2\xi_M}$ be the zeros of the polynomial $P_M(x^2)$. Then the *ansatz* (1.4.16) can be rewritten as

$$\psi(x) = \prod_{i=1}^{M} \left(\frac{x^2}{2} - \xi_i\right) x^p \exp\left\{-\frac{ax^4}{4} - \frac{bx^2}{2}\right\} \qquad (1.4.27)$$

and the spectral problem is reduced to the determination of the numbers $\xi_1, \ldots, \xi_M$ (Ushveridze 1988d, 1989c).

Substituting (1.4.27) into the Schrödinger equation for (1.4.13), we obtain the condition

$$\sum_{i=1}^{M} \frac{1}{\frac{x^2}{2} - \xi_i} \left\{ \sum_{k=1}^{M} \frac{4\xi_i}{\xi_i - \xi_k} + 2p + 1 - 4b\xi_i - 8a\xi_i^2 \right\}$$

$$+ E - (4M + 2p + 1)b - 8a \sum_{i=1}^{M} \xi_i = 0, \qquad (1.4.28)$$

which can be satisfied if and only if the coefficients of the singular terms vanish and the free term is equal to zero. This gives us the following simple expression for the energy:

$$E = (4M + 2p + 1)b + 8a \sum_{i=1}^{M} \xi_i, \qquad (1.4.29)$$

in which $\xi_i, i = 1, \ldots, M$ are numbers satisfying the system of numerical equations

$$\sum_{k=1}^{M} \frac{1}{\xi_i - \xi_k} + \frac{2p+1}{4} \frac{1}{\xi_i} - b - 2a\xi_i = 0, \quad i = 1, \ldots, M. \qquad (1.4.30)$$

Formulas (1.4.27), (1.4.29) and (1.4.30), describing the solutions of the quasi-exactly solvable model (1.4.13), will play a determining role in our consideration (see, for example, sections 1.10 and 1.11).

Thus, we have completed the exposition of our approach to the problem of quasi-exact solvability of the sextic anharmonic oscillator. An interesting feature of this approach is that by its very essence it contains a possibility of generalization: the method developed in it can be used for constructing wide classes of other more complex quasi-exactly solvable models.

In order to demonstrate this fact we use the same language as in section 1.2. Consider the Heisenberg algebra $\mathcal{H}$ and its universal enveloping algebra $\mathcal{U}(\mathcal{H})$. Denote by $\mathcal{U}_k(\mathcal{H})$ the linear subspaces of $\mathcal{U}(\mathcal{H})$ formed by all linear combinations of monomials $(a^+)^l(a^-)^n$ satisfying the constraints $l \leq n + k$. The elements u_k of these subspaces are characterized by the property $u_k \Phi_m \subset \Phi_{m+k}$.

Now, consider the class of operators

$$H = u_0 + u_1(a^+a^- - M) + u_2(a^+a^- - M + 1)(a^+a^- - M) + \ldots$$
$$+ u_k(a^+a^- - M + k - 1)\ldots(a^+a^- - M),$$

$$(1.4.31)$$

in which M and k are given non-negative integers, and $u_0, u_1, u_2, \ldots, u_k$ are certain arbitrarily chosen elements of the spaces $\mathcal{U}_0(\mathcal{H})$, $\mathcal{U}_1(\mathcal{H})$, $\mathcal{U}_2(\mathcal{H}), \ldots, \mathcal{U}_k(\mathcal{H})$. Obviously, all these operators belong to the space $\mathcal{U}_k(\mathcal{H})$. At first glance, this gives us the possibility of writing $H\Phi_m \subset \Phi_{m+k}$. However, this formula is true only if $m \neq M$. Taking $m = M$ we see that operators in brackets cancel the leading basis elements of the space Φ_M ensuring its invariance under the action of H:

$$H\Phi_M \subset \Phi_M. \tag{1.4.32}$$

This enables us to treat the operators (1.4.31) as "hamiltonians" of abstract quasi-exactly solvable models.

Of course, the condition for these models to be physically meaningful restricts the class of admissible "hamiltonians" H. We know that they must be second-order differential operators. Using differential realizations of the generators of the Heisenberg algebra given in formula (1.2.22), one can see that this condition can be satisfied if the degrees of the lowering operator a^- in (1.4.31) do not exceed two. In this case the most general expression for H is

$$H = u_0 + u_1(a^+a^- - M) + u_2(a^+a^- - M + 1)(a^+a^- - M), \tag{1.4.33}$$

where

$$u_0 = A_0(a^+)^2(a^-)^2 + B_0 a^+(a^-)^2 + C_0(a^-)^2 + D_0 a^+ a^- + E_0 a^- + F_0,$$
$$u_1 = \Lambda_1(a^+)^2 a^- + B_1 a^+ a^- + C_1 a^- + D_1 a^+ + E_1,$$
$$u_2 = A_2(a^+)^2 + B_2 a^+ + C_2. \tag{1.4.34}$$

Substituting (1.2.22) into (1.4.33) and (1.4.34) we obtain second-order differential operators of the type (1.2.24), which, after imposing the necessary constraints on functions $A(x)$ and $B(x)$ (see formulas (1.2.22) and (1.2.25)), are reduced to the Schrödinger form. This gives us a rather wide class of one-dimensional quasi-exactly solvable models with potentials expressed in terms of rational, trigonometric, hyperbolic and elliptic functions (Ushveridze 1988c). All these models will be discussed in detail in chapter 4.

1.5 Non-perturbative effects in an explicit form and convergent perturbation theory

In the preceding section we have demonstrated that the sextic anharmonic oscillator with the potential

$$V(x) = b^2 x^2 + a[2bx^4 - 4(M + 2p + 3)x^2] + a^2 x^6 \tag{1.5.1}$$

is quasi-exactly solvable, and has, for any given M and p, $M+1$ exact (algebraic) solutions with the numbers $2K+p$ where $K = 0, 1, \ldots, M$.

Now we discuss some physical properties of the model (1.5.1). It is easily seen that it can be viewed as a perturbed harmonic oscillator with the potential

$$V_0(x) = b^2 x^2. \tag{1.5.2}$$

Here the role of perturbation parameter is played by a. Since the perturbation theory in the parameter a is well defined and easy to construct, we can compare the perturbative results with the exact ones. Such a comparison reveals the presence of a number of non-perturbative effects in the model (1.5.1), i.e., effects which do not appear in any finite order of perturbation theory.

First of all, let us establish the rough boundary between the weak- and strong-coupling regimes. For this we remember that the most essential values of x in the model (1.5.2) are

$$x \sim |b|^{-\frac{1}{2}}. \tag{1.5.3}$$

Comparing the perturbation terms in (1.5.1) with the unperturbed term for these values of x, we find that the effective coupling constant is ab^{-2}. Therefore, for $ab^{-2} \ll 1$ we shall have the weak-coupling regime, and for $ab^{-2} \gtrsim 1$ the strong-coupling regime.

The form of the potential (1.5.1) in the weak-coupling regime strongly depends on the sign of parameter b. If b is positive, the potential (1.5.1) has qualitatively the same form as the harmonic potential (1.5.2) (see figure 1.4) but for negative values of b the form of the potential (1.5.1) becomes more complicated (see figure 1.5). We see that it has now two additional minima lying lower than the minimum at zero. This means that the probability of finding the quantum particle near these minima is larger than the probability of finding it in the vicinity of the origin. Therefore, the physical situation described by the potential $V(x)$ for $b < 0$ differs from the situation described by the harmonic potential $V_0(x)$, irrespective of how weak the coupling is.

In order to see this more explicitly, let us consider the exact ground state wavefunction for the model (1.5.1) found in the preceding section for $M = 0$

$$\psi(x) = \begin{cases} \exp\left(-\frac{ax^4}{4} - \frac{|b|x^2}{2}\right), & b > 0 \\ \exp\left(-\frac{ax^4}{4} + \frac{|b|x^2}{2}\right), & b < 0 \end{cases} \tag{1.5.4}$$

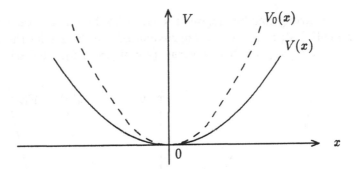

Figure 1.4. The form of the potential (1.5.1) when $b > 0$ and $ab^{-2} < (4M + 2p + 3)^{-1}$.

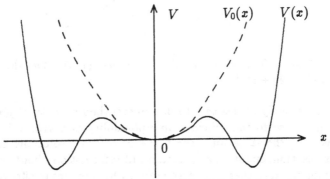

Figure 1.5. The form of the potential (1.5.1) when $b < 0$ and $ab^{-2} < (4M + 2p + 3)^{-1}$.

and compare it with the ground state wavefunction for the harmonic oscillator

$$\psi_0(x) = \exp\left(-\frac{|b|x^2}{2}\right). \tag{1.5.5}$$

We see that in the first case ($b > 0$) the wavefunctions of both models coincide qualitatively with each other: they have the maximum at the origin. In the second case ($b < 0$), the difference between the wavefunctions becomes large: the function $\psi(x)$ has the minimum at the origin, but not the maximum. At the same time it has two maxima lying at the points $x_\pm = \pm\sqrt{\frac{|b|}{a}}$. It is evident that this cannot be seen in any finite order of perturbation theory and, therefore, we have a typical non-perturbative effect.

In the strong-coupling regime the situation becomes more non-trivial. The potential $V(x)$ cannot be approximated any longer by the potential $V_0(x)$, even in a perturbative sense (see figure 1.6). Indeed, the first

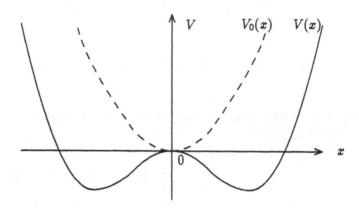

Figure 1.6.   The form of the potential (1.5.1) for the case when $ab^{-2} > (4M + 2p + 3)^{-1}$.

correction already makes the stable unperturbed potential $V_0(x)$ unstable, which gives rise to the principal inapplicability of perturbation theory in this case. At the same time, the presence of exact solutions, found in the previous section, enables us to obtain full information about the system.

Unfortunately, there is one exception which relates to effects of splitting the energy levels in the Z_2-symmetric double-well potential depicted in figure 1.6. This very important non-perturbative effect cannot be studied by means of the exact solutions for the model (1.5.1). The fact is that the levels to be split have opposite parities, while the exact solutions for the model under consideration have the same parities by construction.

Let us now study the behaviour of large-order terms of the perturbation series for the model (1.5.1). Remember that the problem of studying the asymptotics of perturbation expansions was popular in the middle of the seventies, when many people believed that knowledge of such asymptotics could help in constructing the correct summation procedure for a diverging series. At that time, many models with polynomial potentials were studied from this point of view (see e.g. Graffi *et al* 1970, Turchetti 1971, Lipatov 1976, Popov *et al* 1977, 1978, Graffi and Grecchi 1978, Kazakov and Shirkov 1980). The results were one and the same: the perturbation series diverge factorially. The simplest example of such a divergence was considered in section 1.3 for the case of the quartic anharmonic oscillator.

Now consider the situation with the sextic anharmonic oscillator with the potential (1.5.1). This model is especially interesting for us since we have an unique possibility of comparing the results obtained by means of approximative (asymptotic) methods with exact results which follow from the explicit solutions for the model (1.5.1).

First of all, note that Dyson's instability argument does not work for the model (1.5.1) since the change of sign of perturbation parameter a does not change the behaviour of the potential $V(x)$ at infinity: the stable potential remains stable. To study the large-order behaviour of the perturbation series we use other reasonings based on the representation of the (ground state) energy level via a functional integral:

$$E(a) = -\frac{1}{\int dT} \ln \left\{ \int \mathcal{D} \, x(\tau) \exp \left[-\int H(a,\tau) \, d\tau \right] \right\}. \qquad (1.5.6)$$

Here

$$
\begin{aligned}
H(a,\tau) \;=\;& \dot{x}^2(\tau) + b^2 x^2(\tau) + a[2bx^4(\tau) - (4M + 2p + 3)x^4(\tau)] \\
& + a^2 x^6(\tau)
\end{aligned} \qquad (1.5.7)
$$

is a classical hamiltonian of the particle.

Expanding $E(a)$ in powers of a:

$$E(a) = \sum_{N=0}^{\infty} a^N E_N, \qquad (1.5.8)$$

we see that the Nth term of this expansion can be written as

$$E_N = \sum_{n+2m=N} (-1)^{n+m} A_{nm} \qquad (1.5.9)$$

where

$$
\begin{aligned}
A_{nm} \;=\;& \frac{1}{n!m!} \int \mathcal{D} \, x(\tau) \exp \left\{ -\int [\dot{x}^2(\tau) + b^2 x^2(\tau)] \, d\tau \right. \\
& + n \ln \int [2bx^4(\tau) - (4M + 2p + 3)x^2(\tau)] \, d\tau \\
& \left. + m \ln \int x^6(\tau) \, d\tau \right\}. \qquad (1.5.10)
\end{aligned}
$$

For large n and m these functional integrals can be computed by means of the saddle-point method. The details of such computation are well known (see e.g. Bogomolny *et al* 1980) and we shall not repeat them here. Note only that the classical saddle-point trajectory, on which the subintegral expression has a maximum, exists and, therefore, the most relevant values of $x^2(\tau)$ are of the order

$$x^2(\tau) \sim 2n + 3m. \qquad (1.5.11)$$

This leads us to the following approximate expression for A_{nm}:

$$A_{nm} \approx \frac{(2n + 3m)!}{n!m!} \approx N! \qquad (1.5.12)$$

Substituting (1.5.12) into (1.5.9) we see that the Nth term of this expansion is of order $N!$.

Of course, the presence of the factor $(-1)^{n+m}$ with alternate sign may, in principle, cancel the leading contributions of A_{nm} values to the sum (1.5.9). However, the total cancellation seems to be absolutely unrealistic. Miracles do not appear in physics without deep reasons. Therefore, we have come to think that the perturbation series for the model (1.5.1) grows factorially and has (as usual) zero radius of convergence.

Surprising though it is, this conclusion is wrong! As was demonstrated by Ushveridze (1987b, 1988k), the sums (1.5.9) have a standard factorial behaviour only if $M \neq 0, 1, 2, \ldots$. If M is an arbitrary non-negative integer, the amazing cancellation of growing contributions occurs and the series becomes convergent!

At first sight this fact (which is easily confirmed by direct numerical calculation of the perturbation terms) is very paradoxical, since until now there were no examples of models with polynomial potentials allowing a convergent perturbation theory. However, the fact remains and the best way to make sure that it reflects reality is to look at the explicit solutions for the model (1.5.1).

Consider, for example, the solution (1.4.26) corresponding to the case $M = 1$. We see that the energy level E_- (the ground state) considered as an analytic function of the parameter a is regular at the origin and, therefore, its expansion in powers of a must have a finite radius of convergence. The complex a-plane for the level $E_-(a)$ is depicted in figure 1.7. We see that the radius of convergence, defined in usual way as the distance to the nearest singularity, is equal to $\frac{b^2}{4p+2}$. An analogous result can also be obtained also for other non-negative integer values of M for states belonging to exactly calculable parts of the spectrum. At the same time, the energy levels lying beyond these parts have absolutely standard factorially divergent perturbative expansions (Ushveridze 1989c).

Now let us return to figure 1.7. What is the physical meaning of the singular point a_S lying on the left half axis Re a? From expression (1.4.26a)

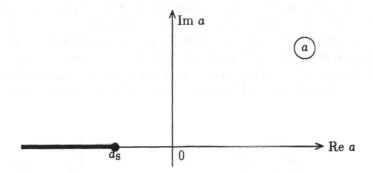

Figure 1.7. Complex a-plane for the function $E_-(\beta)$ defined by the expression (1.4.26). Here $a_S = -(4p+2)^{-1}b^2$.

it is clearly seen that at this point the levels E_- and E_+ (the ground and the second excited states) coincide. At first glance, this coincidence contradicts the familiar theorem about the non-degeneracy of spectra of one-dimensional quantum mechanical systems (Landau and Lifshitz 1977). However, from the second formula (1.4.26b) we can see that, at this crossing point, the corresponding wavefunctions $\psi_-(x)$ and $\psi_+(x)$ also coincide and therefore the geometrical multiplicity of the degenerate eigenvalue remains equal to unity. Thus, there is no contradiction. Note also that the negative values of a are unphysical ones since in this domain the wavefunctions $\psi_\pm(x)$ are not normalizable. From (1.4.26a) it also follows that at the point a_S the functions $E_-(a)$ and $E_+(a)$ have a branch-point singularity of square root type. The corresponding cut is depicted in figure 1.7. This means that by analytic continuation of the level E_- along the closed contour passing round the point a_S, it is possible to obtain the level E_+ and vice versa. In other words, the levels E_- and E_+ are plaited and form a common two-sheeted Riemann surface. Summarizing, we can conclude that the point a_S is none other than the ordinary Bender and Wu singularity (Turbiner and Ushveridze 1987).

Thus, we have obtained an exact and very elegant result which for other models of such a sort could be obtained only after serious technical efforts and, even then, no more than approximately. As is well known, the problem of studying the Bender and Wu singularities is one of the most complex problems in quantum mechanics.

Note that the Bender and Wu singularity for the levels $E_\pm$ in the model (1.5.1) lies outside the region where the perturbation theory is applicable. Therefore, it is a typical strong-coupling non-perturbative effect.

Another non-perturbative effect which is also present in the exact solution (1.4.26a) is related to the behaviour of the energy levels $E_\pm$ as functions of the parameter b when a is small. Here, if one looks at a poor-resolution graph of the two functions $E_\pm(b)$, one receives the impression that the levels intersect and, doing so, exchange quantum numbers. With better resolution it becomes clear that there is no such intersection (see figure 1.8). This effect is well known in nuclear physics, where it is referred

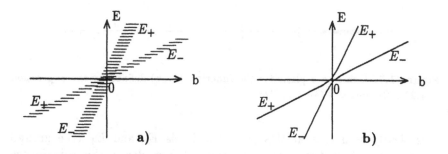

Figure 1.8. The graph of the function $E_\pm(b)$ with a) poor resolution and b) good resolution.

to as the quasi-intersection of levels (see e.g. Solovyev 1981). It cannot be seen in perturbation theory, since its characteristic scale is determined by the relation $b^2 \sim a$. The existence of exact solutions not only allows us to see it, but also gives a simple explanation of it: the quasi-intersection of levels is a manifestation of their actual intersection (plaiting) at a point lying close to the real axis (in the complex plane of the parameter of the problem) (Ushveridze 1988a, 1989c).

Quasi-exactly solvable models possess another rather curious feature. They can be considered as good approximations to exactly non-solvable models. As an example, we consider again the model (1.5.1) and recall that $M + 1$ is the order of the secular equation which determines the algebraic part of the spectrum. If $M = \infty$, the problem of finding the spectrum ceases to be algebraic, and the equation becomes exactly non-solvable. Therefore, to obtain an exactly non-solvable model from (1.5.1), M must be taken to infinity. For the potential to remain finite in this limit, parameters a and b must be dependent on M. For example, if this dependence is determined by the condition

$$b^2 - a(4M + 2p + 3) = \alpha, \qquad (1.5.13)$$

then, for $M \to \infty$, the coefficient of the sextic term in the potential (1.5.1) vanishes as $M^{-2/3}$ and we obtain the exactly non-solvable model of the anharmonic oscillator with potential (1.3.1) (Ushveridze 1988g, 1989c).

Finally, we note that model (1.5.1) can be considered also as a zeroth-order approximation by constructing various perturbative schemes. This assertion refutes the usual opinion that any perturbation theory requires knowledge of the entire spectrum of the non-perturbed hamiltonian. However, such a requirement is necessary only when we work in the energy representation. In this case we must know all energy levels of a non-perturbed Schrödinger equation to construct the needed Green function. If we deal with the coordinate representation it is sufficient to know only the energy level for which we seek the correction. Indeed, according to the well known theorem of mathematical analysis, the knowledge of one (normalizable) solution of the one-dimensional Schrödinger-type equation allows us to build a second, linearly independent solution with the same energy. Now, having two linearly independent solutions one can easily restore the coordinate Green function, the knowledge of which is sufficient to write down explicitly (in quadratures) any term of a perturbation theory for the level under consideration. In case of model (1.5.1) we know explicitly a certain limited part of the spectrum. This gives us the possibility of constructing perturbation theory for all the levels belonging to this part. From a practical point of view the most convenient way to realize such a scheme is to use the so-called "Price–Zeldovich technique", in which instead of the Schrödinger equation, its delinearized Riccatian version is considered (see e.g. Price 1954, Zeldovich 1956, Turbiner 1984).

In order to demonstrate how this (non-standard) perturbation theory works, let us consider the model (1.5.1) for $M = 0$ and $p = 0$, perturbed by the potential $\lambda V_1(x)$ where λ is a perturbation parameter and $V_1(x)$ an arbitrary fixed function. The corresponding Schrödinger equation has the form:

$$\left\{-\frac{\partial^2}{\partial x^2} + (b^2 - 3a)x^2 + 2abx^4 + a^2x^6 + \lambda V_1(x)\right\} \psi(x) = E\psi(x).$$

$$(1.5.14)$$

In this case we know only one solution of the unperturbed problem, which is the ground state with energy $E_0 = b$ and the wavefunction $\psi_0(x) = \exp\left(-\frac{ax^4}{4} - \frac{bx^2}{2}\right)$. Let us now show that this information is quite sufficient to construct the perturbation theory for the ground state in the model (1.5.1).

First of all, we introduce the logarithmic derivative of the wavefunction

$$y(x) = \frac{\psi'(x)}{\psi(x)} \tag{1.5.15}$$

and rewrite equation (1.5.14) in the Riccati form:

$$y'(x) + y^2(x) = (b^2 - 3a)x^2 + 2abx^4 + a^2x^6 + \lambda V_1(x) - E. \quad (1.5.16)$$

Taking

$$E = E_0 + \lambda E_1 + \lambda^2 E_2 + \ldots \quad (1.5.17a)$$

and

$$y(x) = y_0(x) + \lambda y_1(x) + \lambda^2 y_2(x) + \ldots \quad (1.5.17b)$$

(where by $y_0(x)$ we have denoted the unperturbed logarithmic derivative $\psi_0'(x)/\psi_0(x)$) and substituting the expressions (1.5.17) into (1.5.16), we obtain the equations

$$\frac{\partial}{\partial x}\left(\psi_0^2(x)y_n(x)\right) = (V_n(x) - E_n)\,\psi_0^2(x), \quad n = 1, 2, \ldots, \quad (1.5.18)$$

in which

$$V_n(x) = -\sum_{i=1}^{n-1} y_i(x)y_{n-i}(x), \quad \text{for } n \geq 2. \quad (1.5.19)$$

The integration of (1.5.18) from $-\infty$ to $+\infty$ and x gives

$$E_n = \frac{\int\limits_{-\infty}^{\infty} \psi_0^2(x)V_n(x)\,\mathrm{d}x}{\int\limits_{-\infty}^{\infty} \psi_0^2(x)\,\mathrm{d}x} \quad (1.5.20)$$

and

$$y_n(x) = \psi_0^{-2}(x)\int\limits_{-\infty}^{x} [V_n(x') - E_n]\,\psi_0^2(x')\,\mathrm{d}x'. \quad (1.5.21)$$

Taking in formulas (1.5.20) and (1.5.21) $N = 1, 2, \ldots$, we obtain a simple recurrence procedure for determination of all perturbative corrections for E_0 and $\psi_0(x)$.

Note that, according to formula (1.5.20), the first correction for the energy in this non-standard perturbation scheme,

$$E_1 = \frac{\int\limits_{-\infty}^{\infty} V_1(x)\psi_0^2(x)\,\mathrm{d}x}{\int\limits_{-\infty}^{\infty} \psi_0^2(x)\,\mathrm{d}x}, \quad (1.5.22)$$

coincides with the analogous correction in the Rayleigh–Schrödinger perturbation theory (Turbiner 1984).

Concluding this brief review of some properties of model (1.5.1), one can say that it occupies an intermediate position between the exactly solvable models and exactly non-solvable ones, having the features of both the former and the latter simultaneously.

However, the theme is far from being exhausted. In the next sections we show that model (1.5.1) has many other intriguing properties, the study of which leads us to a better understanding of the nature of quasi-exact solvability in quantum mechanics.

1.6 Partial algebraization of the spectral problem. Hidden dynamical symmetries

Let us consider the quasi-exactly solvable model of the sextic anharmonic oscillator discussed in the two previous sections. It is not difficult to verify that its hamiltonian (1.4.13) can be rewritten in the form:

$$H = -4S^0 S^- + 4aS^+ + 4bS^0 - (2M + 4p + 2)S^- + (2M + 2p + 1)b,$$

$$(1.6.1)$$

where S^-, S^0 and S^+ are the operators

$$
\begin{aligned}
S^- &= \frac{1}{2x}\left(\frac{\partial}{\partial x} + ax^3 + bx - \frac{p}{x}\right), \\
S^0 &= \frac{x}{2}\left(\frac{\partial}{\partial x} + ax^3 + bx - \frac{p}{x}\right) - \frac{M}{2}, \\
S^+ &= \frac{x^3}{2}\left(\frac{\partial}{\partial x} + ax^3 + bx - \frac{p}{x}\right) - Mx^2,
\end{aligned}
\qquad (1.6.2)
$$

satisfying the following commutation relations

$$[S^-, S^+] = 2S^0, \quad [S^0, S^\pm] = \pm S^\pm, \qquad (1.6.3)$$

and, thus, forming the $sl(2)$-algebra.

Since we have explicit expressions for the operators $S^\pm$ and S^0 (generators of the algebra $sl(2)$) it is not difficult to construct the representation space in which they act.

First of all, remember that representations of the algebra $sl(2)$ can be defined by the formulas

$$S^-|0\rangle = 0 \qquad (1.6.4)$$

and

$$S^0|0\rangle = -j|0\rangle, \qquad (1.6.5)$$

in which $|0\rangle$ is the so-called lowest-weight vector and $(-j)$ is the corresponding lowest weight. The representation space, which we denote by Φ_j, is formed by the vectors

$$|n\rangle = (S^+)^n|0\rangle, \quad n = 0, 1, 2, \ldots. \qquad (1.6.6)$$

It is known that the dimension of this representation is finite if and only if the lowest weight is a non-positive semi-integer number:

$$j = 0, \frac{1}{2}, 1, \frac{3}{2}, \ldots. \qquad (1.6.7)$$

In this case there exists the highest-weight vector $|2j\rangle$ which is annihilated by the raising operator S^+:

$$S^+|2j\rangle = 0. \qquad (1.6.8)$$

The corresponding highest weight is $(+j)$:

$$S^0|2j\rangle = +j|2j\rangle. \qquad (1.6.9)$$

Then we have

$$\dim \Phi_j = 2j + 1. \qquad (1.6.10)$$

In all other cases the representation space Φ_j is infinite dimensional.

Now it is clear that to construct the representation space for the operators (1.6.2) we must find the lowest-weight vector and then determine the lowest weight.

Substituting the explicit expression (1.6.2) for the generator S^- into (1.6.4) and solving this equation we obtain

$$|0\rangle = x^p \exp\left\{-\frac{ax^4}{4} - \frac{bx^2}{2}\right\}. \qquad (1.6.11)$$

Substituting the found expression for $|0\rangle$ into (1.6.5) we find

$$j = \frac{M}{2}. \qquad (1.6.12)$$

Since the number j is found to be semi-integer, the representation of the algebra $sl(2)$ with the generators (1.6.2) is finite dimensional. Comparing (1.6.10) with (1.6.12) we obtain:

$$\dim \Phi_j = M + 1. \qquad (1.6.13)$$

Acting by the operators $(S^+)^n$, $n = 0, \ldots, 2j = M$ on the function (1.6.11) we obtain the basis vectors

$$|n\rangle = x^{2n+p} \exp\left\{-\frac{ax^4}{4} - \frac{bx^2}{2}\right\}, \quad n = 0, 1, \ldots, M. \qquad (1.6.14)$$

We see that the highest-weight vector $|M\rangle$ is actually annihilated by the raising operator S^+ defined by formula (1.6.2).

Thus, we have constructed the representation space for generators (1.6.3), the most general element of which can be written as

$$\psi(x) = x^p P_M(x^2) \exp\left\{-\frac{ax^4}{4} - \frac{bx^2}{2}\right\}. \qquad (1.6.15)$$

We see that this representation space coincides with the space (1.4.16) in which the Schrödinger equation for the hamiltonian (1.4.13) has exact solutions.

This enables us to give a simple explanation of the phenomenon of quasi-exact solvability of the model (1.4.13): this model is quasi-exactly solvable, since its hamiltonian is expressed in terms of generators of a certain finite-dimensional representation of the algebra $sl(2)$. In this case, the corresponding representation space Φ is automatically invariant under the action of the hamiltonian H. Therefore the spectral problem for H in Φ is formulated correctly and can be solved algebraically. The number of algebraic solutions of the Schrödinger equation is equal to a dimension of the representation, which in the case of the model (1.4.13) is $M + 1$. Since the algebra $sl(2)$ generates the spectrum of the hamiltonian, it plays the role of a hidden dynamical symmetry for the model (1.4.13).

From the physical point of view, formula (1.6.1) means that the hamiltonian of the sextic anharmonic oscillator is equivalent to a spherically non-symmetric quantum top in an external magnetic field.

Finally, note that the explanation of the phenomenon of quasi-exact solvability given above enables us to formulate a rather general method of constructing quasi-exactly solvable problems of quantum mechanics.

Indeed, remember that any $(M + 1)$-dimensional representation of the algebra $sl(2)$ can be realized in the space of Mth-order polynomials $P_M(t)$ (see e.g. Zhelobenko 1965). The corresponding generators S^+, S^0, S^- have in this case the form of the first-order linear differential operators:

$$S^- = \frac{\partial}{\partial t}, \quad S^0 = t\frac{\partial}{\partial t} - \frac{M}{2}, \quad S^+ = t^2\frac{\partial}{\partial t} - Mt. \qquad (1.6.16)$$

Consider the set of arbitrary second-order polynomials in generators (1.6.16):

$$H = C_1 S^+ S^+ + C_2 S^+ S^0 + C_3 S^+ S^-$$

$$+C_4 S^0 S^0 + C_5 S^0 S^- + C_6 S^- S^-$$
$$+C_7 S^+ + C_8 S^0 + C_9 S^-. \tag{1.6.17}$$

The substitution of (1.6.16) into (1.6.17) gives us a class of second-order linear differential operators

$$H = P_4(t)\frac{\partial^2}{\partial t^2} + Q_3(t)\frac{\partial}{\partial t} + R_2(t), \tag{1.6.18}$$

in which $P_4(t)$, $Q_3(t)$ and $R_2(t)$ are certain polynomials of orders four, three and two, respectively. Since these operators act in an $(M+1)$-dimensional space, the spectral problem for them can be solved algebraically. This leads us to a set of second-order linear differential equations of the spectral type having a finite number of exact polynomial solutions.

Generally speaking, the equations obtained are not Schrödinger-type equations and their polynomial solutions are not normalizable. Note however that we always can introduce new variables and make homogeneous transformations of the unknown function. These transformations conserve the linearity and order of the differential equation, but change its concrete form and the normalization properties of its solutions. Therefore they may, in principle, reduce the obtained equations to Schrödinger form. In this case we obtain a class of quasi-exactly solvable models.

In order to demonstrate how this scheme works, let us substitute the generators (1.6.16) into the spectral equation for (1.6.1) and obtain the quasi-exactly solvable model (1.4.13).

The substitution gives the equation:

$$\left\{ -4t\frac{\partial^2}{\partial t^2} + (4at^2 + 4bt - 4p - 2)\frac{\partial}{\partial t} - 4aMt + (2p+1)b \right\} P_M(t)$$
$$= E P_M(t) \tag{1.6.19}$$

which, after introducing a new variable x and a new function $\psi(x)$ by the formulas

$$x = x(t), \tag{1.6.20a}$$
$$\psi(x) = g(t) P_M(t), \tag{1.6.20b}$$

takes the form:

$$\left\{ -4t[x'(t)]^2 \frac{\partial^2}{\partial x^2} + \left[\left(8t\frac{g'(t)}{g(t)} + 4at^2 + 4bt - 4p - 2 \right) x'(t) - 4tx''(t) \right] \frac{\partial}{\partial x} \right.$$
$$+ 4t \left[\frac{g'(t)}{g(t)} \right]' - 4t \left[\frac{g'(t)}{g(t)} \right]^2 - (4at^2 + 4bt - 4p - 2)\frac{g'(t)}{g(t)}$$
$$\left. - 4aMt + (2p+1)b \right\} \psi(x) = E\psi(x). \tag{1.6.21}$$

For equation (1.6.21) to have Schrödinger form, the coefficient of the second derivative must be equal to -1 and the coefficient of the first derivative must vanish. This gives us two equations:

$$4t[x'(t)]^2 = 1 \qquad (1.6.22a)$$

and

$$\left(8t\frac{g'(t)}{g(t)} + 4at^2 + 4bt - 4p - 2\right) x'(t) - 4tx''(t) = 0 \qquad (1.6.22b)$$

with respect to $x(t)$ and $g(t)$. Solving these equations we obtain

$$x(t) = \sqrt{t}, \quad g(t) = t^{\frac{z}{3}} \exp\left\{-\frac{a}{4}t^2 - \frac{b}{2}t\right\}. \qquad (1.6.23)$$

Substituting (1.6.23) into (1.6.21) we obtain the Schrödinger equation with the potential described by formula (1.5.1) and with the solution having the form (1.4.16). This completes the reduction procedure.

Analogously, starting with other bilinear combinations (1.6.17) of the generators (1.6.16) and performing the reduction procedure, one can obtain other one-dimensional quasi-exactly solvable models with potentials expressed via the rational, trigonometric, hyperbolic and elliptic functions.

This method was proposed by Turbiner (1988a) for the one-dimensional case and then generalized by Shifman and Turbiner (1989) to the multi-dimensional case. It is known as the method of partial algebraization (see e.g. Shifman 1989a, Morozov *et al* 1990). In sections 1.16 and 5.1 and appendix A we discuss this method in more detail.

In conclusion, note that the method of partial algebraization leads to the same quasi-exactly solvable models as the method of raising and lowering operators discussed in section 1.4. This follows from the fact that the generators of the algebra $sl(2)$ are easily expressed via the generators of the Heisenberg algebra:

$$S^- = a^-, \quad S^0 = a^+a^- - M/2, \quad S^+ = (a^+)^2a^- - Ma^+, \qquad (1.6.24)$$

and their substitution into (1.6.17) reduces the operators H to the form (1.4.33)–(1.4.34).

1.7   The two-dimensional harmonic oscillator. The separation of variables

In this section we discuss a striking connection between the one-dimensional quasi-exactly solvable model of the sextic anharmonic oscillator, and the exactly solvable model of a two-dimensional harmonic oscillator.

In order to establish this connection, let us consider two identical Schrödinger equations for (1.4.13), rewritten in terms of two different variables x and y:

$$\left\{-\frac{\partial^2}{\partial x^2} + a^2 x^6 + 2abx^4 + [b^2 - a(4M + 2p + 3)]x^2\right\} \psi(x) = E\psi(x),$$

(1.7.1a)

$$\left\{-\frac{\partial^2}{\partial y^2} + a^2 y^6 + 2aby^4 + [b^2 - a(4M + 2p + 3)]y^2\right\} \psi(y) = E\psi(y).$$

(1.7.1b)

Remember that the eigenvalues E and the corresponding eigenfunctions $\psi(x)$ are determined by formulas (1.4.29) and (1.4.27).

Multiplying (1.7.1a) by $\psi(y)$ and (1.7.1b) by $\psi(x)$ and subtracting the results from each other we obtain a single two-dimensional equation

$$\left\{-\frac{\partial^2}{\partial x^2} + \frac{\partial^2}{\partial y^2} \quad + \quad a^2(x^6 - y^6) + 2ab(x^4 - y^4)\right.$$

$$\left. + \quad [b^2 - a(4M + 2p + 3)](x^2 - y^2)\right\}\psi(x)\psi(y) = 0,$$

(1.7.2)

which does not contain the spectral parameter E any longer. Nevertheless, we can interpret (1.7.2) as a spectral equation again, identifying the quantity $b^2 - a(4M + 2p + 3)$ with a new spectral parameter Γ. Dividing (1.7.2) by $x^2 - y^2$ we obtain

$$\left\{-\frac{\frac{\partial^2}{\partial x^2} - \frac{\partial^2}{\partial y^2}}{x^2 - y^2} + a^2(x^4 + y^4 + x^2 y^2) + 2ab(x^2 y^2) + b^2\right\}\psi(x,y) = \Gamma\psi(x,y).$$

(1.7.3)

Evidently, the equation (1.7.3) obtained is exactly solvable by construction. It has an infinite number of exact eigenvalues

$$\Gamma = b^2 - a(4M + 2p + 3)$$

(1.7.4)

and corresponding eigenfunctions

$$\psi(x,y) = (xy)^p \prod_{i=1}^{M} \left(\frac{x^2 y^2}{4} - \frac{x^2 + y^2}{2}\xi_i + \xi_i^2\right)$$

$$\times \exp\left\{-\frac{a(x^4 + y^4)}{4} - \frac{b(x^2 + y^2)}{2}\right\},$$

(1.7.5)

where ξ_i, $i = 1, \ldots, M$ are numbers satisfying equation (1.4.30). We see that the spectrum (1.7.4) is degenerate: for the Mth eigenvalue we have $M + 1$ different linearly independent solutions $\psi(x, y)$.

Equation (1.7.3) and its solutions (1.7.5) can be simplified if we introduce new variables

$$\lambda = \frac{b}{2a} + \frac{x^2 + y^2}{2}, \qquad \mu = ixy, \tag{1.7.6}$$

and new functions

$$\phi(\lambda, \mu) = \psi(x, y). \tag{1.7.7}$$

Then (1.7.3) takes the form

$$\left\{ -\frac{\partial^2}{\partial \lambda^2} - \frac{\partial^2}{\partial \mu^2} + a^2(4\lambda^2 + \mu^2) \right\} \phi(\lambda, \mu) = \Gamma \phi(\lambda, \mu), \tag{1.7.8}$$

where Γ is defined by (1.7.4) and

$$\phi(\lambda, \mu) = \mu^p \prod_{i=1}^{M} \left[\frac{\mu^2}{4} + \left(\lambda - \frac{b}{2a} \right) \xi_i - \xi_i^2 \right] \exp \left(-a\lambda^2 - a\frac{\mu^2}{2} \right). \tag{1.7.9}$$

We have obtained a surprising result: the infinite series of one-dimensional quasi-exactly solvable sextic anharmonic oscillators has turned out to be equivalent to a single two-dimensional exactly solvable model of a harmonic oscillator with correct eigenvalues and eigenfunctions. The last assertion follows from the fact that the function (1.7.9) can always be rewritten in the form

$$\phi(\lambda, \mu) = \sum_{m=0}^{M} c_m H_{M-m}(\lambda) H_{2m+p}(\mu) \exp \left(-a\lambda^2 - a\frac{\mu^2}{2} \right) \tag{1.7.10}$$

where c_m, $m = 0, \ldots, M$ are certain real numbers and $H_n(\lambda)$ are Hermite polynomials.

In order to understand the reason for such amazing equivalence let us look at our derivation from another point of view. Assume that the model of a two-dimensional oscillator (1.7.8) is given and nothing about the existence of quasi-exactly solvable models is known. Trying to separate the variables in equation (1.7.8), we can make sure that this is possible in many coordinate systems, and in particular, in the system defined by formulas (1.7.6). After the separation, we obtain two identical one-dimensional equations

$$\left(\frac{\partial^2}{\partial z^2} - a^2 z^6 - 2abz^4 + E \right) \psi(z) = \Gamma z^2 \psi(z) \tag{1.7.11}$$

(with $z = x$ or $z = y$) containing two spectral parameters of different mathematical meaning. One of them, namely Γ, is a spectral parameter of the initial two-dimensional problem (1.7.8), while the second one, E, is an additional parameter — the so-called separation constant. It is known that in the general case the values of a separation constant depend on the sort of solution. According to formula (1.7.10), the spectrum of the initial equation is degenerate and, therefore, for any admissible value of Γ ($\Gamma = b^2 - a(4M + 2p + 3)$) the separation constant E takes several $(M + 1)$ values. Due to the exact solvability of the initial problem, all these values also can be found exactly. This enables us to treat (1.7.11) as an ordinary spectral equation (with a single spectral parameter E), having $M + 1$ exact solutions for any given M. Identifying the parameter E with the energy, and inserting Γ in the potential, we obtain an infinite set of quasi-exactly solvable Schrödinger-type equations of orders $M + 1 = 1, 2, 3, \ldots$.

It is easy to understand that these reasonings have a quite general character and are applicable to any two-dimensional exactly solvable model allowing separation of variables and having a degenerate spectrum. Thus, they can be considered as a method of constructing quasi-exact solvable models. Hereafter we shall refer to this method as the method of separation of variables. Its many aspects will be discussed in detail in chapter 5.

1.8 Completely integrable quantum systems and quasi-exact solvability

The procedure of going from two identical one-dimensional quasi-exactly solvable equations to a single two-dimensional exactly solvable equation essentially solves the inverse problem of separation of variables. In the preceding section we have identified the energy spectral parameter E with the separation constant and eliminated it by means of this procedure.

However, it is absolutely evident that the same procedure enables us to eliminate the second "potential" spectral parameter Γ. In this case we obtain a two-dimensional exactly solvable model with a non-degenerate spectrum.

In order to obtain the explicit form of this equation, it is sufficient to multiply (1.7.1a) by $y^2\psi(y)$ and (1.7.1b) by $x^2\psi(x)$, and then to subtract one result from the other. Dividing the resulting equation by $y^2 - x^2$ and changing the variables by formula (1.7.6), we obtain

$$\left\{ \frac{b}{a} \left(-\frac{\partial^2}{\partial \mu^2} + a^2 \mu^2 \right) + 2\lambda \left(\frac{\partial^2}{\partial \mu^2} + a^2 \mu^2 \right) - \left(2\mu \frac{\partial}{\partial \mu} + 1 \right) \frac{\partial}{\partial \lambda} \right\} \phi(\lambda, \mu)$$

$$= E\phi(\lambda, \mu). \quad (1.8.1)$$

Equation (1.8.1) is exactly solvable by construction. Its eigenvalues E and eigenfunctions $\phi(\lambda, \mu)$ are determined by formulas (1.4.29) and (1.7.10), respectively.

We shall not try to seek here the physical meaning of this equation. This will be done later. Our aim is to consider equations (1.7.8) and (1.8.1) together from a mathematical point of view.

First of all consider their "hamiltonians"

$$H_\Gamma = \left(-\frac{\partial^2}{\partial \mu^2} + a^2 \mu^2\right) + \left(-\frac{\partial^2}{\partial \lambda^2} + 4a^2 \lambda^2\right) \qquad (1.8.2)$$

and

$$H_E = \frac{b}{a}\left(-\frac{\partial^2}{\partial \mu^2} + a^2 \mu^2\right) + 2\lambda\left(-\frac{\partial^2}{\partial \mu^2} + a^2 \mu^2\right)$$
$$- \left(2\mu\frac{\partial}{\partial \mu} + 1\right)\frac{\partial}{\partial \lambda} \qquad (1.8.3)$$

and note that they can be rewritten in a much simpler form if we use the generators of the Heisenberg algebra $\mathcal{H}$. Since we deal with two-dimensional models we must introduce two groups of such generators A^+, A^-, A^0 and B^+, B^-, B^0 which can be defined as follows:

$$A^+ = \frac{1}{2\sqrt{a}}\left(2a\lambda - \frac{\partial}{\partial \lambda}\right), \quad A^- = \frac{1}{2\sqrt{a}}\left(2a\lambda + \frac{\partial}{\partial \lambda}\right), \quad A^0 = 1,$$
$$(1.8.4a)$$

$$B^+ = \frac{1}{\sqrt{2a}}\left(a\mu - \frac{\partial}{\partial \mu}\right), \quad B^- = \frac{1}{\sqrt{2a}}\left(a\mu + \frac{\partial}{\partial \mu}\right), \quad B^0 = 1.$$
$$(1.8.4b)$$

These generators satisfy the usual commutation relations

$$[A^-, A^+] = A^0, \quad [B^-, B^+] = B^0, \qquad (1.8.5)$$

and representation spaces for them are defined in the same way as in section 1.2.

In terms of these generators the "hamiltonians" (1.8.2) and (1.8.3) take the form

$$H_\Gamma = a\left\{4A^+A^- + 2B^+B^- + 3\right\}, \qquad (1.8.6a)$$
$$H_E = b\left\{2B^+B^- + 1\right\} + 2\sqrt{a}\left\{A^-(B^+)^2 + A^+(B^-)^2\right\}. \quad (1.8.6b)$$

It is not difficult to verify that they commute with each other:

$$[H_\Gamma, H_E] = 0. \qquad (1.8.7)$$

But this means that we have obtained the exactly solvable and completely integrable two-dimensional system! In other words, we have proved that the one-dimensional quasi-exactly solvable model (1.4.13) is equivalent to the completely integrable system with two commuting integrals of motion H_Γ and H_E.

The physical meaning of these integrals is very simple. Their eigenvalues in the representation space of algebra $\mathcal{H} \oplus \mathcal{H}$ are the admissible values of spectral parameters Γ and E entering into the Schrödinger equation (1.7.1). This observation enables us to give a simple group-theoretical explanation of the fact that the model (1.4.13) is quasi-exactly solvable.

Remember that the quasi-exact solvability of the model (1.7.1) follows from the fact that the spectral parameter Γ entering into the potential is degenerate with respect to the energy parameter E. In the language of the completely integrable system (1.8.6) this means that the spectrum of the operator H_Γ is degenerate with respect to the spectrum of the operator H_E. Since the operators H_Γ and H_E commute with each other and thus have a common set of eigenvectors, this is possible if and only if there exists a certain group of symmetry G, under which the operator H_Γ is invariant, while the operator H_E is not. Evidently, the group G acts in the eigenspaces of operator H_Γ, which play in this case the role of spaces of irreducible representations of G. The dimension of each such irreducible representation determines the multiplicity of degeneracy, which, in turn, determines the order of the corresponding quasi-exactly solvable model.

Using explicit expressions (1.8.6) for operators H_Γ and H_E it is not difficult to make sure that such a group actually exists. Its corresponding Lie algebra is formed by the generators

$$J_1 = A^+ A^-, \quad J_2 = B^+ B^-,$$
$$K^+ = A^+(B^-)^2, \quad K^- = A^-(B^+)^2 \tag{1.8.8}$$

and any possible commutators of them. Evidently, this algebra is infinite dimensional, although it can be viewed as finite-dimensional non-linear algebra specified by the commutation relations:

$$[J_1, J_2] = 0,$$
$$[J_1, K^+] = K^+, \quad [J_1, K^-] = K^-,$$
$$[K^+, J_2] = 2K^+, \quad [K^-, J_2] = -2K^-, \tag{1.8.9}$$
$$[K^-, K^+] = J_2^2 - 4J_1 J_2 - J_2 - 2J_1.$$

The representation space for (1.8.9), defined as the linear span of the vectors

$$|n, m\rangle = (A^+)^n (B^+)^m |0, 0\rangle, \quad n, m = 0, 1, 2, \ldots \tag{1.8.10}$$

(where $|0,0\rangle$ is the vector annihilated by operators A^- and B^-), is reducible. In order to construct irreducible subspaces of this space, consider the action of operators (1.8.8) on the basis (1.8.10). We have

$$J_1|n,m\rangle = n|n,m\rangle, \quad J_2|n,m\rangle = m|n,m\rangle,$$
$$K^+|n,m\rangle = m(m-1)|n+1,m-2\rangle, \qquad (1.8.11)$$
$$K^-|n,m\rangle = n|n-1,m+2\rangle,$$

from which it follows that the number

$$l = 2n + m \qquad (1.8.12)$$

is invariant under such action. Therefore the space Φ_l, formed by vectors $|n,m\rangle$ with given quantum number l, is an irreducible representation space for the algebra (1.8.9), and it is finite dimensional. The dimension of this space is equal to the number of solutions of equation (1.8.12) for non-negative integers n and m. The general formula has the form:

$$\dim \Phi_l = \left[\frac{l}{2}\right] + 1. \qquad (1.8.13)$$

Due to the commutativity of operators (1.8.8) with H_Γ, the spaces Φ_l play simultaneously the role of eigenspaces Φ_Γ corresponding to the eigenvalue Γ. Expressing the operator H_Γ in terms of generators J_1 and J_2:

$$H_\Gamma = a(4J_1 + 2J_2 + 3) \qquad (1.8.14)$$

and using formulas (1.8.11) and (1.8.12) we obtain

$$H_\Gamma \Phi_\Gamma = \Gamma \Phi_\Gamma \qquad (1.8.15)$$

where

$$\Gamma = a(2l + 3). \qquad (1.8.16)$$

Comparing (1.8.16) with the old expression (1.7.4) for Γ, we find that

$$l = 2M + p, \qquad (1.8.17)$$

which gives us finally

$$\dim \Phi_\Gamma = M + 1. \qquad (1.8.18)$$

Thus, we have shown that the dimensions of irreducible representations of the algebra (1.8.9) determine the orders of quasi-exactly solvable models (1.4.13).

Finally note that a connection between the quasi-exactly solvable models and completely integrable quantum systems was discovered by Ushveridze (1988d, i) and then studied in the series of papers by Ushveridze (1988f, n, 1989c, e, 1990a, 1992) and Maglaperidze and Ushveridze (1989b, 1990). In sections 1.9 and 1.10 and, especially, in chapter 5 we discuss this connection in more detail.

1.9   Deformation of completely integrable models.   The projection method

In the preceding sections we have shown that the one-dimensional quasi-exactly solvable model (1.4.13) is associated with a completely integrable system of two two-dimensional operators (1.8.6) (integrals of motion), one of which has degenerate spectra.

In this section we will demonstrate that any completely integrable system with similar properties can be considered as a starting point in constructing quasi-exactly solvable models.

Indeed, let H_E and H_Γ be certain commuting operators constructed from the generators of a certain Lie algebra and acting in the corresponding (infinite-dimensional) representation space which we denote by Φ. Assume that the spectral problem for operators H_E and H_Γ in Φ is exactly solvable. Denote by Φ_E and Φ_Γ the eigenspaces of operators H_E and H_Γ corresponding to the eigenvalues E and Γ. Suppose that the spectrum Γ is degenerate with respect to E, so that $\dim \Phi_\Gamma > 1$ while $\dim \Phi_E = 1$.

Consider the operators

$$H = H_E + Q(H_\Gamma - \gamma), \tag{1.9.1}$$

in which Q is an arbitrarily fixed operator acting in the space Φ, but not commuting with H_E and H_Γ, and γ is a parameter. It is quite obvious that for arbitrary (generic) values of γ the spectral problem for H

$$\{H_E + Q(H_\Gamma - \gamma)\}\varphi = E\varphi, \quad \varphi \in \Phi \tag{1.9.2}$$

cannot be solved exactly.

However, if γ coincides with the eigenvalue Γ of the operator H_Γ, the term proportional to Q vanishes for any $\varphi \in \Phi_\Gamma$ and we come to the equation allowing several, $\dim \Phi_\Gamma$, exact solutions. If the spectrum of the operator H_Γ is infinite, we come to an infinite series of quasi-exactly solvable equations.

Now note that generators of algebras forming the operators H_E, H_Γ (and Q) admit, as a rule, differential realizations. Using these realizations we can reduce equation (1.9.2) to a differential form. In other words, we obtain an infinite series of differential quasi-exactly solvable equations. In many cases the form of these equations can be considerably simplified by means of the so-called projection method.

In order to explain the essence of this method we assume for definiteness that H_Γ and H_E are two-dimensional operators of first and second order, respectively.

We denote by Ψ_Γ the space of all functions φ satisfying the equation:

$$H_\Gamma \varphi = \Gamma \varphi. \tag{1.9.3}$$

Here we do not require that solutions of (1.9.3) belong necessarily to the representation space Φ. Therefore Ψ_Γ is wider than the eigenspace Φ_Γ:

$$\Phi_\Gamma \subset \Psi_\Gamma. \tag{1.9.4}$$

Since H_Γ is a first-order differential operator by assumption, the most general solution of the equation (1.9.3) can be written as

$$\varphi = \varphi_\Gamma f(\varphi_0) \tag{1.9.5}$$

where φ_Γ is a certain solution of the inhomogeneous equation (1.9.3), φ_0 is a partial solution of the homogeneous analogue of this equation, and f is an arbitrary analytic function. (The set of such functions is denoted by F).

Denote by F_Γ the space of all such functions $f \in F$ for which the solutions (1.9.5) belong to the eigenspace Φ_Γ. Evidently, $\dim F_\Gamma = \dim \Phi_\Gamma$.

Consider the equation

$$\{H_E + Q(H_\Gamma - \Gamma)\}\varphi = E\varphi, \quad \varphi \in \Psi_\Gamma, \tag{1.9.6}$$

which as we already know has $\dim \Phi_\Gamma$ exact solutions belonging to the space $\Phi_\Gamma \subset \Psi_\Gamma$. Using formulas (1.9.3) and (1.9.5) we can rewrite (1.9.6) in the form:

$$H_E(\Gamma)f(\varphi_0) = Ef(\varphi_0), \quad f \in F, \tag{1.9.7}$$

where

$$H_E(\Gamma) = \varphi_\Gamma^{-1} H_E \varphi_\Gamma. \tag{1.9.8}$$

Obviously, equation (1.9.7) will also have $\dim F_\Gamma$ solutions belonging to the space $F_\Gamma \subset F$. The operator $H_E(\Gamma)$ defined by formula (1.9.8) is, evidently, a second-order differential operator depending explicitly on Γ.

Now note that the functions φ_Γ and φ_0 are functionally independent. This gives us the possibility of introducing the new variables λ and μ by the formulas

$$\lambda = \varphi_0, \quad \mu = \varphi_\Gamma. \tag{1.9.9}$$

Then, functions $f(\varphi_0)$ forming spaces F and F_0 take the form of functions of a single variable λ, while the operator $H_E(\Gamma)$ is, as before, a second-order differential operator in two variables λ and μ. This is possible in one case only, when the projection of $H_E(\Gamma)$ on F is an operator in a single variable λ. Thus we come to the most general expression for $H_E(\Gamma)$:

$$H_E(\Gamma) = h_1(\Gamma, \lambda)\frac{\partial^2}{\partial \lambda^2} + h_2(\Gamma, \lambda)\frac{\partial}{\partial \lambda} + h_3(\Gamma, \lambda)$$

$$+ g_1(\Gamma, \lambda, \mu)\frac{\partial^2}{\partial \mu^2} + \left[g_2(\Gamma, \lambda, \mu)\frac{\partial}{\partial \lambda} + g_3(\Gamma, \lambda, \mu) \right]\frac{\partial}{\partial \mu}. \tag{1.9.10}$$

The substitution of (1.9.10) into (1.9.7) gives us a one-dimensional equation of the form

$$\left\{ h_1(\Gamma, \lambda)\frac{\partial^2}{\partial \lambda^2} + h_2(\Gamma, \lambda)\frac{\partial}{\partial \lambda} + h_3(\Gamma, \lambda) \right\} = E f(\lambda),$$

$$f(\lambda) \in F, \tag{1.9.11}$$

which allows $\dim F_\Gamma$ exact solutions belonging to the space F_Γ. Thus, equation (1.9.11) is a quasi-exactly solvable equation of order $\dim F_\Gamma$. Obviously, any such equation can be reduced to the Schrödinger form, which gives us a series of one-dimensional quasi-exactly solvable models.

The method described above we shall refer below to as the projection method.

In order to demonstrate how this method works, let us consider again the commuting operators H_E and H_Γ defined by parameters (1.8.6) and forming a completely integrable system on the algebra $\mathcal{H} \oplus \mathcal{H}$. According to general prescriptions given above we must reduce the operators H_E and H_Γ to differential form. The most simple way to do this is to use the following differential realizations of generators of Heisenberg algebras:

$$A^+ = z, \quad A^- = \frac{\partial}{\partial z}, \quad A^0 = 1, \tag{1.9.12a}$$

$$B^+ = t, \quad B^- = \frac{\partial}{\partial t}, \quad B^0 = 1. \tag{1.9.12b}$$

Then we obtain:

$$H_\Gamma = a\left\{ 4z\frac{\partial}{\partial z} + 2t\frac{\partial}{\partial t} + 3 \right\}, \tag{1.9.13a}$$

$$H_E = 2\sqrt{a}\left\{t^2\frac{\partial}{\partial z} + z\frac{\partial^2}{\partial t^2}\right\} + b\left(2t\frac{\partial}{\partial t} + 1\right). \qquad (1.9.13\text{b})$$

We see that H_Γ and H_E are first- and second-order differential operators acting in the space of polynomials in two variables z and t. (This is a representation space for the algebra $\mathcal{H} \oplus \mathcal{H}$.)

Taking

$$Q = Q(z,t), \qquad (1.9.14)$$

we obtain the following expression for the operator (1.9.1):

$$H = 2\sqrt{a}\left\{z\frac{\partial^2}{\partial t^2} + t^2\frac{\partial}{\partial z}\right\} + b\left(2t\frac{\partial}{\partial t} + 1\right)$$
$$+ Q(z,t)\left\{a\left[4z\frac{\partial}{\partial z} + 2t\frac{\partial}{\partial t} + 3\right] - \Gamma\right\}. \qquad (1.9.15)$$

The spectral problem for (1.9.15) is, in general, exactly non-solvable. However, if

$$\Gamma = a(4M + 2p + 3), \qquad (1.9.16)$$

it has $M + 1$ exact solutions belonging to the $(M + 1)$-dimensional space Φ_Γ. This space is formed by the functions

$$|n,m\rangle = z^n t^m, \quad 2n + m = 2M + p, \qquad (1.9.17)$$

the most general form of which is

$$\varphi = t^{2M+p} P_M\left(\frac{z}{t^2}\right), \qquad (1.9.18)$$

where P_M are *polynomials of degree M*.

Now let us construct the space Ψ_Γ formed by all solutions of equation (1.9.3). In our case this equation takes the form:

$$\left\{2z\frac{\partial}{\partial z} + t\frac{\partial}{\partial t} - 2M - p\right\}\varphi = 0. \qquad (1.9.19)$$

Evidently, the role of φ_Γ and φ_0 is played by the functions

$$\varphi_\Gamma = t^{2M+p}, \quad \varphi_0 = \frac{z}{t^2} \qquad (1.9.20)$$

and, therefore, according to (1.9.5), the most general solution of (1.9.19) is

$$\varphi = t^{2M+p} f\left(\frac{z}{t^2}\right), \qquad (1.9.21)$$

where f is an *arbitrary function*. We see that the space Ψ_Γ is actually larger than Φ_Γ.

The spectral equation for H in Ψ_Γ can be written as follows:

$$\left\{2\sqrt{a}\left(z\frac{\partial^2}{\partial t^2}+t^2\frac{\partial}{\partial z}\right)+b\left(2t\frac{\partial}{\partial t}+1\right)\right\}t^{2M+p}f\left(\frac{z}{t^2}\right)$$

$$= Et^{2M+p}f\left(\frac{z}{t^2}\right). \qquad (1.9.22)$$

After dividing it by t^{2M+p} and changing variables

$$\lambda=\frac{z}{t^2}, \quad \mu=t, \qquad (1.9.23)$$

it takes the form (1.9.10) with

$$H_E(\Gamma)=\left\{8\sqrt{a}\lambda^3\frac{\partial^2}{\partial\lambda^2}-[4\sqrt{a}(4M+2p+3)\lambda^2+4b\lambda-2\sqrt{a}]\frac{\partial}{\partial\lambda}\right.$$

$$\left.+2\sqrt{a}(2M+p)(2M+p-1)\lambda+b(4M+2p+1)\right\}$$

$$+\{2\sqrt{a}\lambda\mu^2\}\frac{\partial^2}{\partial\mu^2}-\left\{8\sqrt{a}\lambda^2\mu\frac{\partial}{\partial\lambda}-2\sqrt{a}(4M+2p)\lambda\mu-2b\mu\right\}\frac{\partial}{\partial\mu}.$$

$$(1.9.24)$$

By projection of this operator onto the space F of analytic functions of a single variable λ, terms proportional to $\frac{\partial}{\partial\mu}$ and $\frac{\partial^2}{\partial\mu^2}$ vanish and the resulting equation for $H_E(\Gamma)$ takes the form:

$$\left\{8\sqrt{a}\lambda^3\frac{\partial^2}{\partial\lambda^2}-[4\sqrt{a}(4M+2p-3)\lambda^2+4b\lambda-2\sqrt{a}]\frac{\partial}{\partial\lambda}\right.$$

$$\left.+2\sqrt{a}(2M+p)(2M+p-1)\lambda+(4M+2p+1)b\right\}f(\lambda)=Ef(\lambda),$$

$$f(\lambda)\in F. \qquad (1.9.25)$$

This equation allows $M+1$ exact solutions in the space F_Γ of Mth-order polynomials $P_M(\lambda)$ by construction. Thus, we have obtained a one-dimensional quasi-exactly solvable spectral equation of order $M+1$.

After a simple change of the variable λ and homogeneous transformation of the function $f(\lambda)$

$$\lambda=-\frac{1}{2\sqrt{a}x^2},$$

$$f(\lambda)=\lambda^{M+\frac{2p-3}{4}}\exp\left[\frac{1}{16\lambda^2}-\frac{b}{4\sqrt{a}}\frac{1}{\lambda}\right]\psi(x), \qquad (1.9.26)$$

this equation takes the Schrödinger form and we arrive again at the quasi-exactly solvable model (1.4.13).

1.10 Quasi-exact solvabillity and the Gaudin model. Bethe ansatz in quantum mechanics

In this section we will show that the completely integrable quantum system defined by formulas (1.8.6) is none other than the well known Gaudin model (Gaudin 1976, 1983).

The "hamiltonians" of the Gaudin model are constructed from the generators of the so-called Gaudin algebra. This algebra is known as the simplest infinite-dimensional generalization of the algebra $sl(2)$. Its three generators $S^+(\rho), S^-(\rho)$ and $S^0(\rho)$ depend continuously on a certain complex parameter ρ and satisfy the following commutation relations

$$
\begin{aligned}
&[S^-(\rho), S^+(\sigma)] = -\frac{2}{\rho-\sigma}(S^0(\rho) - S^0(\sigma)), \\
&[S^0(\rho), S^\pm(\sigma)] = \pm\frac{1}{\rho-\sigma}(S^\pm(\rho) - S^\pm(\sigma)).
\end{aligned}
\tag{1.10.1}
$$

The representations of algebra (1.10.1) can be obtained by analogy with the $sl(2)$ case. We introduce the notion of the lowest-weight vector $|0\rangle$, which is annihilated by all lowering operators $S^-(\rho)$,

$$
S^-(\rho)|0\rangle = 0,
\tag{1.10.2}
$$

and define the lowest weight as the function $F(\rho)$ whose values are the eigenvalues of the operators $S^0(\rho)$ on $|0\rangle$:

$$
S^0(\rho)|0\rangle = F(\rho)|0\rangle.
\tag{1.10.3}
$$

Then the representation space Φ is defined as the linear span of the vectors:

$$
\begin{aligned}
&|0\rangle, \quad S^+(\rho_1)|0\rangle, \quad S^+(\rho_1)S^+(\rho_2)|0\rangle, \\
&\ldots, S^+(\rho_1)\ldots S^+(\rho_M)|0\rangle, \ldots,
\end{aligned}
\tag{1.10.4}
$$

where $\rho_1, \rho_2, \ldots, \rho_M, \ldots$ are arbitrary numbers. Evidently, the space Φ is, in general, infinite dimensional.

Consider the operators

$$
K(\rho) = S^0(\rho)S^0(\rho) - \frac{1}{2}S^+(\rho)S^-(\rho) - \frac{1}{2}S^-(\rho)S^+(\rho),
\tag{1.10.5}
$$

acting in the representation space Φ and having the same structure as the Casimir operator for the algebra $sl(2)$. It is easy to verify that $K(\rho)$ is not the Casimir operator for the Gaudin algebra since the commutators of $K(\rho)$ with elements $S^{\pm}(\rho), S^0(\rho)$ differ, generally, from zero. Nevertheless, the operators $K(\rho)$ have another very interesting property which is especially important for us: they form a commutative family

$$[K(\rho), K(\sigma)] = 0 \qquad (1.10.6)$$

which gives us the possibility of interpreting them as integrals of motion of certain completely integrable system.

This system is known as the Gaudin model and has many intriguing physical and mathematical properties.

First of all, note that the Gaudin spectral problem

$$K(\rho)\phi = \Lambda(\rho)\phi, \quad \phi \in \Phi \qquad (1.10.7)$$

is exactly solvable in spite of the fact that the representation space Φ in which we seek the solutions is infinite dimensional.

The exact (algebraic) solutions of equation (1.10.7) can be obtained in the framework of the Bethe *ansatz* method. The corresponding *ansatz* has the form

$$\phi = S^+(\xi_1)S^+(\xi_2)\dots S^+(\xi_M)|0\rangle \qquad (1.10.8)$$

where M is an arbitrary non-negative integer, and $\xi_1,\dots,\xi_M$ are certain (unknown) complex numbers. Substituting (1.10.8) into the equation (1.10.7), it is not difficult to show that it actually has solutions of the form (1.10.8) if the numbers ξ_i, $i = 1,\dots,M$ satisfy the following system of numerical equations:

$$\sum_{k=1}^{M}\frac{1}{\xi_i - \xi_k} + F(\xi_i) = 0, \quad i = 1,\dots,M, \qquad (1.10.9)$$

which are known as the Bethe *ansatz* equations. Then, for eigenvalues $\Lambda(\rho)$ we have:

$$\Lambda(\rho) = F'(\rho) + F^2(\rho) + 2\sum_{i=1}^{M}\frac{F(\rho) - F(\xi_i)}{\rho - \xi_i}. \qquad (1.10.10)$$

Now let us show that the Gaudin model is equivalent to model (1.8.6) with two integrals of motion H_{Γ} and H_E.

For this purpose we consider the following differential realizations of generators of the Gaudin algebra:

$$S^+(\rho) = -\frac{\mu^2}{4\rho} + \left(\frac{b}{2a} - \lambda\right) + \rho, \qquad (1.10.11a)$$

$$S^0(\rho) = \frac{\mu\frac{\partial}{\partial\mu} + a\mu^2 + p + \frac{1}{2}}{2\rho} + \left(\frac{\partial}{\partial\lambda} + 2a\lambda - b\right) - 2a\rho, \quad (1.10.11b)$$

$$S^-(\rho) = -\frac{\frac{\partial^2}{\partial\mu^2} + 2(a\mu + p)\frac{\partial}{\partial\mu} + a^2\mu^2 + 2ap\mu + a}{\rho} - 4a\frac{\partial}{\partial\lambda} - 8a^2\lambda.$$
$$(1.10.11c)$$

It is easy to verify that operators (1.10.11) satisfy the commutation relations (1.10.1) and thus, actually form the Gaudin algebra. In order to describe the representation in which these operators act, it is sufficient to solve equations (1.10.2) and (1.10.3), determining the lowest-weight vector and the lowest weight. This gives

$$|0\rangle = \exp\left(-a\lambda^2 - \frac{a}{2}\mu^2\right) \quad (1.10.12)$$

and

$$F(\rho) = \frac{2p+1}{4\rho} - b - 2a\rho. \quad (1.10.13)$$

Substituting the explicit expressions for generators (1.10.11) into (1.10.5), we obtain the following differential expression for $K(\rho)$:

$$K(\rho) = 4a^2\rho^2 + 4ab\rho - \left\{-\frac{\partial^2}{\partial\lambda^2} - \frac{\partial^2}{\partial\mu^2} + a^2(4\lambda^2 + \mu^2)\right\}$$
$$-\frac{1}{2}\left\{\frac{b}{a}\left(-\frac{\partial^2}{\partial\mu^2} + a^2\mu^2\right) + 2\lambda\left(\frac{\partial^2}{\partial\mu^2} + a^2\mu^2\right) - \left(2\mu\frac{\partial}{\partial\mu} + 1\right)\frac{\partial}{\partial\lambda}\right\}\frac{1}{\rho}$$
$$+\frac{(2p+1)(2p-3)}{4\rho^2} \quad (1.10.14)$$

from which it follows that the Gaudin model has in this case two non-trivial integrals of motion which are proportional to the expressions in curly brackets. However, from the results of section 1.8 we know that these integrals of motion are none other than operators H_Γ and H_E for the model (1.8.6).

We have obtained an important result, stating that the completely integrable system (1.8.6) coincides with the Gaudin model if the Gaudin algebra acts in the representation with the lowest weight (1.10.13).

In order to make sure that this coincidence is full, let us examine the Bethe *ansatz* solutions for the model (1.10.5) and compare them with the solutions for model (1.8.6).

Using formulas (1.10.8), (1.10.11a) and (1.10.12) we obtain coordinate expressions for the Bethe vectors

$$\phi = \mu^p \prod_{i=1}^M \left[\frac{\mu^2}{4} + \left(\frac{b}{2a} - \lambda\right)\xi_i - \xi_i^2\right]\exp\left(-a\lambda^2 - \frac{a}{2}\mu^2\right), \quad (1.10.15)$$

which, evidently, coincide with the expressions (1.7.9) for wavefunctions in model (1.8.6).

The numbers ξ_i entering into (1.10.15) and determining the concrete form of the Bethe vectors satisfy equation (1.10.9) in which the function $F(\rho)$ is specified by formula (1.10.13). Thus, we have

$$\sum_{k=1}^{M}{}' \frac{1}{\xi_i - \xi_k} + \frac{2p+1}{4\xi_i} - b - 2a\xi_i = 0, \quad i = 1, \ldots, M. \quad (1.10.16)$$

This system coincides with the system of spectral equations (1.4.30) for model (1.8.6).

The last step is to make sure that we have correct expressions for the eigenvalues of the operators H_Γ and H_E. For this purpose let us consider expression (1.10.10). Substituting the function (1.10.13) in it, we obtain

$$\begin{aligned}
\Lambda(\rho) = {} & 4a^2\rho^2 + hab\rho - a(4M + 2p + 3) \\
& - \left[(4M + 2p + 1)b + 8a \sum_{i=1}^{M} \xi_i\right] \frac{1}{2\rho} + \frac{(2p+1)(2p-3)}{4\rho^2}.
\end{aligned}$$

$$(1.10.17)$$

Comparison of (1.10.17) and (1.10.14) shows that the eigenvalues of operators H_Γ and H_E are actually described by formulas (1.7.4) and (1.4.29) which completes our proof of the coincidence of models (1.10.5) and (1.8.6).

Thus, we have established a deep connection between the quasi-exactly solvable models of the sextic anharmonic oscillator described by the hamiltonian (1.4.13) and completely integrable Gaudin model (1.10.5) based on the algebra $sl(2)$. We see that the spectral equations (1.4.30) for models (1.4.13) can be interpreted as Bethe *ansatz* equations for the Gaudin model; the number M determining the order of quasi-exact solvability is simply a number of elementary "spin" excitations in (1.10.5) and the wavefunction zeros play the role of the "collective coordinates" of these excitations. As was shown by Ushveridze (1988d, f, i, n, 1989c, e, 1990a, 1992) and Maglaperidze and Ushveridze (1989b, 1990), this amazing relationship is not specific only for the model (1.4.13), but takes place for many other quasi-exactly solvable models which will be discussed in detail in chapters 4 and 5.

1.11   The classical multi-particle Coulomb problem and the Schrödinger equation

In preceding sections we have shown that solutions of the quasi-exactly solvable Schrödinger equation with potential

$$V(x) = \left[b^2 - a(4M + 2p + 3)\right] x^2 + 2abx^4 + b^2x^6 \qquad (1.11.1)$$

can be represented in the form

$$\psi(x) = \prod_{j=1}^{M} \left(\frac{x^2}{2} - \xi_j\right) x^p \exp\left\{-\frac{ax^4}{4} - \frac{bx^2}{2}\right\} \qquad (1.11.2)$$

and

$$E = b(4M + 2p + 1) + 8a \sum_{j=1}^{M} \xi_j, \qquad (1.11.3)$$

where the numbers ξ_j satisfy the system of numerical equations

$$\sum_{k=1}^{M}{}' \frac{1}{\xi_j - \xi_k} + \frac{2p+1}{4\xi_j} - b - 2a\xi_j = 0, \quad j = 1,\ldots,M. \qquad (1.11.4)$$

Now let us discuss the properties of equations (1.11.4) in more detail. It is not difficult to understand that system (1.11.4) is equivalent to a system of algebraic equations and therefore its solutions are, generally, complex numbers. Therefore, the numbers ξ_j should also be taken to be complex:

$$\xi_j = \xi_{1j} + i\xi_{2j}. \qquad (1.11.5)$$

Substitution of (1.11.5) into (1.11.4) leads to a system of real equations

$$\sum_{k=1}^{M}{}' \frac{\xi_{1j} - \xi_{1k}}{(\xi_{1j} - \xi_{1k})^2 + (\xi_{2j} - \xi_{2k})^2} + \frac{2p+1}{4} \frac{\xi_{1j}}{\xi_{1j}^2 + \xi_{2j}^2} - b - 2a\xi_{1j} = 0,$$
$$(1.11.6a)$$

$$\sum_{k=1}^{M}{}' \frac{\xi_{2j} - \xi_{2k}}{(\xi_{1j} - \xi_{1k})^2 + (\xi_{2j} - \xi_{2k})^2} + \frac{2p+1}{4} \frac{\xi_{2j}}{\xi_{1j}^2 + \xi_{2j}^2} + 2a\xi_{2j} = 0,$$
$$(1.11.6b)$$

in which $j = 1,\ldots,M$. The system (1.11.6) can be rewritten in a more compact form if we introduce the two-dimensional vectors

$$\vec{\xi_j} = \begin{pmatrix} \xi_{1j} \\ \xi_{2j} \end{pmatrix}. \qquad (1.11.7)$$

Then, instead of (1.11.6) we can write:

$$\sum_{k=1}^{M}{}' \frac{\vec{\xi}_j - \vec{\xi}_k}{|\vec{\xi}_j - \vec{\xi}_k|^2} + \frac{2p+1}{4} \frac{\vec{\xi}_j}{|\vec{\xi}_j|^2} - b\vec{e} - 2a\hat{\sigma}\vec{\xi}_j = 0,$$

$$j = 1, \ldots, M, \qquad (1.11.8)$$

where by $\vec{e}$ and $\hat{\sigma}$ we have denoted the vector and the matrix:

$$\vec{e} = \begin{pmatrix} 1 \\ 0 \end{pmatrix}, \quad \hat{\sigma} = \begin{pmatrix} 1 & 0 \\ 0 & -1 \end{pmatrix}. \qquad (1.11.9)$$

Equation (1.11.8) can be interpreted as the condition for an extremum of the function:

$$W(\vec{\xi_1}, \ldots, \vec{\xi_M}) = -\sum_{i<k}^{M} q_i q_k \ln \left| \vec{\xi_i} - \vec{\xi_k} \right| - \sum_{i=1}^{M} q_i U(\vec{\xi_i}), \qquad (1.11.10)$$

where $q_i = 1$ are unit numbers and

$$U(\vec{\xi}) = -\frac{2p+1}{4} \ln \left| \vec{\xi} \right| + b\vec{e} \cdot \vec{\xi} + a\vec{\xi}\hat{\sigma}\vec{\xi}. \qquad (1.11.11)$$

It is not difficult to see that (1.11.11) is none other than the potential of a two-dimensional (logarithmic) Coulomb system consisting of M particles with coordinates $\vec{\xi_i}$ and unit charges $q_i = 1$ moving in the potential $U(\vec{\xi_i})$ (Ushveridze 1988d, g, Maglaperidze and Ushveridze 1988). Recall that two-dimensional classical electrodynamics is characterized by the logarithmic Coulomb potential.

Potential (1.11.11) is generated by a particle with charge $\frac{2p+1}{4}$ located at the origin and by two particles with charges $\frac{1}{2}(\lambda^2 a \pm \lambda b)$ and coordinates $\pm\lambda$, where λ tends to infinity. This gives the oscillator-type potential in the limit $\lambda = \infty$.

We therefore see that the problem of finding solutions of the system of algebraic equations (1.11.4) is equivalent to the problem of finding equilibrium positions of a system of Coulomb particles in the potential $U(\vec{\xi})$. At first sight this problem does not have solutions since the potential $U(\vec{\xi})$ is unstable. However, the presence of the Z_2-symmetry in the system ($\xi_2 \rightarrow -\xi_2$) leads to the existence of a straight line (coinciding in the present case with the axis ξ_1) on which all forces are longitudinal. The problem of equilibrium of particles moving on this line becomes stable and one dimensional. This gives us the possibility of seeking the solutions of the system (1.11.4) in the real numbers.

Let us now consider the structure of the real ξ_1-axis in more detail. On this axis the potential (1.11.11) essentially simplifies:

$$U(\xi) = -\frac{2p+1}{4}\ln|\xi| + b\xi + a\xi^2. \qquad (1.11.12)$$

Its form is depicted in figure 1.9. What did we gain by introducing the

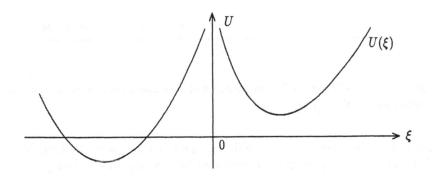

Figure 1.9. The form of the potential (1.11.12) for $b > 0$ and $a > 0$.

analogue system of classical Coulomb particles? The answer is obvious: the rich intuition everybody has in classical mechanics. In particular, looking at the function $U(\xi)$, everybody will immediately say that the equilibrium position does exist. Indeed, the potential wells depicted in figure 1.9 contain M particles with unit charge. The interaction between these particles is of Coulomb type and therefore they repel each other. Furthermore, they interact also with the force centre at the origin which has Coulomb charge $\frac{2p+1}{4}$. At short distances this centre also repels the particles. They cannot run away to infinity because of the attraction term $a\xi^2$ which becomes important at large distances. Hence, equilibrium is necessarily established at some finite distances.

It is not difficult to understand that there are several equilibrium positions, depending on the number of particles situated to the right of the origin. More exactly, there are $M + 1$ possibilities, some of which are depicted in figure 1.10. So we can conclude that the system of equations (1.11.4) has $M + 1$ different solutions. Note that the same result was obtained in preceding sections by different methods.

Now remember that the numbers ξ_i determine the zeros of the wavefunction. From formula (1.11.2) it follows that their positions are

$$x_i = \pm\sqrt{2\xi_i}. \qquad (1.11.13)$$

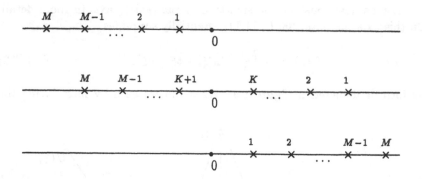

Figure 1.10. Various equilibrium positions of Coulomb particles situated in the external potential (1.11.12).

Remember also that only zeros which are real and lie inside the interval in which the spectral problem is formulated have a physical meaning. They play the role of wavefunction nodes. According to the oscillator theorem, the number of nodes determines the excited state. From (1.11.13) it follows that the wavefunction nodes correspond to the positive values of ξ_i. If K numbers of ξ_i are positive ($0 \leq K \leq M$) the number of nodes is $2K + p$ and thus such a distribution of particles describes the $(2K + p)$th excited state.

We see that in the first case all ξ_i are negative and the wavefunction has p nodes. In the last case, when all particles ξ_i lie on the right semi-axis, the number of states is maximal and equal to $2M + p$.

It is easy to verify that this picture does not contradict formula (1.11.3) in which the values of the energy levels are expressed in terms of the centre of charges of the system of Coulomb particles. Indeed, the more the particles are located in the right well, the more the centre of charges is shifted to the right and the higher is the energy.

Note that all these reasonings are valid only when the number a, determining the behaviour of the system at large distances, is positive.

What happens if $a = 0$? From section 1.5 we know that this case corresponds to a simple harmonic oscillator, which for any given M has only one solution with $2M + p$ nodes. The classical potential (1.11.12) takes in this case the form depicted in figure 1.11. We see that this potential has only one stable well, and therefore for any given M we actually have only one possibility, when all particles are situated in this well and, thus, have positive coordinates.

Now let us assume that the parameter a is small and negative. The

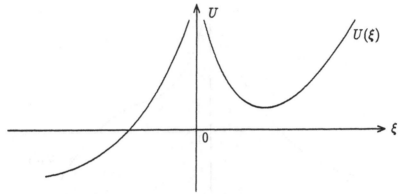

Figure 1.11. The form of the potential (1.11.12) for $b > 0$ and $a = 0$.

classical potential corresponding to this case is depicted in figure 1.12. We

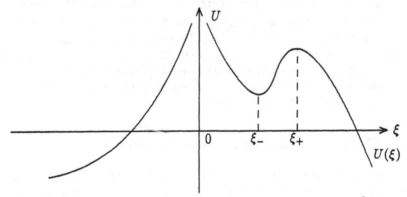

Figure 1.12. The form of the potential (1.11.12) for $b > 0$ and $-\frac{b^2}{4p+2} < a < 0$.

have obtained a potential well of finite depth in which several particles may lie. Consider the simplest case, when only one particle is situated in this well. We see that, unlike the case with $a > 0$, we have an attractive force at infinity which, however, does not destroy the existing equilibrium of the particle, which may lie in the minimum of the potential $U(\xi)$ at the point ξ_- (the stable equilibrium) or in the maximum at the point ξ_+ (the unstable equilibrium). If we now begin to increase the modulus of the negative parameter a, at some instant the classical potential ceases to have extrema, equilibrium of the particle in this potential becomes impossible and it runs away to infinity (see figure 1.13). It is evident that in this case two equilibrium positions — one stable (at the point

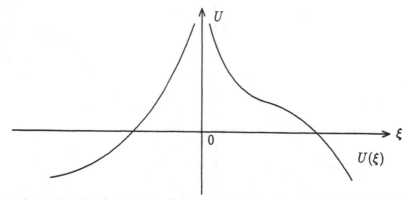

Figure 1.13. The form of the potential (1.11.12) for $b > 0$ and $a < -\frac{b^2}{4p+2}$.

$\xi_- = [-b - \sqrt{b^2 + 2a(2p+1)}]/4a)\dot{\ }$ and one unstable (at the point $\xi_+ = [-b + \sqrt{b^2 + 2a(2p+1)}]/4a)$ — merge at the point when the instability arises. This is the point $a = -\frac{b^2}{4p+2}$. But from the results of section 1.5 we know that this is none other than the position of the Bender and Wu singularity! This enables us to assert that, in the language of the classical Coulomb analogue system, Bender and Wu singularities can be interpreted as the points at which the system of classical Coulomb particles becomes classically unstable (Ushveridze 1988e, g).

1.12   Classical formulation of quantal problems

From the results of the preceding section it follows that the quasi-exactly solvable model of the sextic anharmonic oscillator with hamiltonian (1.4.13) is partially equivalent to a classical Coulomb system. Partially, because we can describe in the classical language only a finite part of the spectrum (consisting of $M + 1$ levels), but not the whole spectrum.

In this connection let us pose the question: what happens if M tends to infinity? From the reasoning given above we know that the larger M is, the larger is the number of states allowing classical interpretation. Therefore taking $M = \infty$ we must obtain a model of which all the energy levels and corresponding wavefunctions can be described in the language of classical electrostatics. In this sense, the limiting quantum mechanical model will be completely equivalent to a classical Coulomb system (Maglaperidze and Ushveridze 1988, 1989b).

In section 1.5 we have already considered the limit $M \to \infty$ for the model (1.4.13). In this limit the system of equations (1.11.4) becomes infinitely complicated and ceases to be algebraic. Therefore, the limiting model is exactly non-solvable. As we know, this is a quartic anharmonic oscillator. It is emphasized that a correct passage to the limit $M \to \infty$ implies a simultaneous change of the parameters a and b determining the form of the hamiltonian (1.4.13). Making, for example, the substitution

$$b = (2\beta M)^{\frac{1}{3}}\Delta, \quad a = \frac{\beta}{2}(2\beta M)^{-\frac{1}{3}}\Delta^{-1} \tag{1.12.1}$$

in which Δ is a function behaving for large M as

$$\Delta = 1 + \frac{1}{3}\frac{\alpha}{(2\beta M)^{\frac{2}{3}}}, \tag{1.12.2}$$

substituting formulas (1.12.1) and (1.12.2) into the potential (1.4.13) and taking $M \to \infty$ we obtain the limiting potential

$$V(x) = ax^2 + \beta x^4. \tag{1.12.3}$$

In order to construct the corresponding (limiting) classical analogue system for (1.12.3), we must substitute expressions (1.12.1) and (1.12.2) into the equations (1.11.4) and then take $M \to \infty$. As a result we obtain the equilibrium conditions for a system consisting of an infinite number of classical Coulomb particles. Remember, however, that equations of classical mechanics are not very convenient for describing the behaviour of such multi-particle systems. Much more suitable in this case are the solid state equations in which the role of the unknown object is played by the particle distribution density.

In order to reduce equations (1.11.4) to a continous form, let us enumerate the coordinates ξ_i of the particles in the following order:

$$\xi_1 < \xi_2 < \ldots < \xi_M \tag{1.12.4}$$

and assume that $M - K$ particles are situated in the left hand well, while the K remaining particles lie in the right hand well. In other words, we consider the spectral equations for the $(2K + p)$th energy level.

Taking

$$\xi_i = \frac{1}{2\beta}(2\beta M)^{\frac{2}{3}}\Delta^2 \xi\left(\frac{i}{M}\right) \tag{1.12.5}$$

we obtain instead of (1.11.4):

$$\sum_{k=1}^{M}{}' \frac{1}{\xi\left(\frac{i}{M}\right) - \xi\left(\frac{k}{M}\right)} \frac{1}{M} + \frac{2p+1}{4M} \frac{1}{\xi\left(\frac{i}{M}\right)} - \Delta^3\left[1 + \frac{1}{2}\xi\left(\frac{i}{M}\right)\right],$$
$$i = 1, \ldots, M. \tag{1.12.6}$$

Correspondingly, expression (1.11.3) for the energy E takes the form:

$$E = 4M\Delta(2\beta M)^{\frac{1}{3}}\left\{1 + \frac{2p+1}{4M} + \frac{1}{2}\sum_{i=1}^{M}\xi\left(\frac{i}{M}\right)\frac{1}{M}\right\}. \qquad (1.12.7)$$

If M is infinitely large, it is convenient to introduce (instead of the discrete variable i and discrete function $\xi\left(\frac{i}{M}\right)$) the continuous variable $\lambda = \frac{i}{M}$ and continuous function $\xi(\lambda)$. Generally speaking, the function $\xi(\lambda)$ is continuous (in the usual sense of this word) only in the intervals $\lambda \in [0, 1-q]$ and $\lambda \in [1-q, 1]$ where $q = \frac{K}{M}$. At the point $\lambda = 1-q$ it has a discontinuity (a jump) caused by the fact that there are no particles in the vicinity of the repelling centre.

Assuming that M tends to infinity, we can replace the sums in (1.12.6) and (1.12.7) by integrals. This leads us to the integral equation for $\xi(\lambda)$:

$$\fint_{0}^{1}\frac{d\mu}{\xi(\lambda) - \xi(\mu)} = 1 + \frac{1}{2}\xi(\lambda) - \frac{2p+1}{4M}\frac{1}{\xi(\lambda)} \qquad (1.12.8)$$

and to the following explicit expression for E:

$$E = 4M(2\beta M)^{\frac{1}{3}}\left\{1 + \frac{1}{2}\int_{0}^{1}\xi(\lambda)\,d\lambda + \frac{2p+1}{4M}\right\}. \qquad (1.12.9)$$

Equation (1.12.8) and expression (1.12.9) can be rewritten in more convenient form if we introduce a new variable ξ and a new function $\rho(\xi)$ by the formulas

$$\xi = \xi(\lambda), \quad \rho(\xi) = \frac{\partial\lambda}{\partial\xi}. \qquad (1.12.10)$$

Owing to the monotonicity of the function $\xi(\lambda)$, the function $\rho(\lambda)$ is non-negative. However, it differs from zero only in the intervals $[A^-, A^+]$ and $[B^-, B^+]$, the ends $A^\pm$ and $B^\pm$ of which are defined as follows:

$$\begin{aligned} A^- &= \xi(0), \quad A^+ = \xi(1 - q - 0), \\ B^- &= \xi(1 - q + 0), \quad B^+ = \xi(1). \end{aligned} \qquad (1.12.11)$$

It is not difficult to see that $\rho(\xi)$ has the meaning of a particle distribution density.

In terms of the function $\rho(\xi)$ equation (1.12.8) takes the form

$$\fint\frac{\rho(\eta)\,d\eta}{\xi - \eta} = 1 + \frac{1}{2}\xi - \frac{2p+1}{M}\frac{1}{\xi}, \qquad (1.12.12)$$

and for the energy E we have:

$$E = 4M(2\beta M)^{\frac{1}{3}} \left\{ 1 + \frac{1}{2} \int \xi \rho(\xi) \, \mathrm{d}\xi + \frac{2p+1}{4M} \right\}. \qquad (1.12.13)$$

Equation (1.12.12) must be supplemented by the requirement that the function $\rho(\xi)$ is non-zero in the intervals $[A^-, A^+]$ and $[B^-, B^+]$ only, and in these intervals satisfies the normalization conditions

$$\int\limits_{A^-}^{A^+} \rho(\xi) \, \mathrm{d}\xi = 1 - q, \qquad \int\limits_{B^-}^{B^+} \rho(\xi) \, \mathrm{d}\xi = q, \qquad (1.12.14)$$

which mean that charges of particles situated in the left and right hand wells are equal to $1 - q$ and q, respectively. In the chosen normalization the total charge of the particles in both wells is equal to unity.

Note that for $M = \infty$, equations (1.12.12) and (1.12.14) can be interpreted as the equilibrium conditions for a charged liquid distributed between two separate wells of the following form:

$$U(\xi) = \begin{cases} \frac{(\xi+2)^2}{4}, & \xi \neq 0, \\ \infty, & \xi = 0. \end{cases} \qquad (1.12.15)$$

The corresponding potential is depicted in figure 1.14. The charge $q = \frac{K}{M}$ of the liquid in the right hand well determines the number of the excited state. According to formula (1.12.13), the energy of this state is expressed in terms of the centre of charge

$$\bar{\xi} = \frac{\int \xi \rho(\xi) \, \mathrm{d}\xi}{\int \rho(\xi) \, \mathrm{d}\xi} = \int \xi \rho(\xi) \, \mathrm{d}\xi \qquad (1.12.16)$$

of all liquid (in both wells). Note also that the behaviour of the function $\rho(\xi)$ in the interval $[A^-, A^+]$ determines the distribution of complex wavefunction zeros, and its behaviour in the interval $[B^-, B^+]$ the distribution of wavefunction nodes.

We shall seek solutions of the equations (1.12.12) and (1.12.14) in the form:

$$\rho(\xi) = \frac{1}{2\mathrm{i}} \mathrm{Dis}\, R(\xi), \qquad (1.12.17)$$

where $R(\xi)$ is a certain real analytic function of the complex parameter ξ having two cuts between the real points A^-, A^+ and B^-, B^+. Obviously, for any such function $R(\xi)$ the corresponding function $\rho(\xi)$ defined by formula

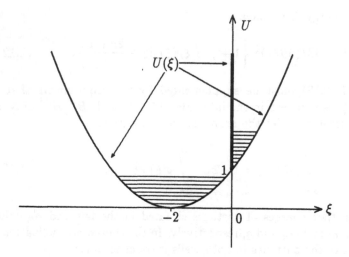

Figure 1.14. The double-well classical potential in which the charged liquid is situated.

(1.12.17) will differ from zero in the intervals $[A^-, A^+]$ and $[B^-, B^+]$ only. This gives us the possibility of writing

$$\fint \frac{\rho(\eta)\, d\eta}{\xi - \eta} = \frac{1}{2} \oint_{C_1} \frac{R(\eta)\, d\eta}{\xi - \eta} + \frac{1}{2} \oint_{C_2} \frac{R(\eta)\, d\eta}{\xi - \eta} = -\frac{1}{2} \oint_C \frac{R(\eta)\, d\eta}{\xi - \eta},$$

$$(1.12.18)$$

where C_1, C_2 and C are the contours shown in figure 1.15. Analogously, one can write:

$$\int \rho(\eta)\, d\eta = \frac{1}{2} \int_{C_1} R(\eta)\, d\eta + \frac{1}{2} \int_{C_2} R(\eta)\, d\eta = -\frac{1}{2} \int_C R(\eta)\, d\eta.$$

$$(1.12.19)$$

From (1.12.17) and (1.12.12) it follows that the function $R(\xi)$ must behave at infinity as $R(\xi) \sim \xi$, $\xi \to \infty$. The simplest function having an analytic structure shown in figure 1.15 is

$$R(\xi) = \frac{i}{2\pi} \sqrt{\frac{(\xi - A^-)(\xi - A^+)(\xi - B^+)}{(\xi - B^-)}}, \qquad (1.12.20)$$

where $A^\pm$ and $B^\pm$ must be considered as unknowns. Substituting (1.12.20) and (1.12.17) into equation (1.12.12) and computing the resulting integrals,

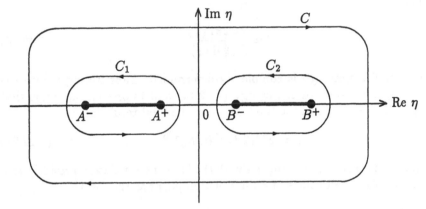

Figure 1.15. The complex η-plane for the function $R(\eta)$ and contours C_1, C_2, and C for the integrals (1.12.18).

we obtain the following conditions

$$B^- = 0, \quad A^+ + A^- + B^+ = -4,$$
$$A^+ A^- + (A^+ + A^-)B^+ = 0. \tag{1.12.21}$$

From this it follows that

$$\rho(\xi) = \frac{1}{2\pi} \mathrm{Re} \sqrt{\frac{e - 4\xi^2 - \xi^3}{\xi}}, \tag{1.12.22}$$

where $e = A^+ A^- B^+$ is a certain unknown parameter. It can be determined from one of the conditions (1.12.14).

Now, consider the case when

$$1 \ll K \ll M. \tag{1.12.23}$$

In this case the charge q is small. Therefore the effective values of coordinate ξ in the second integral (1.12.14) are also small. This enables us to write:

$$q = \int_{B^-}^{B^+} \rho(\xi) \, d\xi \approx \frac{1}{2\pi} \int_0^{\sqrt{e}/2} \sqrt{e - 4\xi^2} \frac{d\xi}{\sqrt{\xi}} = \frac{\Gamma(\frac{3}{2})\Gamma(\frac{1}{4})}{2\sqrt{2}\pi\Gamma(\frac{7}{4})} e^{3/4}, \tag{1.12.24}$$

from which it follows that

$$e = (Cq)^{\frac{4}{3}}, \tag{1.12.25}$$

where

$$C = \frac{2\sqrt{2}\pi\Gamma(\frac{7}{4})}{\Gamma(\frac{3}{2})\Gamma(\frac{1}{4})}. \qquad (1.12.26)$$

Now the function $\rho(\xi)$ is completely determined. Substituting it into the limiting expression for the energy (1.12.13) and taking into account that $Mq = K$, we obtain in the limit $M \to \infty$ the final result for E

$$E = 2^{-1}(2\beta)^{1/3}(CK)^{4/3}, \qquad (1.12.27)$$

which, as is easy to see, coincides with the known semi-classical result. Note also that for small values of ξ the function $\rho(\xi)$ has the form

$$\rho(\xi) = \frac{\sqrt{C}q^{\frac{2}{3}}}{2\pi}\frac{1}{\sqrt{\xi}}, \qquad (1.12.28)$$

which gives us the possibility of restoring the correct semi-classical distribution of wavefunction nodes:

$$x_n \sim n^{\frac{1}{3}}. \qquad (1.12.29)$$

Strictly speaking, equations (1.12.12) and (1.12.13) from which the results (1.12.27) and (1.12.29) were obtained are valid only in the quasi-classical limit when the number K is large (see formula (1.12.23)). In this limit the charge of the liquid situated in the right hand well is not negligibly small; the expression in the curly brackets in (1.12.13) differs from zero and is of order $K^{4/3}/M^{4/3}$. This gives us a true leading term of the semi-classical expansion for E.

However, it is evident that equations (1.12.12) and (1.12.13) cannot always be true since they do not contain any information about the second parameter α entering into the potential (1.12.3). This parameter was simply lost in the derivation of these equations! This happened when we made the transition from the discrete formulas (1.12.6), (1.12.7) to their continuous analogues (1.12.8) and (1.12.9). Indeed, considering the limit $M \to \infty$ we have taken $\Delta = 1$ neglecting the correction which, according to (1.12.2), depends explicitly on α. Besides, replacing the sums entering into (1.12.6) and (1.12.7) by the integrals, we have neglected the corrections to the Euler–McLaurin summation formula. In principle, all these corrections are essential for performing a correct passage to the limit $M \to \infty$. However, if $K \gg 1$ they are relatively small (of order $\frac{1}{K}$), and therefore the leading term of the semi-classical expansion found from equations (1.12.12) and (1.12.13) turned out to be true. Of course, if we want to obtain the next terms of the semi-classical expansion or describe the low excitations in the model (1.12.3) the corrections must be taken into account. In this case, the improved equations (1.12.12) and (1.12.13) become more complicated; however, they can be solved by means of a simple iteration procedure which leads to correct results for the spectrum in the model (1.12.3) (Maglaperidze and Ushveridze 1989b). An example of such a calculation will be considered in detail in chapter 2.

1.13 The Infeld–Hull factorization method and quasi-exact solvability

In this section we show that the model of the quasi-exactly solvable sextic anharmonic oscillator (1.4.13) discussed in the preceding sections can be used as a starting point by constructing new quasi-exactly solvable models, the potentials of which are expressed in terms of rational functions (Shifman 1989b, Ushveridze 1989d, g).

This can be done by means of the so-called Infeld–Hull factorization method (Infeld and Hull 1951), the basic idea of which can be formulated as follows.

Let $V(x)$ be a potential of the quasi-exactly solvable model (1.11.1) for which $M + 1$ explicit solutions E_{2k+p} and $\psi_{2k+p}(x)$, $\quad k = 0, 1, \ldots, M$ are known. We can write:

$$\left\{ -\frac{\partial^2}{\partial x^2} + V(x) \right\} = E_{2k+p}\psi_{2k+p}(x). \tag{1.13.1}$$

Taking $k = 0$ in (1.13.1) we obtain the relation

$$\psi_p''(x) = [V(x) - E_p]\,\psi_p(x), \tag{1.13.2}$$

from which it follows that

$$V(x) - E_p = y_p'(x) + y_p^2(x), \tag{1.13.3}$$

where

$$y_p(x) = \frac{\psi_p'(x)}{\psi_p(x)}. \tag{1.13.4}$$

Substituting (1.13.3) into (1.13.1) and taking into account that

$$-\frac{\partial^2}{\partial x^2} + y_p'(x) + y_p^2(x) = \left[y_p(x) + \frac{\partial}{\partial x} \right] \left[y_p(x) - \frac{\partial}{\partial x} \right], \tag{1.13.5}$$

we can rewrite the equation (1.13.1) in the following factorized form:

$$\left[y_p(x) + \frac{\partial}{\partial x} \right] \left[y_p(x) - \frac{\partial}{\partial x} \right] \psi_{2k+p}(x) = (E_{2k+p} - E_p)\psi_{2k+p}(x). \tag{1.13.6}$$

Now let us act by the operator $y_p(x) - \frac{\partial}{\partial x}$ on (1.13.6). This gives:

$$\left[y_p(x) - \frac{\partial}{\partial x}\right]\left[y_p(x) + \frac{\partial}{\partial x}\right]\left[y_p(x) - \frac{\partial}{\partial x}\right]\psi_{2k+p}(x)$$

$$= (E_{2k+p} - E_p)\left[y_p(x) - \frac{\partial}{\partial x}\right]\psi_{2k+p}(x). \qquad (1.13.7)$$

Using the relation

$$\left[y_p(x) - \frac{\partial}{\partial x}\right]\left[y_p(x) + \frac{\partial}{\partial x}\right] = -\frac{\partial^2}{\partial x^2} - y_p'(x) + y_p^2(x), \qquad (1.13.8)$$

we can rewrite equation (1.13.7) in the following final form:

$$\left[-\frac{\partial^2}{\partial x^2} + \tilde{V}(x)\right]\tilde{\psi}_{2k+p}(x) = E_{2k+p}\tilde{\psi}_{2k+p}(x), \qquad (1.13.9)$$

where

$$\begin{aligned}\tilde{V} &= E_p + y_p^2(x) - y_p'(x)\\ &= V(x) - 2y_p'(x)\\ &= V(x) - 2\frac{\partial^2}{\partial x^2}\ln\psi_p(x) \qquad (1.13.10)\end{aligned}$$

and

$$\begin{aligned}\tilde{\psi}_{2k+p}(x) &= \left[y_p(x) - \frac{\partial}{\partial x}\right]\psi_{2k+p}(x)\\ &= \left[\frac{\partial}{\partial x}\ln\frac{\psi_{2k+p}(x)}{\psi_p(x)}\right]\psi_{2k+p}(x)\\ &= (E_{2k+p} - E_p)\frac{\int\limits_{-\infty}^{x}\psi_{2k+p}(x)\psi_p(x)\,dx}{\psi_p(x)}. \qquad (1.13.11)\end{aligned}$$

From (1.13.11) it follows that $\tilde{\psi}_p(x) = 0$. Therefore, the normalizable functions $\tilde{\psi}_{2k+p}(x)$ correspond to the values $k = 1,\ldots,M$. Thus, we have obtained a new quasi-exactly solvable model with the potential $\tilde{V}(x)$ having only M exact solutions $\tilde{\psi}_{2k+p}(x)$ corresponding to the old eigenvalues E_{2k+p} with $x = 1,\ldots,M$.

The explicit form of this model and its solutions can be obtained by substituting expressions (1.11.1) and (1.11.2) into (1.13.10) and (1.13.11).

This gives:

$$\tilde{V}(x) = a^2 x^6 + 2abx^4 + [b^2 - a(4M + 2p + 3)] + 2b + \frac{2p}{x^2}$$

$$-\sum_{i=1}^{M} \frac{2}{\frac{x^2}{2} - \xi_{0i}} - \sum_{i=1}^{M} \frac{4\xi_{0i}}{\left(\frac{x^2}{2} - \xi_{0i}\right)^2} \qquad (1.13.12)$$

and

$$\tilde{\psi}_{2K+p}(x) = \prod_{i=1}^{M}\left(\frac{x^2}{2} - \xi_{Ki}\right)\left\{\sum_{i=1}^{M} \frac{1}{\frac{x^2}{2} - \xi_{Ki}} - \sum_{i=1}^{M} \frac{1}{\frac{x^2}{2} - \xi_{0i}}\right\}$$

$$\times x^{p+1} \exp\left\{-\frac{ax^4}{4} - \frac{bx^2}{2}\right\}, \qquad (1.13.13)$$

where by ξ_{Ki} and ξ_{0i} we have denoted the number ξ corresponding to the solutions $\psi_{2K+p}(x)$ and $\psi_p(x)$ of the initial quasi-exactly solvable model (1.11.1).

Now let us study model (1.13.12) and the corresponding solution (1.13.13) from a physical point of view.

First of all, remember that all numbers ξ_{0i} are negative since they correspond to the lowest energy level with parity p in model (1.11.1) (the explanation of this fact was given in section 1.11). From the negativity of the numbers ξ_{0i} it follows that the potential (1.13.12) is regular for any $x \neq 0$. If $p = 0$, it is regular even at the point $x = 0$ and has the form of a smooth potential well of an infinite depth (see figure 1.16a). However, if

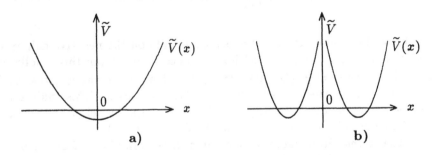

Figure 1.16. The form of the potential (1.13.12) for a) $p = 0$ and b) $p = 1$.

$p = 1$, the potential is singular at the point $x = 0$ and has the form of two separated potential wells as depicted in figure 1.16b. Since the potential barrier at the origin is not penetrable for the particle, we come to two mirror-like Schrödinger problems on the negative and positive half axes.

Now let us discuss the wavefunction properties in the resulting models. From (1.13.13) it follows that all wavefunctions $\tilde{\psi}_{2K+p}(x)$ have poles at points $x = \pm\sqrt{2\xi_{0i}}$. Since all points ξ_{0i} are negative, the wavefunctions are regular on all the real x-axis.

Note that for $p = 0$ the functions (1.13.13) are odd. Therefore they describe only energy levels of odd parity. To obtain a correct numbering of the levels in this case, let us consider the last formula (1.13.11) which becomes in our case

$$\tilde{\psi}_{2K}(x) \sim \frac{1}{\psi_0(x)} \int\limits_{-\infty}^{x} \psi_{2K}(x)\psi_0(x)\, dx. \qquad (1.13.14a)$$

From the sign-definiteness of the ground state wavefunction $\psi_0(x)$ and also from the fact that the function $\psi_{2K}(x)$ has $2K$ nodes it follows immediately that the function $\tilde{\psi}_{2K}(x)$ has only $2K-1$ nodes on the real x-axis and, thus, describes the $(2K-1)$th excitation in the model (1.13.12). This leads us to the assertion that the quasi-exactly solvable sector in the model (1.13.12) with $p = 0$ is formed by the levels with numbers $1, 3, \ldots, 2M - 1$.

When $p = 1$ the wavefunctions (1.13.13) are even, but this fact is not important for us, since only the parts of the wavefunctions defined on the negative (or positive) x-axes have a physical meaning. In this case formula (1.13.11) has the form:

$$\tilde{\psi}_{2K+1}(x) = \frac{1}{\psi_1(x)} \int\limits_{-\infty}^{x} \psi_{2K+1}(x)\psi_1(x)\, dx. \qquad (1.13.14b)$$

We know that the function $\psi_1(x)$ is sign-definite on the negative half axis and that the function $\psi_{2K+1}(x)$ has K nodes there. From this it follows that the function $\tilde{\psi}_{2K+1}(x)$ has exactly $K - 1$ nodes on the negative half axis and, thus, corresponds to the $(K - 1)$th excitation. This enables us to assert that the quasi-exactly solvable sector in the model (1.13.12) with $p = 1$ consists of levels with the numbers $0, 1, \ldots, M - 1$.

The Infeld–Hull factorization procedure described above has a very transparent interpretation within Witten's supersymmetric quantum mechanics (Witten 1982).

Indeed, due to the Z_2 symmetry ($x \rightarrow -x$) the quasi-exactly solvable equation for (1.4.13) can be considered as an equation on the (positive) half axis $x \in [0, \infty]$ with the boundary conditions

$$\psi'(0) = 0, \quad \psi(\infty) = 0, \qquad (1.13.15a)$$

for $p = 0$, and

$$\psi(0) = 0, \quad \psi(\infty) = 0, \tag{1.13.15b}$$

for $p = 1$. In both these cases the Schrödinger problem is well defined and the corresponding hamiltonians are hermitian. In terms of the new boundary conditions the models (1.4.13) can be interpreted as the quasi-exactly solvable models of orders $M + 1$ in which the energy levels with the numbers $0, 1, \ldots, M$ are known exactly.

Following Shifman (1989b) note that the hamiltonian of each of these models can be considered as "one half" of Witten's hamiltonian

$$H = Q_1^2 = Q_2^2, \tag{1.13.16}$$

where Q_1 and Q_2 are supercharges:

$$Q_1 = \sigma_1 \hat{p} + \sigma_2 W(x), \quad Q_2 = \sigma_2 \hat{p} - \sigma_1 W(x). \tag{1.13.17}$$

By reconstructing the missing "other half", we obtain a fully supersymmetric system with "bosonic" and "fermionic" sectors. The hamiltonian acting in the "fermionic" sector — let us call it the daughter hamiltonian — has the same energy eigenvalues as the original one. Moreover, the eigenfunctions of the daughter hamiltonian are obtained from those of the original hamiltonian by applying supercharge. Since $M + 1$ levels of the original hamiltonian are exactly derivable, the same is valid for the daughter system. More exactly, we get explicitly M levels, because the ground state is annihilated by the supercharge, provided that supersymmetry is not broken spontaneously. Here we have considered only the case of the unbroken supersymmetry, since only in this case does the whole construction yield a normalizable wavefunction. As a result we obtain a new quasi-exactly solvable system having M energy levels which can be constructed exactly. As in the initial case there are the ground state, the first excitation, ..., and the $(M - 1)$th excitation.

Obviously, this procedure can be repeated further. In fact, we know the ground state for the daughter hamiltonian, and hence we can construct a new supersymmetric pair of potentials. The potential $\tilde{V}(x)$ will now play the role of the original potential, and starting from this, we shall get the "daughter–daughter" potential $\tilde{\tilde{V}}(x)$, in which we shall know $M - 1$ energy levels, and so on. For a given M this procedure will be exhausted on the Mth step when we come to the model for which only one (the ground state) energy level is known.

1.14   The Gelfand–Levitan equation. Extensions of quasi-exactly solvable problems

In the preceding section we have described the method of expanding the class of quasi-exactly solvable models of type (1.4.13). Obviously, this procedure has a quite general character and can be applied to any quasi-exactly solvable model of one-dimensional quantum mechanics.

Note, however, that models obtained by means of this procedure are "poorer" than the initial model. Indeed, as we have seen, starting with the two-parameter class of quasi-exactly solvable models of order $M + 1$ we obtained a new two-parameter class of models, the order of which is only M. Therefore, on the Mth step this procedure is exhausted.

In this section we will discuss another procedure that gives the possibility of obtaining richer classes of quasi-exactly solvable problems. We show that starting with the L-parameter class of one-dimensional quasi-exactly solvable models of order $M + 1$ it is possible to obtain a new $(L + M + 1)$-parameter class of quasi-exactly solvable models of the same order $M + 1$. This gives us a convenient tool for constructing wide classes of such models, parametrized by an arbitrarily large number of free parameters (Ushveridze 1991a).

This procedure is based on the use of the so-called Gelfand–Levitan equations (Gelfand and Levitan 1951), which until now have been used only for expanding the classes of exactly solvable problems (see e.g. McKean and Trubowitz 1981, Levitan 1987, Pöschl and Trubowitz 1987).

First of all, let us recall some statements of the standard Gelfand–Levitan approach. For this purpose we consider again the model (1.4.13) in which we know exactly $M+1$ wavefunctions $\psi_{2k+p}(x)$ and the corresponding energy levels E_{2k+p}, $x = 0, \ldots, M$. One can write as before

$$\left[-\frac{\partial^2}{\partial x^2} + V(x) \right] \psi_{2k+p}(x) = E_{2k+p}\psi_{2k+p}(x). \tag{1.14.1}$$

The Gelfand–Levitan method allows us to construct a class of new Schrödinger equations

$$\left[-\frac{\partial^2}{\partial x^2} + \widetilde{V}(x) \right] \widetilde{\psi}_{2k+p}(x) = E_{2k+p}\widetilde{\psi}_{2k+p}(x) \tag{1.14.2}$$

with the same spectrum. The daughter potential $\widetilde{V}(x)$ and the functions $\widetilde{\psi}_{2k+p}(x)$ satisfying this equation can be determined from the following simple relations:

$$\widetilde{V}(x) = V(x) + 2\frac{\mathrm{d}}{\mathrm{d}x}K(x,y) \tag{1.14.3}$$

and

$$\tilde{\psi}_{2k+p}(x) = \psi_{2k+p}(x) + \int\limits_{-\infty}^{x} K(x,y)\psi_{2k+p}(y)\, \mathrm{d}y, \qquad (1.14.4)$$

where $K(x,y)$ is a solution of the Gelfand–Levitan equation

$$K(x,y) + Q(x,y) + \int\limits_{-\infty}^{x} K(x,z)Q(z,y)\, \mathrm{d}z = 0 \qquad (1.14.5)$$

with the kernel

$$Q(x,y) = \sum_{n=0}^{\infty} \gamma_n \psi_n(x)\psi_n(y), \qquad (1.14.6)$$

in which γ_n are arbitrary constants. The summation in (1.14.6) is performed over all eigenfunctions of the initial hamiltonian (1.4.13).

Now let us assume that the numbers γ_n differ from zero only for the known levels. In our case these are levels with the numbers $n = 2k + p$, where $k = 0,\ldots,M$. This leads us to the following (degenerate) kernel (1.14.6):

$$Q(x,y) = \sum_{k=0}^{M} c_k \psi_{2k+p}(x)\psi_{2k+p}(y) \qquad (1.14.7)$$

with $c_k = \gamma_{2k+p}$.

Substituting (1.14.7) into the Gelfand–Levitan equation (1.14.5) and seeking the function $K(x,y)$ in the form

$$K(x,y) = \sum_{k=0}^{M} f_k(x)\psi_{2k+p}(y), \qquad (1.14.8)$$

we come to the following system of equations for the coefficient functions $f_k(x)$:

$$f_k(x) + c_k \psi_{2k+p}(x) + c_k \sum_{l=0}^{M} f_l(x) \int\limits_{-\infty}^{x} \psi_{2l+p}(z)\psi_{2k+p}(z)\, \mathrm{d}z,$$

$$x = 0,\ldots,M \quad (1.14.9)$$

which, evidently, can be solved algebraically. The result has the form:

$$K(x,y) = -\sum_{n,m=0}^{M} \left\|\delta_{nm} + c_n \int_{-\infty}^{x} \psi_{2n+p}(z)\psi_{2m+p}(z)\,dz\right\|^{-1}$$

$$\times c_m \psi_{2m+p}(x)\psi_{2n+p}(y). \tag{1.14.10}$$

Substituting (1.14.10) into formulas (1.14.3) and (1.14.4) we obtain

$$\tilde{V}(x) = V(x) - 2\frac{\partial^2}{\partial x^2}\ln\det\left\|\delta_{nm} + c_n \int_{-\infty}^{x} \psi_{2n+p}(z)\psi_{2m+p}(z)\,dz\right\|$$

$$\tag{1.14.11}$$

and

$$\tilde{\psi}_{2n+p}(x) = \psi_{2n+p}(x)$$

$$- \sum_{l,m=0}^{M} \left\|\delta_{nm} + c_n \int_{-\infty}^{x} \psi_{2n+p}(z)\psi_{2m+p}(z)\,dz\right\|^{-1} c_m \psi_{2m+p}(x)$$

$$\times \int_{-\infty}^{x} \psi_{2m+p}(z)\psi_{2n+p}(z)\,dz. \tag{1.14.12}$$

Thus, we have obtained a class of daughter potentials and corresponding solutions in explicit form. From (1.14.11) it follows that this class is parametrized by $M+1$ additional parameters $c_0, c_1, \ldots, c_M$. The largest $((M+1)$-parameter) class arises when all these parameters differ from zero. In the opposite case, when they are all equal to zero, it reduces to the old (initial) model (1.4.13). In the simplest non-trivial case, when only one parameter c_0 differs from zero, formulas (1.14.11) and (1.14.12) take an especially simple form:

$$\tilde{V}(x) = V(x) - 2\frac{\partial^2}{\partial x^2}\ln\left[1 + c_0 \int_{-\infty}^{x} \psi_p^2(z)\,dz\right] \tag{1.14.13}$$

and

$$\tilde{\psi}_{2k+p}(x) = \psi_{2k+p}(x) - \psi_p(x)\frac{c_0 \int_{-\infty}^{x} \psi_p(z)\psi_{2k+p}(z)\,dz}{1 + c_0 \int_{-\infty}^{x} \psi_p^2(z)\,dz}. \tag{1.14.14}$$

In this case we have a three-parameter class of quasi-exactly solvable models and formula (1.14.13) describes a continuous deformation of the initial model (1.4.13) into the other models of this class. We see that for any c_0 satisfying the constraint

$$-\frac{1}{\int\limits_{-\infty}^{\infty} \psi_p^2(z)\, dz} \leq c_0 < \infty \qquad (1.14.15)$$

the wavefunctions (1.14.14) are normalizable and satisfy the needed boundary conditions.

1.15   Summary

Let us now summarize the results obtained above for the sextic anharmonic oscillator with the potential (1.11.1), in which a and b are real continuous parameters, M is an arbitrary non-negative integer and p takes the values 0 and 1.

1. If $a = 0$ this model reduces to a simple harmonic oscillator with the frequency b.

2. If $a \neq 0$ potential (1.11.1) describes a quasi-exactly solvable model of order $M + 1$. For any given M it has $M + 1$ exact solutions corresponding to states with the numbers $2K + p$, $K = 0, 1, \ldots, M$. The spectrum of this model can be expressed in terms of the parameters ξ_i which play the role of the wavefunction zeros and satisfy the system of numerical equations (1.11.4).

3. It is possible to observe explicitly many non-perturbative effects in the model (1.11.1): the convergence of the perturbation series in a, the quasi-crossing of the energy levels, the plaiting of the levels in the complex a-plane, the Bender and Wu singularities, and so on.

4. If $a \sim M^{-\frac{1}{3}}$, $b \sim 4M^{\frac{1}{3}}$ and $M \to \infty$, the model (1.11.1) reduces to the exactly non-solvable model of the quartic anharmonic oscillator.

5. Model (1.11.1) can be used as a zero-order approximation by constructing perturbation theory. For any given perturbation, the terms of the perturbative expansion can be obtained explicitly in quadratures.

6. Model (1.11.1) can be viewed as a spherically non-symmetric quantum top in an external magnetic field. The parameters a, b, M and p determine the strength of this field. Simultaneously, the parameter M determines the spin of the top which is a semi-integer: $s = \frac{M}{2}$. In this

language the order of quasi-exact solvability is equal to the dimension of spin space: $M + 1 = 2s + 1$.

7. Model (1.11.1) can be considered as a result of a separation of variables in the spherically non-symmetric two-dimensional harmonic oscillator with degenerate spectrum. The parameter a determines the frequencies of this oscillator which are equal to a and $2a$. The second parameter b describes the form of the oscillator wavefunction for which separation of variables is possible, and the integer parameters M and p determine the index of the corresponding energy level. The order of quasi-exact solvability, $M + 1$, is equal to the degree of degeneracy of this energy level.

8. Model (1.11.1) is also connected with the completely integrable Gaudin model, which can be solved exactly in the framework of the Bethe *ansatz* method. In this case, parameters a, b and p determine an infinite-dimensional representation of the Gaudin algebra. The parameter M is now the number of elementary excitations in the Gaudin model. The wavefunction zeros ξ_i play the role of the "rapidities" of these excitations and their spectral equations (1.11.4) coincide with the Bethe *ansatz* equations. The order of quasi-exact solvability, $M + 1$, is the dimension of the irreducible representation of the hidden symmetry group for the Gaudin model.

9. Model (1.11.1) is equivalent to a system of M two-dimensional Coulomb particles moving in an external electrostatic field. The parameters a, b and p determine the strength of this field. The non-negative integer M coincides with the number of particles, the wavefunction zeros ξ_i play the role of the coordinates of these particles, and the spectral equations (1.11.4) coincide with the equations for their equilibrium. The order of quasi-exact solvability, $M + 1$, is the number of different equilibrium positions of the particles.

10. The hamiltonian of the model (1.11.1) can be interpreted as a hamiltonian of the bosonic sector in Witten's supersymmetric quantum mechanics. The construction of the fermionic sector is possible and leads to a new quasi-exactly solvable model of order M, the potential of which is expressed in terms of rational functions.

11. The model (1.11.1) can be viewed as a starting point for constructing wide classes of other quasi-exactly solvable models of the same order. This can be done by means of the Gelfand–Levitan equation method.

This list is schematically depicted in figure 1.17. The scheme

Figure 1.17. The problems of theoretical physics appearing by the study of quasi-exactly solvable models of anharmonic oscillators.

clearly demonstrates that model (1.11.1) has extremely rich physical and mathematical properties, can be discussed from various points of view and, beyond any doubt, is a very interesting object to study. Especially important for us were various hidden group-theoretical (Lie-algebraic) properties of model (1.11.1) discussed in sections 1.4, 1.6 and 1.8. The analysis of these properties allowed us to give simple answers to the question: "why is this model quasi-exactly solvable?" In turn, the answers led us to the possibility of formulating simple group-theoretical methods of constructing quasi-exactly solvable models. In the present chapter we formulated only the basic ideas of these methods. In the next chapters we will discuss these methods in more detail and show that they lead to wide classes of various quasi-exactly solvable models, both one dimensional and multi-dimensional.

1.16 Historical comments

Let us make a few remarks about the history of the problem. Following the work of Singh *et al* (1978) devoted to the explicit construction of the ground state solution for the sextic anharmonic oscillator many people started feeling happy even when a single solution of the Schrödinger equation with a non-trivial potential acquired an elementary form. The construction of such solutions was extended by Magyari (1981), Gershenson and Turbiner (1982), Rampal and Datta (1984) and Leach *et al* (1989) to arbitrary polynomials. In this connection we mention also an interesting work of Taylor and Leach (1989) in which two-dimensional models with polynomial potentials are discussed. During the later development the interest in polynomial forces was complemented by studies of various non-polynomial interactions (Flessas 1981, 1982, Znojil 1983, 1984, Blecher and Leach 1987, Hislop *et al* 1990) and strongly singular potentials (Znojil 1982).

The first non-trivial model with two exactly calculable energy levels and with a potential expressed in terms of hyperbolic functions was obtained heuristically by Razavy (1981). A model with similar properties but with a polynomial potential was found by Leach (1984, 1985). Other examples of quantum models admitting several exact solutions and characterized by non-polynomial potentials were discussed by Blecher and Leach (1987), Gallas (1988), Vanden Berghe and De Meyer (1989), Lakhtakia (1989) and Znojil (1990). For more details see the recent paper of Znojil and Leach (1992).

The term "quasi-exact solvability" was introduced in the work of Turbiner and Ushveridze (1986), where two-dimensional quasi-exactly solvable models with degenerate spectra were considered and studied.

The first examples of infinite series of one-dimensional models with arbitrary, arbitrarily large, exactly calculable segments of the spectrum were given by Zaslavsky and Ulyanov (1984) (for the hyperbolic potentials), Bagrov and Vshivtsev (1986) (for the exponential potentials) and Turbiner and Ushveridze (1987) (for the polynomial potentials). Then individual infinite series of models with potentials expressed as powers of exponential, hyperbolic and trigonometric functions were found by Turbiner (1988b). The list of one-dimensional quasi-exactly solvable models was further extended significantly by Ushveridze (1988c, l). In particular, this author found new models with singular trigonometric and hyperbolic potentials, and also a series of models with potentials involving elliptic functions. The existence of a finite series of quasi-exactly solvable models was pointed out in that study. In this connection it is also worth mentioning the papers (Ushveridze 1988k, o, p, 1989d) in which the existence of an infinitely (functionally) large number of quasi-exactly solvable models of not more than fourth order was proved.

The first examples of multi-dimensional quasi-exactly solvable models have been considered by the same author (Ushveridze 1988c, k, m). In these papers, the example of an infinite-dimensional quasi-exactly solvable model was also discussed in detail. Other methods of constructing multi-dimensional quasi-exactly solvable models of limited order are given by Ushveridze (1989c).

The next stage in the history of quasi-exact solvability is characterized by attempts to understand this phenomenon and to formulate general principles allowing the construction and investigation of all possible quasi-exactly solvable models. These attempts led to the development of two fundamentally different approaches, which we shall refer to as the algebraic and analytic approaches.

The algebraic approach, formulated by Turbiner (1988a) for the one-dimensional case and generalized by Shifman and Turbiner (1989) to the case of an arbitrary dimension, is based on the observation that finite-dimensional representations of Lie algebras can be used for generating various quasi-exactly solvable models. Note that the same idea was formulated independently in the paper of Kamran and Olver (1990) which appeared a little later. The essence of this idea can be formulated as follows.

Let I_i be generators of a certain finite-dimensional representation of a Lie algebra $\mathcal{L}$. Then the spectral equation for the operator $H = a_{ik} I^i I^k + b_i I^i$, acting in the corresponding representation space, is finite dimensional and can be solved algebraically for any numbers a_{ik} and b_i. It is known that representations of Lie algebras can be realized in the space of polynomial functions depending on $D = \frac{1}{2}(\dim \mathcal{L} - \operatorname{rank} \mathcal{L})$ variables (Kirillov 1972, Kostant 1977, 1979, Hurt 1983). In this case, the generators of the representations take the form of D-dimensional first-order differential operators. For this reason the spectral equation for H can be interpreted as a certain D-dimensional second-order differential equation. If its reduction to Schrödinger form is possible, we obtain as a result the D-dimensional quasi-exactly solvable model. The number of exactly calculable energy levels in this model is equal to the dimension of the representation. Note also that the spectral operator H can be treated as the hamiltonian of a quantum top, based on the algebra $\mathcal{L}$. Thus, the algebraic approach establishes the connection between quantum tops and quasi-exactly solvable models of quantum mechanics.

There exists only one algebra that leads to one-dimensional quantal problems. This is the algebra $sl(2)$. Its representation with "spin" j has dimension $2j + 1$ and can be realized in the space of polynomials of order $2j$. The corresponding generators have the form (1.6.16). The one-dimensional case was discussed in detail by Turbiner (1988a). (See also the recent paper of González-López *et al* (1993b) in which the problem of normalizability of solutions of one-dimensional quasi-exactly solvable models is discussed.) Later, in the papers of Shifman and Turbiner (Shifman and Turbiner 1989, Shifman 1989a) a number of two-dimensional quasi-exactly solvable models, corresponding to the algebras $sl(3)$, $so(3)$ and $sl(2) \oplus sl(2)$ has been considered. In the same papers, the authors discussed supersymmetric quasi-exactly solvable models, connected with finite-dimensional representations of graded Lie algebras. The exhaustive analysis of all two-dimensional quasi-exactly solvable models is given in the series of papers by González-López *et al* (1991a, b, 1992a, b).

The algebraic approach is attractive primarily because of the simplicity of the idea on which it is based. It is worth stressing that the final formulation of this approach preceded studies by Zaslavsky and Ulyanov (1984), Bagrov and Vshivtsev (1986), Zamolodchikov (1987), who discussed similar ideas. Unfortunately, the algebraic approach is, apparently, not universal, since it cannot be used to describe all quasi-exactly solvable models (Shifman 1989b, c, Ushveridze 1988o, 1992).

The analytic approach was formulated by the present author (Ushveridze 1988c, d, h, 1989c). It is based on the observation that quasi-exactly solvable equations can be viewed as equations with several spectral parameters, some of which are involved in the potential (for example, the parameter M in (1.11.1)), while one plays the role of the "energy" parameter. If the spectra of the "potential" spectral parameters are degenerate with respect to the spectrum of the "energy" parameter, the model is quasi-exactly solvable and its order is equal to the degree of degeneracy. Therefore, the construction of quasi-exactly solvable models reduces to the problem of constructing multi-parameter spectral equations and studying degeneracies in their spectra. This problem can be solved by means of purely analytic methods. It turns out that the mathematical techniques used in solving this problem are very similar to those used in the quantum theory of completely integrable models of magnetic systems based on Lie algebras, and also in the classical multi-particle Coulomb problem, so that three seemingly unrelated branches of quantum and classical physics are seen to be equivalent. The reason is that the allowed values of the "potential" and "energy" spectral parameters of the multi-parameter spectral equation under consideration can be interpreted as the eigenvalues of certain commuting operators with exactly calculable spectra. The degeneracy responsible for quasi-exact solvability is present because of a hidden symmetry in the problem under which the "potential" operators are invariant, while the "energy" operator is not. The full set of "potential" and "energy" operators can be thought of as the integrals of motion of some completely integrable system. In this sense, all the quasi-exactly solvable models arising in the framework of the analytic approach turn out to be equivalent to completely integrable so-called "$sl(2)$ Gaudin magnetic systems", based on infinite-dimensional representations of algebras $sl(2) \oplus \ldots \oplus sl(2)$ and their contractions. These systems permit exact solutions with the help of the algebraic Bethe *ansatz* method. The Bethe equations coincide with the equations determining the spectra of quasi-exactly solvable models. They also coincide with the equilibrium equations for a system of classical two-dimensional Coulomb particles in an external electrostatic field, so that the spectral problem for quasi-exactly solvable models can be posed in purely classical language. If the order of such a model tends to infinity, a non-exactly solvable model arises. Therefore, the equivalence between the problems in non-relativistic quantum mechanics, the theory of completely integrable quantum spin models and classical multi-particle Coulomb systems, discussed for the quasi-exactly solvable case, is also preserved in the non-exactly solvable case.

Further development of the analytic approach led to a more thorough understanding of the problem and stimulated the creation of a new algebraic approach, based on the use of generalized Gaudin models connected with arbitrary simple Lie algebras (Ushveridze 1988d, f, i, n, 1989c, 1990a, 1992, Maglaperidze and Ushveridze 1989a, b, 1990). As in the simplest $sl(2)$ case, these models are completely integrable and their spectral problems can be solved exactly in the framework of the Bethe *ansatz* method.

The connection between the generalized Gaudin models and quasi-exactly solvable ones is caused by the fact that the Hilbert space in which the Gaudin operator acts can be represented as a direct sum of invariant finite-dimensional subspaces. Therefore, the initial Gaudin spectral equation having an infinite exactly calculable spectrum breaks up into an infinite series of equations with a finite number of exactly calculable eigenvectors and eigenvalues. Thus, we obtain an infinite set of quasi-exactly solvable algebraic equations. The next step is to reduce each of these equations to differential form. This can be done by substituting differential realizations of generators of Lie algebras into the Gaudin operator and projecting it on each of the finite-dimensional invariant subspaces of the Hilbert space. It turns out that the projection procedure is equivalent to the procedure of partial separation of variables in a differential version of the Gaudin spectral problem. As a result, we obtain an infinite set of quasi-exactly solvable differential equations, parametrized by multiplets of non-negative integers, enumerating the invariant finite-dimensional subspaces and, simultaneously, playing the role of separation constants. The final step is to rewrite the obtained equations in Schrödinger form. We note that, as in the case of the algebra $sl(2)$, the spectral problems thus obtained allow the reformulation in terms of a classical multi-particle Coulomb problem. However, in contrast to the $sl(2)$ case, the quasi-exactly solvable models connected with the higher Lie algebras turn out to be equivalent to the systems of vector-charged particles.

It is worth stressing that in both the algebraic and analytic approaches, multi-dimensional quasi-exactly solvable equations describe, as a rule, the quantum mechanics on non-trivial curved manifolds. In special cases when the curvature vanishes, we obtain an ordinary quantum mechanics in a flat space. Note also that Lie algebras arise naturally, but in completely different manners, in both approaches. Representations of these algebras used in the author's approach are not finite dimensional as in Turbiner's scheme, but infinite dimensional. The connection between these approaches has been discussed by Maglaperidze and Ushveridze (1989a) and Ushveridze (1990a, 1992).

Finally note that the recent progress in the theory of quasi-exact solvability is associated with the discovery of intriguing parallels between the quasi-exactly solvable problems of quantum mechanics and two-dimensional conformal field theories. These parallels allow one to use quantal methods in conformal field theories and vice versa. The observation of a natural connection between quasi-exactly solvable models in quantum mechanics and conformal field theories was first made in the paper by Morozov *et al* (1990). A new impetus in this line of research is given by two recent studies by Gorsky (1991) and Gorsky and Selivanov (1992). The first paper explicitly demonstrates that the so-called decoupling equations for the conformal blocks in a special class of conformal field theories identically coincide with the quasi-exactly solvable equations for wavefunctions derived by Ushveridze (1989c). Moreover, the computation of the conformal weights is explicitly reduced to that of the corresponding eigenvalues. Finally, probably the most remarkable observation concerns a quantal analogue of the operator product expansion (fusion rules) of conformal field theories. The fact that there should exist a relation between wavefunctions in quasi-exactly solvable models playing the same role as the fusion rules in conformal theories has been conjectured by Morozov *et al* (1990). The issue, however, has not been traced in detail, and the conjectured form of the relation turned out to be unrealistic. Gorsky (1991) has proved that the quasi-exactly solvable analogue of the conformal fusion rules is a set of fusion rules in the parameter space which gives a limiting expression for the wavefunction in a given quasi-exactly solvable model in terms of the corresponding wavefunction of a daughter exactly solvable system. A very detailed review of the results concerning the relationship between quasi-exactly solvable models and conformal field theories has been written by Shifman (1992).

Summarizing, one can conclude that the theory of quasi-exactly solvable models is cross-disciplinary in the full sense of the word. Indeed, in this theory are naturally entwined such branches of mathematical physics as group theory, differential geometry, quantum mechanics, the theory of quantum tops, two-dimensional classical electrostatics, two-dimensional conformal field theory, and also the theory of completely integrable magnetic systems with all its mathematical techniques. It is worth noting that there exists a number of other methods of constructing quasi-exactly solvable models (see e.g. Shifman 1989b, c, Ushveridze 1989c, h, 1991a, b), which, probably, will be included in the coming more general theory. The main aim of this book is to discuss all the methods known at present.

Chapter 2

Simplest analytic methods for constructing quasi-exactly solvable models

2.1 The Lanczos tridiagonalization procedure

In this chapter we discuss some simplest analytic methods of constructing one- and multi-dimensional quasi-exactly solvable models with rational or, more precisely, with quasi-polynomial potentials. We shall call a D-dimensional potential "quasi-polynomial" if it consists of two parts, one of which is an ordinary polynomial in D coordinates x_i^2, while the second is a linear combination of D singular terms x_i^{-2}. Remember that these potentials naturally arise in many problems of quantum mechanics as a result of separation of variables in multi-dimensional toroidal coordinates. In particular, in the one-dimensional case they describe so-called radial Schrödinger equations appearing after the separation of variables in multi-dimensional anharmonic oscillators with spherical symmetry.

We start our discussion with the one-dimensional case. We consider the potential

$$V(x) = \frac{\sigma}{x^2} + \alpha x^2 + \gamma x^6 \qquad (2.1.1)$$

defined on the half-axis $x \in [0, \infty]$ and show that it describes a quasi-exactly solvable model of order $M+1$ if the parameters α, γ and σ satisfy the following condition

$$-\frac{\alpha}{4\gamma} + \sqrt{\tfrac{1}{4} + \sigma} = M + 1, \quad M = 0, 1, 2, \ldots. \qquad (2.1.2)$$

Below, for the sake of convenience, we shall use more suitable parametrization

$$\gamma = a^2, \quad \alpha = -4a\left(s + \tfrac{1}{2} + \mu\right), \quad \sigma = 4\left(s - \tfrac{1}{4}\right)\left(s - \tfrac{3}{4}\right), \quad (2.1.3)$$

in which condition (2.1.2) becomes especially simple:

$$\mu = M, \quad M = 0, 1, 2, \dots. \quad (2.1.4)$$

In this case the hamiltonian for the model (2.1.1) can be written as

$$H = -\frac{\partial^2}{\partial x^2} + a^2 x^6 - 4a\left(s + \tfrac{1}{2} + \mu\right)x^2 + \frac{4(s - \tfrac{1}{4})(s - \tfrac{3}{4})}{x^2}. \quad (2.1.5)$$

The method to be discussed here (Ushveridze 1988k, 1989a) is based on the observation that for any α, γ and σ it is possible to build a basis in the Hilbert space in which the squared hamiltonian (2.1.5), H^2, takes an explicit tridiagonal form. This gives us the possibility of finding conditions for α, γ and σ when two given hermitian conjugated off-diagonal elements of the hamiltonian matrix vanish, after which the matrix takes block-diagonal form, and the model (2.1.1) becomes quasi-exactly solvable.

In order to construct the needed basis let us introduce the trial function

$$\varphi_0(x) = (x^2)^{s - \frac{1}{4}} \exp\left(-\frac{ax^4}{4}\right) \quad (2.1.6)$$

and consider the sequence

$$\varphi_n(x) = (H^2)^n \varphi_0(x), \quad n = 0, 1, 2, \dots, \quad (2.1.7)$$

the terms of which can be represented in the form

$$\varphi_n(x) = P_n\left[\frac{ax^4}{2}\right](x^2)^{s - \frac{1}{4}} \exp\left(-\frac{ax^4}{4}\right), \quad n = 0, 1, 2, \dots, \quad (2.1.8)$$

where $P_n(t)$ are certain polynomials of order n. Note that all functions (2.1.8) reproduce the asymptotic properties of exact wavefunctions in both large- and small-x limits.

Now, let us orthogonalize sequence (2.1.8) using the standard Gram–Schmidt procedure. The nth orthonormalized function, which we denote by $\phi_n(x)$, is a linear combination of the first n functions $\varphi_n(x)$. Hence,

$$\phi_n(x) = Q_n\left[\frac{ax^4}{2}\right](x^2)^{s - \frac{1}{4}} \exp\left(-\frac{ax^4}{4}\right), \quad n = 0, 1, 2, \dots, \quad (2.1.9)$$

where $Q_n(t)$ are certain other polynomials of nth order. Substituting (2.1.9) into the orthonormalization condition

$$\int_0^\infty \phi_n(x)\phi_m(x)\,\mathrm{d}x = \delta_{nm} \qquad (2.1.10)$$

and introducing the new variable

$$t = a\frac{x^4}{2}, \qquad (2.1.11)$$

one can see that the polynomials $Q_n(t)$ are orthogonal with the weight

$$\omega(t) = t^{s-1}\exp(-t), \qquad (2.1.12)$$

and therefore they are the Laguerre polynomials (see e.g. Abramovitz and Stegun 1965). After computing the normalization coefficients we obtain

$$\phi_n(x) = 2\left[\frac{a}{2}\right]^{\frac{s}{2}}\sqrt{\frac{\Gamma(n+1)}{\Gamma(n+c)}}(x^2)^{s-\frac{1}{4}}L_n^{(s-1)}\left[\frac{ax^4}{4}\right]\exp\left(-\frac{ax^4}{4}\right),$$
$$n = 0,1,2,\ldots. \qquad (2.1.13)$$

Our next step is to demonstrate that the squared hamiltonian H^2 has tridiagonal form in the basis (2.1.13). In other words, we must prove that

$$(H^2)_{nm} \equiv \int_0^\infty \phi_n(x)H^2\phi_m(x)\,\mathrm{d}x = 0, \quad \text{if } |n-m| > 1. \qquad (2.1.14)$$

Indeed, from the obvious expansions

$$H^2\phi_n(x) = \sum_{k=0}^n a_{nk}H^2\varphi_k(x) = \sum_{k=0}^n a_{nk}\varphi_{k+1}(x)$$
$$= \sum_{k=0}^n\sum_{m=0}^{k+1} a_{nk}b_{km}\phi_m(x) = \sum_{m=0}^{n+1} c_{nm}\phi_m(x), \quad (2.1.15)$$

in which a_{ik}, b_{ik} and c_{ik} are certain coefficients, it follows that

$$(H^2)_{nm} = 0 \quad \text{for any} \quad n > m+1. \qquad (2.1.16a)$$

Since H^2 is a hermitian operator, we have

$$(H^2)_{nm} = 0 \quad \text{for any} \quad m > n+1 \qquad (2.1.16b)$$

and, thus, assertion (2.1.14) is proved (Lanczos 1950, see also Wilkinson 1965 and Ushveridze 1987a).

The non-zero matrix elements of the operator H^2 can be calculated without difficulties by using well known properties of the Laguerre polynomials (see e.g. Abramowitz and Stegun 1965). The result

$$(H^2)_{nm} = 32a\{(s+n)(2n-\mu)^2 + n(2n-1-\mu)^2\} \quad (2.1.17a)$$

$$(H^2)_{n,n+1} = (H^2)_{n+1,n}$$

$$= -32a\sqrt{(n+1)(n+s)}(2n-\mu)(2n+1-\mu) \quad (2.1.17b)$$

completes the reduction of the operator H^2 to explicit tridiagonal form.

Now note that if μ is a non-negative integer: $\mu = M = 0, 1, 2, \ldots$, then the elements $(H^2)_{n,n+1}$ and $(H^2)_{n+1,n}$ with $n = \left[\frac{M}{2}\right]$ vanish, so that the infinite-dimensional matrix $(H^2)_{nm}$ takes block-diagonal form:

$$H^2 = (H^2)_{\text{fin}} \oplus (H^2)_{\text{inf}}. \quad (2.1.18)$$

Here $(H^2)_{\text{fin}}$ is a finite $\left(\left[\frac{M}{2}\right]+1\right) \times \left(\left[\frac{M}{2}\right]+1\right)$ block which acts in the $\left(\left[\frac{M}{2}\right]+1\right)$-dimensional Hilbert subspace formed by all linear combinations of the basis functions $\phi_0(x), \ldots, \phi_n(x)$, $n = \left[\frac{M}{2}\right]$. We denote by h and $\eta(x)$ the eigenvalues and eigenfunctions of the finite-dimensional matrix $(H^2)_{\text{fin}}$. Then the corresponding eigenvalues and eigenfunctions of the initial operator H can be determined by the formulas

$$E = \pm\sqrt{h} \quad (2.1.19a)$$

and

$$\psi(x) \sim [H \pm \sqrt{h}]\eta(x), \quad (2.1.19b)$$

which give us exact (algebraic) solutions of the initial Schrödinger equation for model (2.1.5).

Now let us consider several concrete examples corresponding to the cases $M = 0, 1, 2, 3$.

1. Let $M = 0$. In this case $(H^2)_{\text{fin}}$ is a 1×1 matrix:

$$(H^2)_{\text{fin}} = \|0\| \quad (2.1.20)$$

and therefore formulas (2.1.19) determine one explicit solution of the Schrödinger equation:

$$E_n = 0, \quad n = 0, \quad (2.1.21a)$$

$$\psi_n(x) = (x^2)^{s-\frac{1}{4}} \exp\left(-\frac{ax^4}{4}\right), \quad n = 0. \quad (2.1.21b)$$

We see that the wavefunction obtained has no nodes and, hence, according to the oscillator theorem, this solution corresponds to the ground state. It is normalizable if $a > 0$ and $s > 0$. The analytically continued eigenvalue forms the univalent Riemann surface.

2. Let $M = 1$. As in the previous case, $(H^2)_{\text{fin}}$ is a 1×1 matrix

$$(H^2)_{\text{fin}} = \|32as\|. \qquad (2.1.22)$$

Using formulas (2.1.19) we obtain two explicit solutions of the Schrödinger equation:

$$E_n = (-1)^{n+1}\sqrt{32as}, \quad n = 0, 1, \qquad (2.1.23a)$$

$$\psi_n(x) = \left[ax^2 - \frac{E_n}{4}\right](x^2)^s - \tfrac{1}{4} \exp\left(-\frac{ax^4}{4}\right), \quad n = 0, 1. \qquad (2.1.23b)$$

According to the oscillator theorem, these solutions correspond to the ground and first excited states, respectively. They are normalizable when $a > 0$ and $s > 0$. The energy levels continued into the complex s-plane are analytic everywhere except for the point $s = 0$ in which they have a square-root-type branch-point singularity. At this point the levels (2.1.23a) are plaited and, consequently, can be treated as two different sheets of a Riemann surface. Wavefunctions (2.1.23b) coincide when $s = 0$.

3. Let $M = 2$. Now $(H^2)_{\text{fin}}$ is a 2×2 matrix of the form:

$$(H^2)_{\text{fin}} = \left\| \begin{matrix} 128as & -64a\sqrt{s} \\ -64a\sqrt{s} & 32a \end{matrix} \right\|. \qquad (2.1.24)$$

In this case we can construct three explicit solutions of the Schrödinger equation:

$$E_n = (n-1)\sqrt{32a(4c+1)}, \quad n = 0, 1, 2, \qquad (2.1.25a)$$

$$\psi_n(x) \sim \left[ax^4 - \frac{E_n}{4}x^2 + \frac{E_n^2}{32a} - 2s - 1\right]$$

$$\times \ (x^2)^s - \tfrac{1}{4} \exp\left(-\frac{ax^4}{4}\right), \quad n = 0, 1, 2. \qquad (2.1.25b)$$

If $a > 0$ and $s > 0$, the obtained solutions are normalizable and describe the ground, first and second excited states. We see that all levels (2.1.25a) continued analytically into the complex s-plane coincide if $s = -\tfrac{1}{4}$. But only two of them, E_0 and E_2, are singular at this point and can be continued into each other. The third level E_1 is regular everywhere. This means that we have two disconnected Riemann surfaces formed by the odd and even levels, and consisting of one and two sheets, respectively. The wavefunctions also coincide at the point $s = -\frac{1}{4}$ in full accordance with the previous case.

4. Let $M = 3$. In this case $(H^2)_{\text{fin}}$ is again the 2×2 matrix. It has the form:

$$(H^2)_{\text{fin}} = \left\| \begin{array}{cc} 288ac & -192a\sqrt{s} \\ -192a\sqrt{s} & 32a(s+5) \end{array} \right\|, \tag{2.1.26}$$

which allows us to obtain four explicit solutions:

$$E_n = (-1)^{\left[\frac{n}{2}\right]+1} \sqrt{32a}$$
$$\times \sqrt{5\left(s+\tfrac{1}{2}\right) + (-1)^{\left[\frac{n+1}{2}\right]} \sqrt{25\left(s+\tfrac{1}{2}\right)^2 - 9s(s+1)}},$$
$$n = 0,1,2,3, \tag{2.1.27a}$$

$$\psi_n(x) \sim (x^2)^{s-\frac{1}{4}} \exp\left(-\frac{ax^4}{4}\right)$$
$$\times \left\{ a^2 x^6 - a\frac{E_n}{4} x^4 + \frac{E_n^2 - 96a(s+1)}{32} x^2 - \frac{E_n^3}{384a} + [7s+5]\frac{E_n}{12} \right\}$$
$$n = 0,1,2,3, \tag{2.1.27b}$$

corresponding to the zeroth, first, second and third energy levels if the normalization conditions $a > 0$ and $s > 0$ hold. We see that energy levels (2.1.27a) are singular at the points $s = 0$, $s = -1$ and $s = -\frac{1}{2} \pm \frac{3}{2}i$, in which the external and internal roots in (2.1.27a) vanish. As in the previous case, the first two singularities lie on the real s-axis, and the second two are located at the complex conjugated points of the complex s-plane. The four functions E_n, $n = 0,1,2,3$ form a common Riemann surface and hence all the levels E_n described by formula (2.1.27a) can be obtained from each other by a simple analytic continuation.

An analogous analysis can be carried out for the next values of M. It is not difficult to understand that for arbitrarily fixed M we have a quasi-exactly solvable model of order $M + 1$, in which we can construct algebraically $M + 1$ first energy levels. The wavefunctions of these levels have the form:

$$\psi(x) = P_M(x^2)(x^2)^{s-1} \exp\left(-\frac{ax^4}{4}\right), \tag{2.1.28}$$

where $P_M(t)$ are certain polynomials of order M.

In conclusion, note that the introduction of the auxiliary operator H^2 is no more than a convenient trick that allows us to observe the appearance of finite-dimensional blocks in the infinite-dimensional hamiltonian matrix. Obviously, we could work immediately with the hamiltonian H.

2.2   The sextic oscillator with a centrifugal barrier

Now let us consider the class of models described by the hamiltonian

$$
\begin{aligned}
H \;=\; & -\frac{\partial^2}{\partial x^2} + \frac{4\left(s - \frac{1}{4}\right)\left(s - \frac{3}{4}\right)}{x^2} + \left[b^2 - 4a\left(s + \frac{1}{2} + M\right)\right]x^2 \\
& + 2abx^4 + a^2 x^6
\end{aligned}
\tag{2.2.1}
$$

and defined on the positive half-axis $x \in [0, \infty]$. These models (which were found by Turbiner (1988b)) differ from those discussed in the preceding section by the presence of an additional parameter b. If $b = 0$ then (2.2.1) reduces to the old hamiltonian (2.1.5).

The model (2.2.1) is quasi-exactly solvable for any values of b. For any given non-negative integer M, it has $M + 1$ solutions which can be found algebraically.

Unfortunately, the method used above for demonstrating the quasi-exact solvability of model (2.1.5) cannot be easily generalized to the case of models (2.2.1). The problems arising are connected with the quartic term in (2.2.1) that does not permit us to reduce this hamiltonian to an explicit tridiagonal form. This assertion can be clarified as follows.

Let us consider the function

$$
\varphi(x) = (x^2)^{s-1} \exp\left(-\frac{ax^4}{4} - \frac{bx^2}{2}\right)
\tag{2.2.2}
$$

which reproduces the asymptotic properties of exact eigenfunctions of H in both the small- and large-x limits and thus, is a most natural generalization of trial function (2.1.6). Repeating the reasoning of the preceding section and acting on $\varphi(x)$ by the operators H^n, $n = 0, 1, 2, \ldots$, we obtain the sequence

$$
\varphi_n(x) = P_n(x^2)\,(x^2)^{s-1} \exp\left(-\frac{ax^4}{4} - \frac{bx^2}{2}\right), \quad n = 0, 1, 2, \ldots
\tag{2.2.3}
$$

the elements of which, after applying the orthonormalization procedure, form a basis in which the hamiltonian H takes tridiagonal form. Obviously this procedure implies a knowledge of the polynomials depending on x^2 and being orthogonal with the weight $(x^4)^{s-1} \exp\left(-\frac{ax^4}{2} - bx^2\right)$ on the positive half-axis $x \in [0, \infty]$. When $b = 0$, explicit expressions for such polynomials are known. They coincide with the classical Laguerre polynomials depending on x^4. The eveness of these polynomials with respect to x^2 makes the Lanczos procedure more sensible for the squared hamiltonian H^2. However, if $b \neq 0$, the orthogonal polynomials can be constructed explicitly for the first several values of n only. General expressions for them are not known. This means that, in this case, explicit tridiagonalization of the hamiltonian matrix is impossible.

In order to prove that model (2.2.1) is quasi-exactly solvable, we use another method (Ushveridze 1988k). Consider the Schrödinger equation

$$H\psi(x) = E\psi(x) \qquad (2.2.4)$$

for (2.2.1) and note that a linear combination of basis functions (2.2.3),

$$\psi(x) = P(x^2) (x^2)^{s-\frac{1}{4}} \exp\left(-\frac{ax^4}{4} - \frac{bx^2}{2}\right), \qquad (2.2.5)$$

can be viewed as an appropriate *ansatz* for this equation. Indeed, subsituting (2.2.5) into (2.2.4) and eliminating the common factor (2.2.2) in (2.2.4), we obtain a new equation for the function $P(x^2)$.

$$QP(x^2) = EP(x^2). \qquad (2.2.6)$$

Here

$$Q = -\left[\frac{\partial^2}{\partial x^2} + \frac{4s-1}{x^2}\right] + 2b\left[x\frac{\partial}{\partial x} + 2s\right] + 2ax^2\left[x\frac{\partial}{\partial x} - 2M\right]. \qquad (2.2.7)$$

Now, let us assume that M is a non-negative integer: $M = 0, 1, 2, \ldots$. In this case the differential spectral equation (2.2.6) can easily be transformed to an algebraic form. Indeed, if $P(x^2)$ is a polynomial in x^2 of order M, then the action of the Q-operator (2.2.7) on $P(x^2)$ gives again a certain polynomial in x^2 of the same order. Therefore, formula (2.2.6) expresses the equality of two polynomials of order M. Considering the coefficients of the polynomial $P(x^2)$ as components of an $(M + 1)$-dimensional vector, one can treat (2.2.6) as an $(M + 1)$-dimensional spectral matrix equation. In general, it has $M + 1$ different solutions. This means that the initial Schrödinger equation (2.2.4) has at least $M+1$ explicit solutions of the form (2.2.5) and thus can be interpreted as a quasi-exactly solvable equation of order $M + 1$.

From formula (2.2.5) it follows that normalization conditions for the solutions hold when $a > 0$ and $c > 0$. The parameter b can be chosen arbitrarily. Note also that these $M + 1$ solutions describe the ground state and first M excitations, since the polynomial $P(x^2)$ cannot have more than M zeros on the half-axis $x \in [0, \infty]$.

Now let us consider three particular cases when M takes the values 0, 1 and 2.

1. $M = 0$. Ground state.

$$E = 4bs, \tag{2.2.8a}$$

$$\psi(x) \sim x^{2(s-\frac{1}{4})} \exp\left(-\frac{ax^4}{4} - \frac{bx^2}{2}\right). \tag{2.2.8b}$$

2. $M = 1$. Ground state and first excitation.

$$E = 4bs + \lambda, \tag{2.2.9a}$$

$$\psi(x) \sim \left[1 - \frac{\lambda}{8s}x^2\right] x^{2(s-\frac{1}{4})} \exp\left(-\frac{ax^4}{4} - \frac{bx^2}{2}\right). \tag{2.2.9b}$$

The parameter λ satisfies the following quadratic equation:

$$\lambda(\lambda - 4b) - 32as = 0. \tag{2.2.10}$$

3. $M = 1$. Ground state and first two excitations.

$$E = 4bs + \lambda, \tag{2.2.11a}$$

$$\psi(x) \sim \left[1 - \frac{\lambda}{8s}x^2 - \frac{a\lambda}{2s(\lambda - 8b)}x^4\right] x^{2(s-\frac{1}{4})} \exp\left(-\frac{ax^4}{4} - \frac{bx^2}{2}\right). \tag{2.2.11b}$$

In this case, the parameter λ satisfies the following cubic equation:

$$\lambda[(\lambda - 4b)(\lambda - 8b) - 32a(2s + 1)] - 64as(\lambda - 8b) = 0. \tag{2.2.12}$$

Analogous explicit formulas can also be obtained for the next values of M. Note that for any fixed M, energy levels belonging to the algebraized part of the spectrum can be written in the form

$$E = 4bs + \lambda, \tag{2.2.13}$$

where λ satisfies a certain algebraic (secular) equation of order $M + 1$:

$$Q_{M+1}(a, b, s; \lambda) = 0. \tag{2.2.14}$$

As in the case of the simpler model (2.1.5), these expressions can be used for studying various spectral singularities in model (2.2.1). Remember that these so-called Bender and Wu singularities appear when two or more energy levels, analytically continued into the complex plane of parameters a, b and s, coincide. The condition of coincidence of $K + 1$ energy levels satisfying equation (2.2.14) can be written as

$$Q_{M+1}(a, b, s; \lambda) = 0,$$

$$\frac{\partial}{\partial \lambda} Q_{M+1}(a, b, s; \lambda) = 0,$$

$$\cdots$$

$$\left(\frac{\partial}{\partial \lambda}\right)^{K+1} Q_{M+1}(a, b, s; \lambda) = 0. \tag{2.2.15}$$

Note that the energy levels in model (2.2.1) depend non-trivially on two combinations of parameters a, b and s only. The third independent combination can easily be eliminated by means of a scale transformation. Therefore, the number K in (2.2.15) cannot exceed two. Solving the system (2.2.15) for $K = 1$ and $K = 2$ we obtain surfaces in the three-dimensional complex plane of parameters a, b and s, on which the double and triple spectral points lie. For example, for $M = 1$, equation (2.2.14) has the form (2.2.10). Taking $K = 1$ we obtain the double-point surface determined by the equation

$$b^2 - 8as = 0. \tag{2.2.16}$$

If $M = 2$, then equation (2.2.14) takes the form (2.2.12). In this case the existence of both double- and triple-point surfaces are possible. Taking, for instance, $K = 2$, we obtain two equations

$$b = 0, \quad s = -\frac{1}{4}, \tag{2.2.17}$$

determining the triple-point line (Ushveridze 1988k).

Spectral equations for model (2.2.1) can also be rewritten in terms of wavefunction zeros. To this end we rewrite the *ansatz* (2.2.3) as:

$$\psi(x) = \prod_{i=1}^{M} \left(\frac{x^2}{2} - \xi_i\right) (x^2)^{s-\frac{1}{4}} \exp\left(-\frac{ax^4}{4} - \frac{bx^2}{2}\right). \tag{2.2.18}$$

Then substituting (2.2.18) into (2.2.4) we obtain the following expression for the energy:

$$E = 4b(M + s) + 8a \sum_{i=1}^{M} \xi_i, \tag{2.2.19}$$

in which the numbers ξ satisfy the system of numerical equations:

$$\sum_{k=1}^{M} \frac{1}{\xi_i - xi_k} + \frac{s}{\xi_i} - b - 2a\xi_i = 0, \quad i = 1, \ldots, M. \qquad (2.2.20)$$

This system can be considered as an equilibrium condition for M Coulomb particles with unit charges and coordinates ξ_i moving in an external electrostatic field with potential

$$U(\xi) = -s \ln |\xi| + b\xi + a\xi^2. \qquad (2.2.21)$$

Here parameters a and b determine the strength of the electrostatic field at large distances and s is the charge of the repelling centre at the origin. This potential consists of two potential wells separated by a potential barrier. The number of particles in the right-hand well determines the number of excitation.

Note that this electrostatic analogue makes it easy to construct trajectories in a complex parameter space, along which energy levels transform into other energy levels. To show this, let us assume that the stability condition $a > 0$, $s > 0$ is satisfied and the initial position of the particles corresponds to the Kth energy level. This means that K particles are located in the right-hand well, and $M - K$ in the left-hand one (see figure 2.1a).

Consider the following trajectory in the (real) space of two parameters c and b.

1. The charge s of the repelling centre decreases to zero $(s \rightarrow 0)$ while parameter b remains constant. Of course this leads to a change of equilibrium positions of the particles (see figure 2.1b).

2. The strength b of the electrostatic field increases or decreases $(b \rightarrow b')$ in such a way that K' particles have positive coordinates and $M - K'$ particles have negative coordinates. Quite obviously, it is always possible to guarantee that $K' \neq K$ (see figure 2.1c).

3. The charge of the repelling centre is restored $(0 \rightarrow s)$ while strength b' remains constant (see figure 2.1d).

4. The original strength of the electrostatic field is restored $(b' \rightarrow b)$ (see figure 2.1e).

The final configuration of the particles obviously corresponds to the K'th energy level. Thus, we see that rather simple trajectories in the two-dimensional space of parameters b and s shown in figure 2.2 realize

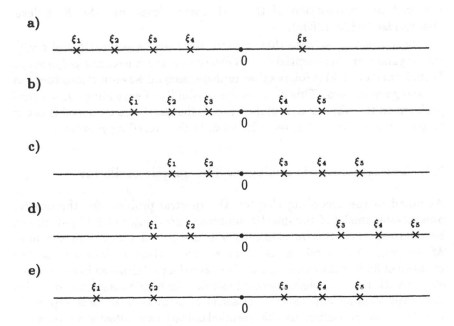

Figure 2.1. The change of dispositions of Coulomb particles by changing the external parameters b and s. Here $M = 5$, $K = 1$ and $K' = 3$.

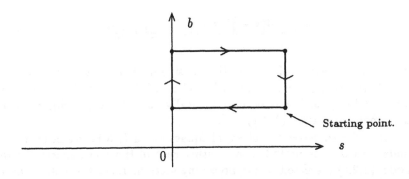

Figure 2.2. The trajectory in the space of parameters b and s realizing analytic continuation of a given energy level into another energy level.

the analytic continuation of the Kth energy level into the K'th level (Ushveridze 1988g, 1989c).

In conclusion, we note that in particular cases, when $s = \frac{1}{4}$ and $s = \frac{3}{4}$, the singular term in potential (2.2.1) disappears and it becomes polynomial. From formula (2.2.5) it follows that in these cases all wavefunctions turn out to be regular at zero. This gives us the possibility of extending the spectral problem on the whole x-axis and we regain the quasi-exactly solvable sextic anharmonic oscillator discussed in detail in the preceding chapter.

2.3   The electrostatic analogue. The quartic oscillator

As noted in the preceding chapter, the spectral problem for the exactly non-solvable model of the quartic anharmonic oscillator (1.3.1) (which can be obtained from the quasi-exactly solvable model (1.4.15) in the limit $M \to \infty$), can be reformulated in a purely classical language as the problem of finding the equilibrium of a charged liquid situated in an external electrostatic field. We demonstrated this fact for high excitations, for which the corresponding equations of classical electrostatics become especially simple. The restriction to this (semi-classical) case allowed us to avoid many difficulties connected with the correct passage to the limit $M \to \infty$. However, in order to assert that the quantum anharmonic oscillator is really equivalent to a classical charged liquid, we must convince ourselves that the classical equations give correct quantum results for low excitations, too.

In this section we will discuss this question considering, as an example, the problem of constructing the ground state in the following exactly non-solvable model

$$V(x) = \frac{4(s - \frac{1}{4})(s - \frac{3}{4})}{x^2} + \alpha x^2 + \beta x^4. \tag{2.3.1}$$

Due to the presence of an additional (singular) term, this model can be considered as a generalization of model (1.3.1). Therefore, any result obtained for (2.3.1) will be automatically valid for the ordinary quartic anharmonic oscillator (1.3.1).

In order to derive the classical equations for (2.3.1), we note that this model can be interpreted as a limiting case of the quasi-exactly solvable model (2.2.5) discussed in the preceding section. Indeed, substituting into (2.2.5)

$$4a = M^{-\frac{1}{3}}(2\beta)^{\frac{2}{3}}\Delta^{-1}(g), \tag{2.3.2a}$$

$$b = M^{\frac{1}{3}}(2\beta)^{\frac{1}{3}}\Delta(g), \tag{2.3.2b}$$

where

$$g = \frac{\alpha}{(2\beta)^{2/3}} \tag{2.3.3}$$

and

$$\Delta(g) = 1 + \frac{g}{3M^{2/3}}, \tag{2.3.4}$$

and taking the limit $M \to \infty$ we obtain model (2.3.1).

Construction of corresponding limiting analogues of the classical equations (2.2.18)–(2.2.20) is a much more non-trivial procedure. Here we reproduce the derivation given in the paper of Maglaperidze and Ushveridze (1989c).

First of all, note that from the technical point of view it is more convenient to deal with a system consisting of $M + 1$ particles. Of course, the replacement of M by $M + 1$ in formulas (2.2.18)–(2.2.20) cannot change the limiting result appearing when $M = \infty$. The equilibrium conditions for the $(M + 1)$-particle Coulomb system are described by equations (2.2.20) which, after using formulas (2.3.2), take the form

$$\sum_{k=0}^{M}{}' \frac{1}{\xi_i - \xi_k} + \frac{s}{\xi_i} - M^{\frac{1}{3}}(2\beta)^{\frac{1}{3}}\Delta(g) - \frac{1}{2}M^{-\frac{1}{3}}(2\beta)^{\frac{2}{3}}\Delta^{-1}(g)\xi_i = 0,$$

$$i = 0, \ldots, M \tag{2.3.5}$$

where the numbers ξ_i are enumerated in increasing order:

$$\xi_0 < \xi_1 < \ldots < \xi_{M-1} < \xi_M. \tag{2.3.6}$$

Taking

$$\xi_i = M^{2/3}(2\beta)^{-\frac{1}{3}}\Delta^2(g)\xi(\tfrac{i}{M}), \tag{2.3.7}$$

we rewrite system (2.3.5) as

$$\sum_{k=0}^{M}{}' \frac{1}{\xi(\frac{i}{M}) - \xi(\frac{k}{M})} \frac{1}{M} - \Delta^3(g)\left[1 + \frac{1}{2}\xi(\tfrac{i}{M})\right] + \frac{s}{M}\frac{1}{\xi(\frac{i}{M})} = 0,$$

$$i = 1, \ldots, M. \tag{2.3.8}$$

This system can be easily reduced to integral form. For this purpose we introduce the continuous variable

$$\lambda = \frac{i}{M} \tag{2.3.9}$$

and the continuous function

$$\xi = \xi(\lambda). \tag{2.3.10}$$

Using the Euler–McLaurin summation formula for (2.3.8) we find:

$$\int_0^{\lambda - \frac{1}{M}} \frac{d\nu}{\xi(\lambda) - \xi(\nu)} + \int_{\lambda + \frac{1}{M}}^1 \frac{d\nu}{\xi(\lambda) - \xi(\nu)}$$

$$+ \frac{1}{2M}\left[\frac{1}{\xi(\lambda) - \xi(0)} + \frac{1}{\xi(\lambda) - \xi(\lambda - \frac{1}{M})} + \frac{1}{\xi(\lambda) - \xi(1)}\right.$$

$$\left. + \frac{1}{\xi(\lambda) - \xi(\lambda + \frac{1}{M})}\right]$$

$$+ \frac{s}{M}\frac{1}{\xi(\lambda)} + \Delta^3(g)\left[1 + \frac{1}{2}\xi(\lambda)\right] = 0. \tag{2.3.11}$$

We have retained only the first two terms in the Euler–McLaurin expansion. The other terms, as we will see later, are non-essential in the limit $M \to \infty$. Expanding (2.3.11) in inverse powers of M and neglecting terms of orders $M^{-2}, M^{-3}, \ldots$, which do not give any contribution to the final result, we obtain

$$\fint_0^1 \frac{d\nu}{\xi(\lambda) - \xi(\nu)} - \Delta^3(g)(1 + \tfrac{\xi(\lambda)}{2})$$

$$+ \frac{1}{2M}\left\{\frac{1}{\xi(\lambda) - \xi(0)} + \frac{1}{\xi(\lambda) - \xi(1)} - \frac{\xi''(\lambda)}{[\xi'(\lambda)]^2} + \frac{2c}{\xi(\lambda)}\right\} = 0. \tag{2.3.12}$$

Introducing the new variable ξ and new function $\rho(\xi)$ by the formulas

$$\xi = \xi(\lambda), \quad \rho(\xi) = \frac{\partial \lambda}{\partial \xi} = \frac{1}{\xi'(\lambda)}, \tag{2.3.13}$$

and taking

$$A^- = \xi(0), \quad A^+ = \xi(1), \tag{2.3.14}$$

we obtain equation

$$\int_{A^-}^{A^+} \frac{\rho(\eta)\,d\eta}{\xi - \eta} - \Delta^3(g)(1 + \tfrac{\xi}{2})$$

$$+ \frac{1}{2M}\left\{\frac{1}{\xi - A^-} + \frac{1}{\xi - A^+} + \frac{2s}{\xi} + \frac{\rho'(\xi)}{\rho(\xi)}\right\} = 0,$$

$$\xi \in [A^-, A^+], \tag{2.3.15}$$

which, obviously, must be supplemented by the normalization condition for $\rho(\eta)$:

$$\int_{A^-}^{A^+} \rho(\xi) \, d\xi = 1. \tag{2.3.16}$$

Thus, we have obtained the correct limiting analogue of the initial equation (2.2.20).

Now, let us consider expressions (2.2.18) and (2.2.19) which, after replacing M by $M + 1$ and using formulas (2.3.2) and (2.3.7) take the form

$$E = 4(2\beta)^{\frac{1}{3}}\Delta(g)M^{\frac{4}{3}}\left[1 + \frac{s+1}{M} + \frac{1}{2}\sum_{i=0}^{M}\xi\left(\frac{i}{M}\right)\frac{1}{M}\right], \tag{2.3.17}$$

$$\begin{aligned}\psi(x) = (x^2)^{s-\frac{1}{4}}\exp\Bigg\{&-M^{\frac{1}{3}}\Delta(g)(2\beta)^{\frac{1}{3}}\frac{x^2}{2}\\&-\frac{1}{4}M^{-\frac{1}{3}}\Delta^{-1}(g)(2\beta)^{\frac{2}{3}}\frac{x^4}{4}\\&+M\sum_{i=0}^{M}\ln\left((2\beta)^{\frac{1}{3}}\frac{x^2}{2} - M^{\frac{2}{3}}\Delta^2\xi(\tfrac{i}{M})\right)\Bigg\}.\end{aligned} \tag{2.3.18}$$

Replacing in (2.3.17) and (2.3.18) the sums by integrals by means of the Euler–McLaurin formula, and rewriting the resulting integral expressions in terms of the variable ξ and function $\rho(\xi)$, we obtain in the large-M limit the expressions

$$E = 4(2\beta)^{\frac{1}{3}}\Delta(g)M^{\frac{3}{4}}\left\{1 + \frac{1}{2}\int_{A^-}^{A^+}\xi\rho(\xi)\, d\xi + \frac{s+1+(A^+ + A^-)/4}{M}\right\}, \tag{2.3.19}$$

and

$$\begin{aligned}\psi(x) = (x^2)^{s-\frac{1}{4}}&\left[(2\beta)^{\frac{1}{3}}\frac{x^2}{2} - M^{2/3}\Delta^2(g)A^-\right]^{\frac{1}{2}}\\&\times\left[(2\beta)^{\frac{1}{3}}\frac{x^2}{2} - M^{2/3}\Delta^2(g)A^+\right]^{\frac{1}{2}}\end{aligned}$$

$$\times \exp\left\{ -(2\beta M)^{\frac{1}{3}}\Delta(g)\frac{x^2}{2} + M\int\limits_{A^-}^{A^+} \ln\left[(2\beta)^{\frac{1}{3}}\frac{x^2}{2} - M^{\frac{2}{3}}\Delta^2(g)\xi\right]\rho(\xi)\,\mathrm{d}\xi\right\},$$

$$(2.3.20)$$

which are the correct limiting analogues of formulas (2.2.19) and (2.2.18).

Now, let us discuss the properties of the obtained expressions in which the role of unknown objects is played by the particle distribution density $\rho(\xi)$ and by the numbers $A^\pm$ determining the ends of the interval in which $\rho(\xi)$ differs from zero.

Let us first take $M = 0$ in (2.3.15) and (2.3.16). This gives us the system of equations

$$\fint\limits_{A^-}^{A^+} \frac{\rho(\eta)}{\xi - \eta}\,\mathrm{d}\eta = 1 + \frac{\xi}{2},\qquad (2.3.21)$$

$$\int\limits_{A^-}^{A^+} \rho(\xi)\,\mathrm{d}\xi = 1,\qquad (2.3.22)$$

which can be interpreted as the equilibrium condition for a charged liquid with charge density $\rho(\xi)$ and with total charge 1. This liquid is situated in an external electrostatic potential of oscillator type $U(\xi) = \frac{1}{4}(\xi + 2)^2$. Corresponding formulas (2.3.19) and (2.3.20) for E and $\psi(x)$ take in this case the form:

$$E = \infty\left\{1 + \frac{1}{2}\int\limits_{A^-}^{A^+} \xi\rho(\xi)\,\mathrm{d}\xi\right\},\qquad (2.3.23)$$

$$\psi(x) = (x^2)^{s-\frac{1}{4}}\exp\left\{-\infty\left[(2\beta)^{\frac{1}{3}}\frac{x^2}{2}\left(1 + \int\limits_{A^-}^{A^+}\frac{\mathrm{d}\xi}{\xi}\rho(\xi)\right) + \ldots\right]\right\}.$$

$$(2.3.24)$$

Multiplying (2.3.21) by $\rho(\xi)$, integrating over ξ and using (2.3.22) we find:

$$1 + \frac{1}{2}\int\limits_{A^-}^{A^+} \xi\rho(\xi)\,\mathrm{d}\xi = 0.\qquad (2.3.25)$$

Taking in (2.3.21) $\xi = 0$ we obtain:

$$1 + \int_{A^-}^{A^+} \frac{d\xi}{\xi} \rho(\xi) = 0. \tag{2.3.26}$$

Substituting (2.3.25) and (2.3.26) into formulas (2.3.23) and (2.3.24) we see that the results for the energy E and wavefunction $\psi(x)$ are expressed in terms of "indefinitenesses" of the type "$\infty \cdot 0$". In order to obtain finite and mathematically sensible expressions for E and $\psi(x)$, these indefinitenesses should be removed. The most natural way to do this is to solve equations (2.3.15) and (2.3.16) assuming that M is arbitrarily large but finite, substitute the results into (2.3.19) and (2.3.20), and only then take the limit $M \to \infty$. Of course, equations (2.3.15) and (2.3.16) cannot be solved exactly, since the corresponding quantum mechanical problem (2.3.1) is exactly non-solvable. However, we can solve these equations by means of an iteration procedure, any step of which, as we will see below, leads to the finite expression for the energy E and wavefunction $\psi(x)$.

As zeroth iteration we choose the function $\rho(\lambda)$ and the numbers $A^{\pm}$ satisfying equations (2.3.21) and (2.3.22). It is not difficult to verify that this function has the form

$$\rho_0(\xi) = \frac{\Delta^3(g)}{2\pi i} \sqrt{(\xi - A_0^-)(\xi - A_0^+)}, \tag{2.3.27}$$

where

$$A_0^- = -4, \quad A_0^+ = 0. \tag{2.3.28}$$

The next iterations $\rho_n(\xi)$ and $A_n^{\pm}$ can be determined from the following recurrence relations:

$$\int_{A_n^-}^{A_n^+} \frac{\rho_n(\eta) \, d\eta}{\xi - \eta} = \Delta^3(g)(1 + \tfrac{\xi}{2})$$

$$+ \frac{1}{2M} \left\{ \frac{1}{\xi - A_{n-1}^-} + \frac{1}{\xi - A_{n-1}^+} + \frac{2s}{\xi} + \frac{\rho_{n-1}'(\xi)}{\rho_{n-1}(\xi)} \right\}, \tag{2.3.29}$$

$$\int_{A_n^-}^{A_n^+} \rho_n(\xi) \, d\xi = 1, \tag{2.3.30}$$

where $n = 1, 2, 3, \ldots$. Substitution of $\rho_n(\xi)$ and $A_n^{\pm}$ obtained from (2.3.29) and (2.3.30) into formulas (2.3.19) and (2.3.20) gives the following expressions for the energy E and wavefunction $\psi(x)$ in the nth approximation:

$$E_n = 4(2\beta)^{\frac{1}{3}}\Delta(g)M^{\frac{3}{4}}$$

$$\times \left\{1 + \frac{1}{2}\int_{A_n^-}^{A_n^+}\xi\rho(\xi)\,\mathrm{d}\xi + \frac{s+1+\frac{1}{4}(A_n^- + A_n^+)}{M}\right\}, \quad (2.3.31)$$

$$\psi_n(x) = (x^2)^{s-\frac{1}{4}}\sqrt{(2\beta)^{\frac{1}{3}}\frac{x^2}{2} - M^{\frac{2}{3}}\Delta^2(g)A_n^-}$$

$$\times \sqrt{(2\beta)^{\frac{1}{3}}\frac{x^2}{2} - M^{\frac{2}{3}}\Delta^2(g)A_n^+}$$

$$\times \exp\left\{-(2\beta M)^{\frac{1}{3}}\Delta(g)\frac{x^2}{2} + M\int_{A_n^-}^{A_n^+}\ln\left[(2\beta)^{\frac{1}{3}}\frac{x^2}{2} - M^{\frac{2}{3}}\Delta^2(g)\xi\right]\rho_n(\xi)\,\mathrm{d}\xi\right\}.$$

$$(2.3.32)$$

Now, let us show that the function $\rho_n(\xi)$ can be represented in the following two equivalent forms:

$$\rho_n(\xi) = \frac{\Delta^3(g)}{2\pi i}\sqrt{(\xi - A_n^-)(\xi - A_n^+)}\frac{\displaystyle\prod_{k=n+2-2^{n+1}}^{n-2^n}(\xi - A_k^-)(\xi - A_k^+)}{\displaystyle\prod_{k=n+1-2^n}^{n-1}(\xi - A_k^-)(\xi - A_k^+)}$$

$$(2.3.33a)$$

or

$$\rho_n(\xi) = \frac{\Delta^3(g)}{2\pi i}\sqrt{(\xi - A_n^-)(\xi - A_n^+)}$$

$$\times \left\{1 + \sum_{k=n+1-2^n}^{n-1}\frac{\Gamma_{nk}^-}{\xi - A_k^-} + \sum_{k=n+1-2^n}^{n-1}\frac{\Gamma_{nk}^+}{\xi - A_k^+}\right\}.$$

$$(2.3.33b)$$

Indeed, substituting (2.3.33a) into the right-hand side of (2.3.29) we obtain:

$$\int_{A_n^-}^{A_n^+}\frac{\rho_n(\eta)\,\mathrm{d}\eta}{\xi - \eta} = \Delta^3(g)\left(1 + \frac{\xi}{2}\right) - \frac{1}{2}\sum_{k=n+1-2^n}^{n-1}\frac{B_{nk}^-}{\xi - A_k^-}$$

$$-\frac{1}{2} \sum_{k=n+1-2^n}^{n-1} \frac{B_{nk}^+}{\xi - A_k^+} = 0, \tag{2.3.34}$$

where $B_{nk}^{\pm}$ are certain numbers (see below). At the same time, substituting (2.3.33b) into the left-hand side of (2.3.34) we see that it takes the same functional structure as the right-hand side. This proves the validity of both representations (2.3.33). Note that the equation appearing after substitution of (2.3.33b) into (2.3.34) is equivalent to the system of numerical equations

$$\Delta^3(g)\Gamma_{nk}^{\pm}\sqrt{(A_k^{\pm} - A_n^-)(A_k^{\pm} - A_n^+)} = B_{nk}^{\pm},$$
$$k = n+1-2^n, \ldots, n-1, \tag{2.3.35a}$$

$$\sum_{k=n+1-2^n}^{n-1} (\Gamma_{nk}^- + \Gamma_{nk}^+) - 2 = \frac{1}{2}(A_n^+ + A_n^-), \tag{2.3.35b}$$

$$\Delta^3(g)\Bigg\{ \sum_{k=n+1-2^n}^{n-1} \left[B_{nk}^- + B_{nk}^+ - \Gamma_{nk}^- A_k^- - \Gamma_{nk}^+ A_k^+ \right.$$
$$\left. + \frac{1}{2}(\Gamma_{nk}^- + \Gamma_{nk}^+)(A_n^- + A_n^+) \right] - \frac{1}{2}A_n^- A_n^+ + \frac{1}{8}(A_n^- + A_n^+)^2 \Bigg\} = 2,$$
$$\tag{2.3.35c}$$

which, obviously, must be supplemented by the values of the numbers $B_{nk}^{\pm}$:

$$B_{nk}^- = \begin{cases} \frac{1}{M}, & n+1-2^n \le k \le n-1-2^{n-1}, \\ -\frac{1}{M}, & n-2^{n-1} \le k \le n-2, \\ \frac{3}{2M}, & k = n-1, \end{cases} \tag{2.3.36a}$$

$$B_{nk}^+ = B_{nk}^- - 2s\delta_{k0}, \tag{2.3.36b}$$

and by the formulas determining the relation between the numbers $A_k^{\pm}$ and $\Gamma_{nk}^{\pm}$:

$$\Gamma_{nk}^- = \frac{\prod_m (A_k^- - A_m^-) \prod_m (A_k^- - A_m^+)}{\prod_l (A_k^- - A_l^+) \prod_l{}' (A_k^- - A_l^-)}, \tag{2.3.37a}$$

$$\Gamma_{nk}^+ = \frac{\prod_m (A_k^+ - A_m^+) \prod_m (A_k^+ - A_m^-)}{\prod_l{}' (A_k^+ - A_l^+) \prod_l (A_k^+ - A_l^-)}, \tag{2.3.37b}$$

(here l takes values from $n + 1 - 2^n$ to $n - 1$ and m from $n + 2 - 2^{n+1}$ to $n - 2^n$).

System (2.3.35)–(2.3.37) is rather cumbersome. For its simplification note that the numbers $B_{nk}^{\pm}$ are of order $\frac{1}{M}$ and, thus, can be represented as

$$B_{nk}^{\pm} = \frac{\beta_{nk}^{\pm}}{M}. \tag{2.3.38}$$

Moreover, the numbers $A_n^{\pm}$ almost coincide with $A_0^{\pm}$ for large M. This gives us the possibility of seeking them in the form

$$A_n^- = -4 + \frac{\alpha_n^-}{M^{2/3}}, \quad A_n^+ = -\frac{\alpha_n^+}{M^{2/3}}. \tag{2.3.39}$$

Analogously, we can use the following appropriate representation for $\Gamma_{nk}^{\pm}$:

$$\Gamma_{nk}^{\pm} = \frac{1}{M^{2/3}} \gamma_{nk}^{\pm}. \tag{2.3.40}$$

Substitution of (2.3.40) into (2.3.35a) gives:

$$\gamma_{nk}^{\pm} = \pm \frac{1}{2} \frac{\beta_{nk}^{\pm}}{\sqrt{\alpha_n^{\pm} - \alpha_k^{\pm}}}. \tag{2.3.41}$$

Now, if we substitute formulas (2.3.4) and (2.3.38)–(2.3.40) into the system (2.3.35)–(2.3.37), eliminate $\gamma_{nk}^{\pm}$ by means of (2.3.41) and take the limit $M \to \infty$, we obtain immediately the system of algebraic equations for $\alpha_k^{\pm}$:

$$\sum_{k=n+1-2^n}^{n-1} \frac{\beta_{nk}^{\pm}}{\sqrt{\alpha_n^{\pm} - \alpha_k^{\pm}}} + \alpha_n^{\pm} - g = 0, \tag{2.3.42a}$$

$$\frac{1}{2} \frac{\beta_{nk}^{\pm}}{\sqrt{\alpha_n^{\pm} - \alpha_k^{\pm}}} = \frac{\displaystyle\prod_{p=n+2-2^n}^{n-2^n} (\alpha_p^{\pm} - \alpha_k^{\pm})}{\displaystyle\prod_{q=n+1-2^n}^{n-1} (\alpha_q^{\pm} - \alpha_k^{\pm})},$$
$$k = n + 1 - 2^n, \ldots, n - 1. \tag{2.3.42b}$$

Here

$$\beta_{nk}^- = \begin{cases} -1, & n + 1 - 2^n \le k \le n - 1 - 2^{n-1} \\ 1, & n - 2^{n-1} \le k \le n - 2, \\ -\frac{3}{2}, & k = n - 1, \end{cases} \tag{2.3.43a}$$

$$\beta_{nk}^+ = \beta_{nk}^- - 2s\delta_{k0}. \tag{2.3.43b}$$

If the numbers $\alpha_k^{\pm}$ with $k = n + 1 - 2^n, \ldots, n - 1$ are known, then the equation (2.3.42a) determines $\alpha_k^{\pm}$, and the system (2.3.42b) determines the numbers $\alpha_k^{\pm}$ with $k = n - 2 - 2^{n+1}, \ldots, n - 2^n$. Therefore, all the $\alpha_k^{\pm}$ with $k = n' + 1 - 2^{n'}, \ldots, n' - 1$ $(n' \equiv n + 1)$ become known.

The energy level E and wavefunction $\psi(x)$ corresponding to this iteration can be obtained if we substitute (2.3.4) and (2.3.38)–(2.3.40) into expressions (2.3.31) and (2.3.32). Taking $M = 0$ we find

$$
E_n = (2\beta)^{\frac{1}{3}} \left\{ \alpha_n^+ \left(\frac{3}{2}\alpha_n^+ - g \right) + \sum_{k=n+1-2^n}^{n-1} \frac{2\beta_{nk}^+ \alpha_k^+}{\sqrt{\alpha_n^+ - \alpha_k^+}} \right\}
$$

$$
- (2\beta)^{\frac{1}{3}} \left\{ \alpha_n^- \left(\frac{3}{2}\alpha_n^- - g \right) + \sum_{k=n+1-2^n}^{n-1} \frac{2\beta_{nk}^- \alpha_k^-}{\sqrt{\alpha_n^- - \alpha_k^-}} \right\},
$$

$$(2.3.44)$$

$$
\psi(x) = (x^2)^{s - \frac{1}{4}} \sqrt{(2\beta)^{\frac{1}{3}} \frac{x^2}{2} + \alpha_n^+} \prod_{k=n+1-2^n}^{n-1} \left[(2\beta)^{\frac{1}{3}} \frac{x^2}{2} + \alpha_k^+ \right]^{\frac{\beta_{nk}^+}{2}}
$$

$$
\times \exp\left\{ -\int_0^{\frac{x^2}{2}} \sqrt{\lambda + \alpha_n^+} \left[1 + \frac{1}{2} \sum_{k=n+1-2^n}^{n-1} \frac{\beta_{nk}^+}{\sqrt{\alpha_n^+ - \alpha_k^+ (\lambda + \alpha_k^+)}} \right] d\lambda \right\}.
$$

$$(2.3.45)$$

Thus, we have obtained closed recursion formulas giving the possibility of determining the nth iteration if the $(n - 1)$th one is known. Consider for example the first two steps of this iteration procedure.

0. Zeroth iteration.

$$
\alpha_0^+ = 0, \quad \alpha_0^- = 0. \tag{2.3.46}
$$

1. First iteration. The numbers α_1^+ and α_1^- can be found from the equations

$$
\frac{\frac{3}{2} + 2s}{\sqrt{\alpha_1^+}} = \alpha_1^+ - g, \qquad \frac{\frac{3}{2}}{\sqrt{\alpha_1^-}} = \alpha_1^- - g. \tag{2.3.47}
$$

Knowledge of the numbers $\alpha_1^{\pm}$ allows us to find the energy E_1 and wavefunction $\psi_1(x)$ in the first approximation.

$$
E_1 = (2\beta)^{\frac{1}{3}} \left\{ \alpha_1^+ \left(\frac{3}{2}\alpha_1^+ - g \right) - \alpha_1^- \left(\frac{3}{2}\alpha_1^- - g \right) \right\}, \tag{2.3.48}
$$

$$\psi_1(x) \sim (x^2)^{s-\frac{1}{4}}\sqrt{(2\beta)^{\frac{1}{3}}\frac{x^2}{2}+\alpha_1^+}$$

$$\times \exp\left\{-\int\limits_0^{\frac{x^2}{2}}\sqrt{\lambda+\alpha_1^+}\left[1-\frac{s-\frac{3}{4}}{\sqrt{\alpha_1^+\lambda}}\right]d\lambda\right\}. \quad (2.3.49)$$

The numbers $\alpha_{-1}^{\pm}$ necessary for constructing the next iteration are determined by the formulas:

$$\alpha_{-1}^+ = -\frac{\frac{3}{4}+s}{\sqrt{\alpha_1^+}}, \quad \alpha_{-1}^- = \frac{\frac{3}{4}}{\sqrt{\alpha_1^-}}. \quad (2.3.50)$$

2. Second iteration. The numbers $\alpha_2^{\pm}$ can be found from the equations

$$\frac{1}{\sqrt{\alpha_2^+-\alpha_{-1}^+}} - \frac{1-2s}{\sqrt{\alpha_2^+}} + \frac{\frac{3}{2}}{\sqrt{\alpha_2^+-\alpha_1^+}} = \alpha_2^+ - g, \quad (2.3.51a)$$

$$\frac{1}{\sqrt{\alpha_2^--\alpha_{-1}^-}} - \frac{1}{\sqrt{\alpha_2^-}} + \frac{\frac{3}{2}}{\sqrt{\alpha_2^--\alpha_1^-}} = \alpha_2^- - g. \quad (2.3.51b)$$

Knowledge of these numbers allows us to compute the energy E_2 and wavefunction $\psi_2(x)$ in the second approximation:

$$E_2 = \left\{\alpha_2^+\left(\tfrac{3}{2}\alpha_2^+ - g\right) - \alpha_2^-\left(\tfrac{3}{2}\alpha_2^- - g\right)\right\}$$

$$-\left\{\frac{2\alpha_{-1}^+}{\sqrt{\alpha_2^+-\alpha_{-1}^+}} + \frac{3\alpha_1^+}{\sqrt{\alpha_2^+-\alpha_1^+}} - \frac{2\alpha_{-1}^-}{\sqrt{\alpha_2^--\alpha_{-1}^-}} - \frac{3\alpha_1^-}{\sqrt{\alpha_2^--\alpha_1^-}}\right\},$$

$$(2.3.52)$$

$$\psi_2(x) = (x^2)^{s-\frac{1}{4}}\frac{\sqrt{(2\beta)^{\frac{1}{3}}\frac{x^2}{2}+\alpha_2^+}}{\sqrt{(2\beta)^{\frac{1}{3}}\frac{x^2}{2}+\alpha_{-1}^+}\left[(2\beta)^{\frac{1}{3}}\frac{x^2}{2}+\alpha_1^+\right]^{\frac{3}{4}}}$$

$$\times \exp\left\{-\int\limits_0^{\frac{x^2}{2}}\sqrt{\lambda+\alpha_2^+}\left[1-\frac{\frac{1}{2}}{\sqrt{\alpha_2^+-\alpha_{-1}^+}(\lambda+\alpha_{-1}^+)} + \frac{\frac{1}{2}-s}{\sqrt{\alpha_2^+\lambda}}\right.\right.$$

$$\left.\left.-\frac{\frac{3}{4}}{\sqrt{\alpha_2^+-\alpha_1^+}(\lambda+\alpha_1^+)}\right]d\lambda\right\}.$$

$$(2.3.53)$$

The subsequent iterations can be constructed analogously.

We see that even these two approximations properly reproduce the asymptotic properties of wavefunctions in both the large- and small-x limits. Moreover, they give correct answers for perturbation theory, which is applicable if the constant g is large. Indeed, solving equations (2.3.47) for large g, substituting the results into (2.3.48), and using (2.3.3) we find:

$$E_1 = 45\sqrt{\alpha} + \dots \qquad (2.3.54)$$

Analogously, solving equations (2.3.50) and (2.3.51) and using formula (2.3.52) we obtain:

$$E_2 = 45\sqrt{\alpha} + 2s(2s+1)\frac{\beta}{\alpha} + \dots \qquad (2.3.55)$$

It is easy to see that expressions (2.3.54) and (2.3.55) reproduce correctly the zeroth and first terms of perturbation theory for the ground state energy level in model (2.3.1). Continuing this procedure we can also obtain correct results for the next terms in perturbation theory.

Finally, let us consider the case when s is large. This is a typical semi-classical situation, since the parameter s can be interpreted as an angular momentum in the radial Schrödinger equation. Solving equations (2.3.47) for large s and substituting the result into (2.3.48) we obtain the following asymptotic expression for the energy

$$E_1 \approx \frac{3}{2}(2s)^{\frac{4}{3}} + \dots \qquad (2.3.56)$$

which, evidently, coincides with the semi-classical results (Maglaperidze and Ushveridze 1989c).

2.4   Higher oscillators with centrifugal barriers

It is well known (Landau and Lifshitz 1977) that the radial analogue of the simple harmonic oscillator (with centrifugal barrier)

$$V(x) = \frac{4(s - \frac{1}{4})(s - \frac{3}{4})}{x^2} + \alpha x^2 \qquad (2.4.1)$$

has an infinite number of exact solutions belonging to the class of functions

$$\psi(x) = \prod_{i=1}^{M} \left(\frac{x^2}{2} - \xi_i \right) (x^2)^{s - \frac{1}{4}} \exp\left(-\frac{\alpha x^2}{2} \right) \qquad (2.4.2)$$

(where $a = \sqrt{\alpha}$). As it was shown in section 2.2, the radial analogue of the sextic anharmonic oscillator (with centrifugal barrier)

$$V(x) = \frac{4(s - \frac{1}{4})(s - \frac{3}{4})}{x^2} + \alpha x^2 + \beta x^4 + \gamma x^6 \qquad (2.4.3)$$

may have a finite number of exact solutions if the parameter α takes certain discrete values. These solutions can be written in the form

$$\psi(x) = \prod_{i=1}^{M} \left(\frac{x^2}{2} - \xi_i \right) (x^2)^{s - \frac{1}{4}} \exp\left(-\frac{ax^4}{4} - \frac{bx^2}{2} \right) \qquad (2.4.4)$$

(where $\alpha = \sqrt{\gamma}$, $b = \frac{\beta}{2\sqrt{\gamma}}$).

We see that in both these cases a correct *ansatz* for wavefunctions consists of three factors. The first factor is a polynomial in x^2, the second is a power function, reproducing the behaviour of wavefunctions at the origin, and the third factor is an exponential. The degrees of the polynomials in this exponential are chosen in such a way as to guarantee correct behaviour of the corresponding potentials at infinity. The fact that the general form of the *ansatz* depends only on the degree of these polynomials suggests that the series of models allowing exact solutions can be extended further. So, it would be quite natural to assume that there exist models described by potentials

$$V(x) = \frac{4(s - \frac{1}{4})(s - \frac{3}{4})}{x^2} + \alpha x^2 + \beta x^4 + \gamma x^6 + \delta x^8 + \epsilon x^{10} \qquad (2.4.5)$$

and having exact solutions of the form

$$\psi(x) = \prod_{i=1}^{M} \left(\frac{x^2}{2} - \xi_i \right) (x^2)^{s - \frac{1}{4}} \exp\left(-\frac{ax^6}{6} - \frac{bx^4}{4} - \frac{dx^2}{2} \right). \qquad (2.4.6)$$

In order to verify this assumption, let us consider the "inverse" Schrödinger equation

$$V(x) = E + [(\ln \psi)']^2 + [\ln \psi]'' \qquad (2.4.7)$$

(in which the potential is expressed via the solution) and substitute *ansatz* (2.4.6) into it. This gives:

$$V(x) = E + \left[\sum_{i=1}^{M} \frac{x}{\frac{x^2}{2} - \xi_i} + \frac{2s - \frac{1}{2}}{x} - ax^5 - bx^3 - cx \right]^2$$
$$+ \left[\sum_{i=1}^{M} \frac{x}{\frac{x^2}{2} - \xi_i} + \frac{2s - \frac{1}{2}}{x} - ax^5 - bx^3 - cx \right]'. \qquad (2.4.8)$$

After trivial transformations, expression (2.4.8) takes the form

$$
\begin{aligned}
V(x) &= \left\{ a^2 x^{10} + 2abx^8 + (b^2 + 2ac)x^6 + [2bc + a(4M + 4s + 4)]x^4 \right. \\
&\quad + [c^2 - b(4M + 4s + 2) - 8a \sum_{i=1}^{M} \xi]x^2 + \left. \frac{4(s - \frac{1}{4})(s - \frac{3}{4})}{x^2} \right\} \\
&\quad + \left\{ E - c(4M + 4s) - 8b \sum_{i=1}^{M} \xi_i - 16a \sum_{i=1}^{M} \xi_i^2 \right\} \\
&\quad + \sum_{i=1}^{M} \frac{1}{\frac{x^2}{2} - \xi_i} \left\{ 4\xi_i \sum_{k=1}^{M}{}' \frac{1}{\xi_i - \xi_k} + 4s - 2c\xi_i - 4b\xi_i^2 - 8a\xi_i^3 \right\}.
\end{aligned}
$$

$$(2.4.9)$$

We see that it consists of two parts. The first part has the same structure as (2.4.5), while the second part contains the constant term and the terms proportional to $\left(\frac{x^2}{2} - \xi_i \right)^{-1}$. In order to guarantee the coincidence of the potential given in (2.4.9) with that in (2.4.5), these (unwanted) terms must vanish. This leads to the system of M equations

$$
\sum_{k=1}^{M}{}' \frac{1}{\xi_i - \xi_k} + \frac{s}{\xi_i} - \frac{c}{2} - b\xi_i - 2a\xi_i^2 = 0,
$$

$$i = 1, \dots, M \qquad (2.4.10)$$

for M unknowns ξ_i, $i = 1, \dots, M$, and to the following expression for the energy:

$$
E = c(4M + 4s) + 8b \sum_{i=1}^{M} \xi_i + 16a \sum_{i=1}^{M} \xi_i^2. \qquad (2.4.11)
$$

Then, potential (2.4.9) takes the form

$$
\begin{aligned}
V(x) &= \frac{4(s - \frac{1}{4})(s - \frac{3}{4})}{x^2} + \left[c^2 - b(4M + 4s + 2) - 8a \sum_{i=1}^{M} \xi_i \right] x^2 \\
&\quad + [2bc + a(4M + 4s + 4)]x^4 + (b^2 + 2ac)x^6 + 2abx^8 + a^2 x^{10}.
\end{aligned}
$$

$$(2.4.12)$$

This results in a new class of quasi-exactly solvable models with corresponding exact solutions described by formulas (2.4.6) and (2.4.11).

We emphasize that there is a principal difference between the quasi-exactly solvable models (2.4.12) and the models (2.2.1) discussed above. This difference lies in the fact that, in contrast to model (2.2.1) the potential of which depends only on the number M, potential (2.4.12) depends also on the numbers ξ_i, $i = 1, \ldots, M$ satisfying the system of numerical equations (2.4.10). In other words, the potential $V(x)$ depends on the sort of solution $\psi(x)$, even when the number M is fixed. But this means that formula (2.4.12) describes a set of quasi-exactly solvable models of unit order.

Fortunately, there are some special cases when the order of quasi-exactly solvable models (2.4.12) may exceed one (Ushveridze 1989c). To show this, consider formula (2.4.12). Note that for an explicit construction of a model of order K, it is necessary that for K solutions $\xi_1^{(k)}, \ldots, \xi_M^{(k)}$, $k = 1, \ldots, K$ of the system (2.4.10) the values of the first-order symmetric polynomials

$$\sigma_1 = \sum_{i=1}^{M} \xi_i \tag{2.4.13}$$

entering into the potential (2.4.12) be made independent of k, and the entire k-dependence be concentrated in the second-order symmetric polynomial

$$\sigma_2 = \sum_{i=1}^{M} [\xi_i]^2 \tag{2.4.14}$$

defining the energy of system (2.4.11). To do this, we multiply system (2.4.10) by ξ_i, sum over i and use the relation

$$\sum_{i,k=1}^{M} {}' \frac{\xi_i^{n+1}}{\xi_i - \xi_k} = -\frac{n+1}{2}\sigma_n + \frac{1}{2}\sum_{l=0}^{n}\sigma_n, \tag{2.4.15}$$

where by σ_n we have denoted the symmetric polynomial of order n:

$$\sigma_n = \sum_{i=1}^{M} [\xi_i]^n. \tag{2.4.16}$$

As a result we obtain a system of equations expressing σ_n with $n > 2$ in terms of σ_n with $n \leq 2$. A different system of conditions on the symmetric polynomials σ_n can be obtained by noting that σ_n with $n > M$ are expressed in terms of σ_n with $n \leq M$. Assuming that $M > 2$, and combining these two systems, we arrive at two algebraic equations of the form

$$\sigma_2^N + f_1\sigma_2^{N-1} + \ldots + f_{N-1}\sigma_2 + f_N = 0, \tag{2.4.17a}$$
$$\sigma_2^L + g_1\sigma_2^{L-1} + \ldots + g_{L-1}\sigma_2 + g_L = 0, \tag{2.4.17b}$$

where

$$N = \left[\frac{M+1}{2}\right], \quad L = \left[\frac{M+2}{2}\right]. \tag{2.4.18}$$

The coefficients of these equations depend explicitly on five quantities: a, b, c, s and σ_1. For each equation (2.4.17) to have at least K different solutions for a fixed set of these quantities, it is necessary that the inequalities

$$N \geq K, \quad L \geq K \tag{2.4.19}$$

be satisfied. This leads to the following restriction on M:

$$M \geq 2K - 1. \tag{2.4.20}$$

Note that if

$$M = 2K - 1, \tag{2.4.21}$$

the degree of both the equations (2.4.17) is the same and equal to K:

$$N = K, \quad L = K. \tag{2.4.22}$$

For these equations to be compatible, it is necessary that the coefficients of identical powers coincide:

$$f_1 = g_1, \quad f_2 = g_2, \quad \ldots, \quad f_K = g_K. \tag{2.4.23}$$

Wee see that K equations (2.4.23) are imposed on the functions depending on five quantities, a, b, c, s and σ_1. For system (2.4.23) to be solvable we must require that

$$K \leq 5. \tag{2.4.24}$$

But this means that we have found the maximal possible order of quasi-exactly solvable models of the type (2.4.12).

As an example, let us construct a second-order ($K = 2$) model with the potential (2.4.12). According to formula (2.4.21), in this case $M = 3$, and therefore we have only three algebraically independent polynomials σ_1, σ_2 and σ_3. Polynomials σ_4 and σ_5 can be expressed in terms of σ_1, σ_2 and σ_3 as

$$
\begin{aligned}
\sigma_4 &= A\sigma_2 + B\sigma_1 + B\sigma_2 + C\sigma_1^2\sigma_2 + D\sigma_1^4 \\
\sigma_5 &= A'\sigma_2\sigma_3 + B'\sigma_1\sigma_2^2 + C'\sigma_1\sigma_4 + D'\sigma_1^2\sigma_3 \\
&\quad + E'\sigma_1^3\sigma_2 + F'\sigma_1^5
\end{aligned}
\tag{2.4.25}
$$

with certain computable coefficients $A, B, \ldots$ and $A', B', \ldots$. To compute the coefficients A and A', take for example, $\xi_1 = +1$, $\xi_2 = +1$, $\xi_3 = -1$, which gives $\sigma_1 = 0$, $\sigma_2 = 6$, $\sigma_3 = -6$, $\sigma_4 = 18$, $\sigma_5 = -30$. Substituting these values into (2.4.25) we find that

$$A = \tfrac{1}{2}, \quad A' = \tfrac{5}{6}. \tag{2.4.26}$$

Analogously one can compute the other coefficients in (2.4.25).

Multiplying equation (2.4.10) by ξ_i, ξ_i^2 and ξ_i^3, summing over i and using formula (2.4.15) we obtain three equations:

$$3(s+1) - \frac{c}{2}\sigma_1 - b\sigma_2 - 2a\sigma_3 = 0,$$

$$(s+2)\sigma_1 - \frac{c}{2}\sigma_2 - b\sigma_3 - 2a\sigma_4 = 0, \tag{2.4.27}$$

$$(s+\tfrac{3}{2})\sigma_2 + \tfrac{1}{2}\sigma_1^2 - \frac{c}{2}\sigma_3 - b\sigma_4 - 2a\sigma_5 = 0.$$

Thus, we have five equations (2.4.25) and (2.4.27) for nine quantities σ_1, σ_2, σ_3, σ_4, σ_5, a, b, c and s. We see that we have a sufficient number of free parameters at our disposal. Therefore, for definiteness we can set

$$\sigma_1 = 0, \quad a = \tfrac{1}{2}, \tag{2.4.28}$$

reducing the system to the form

$$\sigma_2^2 - (2b^2 - c)\sigma_2 + 6b(s+1) = 0, \tag{2.4.29a}$$

$$\sigma_2^2 + \tfrac{3}{b}\left[\tfrac{bc}{2} - \tfrac{3}{2}s - 1\right]\sigma_2 - \frac{9c}{2b}(s+1) = 0. \tag{2.4.29b}$$

Equating the coefficients of identical powers of σ_2, we find that

$$s = \tfrac{8}{27}b^3 - \tfrac{2}{3}, \quad c = -\tfrac{4}{3}b^2. \tag{2.4.30}$$

The potential of the corresponding model has the form

$$V(x) = \frac{\left[\tfrac{16}{27}b^3 - \tfrac{11}{6}\right]\left[\tfrac{16}{27}b^3 - \tfrac{17}{6}\right]}{x^2} + \left[\tfrac{16}{27}b^4 - \tfrac{34}{3}b\right]x^2$$

$$- \left[\tfrac{57}{27}b^3 - \tfrac{20}{3}\right]x^4 - \frac{b^2}{3}x^6 + bx^8 + \tfrac{1}{4}x^{10} \tag{2.4.31}$$

and solutions are given by

$$E = -\tfrac{128}{81}b^5 + \tfrac{8}{9}b^2 \pm \sqrt{b^4 - 2b} \tag{2.4.32}$$

and

$$\psi(x) = \left(\tfrac{x^2}{2} - \xi_1\right)\left(\tfrac{x^2}{2} - \xi_2\right)\left(\tfrac{x^2}{2} - \xi_3\right)$$
$$\times (x^2)^{\frac{8}{27}b^3 - \frac{11}{12}} \exp\left(-\frac{x^6}{12} - \frac{bx^4}{4} + \frac{2b^2x^2}{3}\right), \qquad (2.4.33)$$

where ξ_1, ξ_2 and ξ_3 can be found from the equations

$$\xi_1 + \xi_2 + \xi_3 = 0,$$
$$\xi_1^2 + \xi_2^2 + \xi_3^2 = \tfrac{5}{3}b^2 \pm \sqrt{b^4 - 2b}, \qquad (2.4.34)$$
$$\xi_1^3 + \xi_2^3 + \xi_3^3 = 1 - \tfrac{7}{9}b^3 \mp b\sqrt{b^4 - 2b}.$$

Using the oscillator theorem, it is not difficult to see that the levels obtained in such a way may describe the first and second excitations (Ushveridze 1989c).

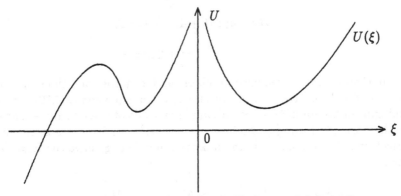

Figure 2.3. The form of the classical potential (2.4.36).

In conclusion, note that the spectral equations for models of the type (2.4.12) also allow a classical interpretation (Maglaperidze and Ushveridze 1988). Indeed, system (2.4.10) can be viewed as the condition for an extremum of the following classical M-particle potential:

$$W(\xi_1, \ldots, \xi_M) = -\sum_{k=1}^{M} \ln|\xi_i - \xi_k| + \sum_{i=1}^{M} U(\xi_i), \qquad (2.4.35)$$

where

$$U(\xi) = -s \ln|\xi| + \frac{c}{2}\xi + \frac{b}{2}\xi^2 + \frac{2a}{3}\xi^3 \qquad (2.4.36)$$

is an external potential of the form depicted in figure 2.3. It is evident that there are such dispositions of M Coulomb particles in the potential (2.4.36) when K of them $(0 \leq K \leq M)$ lie in the right-hand well, but according to the oscillator theorem this means that exact solutions for model (2.4.12) may describe states with the numbers $0, 1, \ldots, M$.

2.5   The electrostatic analogue. The general case

In the preceding section we have shown that the potential of model (2.4.12) depends, in general, on the form of the solution. Nevertheless, the M-dependence of parameters a, b and c entering into the potential (2.4.12) can be chosen in such a way that its dependence on the form of the solution vanishes in the limit $M \rightarrow \infty$. This dependence can be found from the equations

$$c^2 - b(4M + 4s + 2) - 8a \sum_{i=1}^{M} \xi_i = \alpha,$$

$$2bc - a(4M + 4s + 4) = \beta, \qquad (2.5.1)$$

$$b^2 + 2ac = \gamma,$$

where α, β and γ are certain fixed numbers, and from the conditions that the expression for the energy (2.4.11) and the spectral equation (2.4.10) remain meaningful in the limit $M \rightarrow \infty$. A simple analysis of these formulas shows that the correct limiting procedure can be performed if we assume that parameters a, b and c and also the unknown numbers ξ_i have the following orders:

$$a \sim M^{-\frac{1}{2}}, \quad b \sim 1, \quad c \sim M^{\frac{1}{2}}, \quad \xi_i \sim M^{\frac{1}{2}}. \qquad (2.5.2)$$

Then, introducing new parameters A, B, C, and new unknowns $\xi \left(\frac{i}{M}\right)$ by the formulas

$$a = M^{-\frac{1}{2}}A, \quad b = B, \quad c = M^{\frac{1}{2}}C, \quad \xi_i = M^{\frac{1}{2}}\xi \left(\tfrac{i}{M}\right), \qquad (2.5.3)$$

and substituting (2.5.3) into conditions (2.5.2), we obtain

$$B^2 + 2AC \;=\; \gamma,$$

$$2BC - 4A \;=\; O(M^{-\frac{1}{2}}), \qquad (2.5.4)$$

$$C^2 - 4B - 8A \sum_{i=1}^{M} \xi \left(\tfrac{i}{M}\right) \tfrac{1}{M} \;=\; O(M^{-1}),$$

and also

$$\sum_{k=1}^{M} \frac{1}{\xi(\frac{i}{M}) - \xi(\frac{k}{M})} \frac{1}{M} - \frac{C}{2} - B\xi\left(\frac{i}{M}\right) - 2A\xi^2\left(\frac{i}{M}\right) = O(M^{-1}), \quad (2.5.5)$$

$$E = M^{\frac{3}{2}} \left\{ 4C + 8B \sum_{i=1}^{M} \xi\left(\frac{i}{M}\right) \frac{1}{M} + 16A \sum \xi^2\left(\frac{i}{M}\right) \frac{1}{M} + O(M^{-1}) \right\}$$
$$(2.5.6)$$

and

$$V(x) = \frac{4(s - \frac{1}{4})(s - \frac{3}{4})}{x^2} + \alpha x^2 + \beta x^4 + \gamma x^6$$
$$+ O\left(\frac{1}{\sqrt{M}}\right) x^8 + O\left(\frac{1}{M}\right) x^{10}. \quad (2.5.7)$$

If M tends to infinity, the terms with the anharmonicities 8 and 10 disappear and the potential takes the form

$$V(x) = \frac{4(s - \frac{1}{4})(s - \frac{3}{4})}{x^2} + \alpha x^2 + \beta x^4 + \gamma x^6. \quad (2.5.8)$$

We know that, in general, it describes the exactly non-solvable "radial" Schrödinger equation for the sextic anharmonic oscillator.

Introducing the particle distribution density $\rho(\xi)$ and taking $M \to \infty$ in equations (2.5.4) and (2.5.5), we obtain for $\xi \neq 0$ the integral equation

$$\int \frac{\rho(\eta)}{\xi - \eta} \, d\eta - \frac{C}{2} - B\xi - 2A\xi^2 = 0 \quad (2.5.9)$$

supplemented by the normalization condition for $\rho(\xi)$:

$$\int \rho(\xi) \, d\xi = 1, \quad (2.5.10)$$

and by the additional conditions following from the equations (2.5.4):

$$8A \int \xi \rho(\xi) \, d\xi + 4B = C^2,$$
$$BC = 2A, \quad (2.5.11)$$
$$B^2 + 2AC = \gamma.$$

In the limit $M \to \infty$ the expression for the energy becomes

$$E = M^{\frac{3}{2}}\{4C + 8B \int \xi\rho(\xi)\, d\xi + 16A \int \xi^2\rho(\xi)\, d\xi\}. \qquad (2.5.12)$$

It is not difficult to see that the equations obtained describe a classical charged liquid which is distributed between two wells in the potential

$$U(\xi) = \begin{cases} \frac{2A}{3}\xi^3 + \frac{B}{2}\xi^2 + \frac{C}{2}\xi, & \xi \neq 0 \\ \infty, & \xi = 0 \end{cases} \qquad (2.5.13)$$

depicted in figure 2.4. Generally speaking, this potential is not fixed

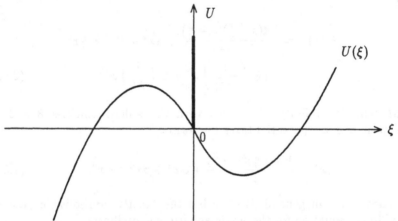

Figure 2.4. The form of the classical potential (2.5.13).

but depends on the particle distribution density (more exactly, on the charge centre $\int \xi\rho(\xi)\, d\xi$) as follows from formulas (2.5.11)). Nevertheless, these equations can be solved exactly and this gives us the possibility of describing the quantum model with the potential (2.5.8) in a purely classical language. Of course, equations (2.5.9)–(2.5.12) correspond only to the zeroth approximation in the iteration scheme described in section 2.3. However, even this rough approximation can be used to describe the spectrum of model (2.5.8) in the semi-classical limit, when the number of excitations K is large: $1 \ll K \ll M$. In this limit, the charge of the liquid which is situated in the right-hand well is not negligibly small; the expression in the curly brackets in (2.5.12) differs from zero, and is of order $K^{\frac{3}{2}}/M^{\frac{3}{2}}$. This gives us a semi-classically correct result for the Kth energy level in model (2.5.8).

Finally, note that we have already two examples of exactly non-solvable models allowing reformulation of the spectral problem in a purely classical language. Similarly, it can be shown that any one-dimensional quantum mechanical model with polynomial potential of degree $2n$ can be obtained as a limiting case of the quasi-exactly solvable model described by potentials of degree $2n+2$ and allowing a classical interpretation. Since any potential can be approximated with arbitrarily high accuracy by polynomial potentials, this means that any one-dimensional problem of quantum mechanics can be formulated in terms of finding the equilibrium of an infinite number of charged classical Coulomb particles in an external electrostatic field (Maglaperidze and Ushveridze 1988).

2.6 The inverse method of separation of variables

From the results of section 2.4 it follows that second-order linear differential equations of the type

$$
\begin{aligned}
\Bigg\{ &-\frac{\partial^2}{\partial x^2} + a^2 x^{10} + 2abx^8 + (b^2 + 2ac)x^6 + [2bc + a(4M + 4s + 4)]x^4 \\
&+ \left[c^2 - b(4M + 4s + 2) - 8a \sum_{i=1}^{M} \xi_i \right] x^2 \\
&+ \left[-c(4M + 4s) - 8b \sum_{i=1}^{M} \xi_i - 16a \sum_{i=1}^{M} \xi_i^2 \right] \\
&+ \frac{4(s - \frac{1}{4})(s - \frac{3}{4})}{x^2} \Bigg\} \psi(x) = 0
\end{aligned}
\tag{2.6.1}
$$

may have exact solutions of the form (2.4.6) if the numbers ξ_i entering simultaneously into (2.6.1) and (2.4.6) satisfy the system of numerical equations (2.4.10). It is not difficult to see that (2.6.1) can be considered as a single exactly solvable spectral equation with three spectral parameters

$$
\Gamma = 2bc - a(4M + 4s + 4),
$$

$$
\Delta = c^2 - b(4M + 4s + 2) - 8a \sum_{i=1}^{M} \xi_i,
\tag{2.6.2}
$$

$$
\Lambda = -c(4M + 4s) - 8b \sum_{i=1}^{M} \xi_i - 16a \sum_{i=1}^{M} \xi_i^2.
$$

One of these parameters, Γ, has a degenerate spectrum. The multiplicity of this degeneracy is equal to a number of solutions of equation (2.4.10) for given M. The presence in (2.6.1) of two spectral parameters Δ and Λ, depending explicitly on ξ_i and, thus, having non-degenerate spectra, prevents the possibility of interpreting (2.6.1) as a quasi-exactly solvable Schrödinger-type equation of sufficiently large order. In fact, if we include one of these parameters in the potential, the latter will depend explicitly on the numbers ξ_i (i.e. on the type of solution) and we arrive at a series of quasi-exactly solvable models of unit order (see section 2.4). On the other hand, we cannot remove both parameters Δ and Λ from the potential since the Schrödinger equation with two "energies" is meaningless. Thus, we see that one of the spectral parameters Δ and Λ is undesired and it would be very temping to except it from the consideration. This can be done as follows.

Let us treat (2.6.1) as a result of the separation of variables in a certain two-dimensional spectral equation. In this case, the unwanted spectral parameter can be identified with a separation constant which (by definition) cannot enter into the initial two-dimensional equation. The latter can, obviously, contain only two spectral parameters, one of which (having non-degenerate spectrum) we identify with the energy, while the second one (with degenerate spectrum) we include in the potential. Such a distribution of remaining spectral parameters gives us a series of two-dimensional quasi-exactly solvable models of an arbitrary, arbitrarily large order.

These models can be constructed by means of the inverse problem of separation of variables (Ushveridze 1988c, d, 1989c, d, e), which has already been discussed in sections 1.7 and 1.8. Following the prescriptions given in these sections, we consider two identical equations (2.6.1) rewritten in terms of two different variables z_1 and z_2.

$$\left\{ -\frac{\partial^2}{\partial z_1^2} + a^2 z_1^{10} + 2ab z_1^8 + (b^2 + 2ac)z_1^6 + \Gamma z_1^4 + \Delta z_1^2 + \Lambda + \frac{4(s-\frac{1}{4})(s-\frac{3}{4})}{z_1^2} \right\}$$

$$\times \, \psi(z_1) = 0, \qquad (2.6.3a)$$

$$\left\{ -\frac{\partial^2}{\partial z_2^2} + a^2 z_2^{10} + 2ab z_2^8 + (b^2 + 2ac)z_2^6 + \Gamma z_2^4 + \Delta z_2^2 + \Lambda + \frac{4(s-\frac{1}{4})(s-\frac{3}{4})}{z_2^2} \right\}$$

$$\times \, \psi(z_2) = 0. \qquad (2.6.3b)$$

Multiplying (2.6.3a) by $\psi(z_2)$, (2.6.3b) by $\psi(z_1)$ and subtracting one result from the other we obtain a simple two-dimensional spectral equation with two spectral parameters Γ and Δ which after dividing by $z_1^2 - z_2^2$ and changing variables

$$x = \frac{z_1^2 + z_2^2}{2}, \quad y = iz_1 z_2, \qquad (2.6.4)$$

takes the form

$$\left\{-\frac{\partial^2}{\partial x^2} - \frac{\partial^2}{\partial y^2} + \frac{4(s - \frac{1}{4})(s - \frac{3}{4})}{y^2} + a^2(16x^4 + 12x^2y^2 + y^4)\right.$$

$$\left. + 2ab(8x^3 + 4y^2x) + (b^2 + 2ac)(4x^2 + y^2) + 2\Gamma x\right\}\psi(x,y)$$

$$= \mathcal{E}\psi(x,y), \qquad (2.6.5)$$

where

$$\mathcal{E} = -\Delta, \qquad (2.6.6)$$

and

$$\psi(x,y) = \psi(z_1)\psi(z_2). \qquad (2.6.7)$$

At first sight, we have obtained a remarkable result: we have constructed a quasi-exactly solvable model of the two-dimensional quartic anharmonic oscillator! However, more careful analysis shows that this assertion is not quite correct. Indeed, let us look at the wavefunction $\psi(x,y)$ describing exact solutions in the model (2.6.5). Substituting expressions (2.4.6) for $\psi(z_1)$ and $\psi(z_2)$ into (2.6.7), and using formulas (2.6.4), we obtain

$$\psi(x,y) = P(x,y)(y^2)^{s - \frac{1}{4}}$$
$$\times \exp\left(-a\left[2x^3 + \tfrac{3}{2}y^2\right] - b[2x^2 + y^2] - cx\right), \quad (2.6.8)$$

where $P(x,y)$ is a certain polynomial. However, it is absolutely obvious that function (2.6.8) is not normalizable!

What does this mean? On the one hand, we see that our idea of using a one-dimensional exactly solvable three-parameter spectral equation to construct a two-dimensional quasi-exactly solvable Schrödinger-type equation was quite successful from a purely mathematical point of view: we have actually obtained such an equation. On the other hand, the exact solutions of this equation turned out to be non-physical ones. Does this mean that the method of the inverse problem of separation of variables cannot be used to construct physically sensible multi-dimensional quasi-exactly solvable models? The answer is no! As we see below, the construction of such models by means of the "inverse procedure" is possible, but it implies the use of more appropriate exactly solvable multi-parameter spectral equations. In order to find the needed form of these equations, we choose the following strategy. First of all, we construct the most general form of two-dimensional Schrödinger-type equations allowing separation of variables. This gives us the possibility of describing simultaneously the class of resulting one-dimensional equations arising after the separation. It is not difficult to understand that it cannot be too large (at least for the Schödinger equations with polynomial potentials) since the number of quantum mechanical models allowing separation of variables is, generally, very small. But this means that the problem of finding exactly solvable multi-parameter spectral equations belonging to this class cannot be too complex.

2.7   The Schrödinger equations with separable variables and quasi-exact solvability

In this section we discuss a general method for building two-dimensional Schrödinger-type equations with polynomial potentials allowing separation of variables (Turbiner and Ushveridze 1988b). The method is based on the observation that any coordinate system in which such a separation is possible can be obtained from the Cartesian system by means of conformal transformations described by a single analytic function. These transformations have the form

$$\xi = f(z) \tag{2.7.1}$$

where

$$z = x + \mathrm{i}y, \quad \xi = \eta + \mathrm{i}\sigma \tag{2.7.2}$$

are complex variables and $f(z)$ is an arbitrary analytic function. Formulas (2.7.1), (2.7.2) describe the transition from Cartesian coordinates (x, y) to new coordinates (η, σ). It is known that the new coordinate system is also orthogonal. The Laplace operator

$$\Delta_{x,y} \equiv \frac{\partial^2}{\partial x^2} + \frac{\partial^2}{\partial y^2} \tag{2.7.3}$$

in this system has the form

$$\Delta_{\eta,\sigma} = F\Delta_{x,y} \tag{2.7.4}$$

where

$$F = |f'(z)|^2. \tag{2.7.5}$$

Therefore, separation of variables (η, σ) in the Laplace equation

$$\Delta_{x,y}\psi = 0 \tag{2.7.6}$$

is always possible.

Now let us consider the Schrödinger equation

$$\{-\Delta_{x,y} + V\}\,\psi = E\psi. \tag{2.7.7}$$

Clearly, the presence of two additional terms V and E in (2.7.7) restricts the set of the coordinate systems (η, σ) in which separation is possible. Our aim is to describe the set of such systems and list the corresponding potentials V. We consider two cases:

(i) The energy E in the Schrödinger equation is fixed (a specific point of the spectrum is being sought),

(ii) The energy is random (a part of the spectrum is being sought).

First case. Without loss of generality, assume that $E = 0$ which corresponds to the choice of reference point. The Schrödinger equation in the varaibles (η, σ) takes the form

$$\left\{-\Delta_{\eta,\sigma} + \frac{V}{F}\right\}\psi = 0. \tag{2.7.8}$$

The following requirement is a condition for the separation of the variables (η, σ) in (2.7.8)

$$V = F\{A(\eta) + B(\sigma)\}. \tag{2.7.9}$$

Here $A(\eta)$ and $B(\sigma)$ are arbitrary functions. Then, after taking

$$\psi = \varphi_a(\eta)\varphi_b(\sigma) \tag{2.7.10}$$

the problem of solving the initial Schrödinger equation (2.7.7) is reduced to solving two independent one-dimensional equations

$$\left\{\frac{\partial^2}{\partial\eta^2} + A(\eta)\right\}\varphi_a(\eta) = 0, \quad \left\{\frac{\partial^2}{\partial\sigma^2} + B(\sigma)\right\}\varphi_b(\sigma) = 0. \tag{2.7.11}$$

Formulas (2.7.9)–(2.7.11) give us the most general form of two-dimensional Schrödinger equations allowing separation of variables for a certain fixed state.

Second case. Let us now discuss the case when the energy E in (2.7.7) may take arbitrary non-zero values. Using the conformal transformation (2.7.1) we obtain the equation

$$\left(-\Delta_{\eta,\sigma} + \frac{V}{F} - \frac{E}{F}\right)\psi = 0. \tag{2.7.12}$$

It is clear that condition (2.7.9) is not alone sufficient for the variables (η, σ) to be separable in (2.7.12), and it is necessary to require in addition that

$$\frac{1}{F} = a(\eta) + b(\sigma), \qquad (2.7.13)$$

where $a(\eta)$ and $b(\sigma)$ are certain arbitrary functions. As a result, we obtain instead of (2.7.11) the following system of equations

$$\left[-\frac{\partial^2}{\partial \eta^2} + A(\eta)\right]\varphi_a(\eta) = [Ea(\eta) + \Gamma]\varphi_a(\eta), \qquad (2.7.14a)$$

$$\left[-\frac{\partial^2}{\partial \sigma^2} + B(\sigma)\right]\varphi_b(\sigma) = [Eb(\sigma) + \Gamma]\varphi_b(\sigma), \qquad (2.7.14b)$$

in which E is the spectral parameter of the intial Schrödinger equation, while Γ is the separation constant which plays the role of an additional spectral parameter in the joint system (2.7.14).

Now consider the additional condition (2.7.13) in more detail. Obviously, it can be rewritten as

$$\frac{\partial^2}{\partial \eta \, \partial \sigma} F^{-1} = \left(\frac{\partial^2}{\partial f^2} - \frac{\partial^2}{\partial f^{*2}}\right)\frac{1}{f'(z)f^{*'}(z)} = 0 \qquad (2.7.15)$$

or, equivalently, as

$$f'(z)\frac{\partial^2}{\partial f^2}\frac{1}{f'(z)} = f^{*'}(z)\frac{\partial^2}{\partial f^{*2}}\frac{1}{f^{*'}(z)}. \qquad (2.7.16)$$

Substituting the identities

$$\frac{\partial^2}{\partial f^2} \equiv \frac{1}{[f'(z)]}\frac{\partial^2}{\partial z^2} - \frac{f''(z)}{[f'(z)]^3}\frac{\partial}{\partial z} \qquad (2.7.17a)$$

$$\frac{\partial^2}{\partial f^{*2}} \equiv \frac{1}{[f'(z^*)]}\frac{\partial^2}{\partial z^{*2}} - \frac{f''(z^*)}{[f'(z^*)]^3}\frac{\partial}{\partial z^*} \qquad (2.7.17b)$$

into (2.7.16) and introducing the new function

$$t(z) = \frac{1}{f'(z)}, \qquad (2.7.18)$$

we obtain for (2.7.16)

$$t(z)t''(z) + [t'(z)]^2 = t(z^*)t''(z^*) + [t'(z^*)]^2. \qquad (2.7.19)$$

Since the left- and right-hand sides of this equation have the same form but depend on different variables z and z^*, equality (2.7.19) holds if and only if both these sides are equal to the same constant:

$$t(z)t''(z) + [t'(z)]^2 = c. \qquad (2.7.20)$$

This equation can be solved without difficulty. Indeed, taking

$$t(z) = t, \quad t'(z) = y(t), \quad t''(z) = y'(t)y(t), \qquad (2.7.21)$$

we obtain an ordinary linear differential equation for $y^2(t)$

$$\frac{\partial}{\partial \ln t^2} y^2(t) + y^2(t) = c, \qquad (2.7.22)$$

the most general solution of which is

$$y^2(t) = c + \frac{c'}{t^2}. \qquad (2.7.23)$$

Using formulas (2.7.18) and (2.7.21) we find the following final expression for $f(z)$:

$$f(z) = c_1 \operatorname{arcsh}(c_2 z + c_3) + c_4 \qquad (2.7.24)$$

in which c_1, c_2, c_3 and c_4 are arbitrary constants. Thus, we have found the most general form of the function $f(z)$ for which the additional condition of separation (2.7.13) can be satisfied.

Let us now consider the two most important limiting cases of formula (2.7.14):

$$f(z) = \sqrt{z} \qquad (2.7.25)$$

and

$$f(z) = \operatorname{arcsh} z, \qquad (2.7.26)$$

and note that all other possible cases can be derived from (2.7.25) and (2.7.26) by means of arbitrary translations, dilatations and rotations in the spaces of variables $z = (x, y)$ and $\xi = (\eta, \sigma)$.

1. Let $f(z) = \sqrt{z}$. Then, obviously,

$$\eta = \sqrt{(r + x)/2}, \quad \sigma = \sqrt{(r - x)/2}, \qquad (2.7.27)$$

where $r = \sqrt{x^2 + y^2}$ is the radius in the (x, y)-plane, and

$$F = \frac{1}{4r} = \frac{1}{4(\eta^2 + \sigma^2)}. \tag{2.7.28}$$

Thus, from (2.7.13) it follows that

$$a(\eta) = 4\eta^2, \quad b(\sigma) = 4\sigma^2. \tag{2.7.29}$$

The explicit form of the functions $A(\eta)$ and $B(\sigma)$ can be found from the condition of quasi-polynomiality of the original potential V in the variables x and y (see formula (2.7.9)). It is clear that the most general form of the functions in this case is

$$A(\eta) = \sum_{k=-1}^{N} c_k \eta^{2k}, \quad B(\sigma) = -\sum_{k=-1}^{N} (-1)^k c_k \sigma^{2k}, \tag{2.7.30}$$

where the c_k in both formulas (2.7.30) are the same coefficients which may be chosen arbitrarily. Substituting (2.7.30) into equations (2.7.14) it is not difficult to see that after replacing σ^2 by $-\eta^2$ in the second equation, it takes the same form as the first one:

$$\left\{ -\frac{\partial^2}{\partial\eta^2} + \sum_{k=-1}^{N} c_k \eta^{2k} - E\eta^2 \right\} \varphi(\eta) = \Gamma\varphi(\eta). \tag{2.7.31}$$

(Here we have taken $\varphi_a(\eta) = \varphi_b(i\eta) = \varphi(\eta)$.) However, (2.7.31) is simply the Schrödinger equation for quasi-exactly solvable models with even quasi-polynomial potentials of orders $2N$. Such models have been discussed in the preceding sections of this chapter. Consider two particular cases:

i. $N = 3$. From the results of section 2.2 we know that in this case equation (2.7.31) may have several exact solutions if parameter E takes specific values. Thus, (2.7.31) can be viewed as an equation with two spectral parameters E and Γ. By transition to the two-dimensional case, the second spectral parameter Γ, which plays the role of the separation constant, disappears and we arrive at the Schrödinger equation with a single spectral parameter E. Obviously, this equation is exactly solvable by construction. Its potential can be recovered from formula (2.7.9). Computing it, we see that the obtained exactly solvable model is a two-dimensional spherically non-symmetric harmonic oscillator with centrifugal barrier.

ii. $N = 5$. According to the results of section 2.4, equation (2.7.31) with $N = 5$ also allows exact solutions, but now it can be viewed as an equation with three spectral parameters: E, Γ and c_2. We know that the spectrum of parameter c_2 is degenerate and therefore, after transition to a two-dimensional case, we obtain the quasi-exactly solvable equation with two spectral parameters: E, which plays the role of the energy, and c_2, which is included in the potential. The potential of this model is described by a quasi-polynomial of order four. Unfortunately, as noted in section 2.6, the wavefunctions corresponding to admissible values of these parameters are not normalizable, and therefore the obtained quasi-exactly solvable model is not of any physical interest.

Note that the next values of N ($N = 7, 9, \ldots$) also lead to equations (2.7.31) allowing exact solutions. However, in these cases equations (2.7.31) contain more than three spectral parameters and consequently, the elimination of one of them by transitions to a two-dimensional case cannot give any interesting quasi-exactly solvable model of sufficiently high order.

Now let us turn to the second limiting case described by formula (2.7.26).

2. Let $f(z) = \text{arcsh}\, z$. Then we have:

$$\eta = \text{arcsh}\, \frac{r_+ + r_-}{2}, \quad \sigma = \arcsin \frac{r_+ - r_-}{2}, \qquad (2.7.32)$$

where by $r_\pm$ we denoted the expressions

$$r_\pm = \sqrt{(x \pm 1)^2 + y^2}. \qquad (2.7.33)$$

We also have

$$F = \frac{1}{|1 + z^2|} = \frac{1}{\text{ch}^2 \eta - \sin^2 \sigma}, \qquad (2.7.34)$$

from which it follows that

$$a(\eta) = \text{ch}^2 \eta, \quad b(\sigma) = -\sin^2 \sigma. \qquad (2.7.35)$$

Obviously, the unique form of functions $A(\eta)$ and $B(\sigma)$ guaranteeing quasi-polynomiality of the potential V is

$$A(\eta) = -\frac{d_1}{\text{ch}^2 \eta} + \frac{d_2}{\text{sh}^2 \eta} + \sum_{n=2}^{2N} e_n \, \text{ch}^{2n} \eta, \qquad (2.7.36a)$$

$$B(\sigma) = \frac{d_1}{\sin^2 \eta} + \frac{d_2}{\cos^2 \eta} - \sum_{n=2}^{2N} e_n \sin^{2n} \sigma, \qquad (2.7.36b)$$

where d_1, d_2 and e_n are arbitrary constants. Substitution of (2.7.36) into system (2.7.14) gives

$$\left\{ -\frac{\partial^2}{\partial \eta^2} - \frac{d_1}{\text{ch}^2 \eta} + \frac{d_2}{\text{sh}^2 \eta} + \sum_{n=1}^{2N} e_n \, \text{ch}^{2n} \eta - E \, \text{ch}^2 \eta - \Gamma \right\} \varphi_a(\eta) = 0,$$

$$\text{(2.7.37a)}$$

$$\left\{ -\frac{\partial^2}{\partial \sigma^2} + \frac{d_1}{\sin^2 \sigma} + \frac{d_2}{\cos^2 \sigma} - \sum_{n=1}^{2N} e_n \sin^{2n} \sigma + E \sin^2 \sigma - \Gamma \right\} \varphi_b(\sigma) = 0.$$

$$\text{(2.7.37b)}$$

Taking

$$\lambda = \text{ch}^2 \eta, \quad \varphi_a(\eta) = [\lambda(\lambda - 1)]^{\frac{1}{4}} \varphi(\lambda), \tag{2.7.38a}$$

$$\lambda = \sin^2 \sigma, \quad \varphi_b(\sigma) = [\lambda(\lambda - 1)]^{\frac{1}{4}} \varphi(\lambda), \tag{2.7.38b}$$

we can see that (2.7.37a) and (2.7.37b) are simply two different forms of a single one-dimensional equation which, after introducing the notations

$$e_1 = -E, \quad e_0 = -\Gamma - d_1 - d_2 + \tfrac{1}{2}, \tag{2.7.39}$$

can be written as

$$\left\{ -\frac{\partial^2}{\partial \lambda^2} + \frac{d_1 - \frac{3}{4}}{4\lambda^2} + \frac{d_2 - \frac{3}{4}}{4(\lambda - 1)^2} + \frac{\sum_{n=0}^{2N} e_n \lambda^n}{4\lambda(\lambda - 1)} \right\} \varphi(\lambda) = 0. \tag{2.7.40}$$

Evidently, (2.7.40) is equivalent to (2.7.37a) if $1 \leq \lambda \leq \infty$, and to (2.7.37b) if $0 \leq \lambda \leq 1$.

Thus, we see that the problem of constructing two-dimensional quasi-exactly solvable Schrödinger problems with quasi-polynomial potentials is reduced to the problem of finding equations belonging to the class (2.7.40) and admitting the interpretation as exactly solvable equations with three spectral parameters.

In order to find such equations note that (2.7.40) is an equation with three singular points located at 0, 1 and ∞. Consequently, the behaviour of the function $\varphi(\lambda)$ in the vicinities of these points cannot be arbitrary. It is determined by the formulas

$$\varphi(\lambda) \sim \begin{cases} \lambda^{s_1}, & \lambda \to 0 \\ (\lambda - 1)^{s_2}, & \lambda \to 1 \\ \exp\{-Q_N(\lambda)\}, & \lambda \to \infty \end{cases} \tag{2.7.41}$$

in which the numbers s_1 and s_2 satisfy the equations

$$4(s_1 - \tfrac{1}{4})(s_1 - \tfrac{3}{4}) = d_1, \quad 4(s_2 - \tfrac{1}{4})(s_2 - \tfrac{3}{4}) = d_2, \tag{2.7.42}$$

and coefficients of the polynomial $Q_N(\lambda)$ are expressed via parameters $e_{N+1}, \ldots, e_{2N}$. For example, if

$$Q_N(\lambda) = \frac{q_N}{2N}\lambda + \frac{q_{N-1}}{2(N-1)} + \ldots \qquad (2.7.43)$$

then

$$q_N^2 = e_N, \quad N \geq 1, \qquad (2.7.44a)$$
$$2q_N q_{N-1} - q_N^2 = e_{N-1}, \quad N \geq 2, \qquad (2.7.44b)$$
$$\ldots$$

The simplest expression reproducing all the asymptotic properties (2.7.41) can be written as

$$\varphi(\lambda) = P(\lambda)\lambda^{s_1}(\lambda - 1)^{s_2}\exp\{-Q(\lambda)\} \qquad (2.7.45)$$

where $P(\lambda)$ is a certain (arbitrary) polynomial. Surprisingly enough, (2.7.45) turns out to be a correct *ansatz* for equation (2.7.40). Substituting (2.7.45) into (2.7.40) we can make sure that this equation can actually be satisfied if the numbers $e_0, e_1, \ldots, e_N$, playing here the role of the spectral parameters, take certain discrete values. The number of these parameters is $N+1$ and, therefore, in order to obtain the needed three-parameter spectral equation we must take $N = 2$. Later we discuss the case $N = 2$ in detail and show that it actually leads to physically sensible two-dimensional quasi-exactly solvable models with quasi-polynomial potentials. However, first it is very helpful to consider the case $N = 1$, which is especially interesting from the point of view of one-dimensional quantum mechanics.

1. $N = 1$. In this case *ansatz* (2.7.45) can be written in the form

$$\varphi(\lambda) = \prod_{i=1}^{M}(\lambda - \xi_i)\lambda^{s_1}(\lambda - 1)^{s_2}\exp\left(-\frac{q_1}{2}\lambda\right), \qquad (2.7.46)$$

where $\xi_1, \ldots, \xi_M$ are certain unknown parameters determining the zeros of the polynomial $P(\lambda)$. Substituting (2.7.46) into (2.7.40) and equating similar terms we find explicit expressions for the two spectral parameters e_0 and e_1

$$e_0 = 8s_1 s_2 + 8s_1\left(\frac{q_1}{2} + \sum_{i=1}^{M}\frac{1}{\xi_i}\right), \qquad (2.7.47a)$$

$$e_1 = -q_1^2 - 4q_1(s_1 + s_2 + M), \qquad (2.7.47b)$$

in which the ξ_i satify the system of equations:

$$\sum_{k=1}^{M}{}' \frac{1}{\xi_i - \xi_k} + \frac{s_1}{\xi_i} + \frac{s_2}{\xi_i - 1} - \frac{q_1}{2} = 0, \quad i = 1, \ldots, M. \qquad (2.7.48)$$

We see that the spectrum of the parameter e_1 is degenerate with respect to the spectrum of e_0, and therefore, equation (2.7.40) can be interpreted as a quasi-exactly solvable equation with a single spectral parameter e_0. To reduce the obtained equation to Schrödinger form, we must identify the non-degenerate spectral parameter e_0 with the energy and include the second degenerate one, e_1, in the potential. Remembering that e_0 is connected with Γ and e_1 with E and making transformations (2.7.38a) and (2.7.38b), we obtain two new one-dimensional quasi-exactly solvable models with hyperbolic and trigonometric potentials (Ushveridze 1988c):

$$V_a(\eta) = -\frac{4(s_1 - \frac{1}{4})(s_2 - \frac{3}{4})}{\text{ch}^2\,\eta} + \frac{4(s_2 - \frac{1}{4})(s_1 - \frac{3}{4})}{\text{sh}^2\,\eta}$$
$$-[q_1^2 + 4q_1(s_1 + s_2 + M)]\,\text{ch}^2\,\eta + q_1^2\,\text{ch}^4\,\eta, \qquad (2.7.49a)$$

$$V_b(\sigma) = -\frac{4(s_1 - \frac{1}{4})(s_2 - \frac{3}{4})}{\sin^2\sigma} + \frac{4(s_2 - \frac{1}{4})(s_1 - \frac{3}{4})}{\cos^2\sigma}$$
$$-[q_1^2 + 4q_1(s_1 + s_2 + M)]\sin^2\sigma + q_1^2\sin^4\sigma. \qquad (2.7.49b)$$

The (exact) wavefunctions for these models have the form:

$$\varphi_a(\eta) = (\text{ch}^2\,\eta)^{s_1 - \frac{1}{4}}(\text{sh}^2\,\eta)^{s_2\frac{1}{4}}e^{-\frac{q_1}{2}\,\text{ch}^2\,\eta}\prod_{i=1}^{M}(\text{ch}^2\,\eta - \xi_i), \qquad (2.7.50a)$$

$$\varphi_b(\sigma) = (\sin^2\sigma)^{s_1 - \frac{1}{4}}(\cos^2\sigma)^{s_2\frac{1}{4}}e^{-\frac{q_1}{2}\sin^2\sigma}\prod_{i=1}^{M}(\sin^2\sigma - \xi_i) \qquad (2.7.50b)$$

and the corresponding energy levels are described by the formula

$$\Gamma = -4(s_1 + s_2 - \tfrac{1}{2})^2 - 8s_1\left(\frac{q_1}{2} + \sum_{i=1}^{M}\frac{1}{\xi_i}\right) \qquad (2.7.51)$$

in which the numbers $\xi_1, \ldots, \xi_M$ satisfy equation (2.7.48).

We see that model (2.7.49a) is defined on the positive half-axis $\eta \in [0, \infty]$, and the model (2.7.49b) on the finite interval $\sigma \in [0, \frac{\pi}{2}]$. The potentials of these models are depicted in figure 2.5.

The wavefunctions of the first model are normalizable in the interval $\eta \in [0, \infty]$ if $s > 2$ and $q_1 > 0$, and the wavefunctions of the second

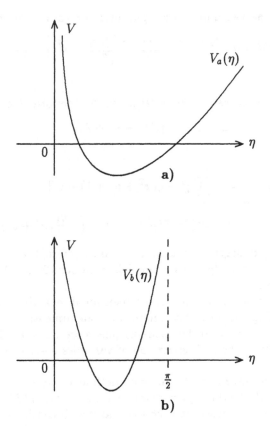

Figure 2.5. The form of the potentials a) (2.7.49a) and b) (2.7.49b).

model satisfy the normalization condition in the interval $\sigma \in [0, \frac{\pi}{2}]$ when $s_1 > 0$ and $s_2 > 0$. It is remarkable that the spectra of both these models coincide, which is a trivial consequence of the fact that they were obtained from the single two-parameter spectral equation (2.7.40). Note also that the models (2.7.49a) and (2.7.49b) can be interpreted as generalizations of the well known hyperbolic and trigonometric Pöschel–Teller potential wells (see, e.g., Flügge 1971).

Concluding the discussion of the case $N = 1$, we emphasize that the two-dimensional model, from which models (2.7.49a) and (2.7.49b) were obtained after the separation, is simply the ordinary harmonic oscillator with potential barriers along the x- and y-axes. Indeed, using formulas (2.7.9), (2.7.13), (2.7.35), (2.7.36), (2.7.32) and (2.7.33), we can reconstruct the corresponding two-dimensional potential which has the form

$$V(x,y) = -\frac{4(s_1 - \frac{1}{4})(s_2 - \frac{3}{4})}{x^2} + \frac{4(s_2 - \frac{1}{4})(s_1 - \frac{3}{4})}{y^2} + q_1^2(x^2 + y^2 + 1)$$

$$(2.7.52)$$

and leads to a degenerate system described by the formulas

$$E = q_1^2 + 4q_1(s_1 + s_1 + M)$$

$$(2.7.53)$$

and

$$\psi(x,y) = \prod_{i=1}^{M}[\xi_i^2 - \xi(x^2 + y^2 + 1) + x^2]$$

$$\times \; (x^2)^{s_1 - \frac{1}{4}}(y^2)^{s_2 - \frac{1}{4}} \exp\left(-\frac{q_1}{2}(x^2 + y^2)\right). \quad (2.7.54)$$

The multiplicity of degeneracy is $M+1$, and therefore, the order of the one-dimensional quasi-exactly solvable models obtained (2.7.49a) and (2.7.49b) is also equal to $M + 1$.

The transition from the two-dimensional exactly solvable model (2.7.52) with degenerate spectrum to two one-dimensional quasi-exactly solvable models of finite order gives us one more example demonstrating how the (direct) method of separation of variables described in section 1.7 works.

Now, let us return to the inverse method of separation of variables and, starting with the three-parameter spectral equation (2.7.40) appearing when $N = 2$, try to construct a non-trivial two-dimensional quasi-exactly solvable model.

2. $N = 2$. In this case, *ansatz* (2.7.45) becomes

$$\varphi(\lambda) = \prod_{i=1}^{M}(\lambda - \xi_i)\lambda^{s_1}(\lambda - 1)^{s_2} \exp\left(-\frac{q_2}{4}\lambda^2 - \frac{q_1}{2}\lambda\right). \quad (2.7.55)$$

Substituting (2.7.55) into (2.7.50) and equating similar terms, we find the following expressions for the three spectral parameters e_0, e_1 and e_2:

$$e_0 = 8s_1 s_2 + 8s_1\left(\frac{q_1}{2} + \sum_{i=1}^{M}\frac{1}{\xi_i}\right), \quad (2.7.56a)$$

$$e_1 = 4q_2(s_1 + M + \tfrac{1}{2}) - 4q_1(s_1 + s_2 + M) - q_1^2 - 4q_2\sum_{i=1}^{M}\xi_i, \quad (2.7.56b)$$

$$e_2 = q_1^2 - 2q_1 q_2 - 2q_2(2s_1 + 2s_2 + 2M + 1), \quad (2.7.56c)$$

in which the ξ_i satisfy the system of M numerical equations

$$\sum_{k=1}^{M} \frac{1}{\xi_i - \xi_k} + \frac{s_1}{\xi_i} + \frac{s_2}{\xi_i - 1} - \frac{q_1}{2} - \frac{q_2}{2}\xi_i = 0,$$

$$i = 1, \ldots, M. \tag{2.7.57}$$

The remaining parameters e_3, e_4 and d_1, d_2 are determined by formulas (2.7.42) and (2.7.44). Since all parameters entering into the expressions for $A(\eta)$ and $B(\sigma)$ are known, we can easily recover the form of the corresponding two-dimensional potential. Using formulas (2.7.9), (2.7.13), (2.7.35), (2.7.36), (2.7.32) and (2.7.33) we find

$$
\begin{aligned}
V(x,y) \;=\; & \frac{4(s_1 - \frac{1}{4})(s_1 - \frac{3}{4})}{x^2} + \frac{4(s_2 - \frac{1}{4})(s_2 - \frac{3}{4})}{y^2} \\
& + \{q_1^2 - 2q_2(2s_1 + 2s_2 + 2M + 1)\} \\
& + \{[q_1^2 - 2q_2(2s_1 + 2s_2 + 2M + 1)]r^2 + [2q_1q_2 + q_2^2]y^2\} \\
& + \{2q_1q_2 r^2 + 2q_2^2 y^2\}r^2 + q_2^2 r^6.
\end{aligned}
\tag{2.7.58}
$$

Analogously, we can obtain the explicit form of wavefunctions satisfying the Schrödinger equation for (2.7.58). Substituting (2.7.55) and (2.7.38) into (2.7.10) and using (2.7.32), (2.7.33) we get

$$
\begin{aligned}
\psi(x,y) \;=\; & \prod_{i=1}^{M} [\xi_i^2 - \xi_i(r^2 + 1) + x^2](x^2)^{s_1 - \frac{1}{4}}(y^2)^{s_2 - \frac{1}{4}} \\
& \times \exp\left\{-\frac{q_1}{2}(r^2 + 1) - \frac{q_2}{4}[(r^2 + 1)^2 - 2x^2]\right\}. \tag{2.7.59}
\end{aligned}
$$

The corresponding energy levels can be obtained from (2.7.39) and (2.7.56b):

$$E = q_1^2 + 4q_1(s_1 + s_2 + M) - 4q_2(s_2 + M + \tfrac{1}{2}) + 4q_2 \sum_{i=1}^{M}\xi_i. \tag{2.7.60}$$

(Remember that the numbers ξ_i can be found from (2.7.57).)

Is is easy to see that model (2.7.58) is defined in the domain $x \geq 0$, $y \geq 0$, and the wavefunctions in (2.7.59) are normalizable in this domain if $s_1 > 0$, $s_2 > 0$ and $q_2 > 0$. Thus, we have constructed a physically sensible two-dimensional quasi-exactly solvable model, the potential of which is described by a spherically non-symmetric quasi-polynomial of order six. The non-negative integer M entering into the potential determines the order of this model, which is equal to the number of solutions of system (2.7.57).

The simplest way to compute this number is to use the classical electrostatic analogue of equations (2.7.57).

Indeed, it is not difficult to understand that system (2.7.57) describes the equilibrium condition for M classical Coulomb particles with unit charges and coordinates ξ_i situated in an external electrostatic field with the potential

$$U(\xi) = -s_1 \ln |\xi| - s_2 \ln |\xi - 1| + \frac{q_1}{2}\xi + \frac{q_2}{4}\xi^2. \qquad (2.7.61)$$

This potential, consisting of three wells separated by singular potential barriers at the points $\xi = 0$ and $\xi = 1$, has the form depicted in figure 2.6.

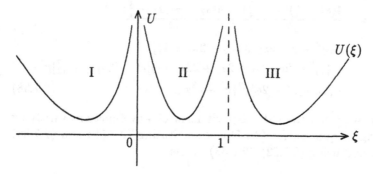

Figure 2.6. The form of the potential (2.7.61).

It is quite obvious that there are $\frac{(M+1)(M+2)}{2}$ different ways to distribute M Coulomb particles between these three wells and for any such distribution a classical stable equilibrium is possible. But this means that the number of solutions for equation (2.7.57) is $\frac{(M+1)(M+2)}{2}$ and this is also the order of the constructed quasi-exactly solvable model (2.7.58).

Note that, as in the one-dimensional case, the numbers ξ_i determine the wavefunction zeros. This follows from expression (2.7.55) and formulas (2.7.38) and (2.7.10). The equations for these zeros, which in the two-dimensional case are located along the lines, can be obtained from the condition $\lambda = \xi_i$ and formulas (2.7.38), (2.7.32) and (2.7.33):

$$\frac{1}{2}\left(\sqrt{(x+1)^2 + y^2} + \sqrt{(x-1)^2 + y^2}\right) = \xi_j, \qquad (2.7.62a)$$

$$\frac{1}{2}\left(\sqrt{(x+1)^2 + y^2} - \sqrt{(x-1)^2 + y^2}\right) = \xi_k. \qquad (2.7.62b)$$

Obviously, these equations allow real solutions if and only if the numbers ξ_j in (2.7.62a) exceed one: $\xi_i \in [1, \infty]$, while the numbers ξ_k in (2.7.62b) are less than one but positive: $\xi_k \in [0,1]$. In this case we can speak of wavefunction nodal lines which may have the form of confocal ellipses (2.7.62a) or hyperbolas (2.7.62b), see figure 2.7.

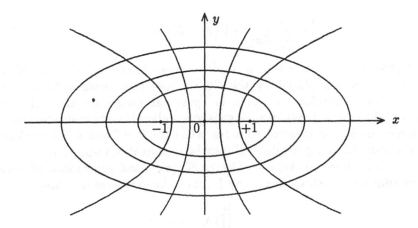

Figure 2.7. The nodal structure of wavefunctions in the model (2.7.58).

The number of elliptic nodal lines is equal to the number of particles situated in the third well (see figure 2.7), and the number of hyperbolic nodal lines is determined by the number of particles situated in the second well. The particles situated in the first well determine the complex zeros of the wavefunction. The case in which all particles are concentrated in the first well corresponds to the ground state.

2.8 Multi-dimensional quasi-exactly solvable models

The method of constructing two-dimensional quasi-exactly solvable models discussed in the previous section can be generalized easily to the multi-dimensional case (Ushveridze 1988b). As stated before the procedure includes three stages. First of all, we choose the coordinate system; then construct a most general class of Schrödinger equations (with quasi-polynomial potentials) allowing separation of variables in this system; and finally, we find the conditions for which the one-dimensional equations appearing after the separation can be interpreted as multi-parameter exactly solvable spectral equations. Note that in the $\mathcal{D}$-dimensional case, these equations must contain $\mathcal{D} + 1$ spectral parameters: $\mathcal{D} + 1$ of them must be identified with the separation constants, one must play the role of the energy in the initial $\mathcal{D}$-dimensional Schrödinger problem, and one additional parameter with degenerate spectrum must be included in the potential.

In order to choose the most suitable coordinates in which the separation of variables in the multi-dimensional Schrödinger equation

$$\left(-\sum_{\alpha=1}^{\mathcal{D}} \frac{\partial^2}{\partial x_\alpha^2} + V(x) - E \right) \psi(\vec{x}) = 0 \qquad (2.8.1)$$

is possible, remember that in the two-dimensional case, the role of such coordinates was played by the elliptic ones (see formulas (2.7.32) and (2.7.33)). Therefore, it would be quite natural to assume that in the multi-dimensional case an analogous role is played by the multi-dimensional generalizations of these coordinates, which for $\mathcal{D} = 3$ and $\mathcal{D} > 3$ are known as ellipsoidal and generalized ellipsoidal coordinates, respectively.

The connection between Cartesian, $\{x_\alpha\}$, and generalized ellipsoidal coordinates, which we denote by $\{\lambda_i\}$, can be expressed as follows:

$$x_\alpha^2 = \frac{\prod\limits_{i=1}^{\mathcal{D}} (\lambda_i - a_\alpha)}{\prod\limits_{\beta=1}^{\mathcal{D}} {}' (a_\beta - a_\alpha)}. \qquad (2.8.2)$$

Here a_α are constants determining the intervals in which the λ_i-coordinates change,

$$a_1 < \lambda_1 < a_2 < \ldots < a_{\mathcal{D}} < \lambda_{\mathcal{D}} < \infty. \qquad (2.8.3)$$

In the new variables the Schrödinger equation (2.8.1) takes the form

$$\left\{ -\sum_{i=1}^{\mathcal{D}} \frac{4(Q(\lambda_i))^{\frac{1}{2}}}{\prod\limits_{k=1}^{\mathcal{D}} {}' (\lambda_i - \lambda_k)} \frac{\partial}{\partial \lambda_i} \left[(Q(\lambda_i))^{\frac{1}{2}} \frac{\partial}{\partial \lambda_i} \right] + V - E \right\} \psi = 0, \quad (2.8.4)$$

where

$$Q(\lambda) = \prod_{\alpha=1}^{\mathcal{D}} (\lambda - a_\alpha). \qquad (2.8.5)$$

Many formulas of this section can be considerably simplified if we introduce the following notation:

$$\langle F(\lambda) \rangle \equiv \sum_{i=1}^{\mathcal{D}} \frac{F(\lambda_i)}{\prod\limits_{k=1}^{\mathcal{D}} {}' (\lambda_i - \lambda_k)}, \qquad (2.8.6)$$

in which $F(\lambda_i)$ is an arbitrary expression containing λ_i. For example, equation (2.8.4) can be rewritten as follows:

$$\left\{-4\langle\sqrt{Q(\lambda)}\frac{\partial}{\partial\lambda}\sqrt{Q(\lambda)}\frac{\partial}{\partial\lambda}\rangle + V - E\right\}\psi = 0. \tag{2.8.7}$$

Now, let us formulate and prove some assertions which allow us to investigate the possibility of separation of variables in equation (2.8.7).

Statement 2.1. *Let*

$$f_L(\lambda) \equiv \langle\lambda^{\mathcal{D}-1+L}\rangle. \tag{2.8.8}$$

Then for any $-(\mathcal{D}-1) \leq L < 0$ *the function* $f_L(\lambda)$ *is identically zero, and for* $L \geq 0$ *it is an Lth-order polynomial of the form*

$$f_L(\lambda) = \sum_{1l_1+\ldots+Ll_L=L} f_{l_1\ldots l_L}\sigma_1^{l_1}(\lambda)\cdot\ldots\cdot\sigma_L^{l_L}(\lambda), \tag{2.8.9}$$

Here $f_{l_1\ldots l_L}$ *are certain computable constants and*

$$\sigma_n(\lambda) = \begin{cases} \displaystyle\sum_{k_1<\ldots<k_n}\lambda_{k_1}\ldots\lambda_{k_n} & 1 \leq n \leq \mathcal{D} \\ 1 & n = 0 \\ 0 & n > \mathcal{D} \end{cases} \tag{2.8.10}$$

are elementary symmetric polynomials.

Proof. Let I and J be arbitrary fixed numbers, $I \neq J$. Introducing the function

$$U_{IJ} \equiv \lambda^{\mathcal{D}-1+L}\left\{\prod_{i\neq I,i\neq J}^{\mathcal{D}}(\lambda - \lambda_i)\right\}^{-1}, \tag{2.8.11}$$

we can rewrite formula (2.8.8) in the following form:

$$\begin{aligned} f_L(\lambda) &= \frac{U_{IJ}(\lambda_I) - U_{IJ}(\lambda_J)}{\lambda_I - \lambda_J} \\ &+ \sum_{i=1}^{\mathcal{D}}\lambda_i^{\mathcal{D}-1+L}\left\{\prod_{k=1}^{\mathcal{D}}{}'(\lambda_i - \lambda_k)\right\}^{-1} \quad i \neq I, i \neq J. \end{aligned} \tag{2.8.12}$$

From the representation (2.8.12) it follows that the function $f_L(\lambda)$ remains finite if λ_I tends to λ_J. The fact that the finiteness of $f_L(\lambda)$ takes place for any I and J implies that this function is regular everywhere.

On the other hand, from definition (2.8.8) it follows that the function $f_L(\lambda)$ can be represented as a fraction whose numerator is a homogenous polynomial of $[\mathcal{D}(\mathcal{D}-1)+L]$th order. The denominator of this fraction is also a homogenous polynomoial of $[\mathcal{D}(\mathcal{D}-1)]$th order and has the form $\prod_{i<k}(\lambda_i-\lambda_k)^2$. We can see that it vanishes if $\lambda_i=\lambda_k$. The fact that function $f_L(\lambda)$ has no poles means that the numerator in the fraction can be divided by the denominator without remainder. Hence, $f_L(\lambda)=0$ if $-(\mathcal{D}-1)\leq L<0$, and $f_L(\lambda)$ is the Lth-order polynomial when $L\geq0$. Since $f_L(\lambda)$ is invariant under all transformations of the permutation group, it can be expressed in terms of the elementary symmetric polynomials defined by (2.8.10). Thus, the statement is proved.

It is not difficult to calculate the first four functions $f_L(\lambda)$. They are

$$
\begin{aligned}
f_0(\lambda) &= 1, \\
f_1(\lambda) &= \sigma_1(\lambda), \\
f_2(\lambda) &= \sigma_1^2(\lambda) - \sigma_2(\lambda), \\
f_3(\lambda) &= \sigma_1^3(\lambda) - 2\sigma_1(\lambda)\sigma_2(\lambda) + \sigma_3(\lambda).
\end{aligned}
\tag{2.8.13}
$$

Statement 2.2. *The identity*

$$
\langle\frac{1}{\lambda-a_\alpha}\rangle = \frac{1}{\displaystyle\prod_{k=1}^{\mathcal{D}}{}'(a_\alpha-a_\beta)}\frac{1}{x_\alpha^2}
\tag{2.8.14}
$$

holds for any a_α and any x_α given by formula (2.8.2).

Proof. Consider the identity

$$
\frac{1}{(\mu-a_\alpha)\displaystyle\prod_{i=1}^{\mathcal{D}}(\mu-\lambda_i)} = \frac{1}{\displaystyle\prod_{i=1}^{\mathcal{D}}(a_\alpha-\lambda_i)}\frac{1}{\mu-a_\alpha} + \langle\frac{1}{(\lambda-a_\alpha)(\mu-\lambda)}\rangle.
\tag{2.8.15}
$$

Multiplying both sides of (2.8.15) by μ and taking $\mu\to\infty$, we obtain the equality

$$
\langle\frac{1}{\lambda-a_\alpha}\rangle = -\frac{1}{\displaystyle\prod_{i=1}^{\mathcal{D}}(a_\alpha-\lambda_i)}
\tag{2.8.16}
$$

which, after taking into account definition (2.8.2), reduces to (2.8.14). The statement is proved.

Statement 2.3. *Let us assume that the potential V in λ-coordinates has the form*

$$V = \left\langle \sum_{\alpha=1}^{\mathcal{D}} \frac{p_\alpha}{\lambda - a_\alpha} \right\rangle + \left\langle \lambda^{\mathcal{D}} P_{2N}(\lambda) \right\rangle, \qquad (2.8.17)$$

where

$$P_{2N}(\lambda) = \sum_{n=0}^{2N} P_n \lambda^n. \qquad (2.8.18)$$

Then, in the initial x-coordinates it is a quasi-polynomial.

Proof. From definition (2.8.2) it follows that

$$\sum_{n=1}^{\mathcal{D}} (-1)^n a_\alpha^n \sigma_n(\lambda) = x_\alpha^2 \prod_{\beta=1}^{\mathcal{D}} (a_\beta - a_\alpha). \qquad (2.8.19)$$

These relations can be considered as the linear equations for $\sigma_n(\lambda)$. Solving the linear system (2.8.19) we obtain

$$\sigma_n(\lambda) = \sigma_n(a) + \sum_{\alpha=1}^{\mathcal{D}} \sigma_{n-1}^{(\alpha)}(a) x_\alpha^2 \qquad (2.8.20)$$

where $\sigma_n^{(\alpha)}(a)$ are defined as nth-order symmetric polynomials of $a_1, a_2, \ldots, a_{\alpha-1}, a_{\alpha+1}, \ldots, a_{\mathcal{D}}$. According to statement 1 the second term in (2.8.17) depends polynomially on $\sigma_n(\lambda)$ and, hence, it is the $(2N+1)$th-order polynomial of $x_1^2, \ldots, x_{\mathcal{D}}^2$. At the same time, from statement 2 it follows that the first term in (2.8.17) is a linear combination of the terms $\frac{1}{x_1^2}, \ldots, \frac{1}{x_{\mathcal{D}}^2}$. Thus, the statement is proved.

Statement 2.4. *If the potential V in λ-coordinates has the form (2.8.17), the Schrödinger equation (2.8.7) allows separation of variables.*

Proof. According to statement 1, the constant E (the energy) can be represented as

$$E \equiv \left\langle E\lambda^{\mathcal{D}-1} + \sum_{n=0}^{\mathcal{D}-2} \Gamma_n \lambda^n \right\rangle. \qquad (2.8.21)$$

Substituting (2.8.17) and (2.8.21) into (2.8.7) and taking

$$\psi = \psi_1(\lambda_1)\ldots\psi_{\mathcal{D}}(\lambda_{\mathcal{D}}), \tag{2.8.22}$$

we obtain the system of one-dimensional spectral equations

$$\left\{ -4Q(\lambda_i)\frac{\partial^2}{\partial\lambda_i^2} - 2Q'(\lambda_i)\frac{\partial}{\partial\lambda_i} + \sum_{\alpha=1}^{\mathcal{D}}\frac{p_\alpha}{\lambda_i - a_\alpha} + \lambda_i^{\mathcal{D}}P_{2N}(\lambda_i)\right\}\psi_i(\lambda_i)$$

$$= \left(E\lambda_i^{\mathcal{D}-1} + \sum_{n=0}^{\mathcal{D}-2}\Gamma_n\lambda_i^n\right)\psi_i(\lambda_i), \quad i = 1,\ldots,\mathcal{D}. \tag{2.8.23}$$

Here the energy E is the spectral parameter of the initial problem, and Γ_n are the separation constants playing the role of the auxiliary spectral parameters of system (2.8.23). Thus, the statement is proved.

Taking into account formula (2.8.18) and using (2.8.8) and (2.8.17) we find

$$V = \sum_{\alpha=1}^{\mathcal{D}} p_\alpha\left\langle\frac{1}{\lambda - a_\alpha}\right\rangle + \sum_{n=0}^{2N} P_n f_{n+1}(\lambda). \tag{2.8.24}$$

Substituting (2.8.13), (2.8.20) and (2.8.14) into (2.8.24), we obtain the explicit form of the potentials in the x-representation, allowing the separation of variables in generalized ellipsoidal coordinates.

The next step in our program is to study equations (2.8.23) from the point of view of their exact solvability.

First of all, note that all equations (2.8.23) essentially coincide and the only difference between them is that variables λ_i belong to different intervals (see formula (2.8.3)). This gives us the possibility of identifying the variables λ_i and functions $\psi_i(\lambda_i)$ by taking

$$\lambda_i = \lambda, \quad \psi_i(\lambda_i) = \psi(\lambda) \tag{2.8.25}$$

and consider, instead of system (2.8.23), a single equation of the following form:

$$\left\{ -4\prod_{\alpha=1}^{\mathcal{D}}(\lambda - a_\alpha)\left[\frac{\partial^2}{\partial\lambda^2} + \frac{1}{2}\left(\sum_{\alpha=1}^{\mathcal{D}}\frac{1}{\lambda - a_\alpha}\right)\frac{\partial}{\partial\lambda}\right]\right.$$

$$\left. + \sum_{\alpha=1}^{\mathcal{D}}\frac{p_\alpha}{x - a_\alpha} + \lambda^{\mathcal{D}}P_{2N}(\lambda)\right\}$$

$$= \left(E\lambda^{\mathcal{D}-1} + \sum_{n=0}^{\mathcal{D}-2}\Gamma_n\lambda^n\right)\psi(\lambda). \tag{2.8.26}$$

It is not difficult to see that (2.8.26) is an equation with $\mathcal{D}+1$ singular points located at a_α, $\alpha = 1, \ldots, \mathcal{D}$ and ∞. Therefore, the behaviour of the function $\psi(\lambda)$ in the vicinities of these points cannot be arbitrary. We can express it by the formula

$$\psi(\lambda) = \begin{cases} (\lambda - a_\alpha)^{s_\alpha - \frac{1}{4}}, & \lambda \to a_\alpha \\ \exp\left\{ -\frac{1}{2} \int A_N(\lambda)\, d\lambda \right\}, & \lambda \to \infty \end{cases} \tag{2.8.27}$$

in which s_α are certain computable parameters (which can be expressed via the numbers p_α), and $A_N(\lambda)$ is an Nth-order polynomial whose $N + 1$ coefficients are connected with the leading $N + 1$ coefficients of the polynomial $P_{2N}(\lambda)$.

The simplest functions $\psi(x)$ reproducing all the asymptotic properties (2.8.27) can be written as follows:

$$\psi(\lambda) = \prod_{i=1}^{M}(\lambda - \xi_i) \prod_{\alpha=1}^{\mathcal{D}}(\lambda - a_\alpha)^{s_\alpha - \frac{1}{4}} \exp\left\{ -\frac{1}{2} \int A_N(\lambda)\, d\lambda \right\}. \tag{2.8.28}$$

Substituting (2.8.28) into equation (2.8.26), we see that (2.8.28) is a correct *ansatz* for this equation! More precisely, for (2.8.26) to be satisfied, the numbers ξ_i entering into the expression (2.8.28) must satisfy the constraints

$$\sum_{k=1}^{M} \frac{1}{\xi_i - \xi_k} + \sum_{\alpha=1}^{\mathcal{D}} \frac{s_\alpha}{\xi_i - a_\alpha} - \frac{1}{2} A_N(\xi_i) = 0, \quad i = 1, \ldots, M. \tag{2.8.29}$$

At the same time, parameters E and Γ_n entering into the right-hand side of (2.8.26), as well as the first N coefficients of the polynomial $P_{2N}(\lambda)$, depend on the numbers ξ_i. Thus, we can assert that (2.8.26) takes the form of an exactly solvable multi-parameter spectral equation. For given N and $\mathcal{D}$, the total number of spectral parameters is $\mathcal{D} + N$.

Note that for $N = 0$ we have only $\mathcal{D}$ spectral parameters E and Γ_n, $n = 0, \ldots, \mathcal{D} - 2$. We know that after transition to the $\mathcal{D}$-dimensional case, the parameters Γ_n (the separation constants) disappear, and we come to an equation with a single spectral parameter E. Obviously, this equation is exactly solvable by construction. Using formula (2.8.24) it is not difficult to show that it is the Schrödinger equation for the $\mathcal{D}$-dimensional harmonic oscillator with potential barriers along the x_α-axes.

Now, let us consider the case with $N = 1$. This case is especially interesting for us since, as we will see below, it leads to multi-dimensional quasi-exactly solvable models of arbitrarily large order.

The polynomial $P_{2N}(\lambda)$ entering into (2.8.26) takes for $N = 1$ the form

$$P_2(\lambda) = P_2\lambda^2 + P_1\lambda + P_0, \tag{2.8.30}$$

and for $A_N(\lambda)$ in (2.8.28) we have

$$A_1(\lambda) = A\lambda + B. \tag{2.8.31}$$

Substituting (2.8.28) into (2.8.26) and equating similar terms, we obtain the following expressions for the coefficients of equation (2.8.26):

$$p_\alpha = 4(s_\alpha - \tfrac{1}{4})(s_\alpha - \tfrac{3}{4}) \prod_{\beta=1}^{D}{}' (a_\alpha - a_\beta), \quad \alpha = 1,\ldots,D, \tag{2.8.32}$$

$$P_2 = A^2, \tag{2.8.33a}$$

$$P_1 = 2AB - A^2\sigma_1(a), \tag{2.8.33b}$$

$$P_0 = B^2 - 2AB\sigma_1(a) + A^2\sigma_2(a) - 4A\left(M + \tfrac{1}{2} + \sum_{\alpha=1}^{D} a_\alpha\right), \tag{2.8.33c}$$

$$E = B^2\sigma_1(a) - 2AB\sigma_2(a) + A^2\sigma_3(a) + 4B\left(M + \sum_{\alpha=1}^{D} s_\alpha\right)$$
$$+ 4A\left[\sum_{\alpha=1}^{D} a_\alpha s_\alpha - \sigma_1(a)\left(M + \tfrac{1}{2} + \sum_{\alpha=1}^{D} s_\alpha\right) + \sum_{i=1}^{M} \xi_i\right]. \tag{2.8.34}$$

The values of the separation constants Γ_n are not of interest to us and, therefore, we shall not write here the explicit expression for them. Note, however, that these constants, as well as the energy E, depend explicitly on ξ_i.

Thus, we see that equation (2.8.26) contains $D+1$ spectral parameters. They are $D-1$ separation constants Γ_n, the energy E of the multi-dimensional Schrödinger equation, and an additional parameter P_0 which plays the role of a free term in the polynomial (2.8.30) and, therefore, must be included in the potential. From formula (2.8.33c) it follows that this parameter depends on the number M, but not on the numbers ξ_i. This means that the spectrum of P_0 is degenerate with respect to the spectra of other parameters E and Γ_n. The multiplicity of this degeneracy is equal to the number of various non-equivalent solutions of the system (2.8.29)

$$\sum_{k=1}^{M}{}' \frac{1}{\xi_i - \xi_k} + \sum_{\alpha=1}^{D} \frac{s_\alpha}{\xi_i - a_\alpha} - \frac{B}{2} - \frac{A}{4}\xi_i = 0,$$
$$i = 1,\ldots,M, \tag{2.8.35}$$

which, as before, can be interpreted as the equilibrium condition for M Coulomb particles moving in an external electrostatic field. In this case the corresponding classical potential has the form of $\mathcal{D} + 2$ potential wells separated by impenetrable barriers. Therefore, the number of non-equivalent distributions for M particles between these $\mathcal{D} + 2$ wells determines the number of solutions of the system (2.8.35), which, obviously, is equal to $\frac{(M+\mathcal{D})!}{M!\mathcal{D}!}$. Summarizing, we can conclude that the $\mathcal{D}$-dimensional Schrödinger equation corresponding to equation (2.8.26) is quasi-exactly solvable and its order for any given M is $\frac{(M+\mathcal{D})!}{M!\mathcal{D}!}$ (Ushveridze 1988d, k).

Thus, we have obtained a new infinite series of multi-dimensional quasi-exactly solvable models. The form of their potentials in x-coordinates can easily be recovered from (2.8.24) by means of formulas (2.8.13), (2.8.14) and (2.8.20). Introducing the notation

$$B_\alpha \equiv Aa_\alpha + B, \quad \alpha = 1, \ldots, \mathcal{D}, \qquad (2.8.36)$$

we can write

$$
\begin{aligned}
V(x_1, \ldots, x_\mathcal{D}) = {} & A^2 \left(\sum_{\alpha=1}^{\mathcal{D}} x_\alpha^2 \right)^3 + 2A \left(\sum_{\alpha=1}^{\mathcal{D}} B_\alpha x_\alpha^2 \right) \left(\sum_{\alpha=1}^{\mathcal{D}} x_\alpha^2 \right) \\
& + \sum_{\alpha=1}^{\mathcal{D}} \left[B_\alpha^2 - 4A \left(M + \tfrac{1}{2} + \sum_{\alpha=1}^{\mathcal{D}} s_\alpha \right) \right] x_\alpha^2 \\
& + \sum_{\alpha=1}^{\mathcal{D}} \frac{4(s_\alpha - \tfrac{1}{4})(s_\alpha - \tfrac{3}{4})}{x_\alpha^2}.
\end{aligned}
\qquad (2.8.37)
$$

Here we have neglected the constant term equal to $B^2 \sigma_1(a) - 2AB\sigma_2(a) + A^2\sigma_3(a) - 4A\sigma_1(a)\left(M + \tfrac{1}{2} + \sum_{\alpha=1}^{\mathcal{D}} s_\alpha \right)$. Therefore, in order to obtain correct expressions for the energies we must subtract the same term from (2.8.34). This gives:

$$E = 4 \sum_{\alpha=1}^{\mathcal{D}} B_\alpha s_\alpha + 4BM + 4A \sum_{i=1}^{M} \xi_i. \qquad (2.8.38)$$

The x-coordinate form of wavefunctions corresponding to (2.8.38) can be found from (2.8.28), (2.8.25) and (2.8.22) by means of formulas (2.8.2) and (2.8.20). It is

$$\psi(x_1, \ldots, x_\mathcal{D}) = \prod_{i=1}^{M} \left\{ \sum_{n=0}^{\mathcal{D}} (-1)^n \xi_i^{\mathcal{D}-n} \left[\sigma_n(a) + \sum_{\alpha=1}^{\mathcal{D}} \sigma_{n-1}^{(\alpha)}(a) x_\alpha^2 \right] \right\}$$

$$\times \prod_{\alpha=1}^{\mathcal{D}} (x_\alpha^2)^{s_\alpha - \frac{1}{4}} \exp \left\{ -\frac{A}{4} \left(\sum_{\alpha=1}^{\mathcal{D}} x_\alpha^2 \right)^2 - \frac{1}{2} \left(\sum_{\alpha=1}^{\mathcal{D}} B_\alpha x_\alpha^2 \right) \right\}.$$

$$(2.8.39)$$

The condition of normalizability of these wavefunctions in the domain $x_\alpha \geq 0$, $\alpha = 1, \ldots, \mathcal{D}$ has the form

$$A > 0, \quad s_\alpha > 0, \quad \alpha = 1, \ldots, \mathcal{D}. \qquad (2.8.40)$$

Note that the fact of quasi-exact solvability of model (2.8.37) can be proved immediately in the x-representation (Ushveridze 1988k). Indeed, denoting the first (polynomial) factor in (2.8.39) by $P_M(x_1^2, \ldots, x_{\mathcal{D}}^2)$ and substituting the expression (2.8.37) into the Schrödinger equation for (2.8.37), we obtain a new equation

$$\hat{Q} P(x_1^2, \ldots, x_{\mathcal{D}}^2) = E P_M(x_1^2, \ldots, x_{\mathcal{D}}^2), \qquad (2.8.41)$$

in which

$$\hat{Q} = -\sum_{\alpha=1}^{\mathcal{D}} \left(\frac{\partial^2}{\partial x_\alpha^2} + \frac{4s_\alpha - 1}{x_\alpha} \frac{\partial}{\partial x_\alpha} \right) + 2 \sum_{\alpha=1}^{\mathcal{D}} B_\alpha \left(x_\alpha \frac{\partial}{\partial x_\alpha} + 2s_\alpha \right)$$

$$+ 2A \left(\sum_{\alpha=1}^{\mathcal{D}} x_\alpha^2 \right) \sum_{\alpha=1}^{\mathcal{D}} \left(x_\alpha \frac{\partial}{\partial x_\alpha} - M \right). \qquad (2.8.42)$$

If M is a non-negative integer, the operator $\hat{Q}$ acts in the finite-dimensional space of polynomials $P_M(x_1^2, \ldots, x_{\mathcal{D}}^2)$ of order M (the order is defined with respect to the variables $x_1^2, \ldots, x_{\mathcal{D}}^2$). Therefore, the spectral differential equation (2.8.42) can be treated as a finite-dimensional matrix equation, in which the role of the eigenvalues and eigenvectors is played by the energy E and coefficients of the polynomials $P_M(x_1^2, \ldots, x_{\mathcal{D}}^2)$. This proves the quasi-exact solvability of the model (2.8.37). A dimensionality of the corresponding matrix equation is determined by the number of non-equivalent terms in the polynomial $P_M(x_1^2, \ldots, x_{\mathcal{D}}^2)$. It is not difficult to verify that this number is $\frac{(M+\mathcal{D})!}{M!\mathcal{D}!}$. Hence, the order of the quasi-exactly solvable model under consideration must also be equal to $\frac{(M+\mathcal{D})!}{M!+\mathcal{D}!}$.

Now, let us consider two concrete examples corresponding to the simplest cases $M = 0$ and $M = 1$.

1. $M = 0$. The Schrödinger equation has one explicit solution

$$E = 4 \sum_{\alpha=1}^{\mathcal{D}} B_\alpha s_\alpha, \qquad (2.8.43)$$

$$\psi(x_1, \ldots, x_{\mathcal{D}}) \sim \prod_{\alpha=1}^{\mathcal{D}} (x_\alpha^2)^{s_\alpha - \frac{1}{4}}$$

$$\times \exp\left\{ -\frac{A}{4} \left(\sum_{\alpha=1}^{\mathcal{D}} x_\alpha^2 \right)^2 - \frac{1}{2} \left(\sum_{\alpha=1}^{\mathcal{D}} B_\alpha x_\alpha^2 \right) \right\}, \qquad (2.8.44)$$

describing the ground state because of the absence of nodes.

2. $M = 1$. In this case the Schrödinger equation has $\mathcal{D}+1$ explicit solutions which can be represented in the form

$$E = 4 \sum_{\alpha=1}^{\mathcal{D}} B_\alpha s_\alpha + \eta, \qquad (2.8.45)$$

$$\psi(x_1, \ldots, x_{\mathcal{D}}) \sim \left[1 + \sum_{\alpha=1}^{\mathcal{D}} \frac{4A x_\alpha^2}{4B_\alpha - \eta} \right] \prod_{\alpha=1}^{\mathcal{D}} (x_\alpha^2)^{s_\alpha - \frac{1}{4}}$$

$$\times \exp\left\{ -\frac{A}{4} \left(\sum_{\alpha=1}^{\mathcal{D}} x_\alpha^2 \right)^2 - \frac{1}{2} \left(\sum_{\alpha=1}^{\mathcal{D}} B_\alpha x_\alpha^2 \right) \right\} \qquad (2.8.46)$$

where by η we denote the parameter $4B + 4A\xi_1$ satisfying the algebraic equation of order $\mathcal{D} + 1$:

$$\eta + 32A \sum_{\alpha=1}^{\mathcal{D}} \frac{s_\alpha}{4B_\alpha - \eta} = 0. \qquad (2.8.47)$$

Let us assume, for definiteness, that all parameters B_α are real numbers, differing from each other and ordered as

$$B_1 < B_2 < \ldots < B_{\mathcal{D}-1} < B_{\mathcal{D}}. \qquad (2.8.48)$$

Analysing equation (2.8.47) graphically, we obtain the result that all its roots η_i are real and lie in the intervals

$$\eta_0 < 4B_1 < \eta_1 < 4B_2 < \ldots < 4B_{\mathcal{D}-1} < \eta_{\mathcal{D}-1} < 4B_{\mathcal{D}} < \eta_{\mathcal{D}}. \quad (2.8.49)$$

We see that the root η_0 corresponds to the ground state, since the wavefunction (2.8.46) has no nodes for $\eta = \eta_0$. The other functions (2.8.4) corresponding to the roots $\eta_1, \ldots, \eta_{\mathcal{D}}$ have nodes and, therefore, describe excited states. From formulas (2.8.46) and (2.8.42) it follows that the nodal surfaces of these wavefunctions have the form of various confocal quadriques with signatures from $(- - \ldots - +)$ up to $(+ + \ldots + +)$, in full accordance with the fact that the Schrödinger equation for the model (2.8.37) admits total separation of variables in $\mathcal{D}$-dimensional ellipsoidal coordinates. This means, in particular, that nodal surfaces for any wavefunction in model (2.8.37) must have the form of certain confocal quadriques.

To conclude this section, we note that whereas the problem of finding the wavefunction nodes for one-dimensional quantum systems is related to solving the problem of the equilibrium of charged particles in an external field, the problem of finding the nodal lines or surfaces for systems of dimension $\mathcal{D} \geq 2$ turns out to be related to the problem of the equilibrium of classical charged strings or membranes in an external field (Ushveridze 1989c). For model (2.8.37) this fact is obvious. It follows immediately from equations (2.8.35) which can be interpreted as conditions of equilibrium of M charged direct nodal lines $(\mathcal{D} = 2)$ or flat nodal surfaces $(\mathcal{D} \geq 3)$ interacting according to the laws of $(\mathcal{D} + 1)$-dimensional subspace. In x-coordinates we have a similar picture, but the nodal lines and surfaces become curved taking the form of confocal quadriques. Note that such a classical interpretation of wavefunction nodes is possible even in the case when separation of variables in the multi-dimensional Schrödinger equation is not possible. In fact, let us consider the problem of constructing a $\mathcal{D}$-dimensional Schrödinger equation

$$[-\Delta + V(\vec{x})]\psi(\vec{x}) = E\psi(\vec{x}), \tag{2.8.50}$$

which is exactly solvable for any one state. It is easily seen that for the choice

$$V(\vec{x}) = E + \Delta\psi(\vec{x})/\psi(\vec{x}) \tag{2.8.51}$$

where $\psi(\vec{x})$ is a smooth function, the Schrödinger equation has the function $\psi(\vec{x})$ itself as a formal solution. The requirement that the potential $V(\vec{x})$ be smooth imposes a number of constraints on the admissible form of the nodal surfaces of $\psi(\vec{x})$. To derive these we assume that the M nodal surfaces of $\psi(\vec{x})$ are described by the equations

$$\vec{x}_i = \vec{x}_i(\vec{t}), \quad \vec{x}_i \in R_{\mathcal{D}}, \quad \vec{t} \in R_{\mathcal{D}-1}, \quad i = 1, \ldots, M. \tag{2.8.52}$$

Then, the wavefunction $\psi(\vec{x})$ (up to a sign) can be written as

$$\psi(\vec{x}) = \exp\left\{-\sum_{i=1}^{M} \int \frac{\sigma[\vec{x}_i(\vec{t})] \, \mathrm{d}^{\mathcal{D}-1}\vec{t}}{|\vec{x}_i - \vec{x}_i(\vec{t})|^{\mathcal{D}-1}}\right\} \exp\phi(\vec{x}) \tag{2.8.53}$$

where $\sigma[\vec{x}_i(\vec{t})] \, \mathrm{d}^{\mathcal{D}-1}\vec{t}$ is an element of the nodal surface and $F(\vec{x})$ is a smooth function. The substitution of (2.8.53) into (2.8.51) and the requirement that $V(\vec{x})$ be smooth lead to the following system of integral equations in $\vec{x}_i(\vec{t})$:

$$
\vec{n}[\vec{x}_i(\vec{t})] \left\{ \int \frac{[\vec{x}_i(\vec{t}) - \vec{x}_i(\vec{t}')]\sigma[\vec{x}_i(\vec{t}')] \, \mathrm{d}^{\mathcal{D}-1}\vec{t}'}{|\vec{x}_i(\vec{t}) - \vec{x}_i(\vec{t}')|^{\mathcal{D}+1}} \right.
$$

$$
\left. + \sum_{k=1}^{M}{}' \int \frac{[\vec{x}_i(\vec{t}) - \vec{x}_k(\vec{t}')]\sigma[\vec{x}_k(\vec{t}')] \, \mathrm{d}^{\mathcal{D}-1}\vec{t}'}{|\vec{x}_i(\vec{t}) - \vec{x}_k(\vec{t})|^{\mathcal{D}+1}} + \vec{\nabla}\phi[\vec{x}_i(\vec{t})] \right\} = 0,
$$

$$
i = 1, \dots, M \qquad (2.8.54)
$$

where $\vec{n}[\vec{x}(\vec{t})]$ is a normal to the surface $\vec{x}(\vec{t})$. It is easy to see that for $\mathcal{D} = 1$ the first term in (2.8.54) vanishes, while the remainder of the equation degenerates into an equation describing the equilibrium of Coulomb particles with coordinates x_i in an external electrostatic field $E(x)$. For $\mathcal{D} > 1$ the resulting system can be interpreted as the equilibrium condition for M absolutely inelastic massless charged strings ($\mathcal{D} = 2$) or membranes ($\mathcal{D} \geq 3$) interacting according to the laws of ($\mathcal{D}+1$)-dimensional electrostatics in a $\mathcal{D}$-dimensional subspace. The charge is distributed uniformly along the strings (surfaces of the membranes) with unit density. Equation (2.8.54) expresses the condition that the normal component of the force acting on each element of a string (membrane) due to the other strings (membranes) and also the external potential $\phi(\vec{x})$, vanish. This electrostatic analogue allows us to understand the features of the nodal surfaces and the singularities associated with their relative location, and also to follow the variation in the model-surface shape as the potential is varied.

2.9   The "field-theoretical" case

When all parameters s_i are equal to $\frac{1}{4}$, the potential barriers along the x_α-axes disappear and the potential (2.8.37) becomes polynomial. Correspondingly, the wavefunctions (2.8.39) become regular everywhere. This gives us the possibility of defining model (2.8.37) in the entire $\mathcal{D}$-dimensional space. If $M = 1$ the hamiltonian of this model takes the form

$$
H = -\sum_{\alpha=1}^{\mathcal{D}} \frac{\partial^2}{\partial x_\alpha^2} + A^2 \left[\sum_{\alpha=1}^{\mathcal{D}} x_\alpha^2\right] \left[\sum_{\alpha=1}^{\mathcal{D}} x_\alpha^2\right] + 2A \left[\sum_{\alpha=1}^{\mathcal{D}} B_\alpha x_\alpha^2\right] \left[\sum_{\alpha=1}^{\mathcal{D}} x_\alpha^2\right]
$$

$$
+ \sum_{\alpha=1}^{\mathcal{D}} [B_\alpha^2 - A(\mathcal{D} + 6)]x_\alpha^2, \qquad (2.9.1)
$$

and the corresponding Schrödinger equation for (2.9.1) has $\mathcal{D} + 1$ exact solutions

$$E = \sum_{\alpha=1}^{\mathcal{D}} B_\alpha + \eta, \qquad (2.9.2)$$

$$\psi(x_1, \ldots, x_{\mathcal{D}}) = \left[1 + \sum_{\alpha=1}^{\mathcal{D}} \frac{4Ax_\alpha^2}{4B_\alpha - \eta} \right]$$

$$\times \exp \left\{ -\frac{A}{4} \left(\sum_{\alpha=1}^{\mathcal{D}} x_\alpha^2 \right)^2 - \tfrac{1}{2} \sum_{\alpha=1}^{\mathcal{D}} B_\alpha x_\alpha^2 \right\},$$

$$(2.9.3)$$

expressed via the auxiliary spectral parameter η satisfying the algebraic equation

$$\eta + \sum_{\alpha=1}^{\mathcal{D}} \frac{8A}{4B\alpha - \eta} = 0 \qquad (2.9.4)$$

of order $\mathcal{D} + 1$.

The model (2.9.1) is interesting because in the limit $\mathcal{D} \to \infty$ it can be transformed into a non-local, non-relativistic quasi-exactly solvable model of field theory with gapless excitations (Ushveridze 1988d, k).

In order to perform this transformation we introduce new parameters

$$\Delta = \sqrt{A}, \quad b_\alpha = \frac{B_\alpha}{\sqrt{A}}, \quad \alpha = 1, \ldots, \mathcal{D},$$

$$\mathcal{E} = \frac{E}{\sqrt{A}}, \quad \epsilon = \frac{\eta}{4\sqrt{A}}, \quad L = \frac{\mathcal{D}\sqrt{A}}{2}, \qquad (2.9.5)$$

and rewrite the Schrödinger equation (2.9.1) in the form

$$\left\{ -\sum_{\alpha=1}^{\mathcal{D}} \left[\frac{1}{\Delta} \frac{\partial}{\partial x_\alpha} \right]^2 \Delta + \left(\sum_{\alpha=1}^{\mathcal{D}} x_\alpha^2 \Delta \right)^3 + 2 \left(\sum_{\alpha=1}^{\mathcal{D}} b_\alpha x_\alpha^2 \Delta \right) \left(\sum_{\alpha=1}^{\mathcal{D}} x_\alpha^2 \Delta \right) \right.$$

$$\left. + \sum_{\alpha=1}^{\mathcal{D}} \left[b_\alpha^2 - \frac{2L}{\Delta} - 6 \right] x_\alpha^2 \Delta \right\} \psi(x_1, \ldots, x_{\mathcal{D}}) = \mathcal{E} \psi(x_1, \ldots, x_{\mathcal{D}}).$$

$$(2.9.6)$$

Making the replacement (2.9.5) in formulas (2.9.3) and (2.9.3) we obtain:

$$\mathcal{E} = \frac{1}{\Delta}\left[\sum_{\alpha=1}^{D} b_\alpha \Delta\right] + \sigma \qquad (2.9.7)$$

$$\psi(x_1,\ldots,x_D) = \left[1 + \sum_{\alpha=1}^{D} \frac{x_\alpha^2}{b_\alpha - \epsilon}\Delta\right]$$

$$\times \exp\left\{-\frac{1}{4}\left[\sum_{\alpha=1}^{D} x_\alpha^2\Delta\right]^2 - \frac{1}{2}\left[\sum_{\alpha=1}^{D} b_\alpha x_\alpha^2\Delta\right]\right\},$$

$$(2.9.8)$$

where ϵ satisfies the equation

$$2\epsilon\Delta = \sum_{\alpha=1}^{D} \frac{\Delta}{\epsilon - b_\alpha}. \qquad (2.9.9)$$

Now, taking $\mathcal{D} \to \infty$, $\Delta \to 0$, $\Delta\mathcal{D} = 2L = \text{constant}$, and

$$\left(\alpha - \frac{\mathcal{D}}{2}\right)\Delta = t,$$

$$\Delta^{-1} = \delta(0),$$

$$x_\alpha \to x(t), \qquad b_\alpha = b(t),$$

$$\frac{1}{\Delta}\frac{\partial}{\partial x_\alpha} \to \frac{\delta}{\delta x(t)}, \qquad \sum_{\alpha=1}^{D}[\ldots]\Delta \to \int_{-L}^{L}[\ldots]\,dt,$$

$$\psi(x_1,\ldots,x_D) \to \Psi[(x(t)], \qquad (2.9.10)$$

we obtain, instead of (2.9.6), the Schrödinger equation in functional derivatives

$$\left\{-\int_{-L}^{L}\left[\frac{\delta}{\delta x(t)}\right]2\,dt + \left[\int_{-L}^{L} x^2(t)\,dt\right]^3 + 2\left[\int_{-L}^{L} b(t)x^2(t)\,dt\right]\left[\int_{-L}^{L} x^2(t)\,dt\right]\right.$$

$$\left. + \int_{-L}^{L}[b^2(t) - 6]x^2(t)\,dt - 2L\delta(0)\int_{-L}^{L} x^2(t)\,dt\right\}\Psi[x(t)] = \mathcal{E}\Psi[x(t)]$$

$$(2.9.11)$$

which describes a certain non-local, non-relativistic and inhomogeneous "field theory" of rather exotic form. The last term in (2.9.11), which is proportional to $\delta(0)$, can be treated as a counterterm cancelling ultraviolet divergencies in the model. Solutions of equation (2.9.11) have the form

$$\mathcal{E} = \delta(0) \int_{-L}^{L} b(t) \, dt + \epsilon, \qquad (2.9.12)$$

$$\Psi[x(t)] = \left[1 - \int_{-L}^{L} \frac{x^2(t) \, dt}{\epsilon - b(t)}\right] \exp\left\{-\frac{1}{4}\left(\int_{-L}^{L} x^2(t) \, dt\right)^2 - \frac{1}{2}\int b(t) x^2(t) \, dt\right\}$$

$$(2.9.13)$$

where ϵ must be considered as an arbitrary real number belonging to the interval $[\min b(t), \max b(t)]$. Note that, owing to condition (2.8.49), the function $b(t)$ is monotonic and, therefore, the density of states can be determined as

$$\rho(\epsilon) = \frac{1}{b'(b^{-1}(\epsilon))}, \qquad (2.9.14)$$

where by $b^{-1}(\epsilon)$ we denote the solution of equation $b(t) = \epsilon$ with respect to t.

Formula (2.9.13) requires some comments, since the subintegral expression in the first factor in (2.9.13) is singular. A most natural way to define the "dangerous" integral is to treat it in the sense of the Cauchy principal value

$$\int_{-L}^{L} \frac{x^2(t) \, dt}{\epsilon - b(t)} \equiv \fint_{-L}^{L} \frac{x^2(t) \, dt}{\epsilon - b(t)}$$

$$= \lim_{\delta^- \to 0} \int_{-L}^{b^{-1}(\epsilon)-\delta^-} \frac{x^2(t) \, dt}{\epsilon - b(t)} + \lim_{\delta^+ \to 0} \int_{b^{-1}(\epsilon)+\delta^+}^{L} \frac{x^2(t) \, dt}{\epsilon - b(t)}.$$

$$(2.9.15)$$

From this definition it follows that the singularity at $t = b^{-1}(\epsilon)$ is integrable, but the result depends essentially on the ratio δ^+/δ^- in the limit when both the regularization parameters δ^+ and δ^- tend to zero.

In order to eliminate this ambiguity it is sufficient to take into account equation (2.9.9) which in the limit takes the form

$$\int_{-L}^{L} \frac{dt}{\epsilon - b(t)} = 0.$$

(2.9.16)

We see that the singular factors in the subintegral expressions in (2.9.15) and (2.9.16) coincide, which allows us to define the integral (2.9.16) by analogy with (2.9.15) as

$$\int_{-L}^{L} \frac{dt}{\epsilon - b(t)} \equiv \fint_{-L}^{L} \frac{dt}{\epsilon - b(t)}$$

$$= \lim_{\delta^- \to 0} \int_{-L}^{b^{-1}(\epsilon)-\delta^-} \frac{dt}{\epsilon - b(t)} + \lim_{\delta^+ \to 0} \int_{b^{-1}(\epsilon)+\delta^+}^{L} \frac{dt}{\epsilon - b(t)}.$$

(2.9.17)

Relation (2.9.17) can be interpreted as an equation for the ratio δ^+/δ^-. For example, if $b(t)$ is the linear function $b(t) = t$, equation (2.9.17) gives $\delta^+/\delta^- = (L + \epsilon)/(L - \epsilon)$. Substituting the resulting value of δ^+/δ^- into (2.9.15) we complete the definition of expression (2.9.13).

The model described by the Schrödinger equation (2.9.11) is quasi-exactly solvable, since only one branch of excitations can explicitly be found in this model. This branch, parametrized by the number ϵ, describes a part of the continuous spectrum $\epsilon \in [\min b(t), \max b(t)]$ and includes the ground state, $\epsilon = \min b(t)$. Therefore, we deal with the model with gapless excitations. The infinite additive term in (2.9.11) is irrelevant. It plays the role of an additive vacuum energy.

Concluding this section we note that the analogous quasi-exactly solvable "field theoretical" models resulting in more complicated expressions for wavefunctionals and spectral densities can be obtained if we start with the multi-dimensional quasi-exactly solvable models (2.8.37) with $M > 1$.

2.10 Other examples of multi-dimensional quasi-exactly solvable models

Concluding our discussion of methods of constructing multi-dimensional quasi-exactly solvable Schrödinger equations, we will mention here another simple method (Ushveridze 1991b) which differs essentially from those already described in the preceding sections.

The idea of this method can be formulated as follows. Let us consider the following simultaneous *ansatz* for the potential

$$
V(x_1,\ldots,x_\mathcal{D}) = 2\delta(\delta-1)\sum_{i<k}^{\mathcal{D}}\frac{1}{(x_i-x_k)^2}
$$
$$
-\frac{\mathcal{D}[(\mathcal{D}-1)(\mathcal{D}\delta+1)-1][(\mathcal{D}-1)(\mathcal{D}\delta+1)-3]}{4}\frac{1}{\sum_{i<k}^{\mathcal{D}}(x_i-x_k)^2}
$$
$$
+\mathcal{D}W\left(\sqrt{\sum_{i<k}^{\mathcal{D}}(x_i-x_k)^2}\right) \tag{2.10.1}
$$

and corresponding wavefunctions

$$
\psi(x_1,\ldots,x_\mathcal{D}) = \left[\prod_{i<x}^{\mathcal{D}}(x_i-x_k)\right]^{\delta}\left[\sum_{i<k}^{\mathcal{D}}(x_i-x_k)^2\right]^{-\frac{(\mathcal{D}-1)(\mathcal{D}\delta+1)-1}{4}}
$$
$$
\times\;\varphi\left(\sqrt{\sum_{i<k}^{\mathcal{D}}(x_i-x_k)^2}\right). \tag{2.10.2}
$$

Here, δ is an arbitrary numerical parameter, and W and φ are unknown functions. Substituting expressions (2.10.1) and (2.10.2) into the Schrödinger equation (2.8.1) and introducing a new variable

$$
\rho = \sqrt{\sum_{i<k}^{\mathcal{D}}(x_i-x_k)^2}, \tag{2.10.3}
$$

we obtain

$$
\left[-\frac{\partial^2}{\partial\rho^2}+W(\rho)\right]\varphi(\rho)=e\varphi(\rho), \tag{2.10.4}
$$

where $e = E/\mathcal{D}$.

Therefore, if equation (2.10.4) is exactly or quasi-exactly solvable, the initial $\mathcal{D}$-dimensional Schrödinger equation with potential (2.10.1) will also be exactly or quasi-exactly solvable. This observation completes the formulation of the method.

Now, let us consider a concrete example. Let

$$
W(\rho) = a^2\rho^6 - 4a(s+M+\tfrac{1}{2})\rho^2 + \frac{4(s-\tfrac{1}{4})(s-\tfrac{3}{4})}{\rho^2}. \tag{2.10.5}
$$

From the results of section 2.1 we know that model (2.10.5) is quasi-exactly solvable and has $M + 1$ exact solutions belonging to the class of functions

$$\varphi(\rho) = P_M(\rho^2)(\rho^2)^{s - \frac{1}{4}} \exp\left(-\frac{a\rho^4}{4}\right), \qquad (2.10.6)$$

where $P_M(t)$ are Mth-order polynomials. Using formulas (2.10.1) and (2.10.2) and taking for definiteness

$$\delta = \frac{1}{\mathcal{D}}\left[\frac{4s}{\mathcal{D} - 1} - 1\right], \qquad (2.10.7)$$

we obtain a $\mathcal{D}$-dimensional quasi-exactly solvable model with potential

$$V(x_1, \ldots, x_{\mathcal{D}}) = \frac{2(4s + 1 - \mathcal{D})(4s + 1 - \mathcal{D}^2)}{\mathcal{D}^2(\mathcal{D} - 1)^2} \sum_{i<k}^{\mathcal{D}} \frac{1}{(x_i - x_k)^2}$$
$$-4a(s + M + \tfrac{1}{2}) \sum_{i_k}^{\mathcal{D}} (x_i - x_k)^2 + a^2 \left[\sum_{i<k}^{\mathcal{D}} (x_i - x_k)^2\right]^3 \qquad (2.10.8)$$

having $M + 1$ exact solutions of the form

$$\psi(x_1, \ldots, x_{\mathcal{D}}) = P_M \left[\sum_{i<k}^{\mathcal{D}} (x_i - x_k)^2\right]$$
$$\times \left[\prod_{i<k}^{\mathcal{D}} (x_i - x_k)\right]^{\frac{1}{\mathcal{D}}\left[\frac{4s}{\mathcal{D}-1} - 1\right]}$$
$$\times \exp\left\{-\frac{a}{4}\left[\sum_{i<k}^{\mathcal{D}} (x_i - x_k)^2\right]^2\right\}. \qquad (2.10.9)$$

Quite obviously, starting with the other one-dimensional quasi-exactly solvable models discussed in this chapter (for example, with models (2.7.49a) and (2.7.49b)) one can obtain more complicated multi-dimensional quasi-exactly solvable Schrödinger equations with potentials expressed in terms of rational, hyperbolic or trigonometric functions. Note also that models of such a sort can be viewed as evident generalizations of the multi-particle exactly solvable models discussed by Olshanetsky and Perelomov (1983).

Chapter 3

The inverse method of separation of variables

3.1 Multi-parameter spectral equations and their properties

In this chapter we discuss a regular analytic method of constructing exactly solvable models of quantum mechanics. This method, proposed by Ushveridze (1988c, d) and then generalized by Ushveridze (1989c), is based on the use of one-dimensional exactly solvable equations with several spectral parameters. We will demonstrate that any such equation can be reduced to an exactly solvable Schrödinger-type equation on a multi-dimensional, in general, curved manifold. The transition to the multi-dimensional case can be realized by means of the inverse method of separation of variables which we discussed briefly in the preceding chapters. Here we will give the most general formulation of this method and describe a convenient procedure of constructing wide classes of exactly solvable multi-parameter spectral equations (MPS equations). The corresponding classes of exactly solvable Schrödinger-type equations will be discussed in the concluding sections of this chapter.

Let $\mathcal{V}$ be an infinite-dimensional complex linear vector space and $\mathcal{W}$ be a certain Hilbert subspace of $\mathcal{V}$. Denote by $\mathcal{X}_i^0$ and $\mathcal{X}_i^\alpha$, $\alpha = 1, \ldots, \mathcal{D}$, $i = 1, \ldots, \mathcal{D}$ linear operators acting from $\mathcal{V}$ to $\mathcal{V}$. Consider the system of equations

$$\mathcal{X}_i^0 \phi = \left(\sum_{\alpha=1}^{\mathcal{D}} \mathcal{X}_i^\alpha \varepsilon_\alpha \right) \phi, \quad \phi \in \mathcal{W}, \quad i = 1, \ldots, \mathcal{D}, \tag{3.1.1}$$

which we shall refer to as "multi-parameter spectral equations" or MPS equations. The role of spectral parameters in (3.1.1) is played by the numbers ε_α, $\alpha = 1, \ldots, \mathcal{D}$. The problem is to find all the vectors

$\varepsilon = (\varepsilon_1, \ldots, \varepsilon_{\mathcal{D}})$ for which the system (3.1.1) has non-zero solutions in $\mathcal{W}$. The set of all such vectors we shall call the spectrum of equation (3.1.1). Depending on the concrete choice of the space $\mathcal{W}$ and the operators $\mathcal{X}_i^0$ and $\mathcal{X}_i^\alpha$ this spectrum may be continuous, discrete (infinite or finite) or empty. Below, we will concentrate on the cases in which the spectrum is infinite and discrete.

The system (3.1.1) can be considered as an evident generalization of the ordinary spectral equations arising when $\mathcal{D} = 1$. Here we shall discuss the general case when $\mathcal{D} \geq 1$.

Statement 3.1. *Let the matrix operator*

$$\hat{X} = \{\mathcal{X}_i^\alpha\}_{i,\alpha=1}^{\mathcal{D}} \tag{3.1.2}$$

entering into the system (3.1.1) be invertible. Then it is possible to construct the one-parameter family of operators

$$L(\lambda) = \sigma^0(\lambda) + \sum_{i,\alpha=1}^{\mathcal{D}} \sigma^\alpha(\lambda)(\hat{X}^{-1})_\alpha^i \mathcal{X}_i^0, \tag{3.1.3}$$

in which $\sigma^0(\lambda)$ and $\sigma^\alpha(\lambda)$ are arbitrary functions of λ. These operators act in the space $\mathcal{W}$ and the spectral problem for them

$$L(\lambda)\varphi = E(\lambda)\varphi, \quad \varphi \in \mathcal{W}, \tag{3.1.4}$$

has the following solutions

$$E(\lambda) = \sigma^0(\lambda) + \sum_{\alpha=1}^{\mathcal{D}} \sigma^\alpha(\lambda)\varepsilon_\alpha, \quad \varphi = \phi, \tag{3.1.5}$$

where ε_α, $\alpha = 1 \ldots \mathcal{D}$ and ϕ are solutions of the initial system (3.1.1).

Proof. Let us introduce the operators L_α, $\alpha = 1, \ldots, \mathcal{D}$ satisfying the system of equations

$$\mathcal{X}_i^0 = \sum_{\alpha=1}^{\mathcal{D}} \mathcal{X}_i^\alpha L_\alpha, \quad i = 1, \ldots, \mathcal{D}. \tag{3.1.6}$$

The invertibility of (3.1.2) enables us to solve system (3.1.6) for L_α:

$$L_\alpha = \sum_{i=1}^{\mathcal{D}} (\hat{X}^{-1})_\alpha^i \mathcal{X}_i^0, \quad \alpha = 1, \ldots, \mathcal{D}. \tag{3.1.7}$$

Applying the inverse operator $\hat{X}^{-1}$ to both sides of (3.1.1) and using (3.1.7), we obtain $\mathcal{D}$ independent one-parameter spectral equations

$$L_\alpha \phi = \varepsilon_\alpha \phi, \quad \alpha = 1, \ldots, \mathcal{D}. \tag{3.1.8}$$

We see that the admissible values of spectral parameters ε_α and corresponding solutions ϕ of the initial MPS equations (3.1.1) are the eigenvalues and eigenvectors of the operators L_α. Therefore, the eigenvalues and eigenvectors of the operators

$$L(\lambda) = \sigma^0(\lambda) + \sum_{\alpha=1}^{\mathcal{D}} \sigma^\alpha(\lambda) L_\alpha \tag{3.1.9}$$

are described by formulas (3.1.5). This proves the statement.

We note that the eigenvectors of the operators $L(\lambda)$ do not depend on λ. This is a typical situation for the commuting operators $L(\lambda)$. However, from the explicit expression (3.1.3) it is not evident that $L(\lambda)$ really form a commuting family. The necessary constraints on $\mathcal{X}_i^0$ and $\mathcal{X}_i^\alpha$ guaranteeing the commutativity of operators $L(\lambda)$ are formulated in the following statement.

Statement 3.2. *Let*

$$[\mathcal{X}_i^0, \mathcal{X}_k^0] = 0, \quad \textit{for any } i \textit{ and } k, \tag{3.1.10}$$
$$[\mathcal{X}_i^\alpha, \mathcal{X}_k^\beta] = 0, \quad \textit{for any } i, k \textit{ and } \alpha, \beta, \tag{3.1.11}$$

and

$$[\mathcal{X}_i^0, \mathcal{X}_k^\alpha] = 0, \quad \textit{for any } \alpha \textit{ and } i \neq k. \tag{3.1.12}$$

Then the operators $L(\lambda)$ commute with each other:

$$[L(\lambda), L(\mu)] = 0, \tag{3.1.13}$$

for any λ and μ.

Proof. Taking into account equations (3.1.6), we consider the following chain of equalities:

$$0 = [\mathcal{X}_i^0, \mathcal{X}_i^0] = \sum_{\alpha,\beta=1}^{\mathcal{D}} [\mathcal{X}_i^\alpha L_\alpha, \mathcal{X}_i^\beta L_\beta]$$

$$= \sum_{\alpha,\beta=1}^{\mathcal{D}} \mathcal{X}_i^\alpha [L_\alpha, \mathcal{X}_i^\beta] L_\beta + \sum_{\alpha,\beta=1}^{\mathcal{D}} \mathcal{X}_i^\beta [\mathcal{X}_i^\alpha, L_\beta] L_\alpha$$

$$+ \sum_{\alpha,\beta=1}^{\mathcal{D}} [\mathcal{X}_i^\alpha, \mathcal{X}_i^\beta] L_\alpha L_\beta + \sum_{\alpha,\beta=1}^{\mathcal{D}} \mathcal{X}_i^\alpha \mathcal{X}_i^\beta [L_\alpha, L_\beta]. \qquad (3.1.14)$$

We see that the first two terms on the right-hand side of (3.1.14) cancel and the third term vanishes because of the commutation relation $[\mathcal{X}_i^\alpha, \mathcal{X}_i^\beta] = 0$ which follows from (3.1.11). Thus, we obtain

$$\sum_{\alpha,\beta=1}^{\mathcal{D}} \mathcal{X}_i^\alpha \mathcal{X}_i^\beta [L_\alpha, L_\beta] = 0. \qquad (3.1.15)$$

Now, using the conditions (3.1.10) and (3.1.12), we can consider an analogous chain for $i \neq k$:

$$0 = [\mathcal{X}_i^0, \mathcal{X}_k^0] = \sum_{\alpha=1}^{\mathcal{D}} [\mathcal{X}_i^\alpha L_\alpha, \mathcal{X}_k^0]$$

$$= \sum_{\alpha=1}^{\mathcal{D}} \mathcal{X}_i^\alpha [L_\alpha, \mathcal{X}_k^0] + \sum_{\alpha=1}^{\mathcal{D}} [\mathcal{X}_i^\alpha, \mathcal{X}_k^0] L_\alpha$$

$$= \sum_{\alpha,\beta=1}^{\mathcal{D}} \mathcal{X}_i^\alpha [L_\alpha, \mathcal{X}_k^\beta L_\beta]$$

$$= \sum_{\alpha,\beta=1}^{\mathcal{D}} \mathcal{X}_i^\alpha \mathcal{X}_k^\beta [L_\alpha, L_\beta] + \sum_{\alpha,\beta=1}^{\mathcal{D}} \mathcal{X}_i^\alpha [L_\alpha, \mathcal{X}_k^\beta] L_\beta$$

$$= \sum_{\alpha,\beta=1}^{\mathcal{D}} \mathcal{X}_i^\alpha \mathcal{X}_k^\beta [L_\alpha, L_\beta] + \sum_{\alpha,\beta=1}^{\mathcal{D}} [\mathcal{X}_i^\alpha L_\alpha, \mathcal{X}_k^\beta] L_\beta$$

$$= \sum_{\alpha,\beta=1}^{\mathcal{D}} \mathcal{X}_i^\alpha \mathcal{X}_k^\beta [L_\alpha, L_\beta] + \sum_{\beta=1}^{\mathcal{D}} [\mathcal{X}_i^0, \mathcal{X}_k^\beta] L_\beta$$

$$= \sum_{\alpha,\beta=1}^{\mathcal{D}} \mathcal{X}_i^\alpha \mathcal{X}_k^\beta [L_\alpha, L_\beta]. \qquad (3.1.16)$$

Combining (3.1.15) and (3.1.16), we obtain

$$\sum_{\alpha,\beta=1}^{\mathcal{D}} \mathcal{X}_i^\alpha \mathcal{X}_k^\beta [L_\alpha, L_\beta] = 0, \quad \text{for any } i \text{ and } k. \qquad (3.1.17)$$

Taking into account the invertibility of the operator (3.1.2) we obtain the relations

$$[L_\alpha, L_\beta] = 0, \quad \text{for any } \alpha \text{ and } \beta, \qquad (3.1.18)$$

guaranteeing the commutativity of the operators $L(\lambda)$. This proves the statement.

Let us now introduce more general operators

$$H(\lambda) = U L(\lambda) U^{-1} \qquad (3.1.19)$$

connected with the operators $L(\lambda)$ by a simple homogenous transformation. Since these operators commute with each other:

$$[H(\lambda), H(\mu)] = 0, \qquad (3.1.20)$$

it would be very temping to try to interpret $H(\lambda)$ as the "hamiltonians" or, more accurately, as the "integrals of motion" of a certain completely integrable quantum system. For this, they must be hermitian operators in the Hilbert space $\mathcal{W}$. Requiring that

$$H^+(\lambda) = H(\lambda) \qquad (3.1.21)$$

and substituting the expression (3.1.19) into (3.1.21), we obtain the equation

$$L^+(\lambda) U^+ U = U^+ U L(\lambda) \qquad (3.1.22)$$

for U which, after using the expression (3.1.9) for $L(\lambda)$, can be rewritten in the form

$$L_\alpha^+ U^+ U = U^+ U L_\alpha, \quad \alpha = 1, \dots, \mathcal{D}. \qquad (3.1.23)$$

Obviously, the system (3.1.23) is overdetermined and therefore, in general, it is not solvable.

Fortunately, there exists a special case when construction of solutions of the system (3.1.23) becomes possible. This case is realized when the operators $\mathcal{X}_i^0$ and $\mathcal{X}_i^\alpha$ are hermitian operators in $\mathcal{W}$.

Statement 3.3. *Let $\mathcal{X}_i^0$ and $\mathcal{X}_i^\alpha, i, \alpha = 1, \dots, \mathcal{D}$, be hermitian operators in $\mathcal{W}$ and let all the conditions of statement 3.2 be satisfied. Then system (3.1.23) is solvable and its solution has the form*

$$U = (\det \hat{X})^{\frac{1}{2}}. \qquad (3.1.24)$$

Proof. First of all, remember that the components $\mathcal{X}_i^\alpha$ of the matrix operator $\hat{X}$ are assumed to be commuting with each other and thus can be considered as "c-numbers" by performing various algebraic operations with the matrix $\hat{X}$. In particular, one can assert that the components of the inverse operator $\hat{X}^{-1}$ have the form

$$(\hat{X}^{-1})_\alpha^i = \frac{x_\alpha^i}{\det \hat{X}}, \qquad (3.1.25)$$

where the determinant in the denominator is understood in a usual matrix sense

$$\det \hat{X} = \sum_{\alpha_1 \ldots \alpha_{\mathcal{D}} = 1}^{\mathcal{D}} \epsilon_{\alpha_1 \ldots \alpha_{\mathcal{D}}} \mathcal{X}_1^{\alpha_1} \ldots \mathcal{X}_{\mathcal{D}}^{\alpha_{\mathcal{D}}}, \qquad (3.1.26)$$

and x_α^i are cofactors of elements $\mathcal{X}_i^\alpha$:

$$x_\alpha^i = \frac{\partial}{\partial \mathcal{X}_i^\alpha} \det \hat{X}. \qquad (3.1.27)$$

Using the fact that operator $\mathcal{X}_i^0$ commutes with all operators $\mathcal{X}_k^\alpha$ with $k \neq i$ and observing the absence of $\mathcal{X}_i^\alpha$ in expression (3.1.27), we can write

$$[x_\alpha^i, \mathcal{X}_i^0] = 0, \quad \text{for any } \alpha \text{ and } i. \qquad (3.1.28)$$

Note also that, due to the commutativity of the operators $\mathcal{X}_i^\alpha$ and their hermiticity in $\mathcal{W}$, operators (3.1.26) and (3.1.27) are hermitian in $\mathcal{W}$:

$$(\det \hat{X})^+ = \det \hat{X}, \qquad (3.1.29)$$
$$(x_\alpha^i)^+ = x_\alpha^i. \qquad (3.1.30)$$

Assuming that $\sigma^0(\lambda)$ and $\sigma^\alpha(\lambda)$, $\alpha = 1, \ldots, \mathcal{D}$, are real functions and using (3.1.3) and (3.1.25), we can rewrite the system (3.1.23) in the form

$$\left[\det^{-1} \hat{X} \sum_{i=1}^{\mathcal{D}} x_\alpha^i \mathcal{X}_i^0\right]^+ U^+ U = U^+ U \left[\det^{-1} \hat{X} \sum_{i=1}^{\mathcal{D}} x_\alpha^i \mathcal{X}_i^0\right]. \qquad (3.1.31)$$

Taking into account formulas (3.1.28), (3.1.29) and (3.1.30), and remembering that $\mathcal{X}_i^0$ are hermitian operators by asumption, we obtain

$$\left[\sum_{i=1}^{\mathcal{D}} x_\alpha^i \mathcal{X}_i^0\right] \left\{\det^{-1} \hat{X} U^+ U\right\} = \left\{U^+ U \det^{-1} \hat{X}\right\} \left[\sum_{i=1}^{\mathcal{D}} x_\alpha^i \mathcal{X}_i^0\right]. \qquad (3.1.32)$$

In order to satisfy this equation it is sufficient to take

$$U^+ U = \det \hat{X}, \qquad (3.1.33)$$

and this gives us the final expression for U which coincides with (3.1.24) provided that U is a hermitian operator. This proves the statement.

Thus, we have constructed the operators

$$H(\lambda) = (\det \hat{X})^{\frac{1}{2}} L(\lambda)(\det \hat{X})^{-\frac{1}{2}} \qquad (3.1.34)$$

acting in the space

$$\mathcal{W}' = (\det \hat{X})^{\frac{1}{2}} \mathcal{W}, \qquad (3.1.35)$$

being hermitian with respect to the scalar product in $\mathcal{W}$ and commuting with each other for any values of λ.

The solutions of the spectral problem for these operators

$$H(\lambda)\psi = E(\lambda)\psi, \quad \psi \in \mathcal{W}' \qquad (3.1.36)$$

have the form

$$E(\lambda) = \sigma^0(\lambda) + \sum_{\alpha=1}^{\mathcal{D}} \sigma^\alpha(\lambda)\varepsilon_\alpha, \qquad (3.1.37a)$$

$$\psi = (\det \hat{X})^{\frac{1}{2}}\phi, \qquad (3.1.37b)$$

where ε_α and ϕ are solutions of the initial system of $\mathcal{D}$ MPS equations (3.1.1).

Of course, we can consider a more general class of equations of the type (3.1.36) having the same spectrum and the same properties of the corresponding "hamiltonians": the hermiticity and commutativity. These equations are connected with (3.1.36) by the homogenous transformations and have the form

$$H_T(\lambda)\psi_T = E(\lambda)\psi_T, \quad \psi_T \in \mathcal{W}_T, \qquad (3.1.38)$$

where

$$H_T(\lambda) = TH(\lambda)T^{-1} \qquad (3.1.39)$$

and

$$\mathcal{W}_T = T\mathcal{W}'. \qquad (3.1.40)$$

Here T is an arbitrary non-singular operator from $\mathcal{V}$ to $\mathcal{V}$. We denote by $\langle \, , \, \rangle$ the scalar product in the initial Hilbert space $\mathcal{W}$. It is not difficult to show that the operators $H_T(\lambda)$ are hermitian with respect to a new scalar product $\langle\ ,\ \rangle_T$ defined as

$$\langle\ldots,\ldots\rangle_T = \langle T^{-1}\ldots T^{-1}\ldots\rangle. \qquad (3.1.41)$$

This follows from the chain

$$
\begin{aligned}
\langle\psi_2, H_T(\lambda)\psi_1\rangle_T &= \langle T^{-1}\psi_2, T^{-1}H_T(\lambda)\psi_1\rangle \\
= \langle T^{-1}\psi_2, H(\lambda)T^{-1}\psi_1\rangle &= \langle H(\lambda)T^{-1}\psi_2, T^{-1}\psi_1\rangle \\
= \langle T^{-1}H_T(\lambda)\psi_2, T^{-1}\psi_1\rangle &= \langle H_T(\lambda)\psi_2, \psi_1\rangle_T,
\end{aligned}
\qquad (3.1.42)
$$

which holds for any $\psi_1, \psi_2 \in \mathcal{W}_T$. Solutions of (3.1.38) are connected with solutions of (3.1.36) by the formula

$$\psi_T = T\psi. \qquad (3.1.43)$$

Note that the normalization properties of these solutions do not depend on a concrete choice of T:

$$\|\psi_T\|_T \equiv \sqrt{\langle\psi_T, \psi_T\rangle_T} = \sqrt{\langle T^{-1}\psi_T, T^{-1}\psi_T\rangle} = \sqrt{\langle\psi, \psi\rangle} \equiv \|\psi\|. \quad (3.1.44)$$

We note also that the spectra of equations (3.1.36) and (3.1.38) coincide.

Thus, we see that, starting with the system of $\mathcal{D}$ MPS equations (3.1.1) satisfying the conditions listed in statements 3.1–3.3, it is possible to construct a family of one-parameter spectral problems for the operators $H_T(\lambda)$ having hermitian symmetry and commuting with each other. Solutions of these spectral problems are expressed in terms of solutions of the initial system (3.1.1). However, if we actually want to interpret these operators as the hamiltonians of certain quantum mechanical systems, we must require that their spectra be bounded from below. In other words, the existence of the ground state is needed.

In order to derive the necessary conditions for this, we consider the initial system (3.1.1) from which it follows that

$$\langle\phi, \mathcal{X}_i^0\phi\rangle = \sum_{\alpha=1}^{\mathcal{D}}\langle\phi, \mathcal{X}_i^\alpha\phi\rangle\varepsilon_\alpha, \quad i = 1,\ldots,\mathcal{D}. \qquad (3.1.45)$$

Solving system (3.1.45) with respect to ε_α, we obtain

$$\varepsilon_\alpha = \sum_{i=1}^{\mathcal{D}}\|\langle\phi, \mathcal{X}_i^\alpha\phi\rangle\|^{-1}\langle\phi, \mathcal{X}_i^0\phi\rangle. \qquad (3.1.46)$$

Substitution of (3.1.46) into formula (3.1.37) for $E(\lambda)$ gives

$$E(\lambda) = \sigma^0(\lambda) + \sum_{i,\alpha=1}^{\mathcal{D}} \sigma^\alpha(\lambda) \|\langle \phi, \mathcal{X}_i^\alpha \phi \rangle\|^{-1} \langle \phi, \mathcal{X}_i^0 \phi \rangle. \qquad (3.1.47)$$

Let us now assume that the operators $\mathcal{X}_i^0$ are positive definite in $\mathcal{W}$, and the inverse matrix $\|\langle \phi, \mathcal{X}_i^\alpha \phi \rangle\|^{-1}$ contains a certain number of rows (or their linear combinations) which are positive definite in $\mathcal{W}$. In this case it is always possible to choose the functions $\sigma^\alpha(\lambda)$, $\alpha = 1,\ldots,\mathcal{D}$ in such a way as to guarantee the positive definiteness of the second term in (3.1.47). Then, the needed inequality for $E(\lambda)$ takes the form

$$E(\lambda) \geq \sigma^0(\lambda). \qquad (3.1.48)$$

Collecting the results obtained above we can formulate the following important statement.

Statement 3.4. *Let $\mathcal{V}$ be an infinite-dimensional complex linear vector space and $\mathcal{W}$ be its Hilbert subspace with scalar product $\langle \ , \ \rangle$. Let $\mathcal{X}_i^0$ and $\mathcal{X}_i^\alpha$, $\alpha = 1,\ldots,\mathcal{D}$, $i = 1,\ldots,\mathcal{D}$ be linear operators in $\mathcal{V}$ satisfying the following conditions:*

1. Operators $\mathcal{X}_i^0$, $i = 1,\ldots,\mathcal{D}$ commute with each other.

2. Operators $\mathcal{X}_i^\alpha$, $i,\alpha = 1,\ldots,\mathcal{D}$ commute with each other.

3. Operators $\mathcal{X}_i^0$ and $\mathcal{X}_k^\alpha$, $i,k,\alpha = 1,\ldots,\mathcal{D}$ commute with each other if and only if $i \neq k$.

4. Operator matrix $\hat{X} = \{\mathcal{X}_i^\alpha\}_{i,\alpha=1}^{\mathcal{D}}$ is invertible.

5. Operators $\mathcal{X}_i^0$ and $\mathcal{X}_i^\alpha$, $i,\alpha = 1,\ldots,\mathcal{D}$ are hermitian in the space $\mathcal{W}$ with respect to the scalar product $\langle \ , \ \rangle$.

Let T be an arbitrary non-singular operator in $\mathcal{V}$ and $\sigma^0(\lambda)$ and $\sigma^\alpha(\lambda)$, $\alpha = 1,\ldots,\mathcal{D}$ be certain real functions.
Then the operators $H_T(\lambda)$ acting in $\mathcal{V}$ and defined by the formula

$$H_T(\lambda) = \sigma^0(\lambda)$$

$$+ T(\det \hat{X})^{\frac{1}{2}} \sum_{i,\alpha=1}^{\mathcal{D}} \sigma^\alpha(\lambda)(\hat{X}^{-1})_\alpha^i \mathcal{X}_i^0 (\det \hat{X})^{-\frac{1}{2}} T^{-1} \qquad (3.1.49)$$

commute with each other for any λ and μ:

$$[H_T(\lambda), H_T(\mu)] = 0. \qquad (3.1.50)$$

They are hermitian operators in the Hilbert space

$$\mathcal{W}_T \equiv T(\det \hat{X})^{\frac{1}{2}} \mathcal{W} \tag{3.1.51}$$

with the scalar product

$$\langle \ldots, \ldots \rangle_T \equiv \langle T^{-1} \ldots, T^{-1} \ldots \rangle. \tag{3.1.52}$$

Their spectral problem

$$H_T(\lambda)\psi_t = E(\lambda)\psi_T, \quad \psi_T \in \mathcal{W}_T, \tag{3.1.53}$$

has the following solutions

$$E(\lambda) = \sigma^0(\lambda) + \sum_{\alpha=1}^{\mathcal{D}} \sigma^\alpha(\lambda)\varepsilon_\alpha, \quad \psi_T = T(\det \hat{X})^{\frac{1}{2}}\phi, \tag{3.1.54}$$

where ε_α, $\alpha = 1, \ldots, \mathcal{D}$ and ϕ satisfy the system of $\mathcal{D}$ MPS equations

$$\mathcal{X}_i^0 \phi = \left(\sum_{\alpha=1}^{\mathcal{D}} \mathcal{X}_i^\alpha \varepsilon_\alpha\right)\phi, \quad \phi \in \mathcal{W}, \quad i = 1, \ldots, \mathcal{D}. \tag{3.1.55}$$

Assume also that the operators $\mathcal{X}_i^0$ and $\mathcal{X}_i^\alpha$ satisfy the following additional conditions:

6. Operators $\mathcal{X}_i^0$, $i = 1, \ldots, \mathcal{D}$ are positive definite in $\mathcal{W}$.

7. The inverse matrix $\|\langle \varphi, \mathcal{X}_i^\alpha \phi \rangle\|^{-1}$ contains a non-zero number of rows (or their linear combinations) which are positive definite in $\mathcal{W}$.

Then, the functions $\sigma^\alpha(\lambda)$, $\alpha = 1, \ldots, \mathcal{D}$ can be chosen in such a way as to guarantee the boundedness of the spectrum of equation (3.1.53) from below

$$E(\lambda) \geq \sigma_0(\lambda). \tag{3.1.56}$$

(This is the end of the statement).

This statement gives us a method of reducing the systems of MPS equations of the type (3.1.55) (with the operators satisfying the conditions 1–7 of the statement) to the class of one-parameter spectral equations (3.1.53) admitting a physical interpretation: they can be considered as Schrödinger-type equations for a certain completely integrable and stable quantum system.

Let us now discuss the case when the initial MPS equations differ from (3.1.55) by a certain equivalence transformation.

Statement 3.5. *Let operators $\mathcal{X}_i^0$ and $\mathcal{X}_i^\alpha$, $\alpha = 1, \ldots, \mathcal{D}, i = 1, \ldots, \mathcal{D}$ entering into the system (3.1.55) be replaced by operators $\mathcal{X}_i'^0$ and $\mathcal{X}_i'^\alpha$, $\alpha = 1, \ldots, \mathcal{D}, i = 1, \ldots, \mathcal{D}$ of the form*

$$\mathcal{X}_i'^0 = B_i A_i \mathcal{X}_i^0 B_i^{-1}, \tag{3.1.57a}$$

$$\mathcal{X}_i'^\alpha = B_i A_i \mathcal{X}_i^\alpha B_i^{-1} \tag{3.1.57b}$$

where A_i and B_i are arbitrary fixed non-singular operators acting in the space V and satisfying the following conditions:

1. Operators A_i commute with any operators $A_k, B_k, \mathcal{X}_k^0$ and $\mathcal{X}_k^\alpha$ having different indices $k \neq i$.

2. Operators B_i commute with any operators $A_k, B_k, \mathcal{X}_k^0$ and $\mathcal{X}_k^\alpha$ having different indices $k \neq i$.

Then, the resulting class of one-parameter spectral equations connected with the transformed MPS *equations (3.1.55) is the same as in the untransformed case.*

Proof. First of all, note that the components of transformed vector and matrix operators $\mathcal{X}_i^0$ and $\mathcal{X}_i^\alpha$ can be represented in the form

$$\mathcal{X}_i'^0 = B \left(\sum_{k=1}^{\mathcal{D}} A_i^k \mathcal{X}_k^0 \right) B^{-1}, \tag{3.1.58a}$$

$$\mathcal{X}_i'^\alpha = B \left(\sum_{k=1}^{\mathcal{D}} A_i^k \mathcal{X}_k^\alpha \right) B^{-1}, \tag{3.1.58b}$$

where

$$B = B_1 B_2 \ldots B_{\mathcal{D}}, \tag{3.1.59}$$

and A is a diagonal matrix operator with the components

$$A_i^k = A_i \delta_i^k. \tag{3.1.60}$$

Introducing the notation

$$\hat{A} = \{A_i^k\}_{i,k=1}^{\mathcal{D}}, \tag{3.1.61}$$

one can write

$$(\hat{X}'^{-1})_\alpha^i = B \left(\sum_{k=1}^{\mathcal{D}} (\hat{X}^{-1})_\alpha^k (\hat{A}^{-1})_k^i \right) B^{-1}. \tag{3.1.62}$$

Substituting formulas (3.1.62) and (3.1.58) into (3.1.49), we obtain the following expression for transformed "hamiltonians":

$$H'_T(\lambda) = \sigma^0(\lambda)$$

$$+ T(\det \hat{X}')^{\frac{1}{2}} \mathcal{B} \left\{ \sum_{i,\alpha=1}^{\mathcal{D}} \sigma^\alpha(\lambda)(\hat{X}^{-1})^i_\alpha \mathcal{X}^0_i \right\} \mathcal{B}^{-1}(\det \hat{X}')^{-\frac{1}{2}} T^{-1}.$$

$$(3.1.63)$$

Taking

$$T' \equiv T(\det \hat{X}')^{\frac{1}{2}} \mathcal{B}(\det \hat{X})^{-\frac{1}{2}} \qquad (3.1.64)$$

we find that

$$H'_T(\lambda) = H_{T'}(\lambda), \qquad (3.1.65)$$

where

$$H_{T'}(\lambda) = \sigma^0(\lambda) + T'(\det \hat{X})^{\frac{1}{2}} \left\{ \sum_{i,\alpha=1}^{\mathcal{D}} \sigma^\alpha(\lambda)(\hat{X}^{-1})^i_\alpha \right\} (\det \hat{X})^{-\frac{1}{2}} T'^{-1}.$$

$$(3.1.66)$$

Comparing (3.1.66) with (3.1.49) we see that the classes of the two one-parameter spectral problems coincide. This proves the statement.

3.2   The method. General formulation

At first sight, the problem of constructing $\mathcal{D}$ different MPS equations entering into the system (3.1.55) and having identical sets of solutions seems to be extremely complicated. Indeed, on the one hand, these equations cannot coincide, since such a coincidence would contradict the condition of invertability of the matrix operator $\hat{X}$. On the other hand, if equations (3.1.55) differ from each other, it is absolutely unclear how the coincidence of their solutions can be attained.

Fortunately, this is only an apparent difficulty. Its origin lies in the fact that we have not taken into account the internal symmetry of the system (3.1.55). In order to reveal this symmetry, remember that operators $\mathcal{X}^0_i$ and $\mathcal{X}^\alpha_i$ entering into the ith equation (3.1.55) commute with analogous operators $\mathcal{X}^0_k$ and $\mathcal{X}^\alpha_k$ entering into the kth equation. This means that all MPS equations forming the system (3.1.55) are completely disconnected from each other and the problems of finding their solutions can be solved separately. This gives us the possibility of realizing one elegant trick simplifying the problem of constructing solutions of system (3.1.55).

The basic idea of this trick can be formulated as follows. Let us assume that the Hilbert space W is a direct product of $\mathcal{D}$ different Hilbert spaces W_i, $i = 1, \ldots, \mathcal{D}$. Assume also that operators $\mathcal{X}_i^0$ and $\mathcal{X}_i^\alpha$ for any given i act non-trivially in the ith space W_i only, while in all other spaces W_k with $k \neq i$ they act as unit operators.

Consider the system of $\mathcal{D}$ different MPS equations in $\mathcal{D}$ different Hilbert spaces W_i:

$$\mathcal{X}_i^0 \varphi_i = \left(\sum_{\alpha=1}^{\mathcal{D}} \mathcal{X}_i^\alpha \epsilon_\alpha \right) \varphi_i, \quad \varphi_i \in W_i, \quad i = 1, \ldots, \mathcal{D}. \qquad (3.2.1)$$

It is obvious that, in general, solutions φ of this system will differ from each other. On the other hand, we can multiply the solution $\varphi \in W_i$ of the ith equation (3.2.1) by the direct product $\bigotimes_{k \neq i}^{\mathcal{D}}{}' \varphi_k \in \bigotimes_{k \neq i}^{\mathcal{D}} \varphi_k$ of solutions of all remaining equations. Due to the triviality of operators $\mathcal{X}_i^0$ and $\mathcal{X}_i^\alpha$ in the spaces W_k with $k \neq i$ and homogeneity of equations (3.2.1) we obtain again a solution of the same ith equation. But now it will have the form $\phi = \bigotimes_{k=1}^{\mathcal{D}} \varphi_i \in \bigotimes_{k=1}^{\mathcal{D}} W_k \equiv W$, which does not depend explicitly on the number i! This gives us a reason to replace the vectors $\varphi_i \in W_i$ in (3.2.1) by $\phi \in W$, which leads us immediately to system (3.1.55). Thus, we see that the system of different MPS equations having different solutions in different Hilbert spaces W_i, $i = 1, \ldots, \mathcal{D}$ can easily be reduced to the system of different MPS equations with identical solutions in the space $W = W_1 \otimes \ldots \otimes W_{\mathcal{D}}$.

Obviously, the system (3.2.1) is mathematically simpler than system (3.1.55), since it is free from the requirements of coincidence of its solutions. This enables us to consider this system as a starting point in constructing Schrödinger-type equations for stable completely integrable quantum models.

In order to reformulate the assertions of statements 3.4 and 3.5 in terms of the system (3.2.1) we must describe the properties of operators $\mathcal{X}_i^0$ and $\mathcal{X}_i^\alpha$ in the Hilbert spaces W_i. But for this we need to know the connection between the scalar products in the spaces W_i and W. There are many ways to establish such a connection. We choose the simplest way, which can be formulated as follows.

Let $\phi^{(1)}$ and $\phi^{(2)}$ be two arbitrary elements of the space W which, evidently, admit the expansions:

$$\phi^{(1)} = \sum_{\varphi_1^{(1)}, \ldots, \varphi_{\mathcal{D}}^{(1)}} \varphi_1^{(1)} \otimes \ldots \otimes \varphi_{\mathcal{D}}^{(1)}, \qquad (3.2.2a)$$

$$\phi^{(2)} = \sum_{\varphi_1^{(2)}, \dots, \varphi_D^{(2)}} \varphi_1^{(2)} \otimes \dots \otimes \varphi_D^{(2)}. \tag{3.2.2b}$$

The summation in (3.2.2) is performed over certain bases $\{\varphi_1\}, \dots, \{\varphi_D\}$ in the spaces $\mathcal{W}_1, \dots, \mathcal{W}_D$. Denote the scalar products in $\mathcal{W}_i$ by $(\ ,\ )_i$, $i = 1, \dots, D$. Then, the scalar product $\langle\ ,\ \rangle$ in $\mathcal{W} = \mathcal{W}_1 \otimes \dots \otimes \mathcal{W}_D$ can be defined as

$$\left\langle \phi^{(1)}, \phi^{(2)} \right\rangle \equiv \sum_{\varphi_1^{(1)}, \dots, \varphi_D^{(1)}} \sum_{\varphi_1^{(2)}, \dots, \varphi_D^{(2)}} (\varphi_1^{(1)}, \varphi_1^{(2)})_1 \dots (\varphi_D^{(1)}, \varphi_D^{(2)})_D,$$

$$\tag{3.2.3}$$

or, in brief, as

$$\langle\ ,\ \rangle = \prod_{i=1}^{D} (\ ,\ )_i. \tag{3.2.4}$$

Now we can formulate the analogues of statements 3.4 and 3.5 in conformity to the system (3.2.1).

Statement 3.6 (analogue of statment 3.4). *Let $\mathcal{V}_i$ be infinite-dimensional complex vector spaces and $\mathcal{W}_i$ be their Hilbert subspaces with the scalar products $(\ ,\ )_i$, $i = 1, \dots, D$. Let X_i and X_i^α, $\alpha = 1, \dots, D$ be linear operators in the spaces $\mathcal{V}_i$ satisfying the following conditions:*

1. For any fixed index i the operators X_i^α, $\alpha = 1, \dots, D$ commute with each other.

2. The columns of the matrix X_i^α are linearly independent.

3. For any fixed i the operators X_i^0 and X_i^α, $\alpha = 1, \dots, D$ are hermitian in the corresponding space $\mathcal{W}_i$.

4. Operators X_i^0 are positive definite in $\mathcal{W}_i$: $(\varphi_i, X_i^0, \varphi_i)_i > 0$ for any non-zero $\varphi_i \in \mathcal{W}_i$.

5. The matrix $\|(\varphi_i, X_i^\alpha, \varphi_i)_i\|^{-1}$ contains a non-zero number of rows (or their linear combinations) which are positive definite for any $\varphi_i \in \mathcal{W}_i$.

Also let

$$\mathcal{W} \equiv \bigotimes_{i=1}^{D} \mathcal{W}_i \subset \bigotimes_{i=1}^{D} \mathcal{V}_i \equiv \mathcal{V} \tag{3.2.5}$$

be a Hilbert space with scalar product defined by formula (3.2.4).

Then, the operators

$$\mathcal{X}_i^0 = I_1 \otimes \ldots \otimes I_{i-1} \otimes X_i^0 \otimes I_{i+1} \otimes \ldots \otimes I_{\mathcal{D}} \qquad (3.2.6a)$$

$$\mathcal{X}_i^\alpha = I_1 \otimes \ldots \otimes I_{i-1} \otimes X_i^\alpha \otimes I_{i+1} \otimes \ldots \otimes I_{\mathcal{D}} \qquad (3.2.6b)$$

(in which I_i are the unit operators in $\mathcal{W}_i$) satisfy all conditions 1–7 of statement 3.4 and thus describe the classes of completely integrable quantum systems, the properties of which are listed in formulas (3.1.49)–(3.1.55). We only note that solutions of the system (3.1.54) now have the form

$$\phi = \bigotimes_{i=1}^{\mathcal{D}} \varphi_i, \qquad (3.2.7)$$

where $\varphi_i \in \mathcal{W}_i$, $i = 1, \ldots, \mathcal{D}$ are solutions of the simplified system

$$X_i^0 \varphi_i = \left(\sum_{\alpha=1}^{\mathcal{D}} X_i^\alpha \epsilon_\alpha \right) \varphi_i, \quad \varphi_i \in \mathcal{W}_i, \quad i = 1, \ldots, \mathcal{D}. \qquad (3.2.8)$$

(This is the end of the statement.)

Proof. Properties 1–3 of operators $\mathcal{X}_i^0$ and $\mathcal{X}_i^\alpha$ listed in statement 3.4 follow immediately from the definitions (3.2.6) of these operators, and also from condition 1 of the present statement. Analogously, property 4 of the matrix operator $\hat{X}$ follows from formula (3.2.6) and condition 2. In order to prove property 5 (hermitian symmetry of operators $\mathcal{X}_i^\alpha$, $\alpha = 0, 1, \ldots, \mathcal{D}$), we consider two arbitrarily chosen elements of the space $\mathcal{W}$ written in the form

$$\phi^{(1)} = \bigotimes_{k=1}^{\mathcal{D}} \varphi_k^{(1)}, \quad \phi^{(2)} = \bigotimes_{k=1}^{\mathcal{D}} \varphi_k^{(2)} \qquad (3.2.9)$$

where $\varphi_k^{(1)}$ and $\varphi_k^{(2)}$ are arbitrary elements of the spaces $\mathcal{W}_k$, $k = 1, \ldots, \mathcal{D}$. Then, using formulas (3.2.3) and (3.2.6) and condition 3 of the present statement, we obtain:

$$\left\langle \phi^{(1)}, \mathcal{X}_i^\alpha \phi^{(2)} \right\rangle \equiv \left\langle \bigotimes_{k=1}^{\mathcal{D}} \varphi_k^{(1)}, \mathcal{X}_i^\alpha \bigotimes_{k=1}^{\mathcal{D}} \varphi_k^{(2)} \right\rangle$$

$$= \left\langle \bigotimes_{k=1}^{\mathcal{D}} \varphi_k^{(1)}, \bigotimes_{k<i} \varphi_k^{(2)} \otimes X_i^\alpha \varphi_i^{(2)} \otimes \bigotimes_{k>i} \varphi_k^{(2)} \right\rangle$$

$$= \left\{ \prod_{k \neq i}^{\mathcal{D}} (\varphi_k^{(1)}, \varphi_k^{(2)})_k \right\} (\varphi_i^{(1)}, X_i^\alpha \varphi_i^{(2)})_i$$

$$= \left\{ \prod_{k \neq i}^{\mathcal{D}} (\varphi_k^{(1)}, \varphi_k^{(2)})_k \right\} (X_i^\alpha \varphi_i^{(1)}, \varphi_i^{(2)})_i$$

$$= \left\langle \bigotimes_{k<i} \varphi_k^{(1)} \otimes X_i^\alpha \varphi_i^{(1)} \otimes \bigotimes_{k>i} \varphi_k^{(1)}, \bigotimes_{k=1}^{\mathcal{D}} \varphi_k^{(2)} \right\rangle$$

$$= \left\langle \mathcal{X}_i^\alpha \bigotimes_{k=1}^{\mathcal{D}} \varphi_i^{(1)}, \bigotimes_{k=1}^{\mathcal{D}} \varphi_i^{(2)} \right\rangle \equiv \left\langle \mathcal{X}_i^\alpha \phi^{(1)}, \phi^{(2)} \right\rangle. \tag{3.2.10}$$

Since any two elements $\phi^{(1)}$ and $\phi^{(2)}$ of the space $\mathcal{W}$ admit the expansion (3.2.2), the equality

$$\langle \phi^{(1)}, \mathcal{X}_i^\alpha \phi^{(2)} \rangle = \langle \mathcal{X}_i^\alpha \phi^{(1)}, \phi^{(2)} \rangle \tag{3.2.11}$$

holds for any $\phi^{(1)}, \phi^{(2)} \in \mathcal{W}$.

The additional properties 6 and 7 of the operators $\mathcal{X}_i^0$ and $\mathcal{X}_i^\alpha$ follow from conditions 4 and 5 of the present statement and also from the chain (3.2.10). Indeed, for any $\phi = \bigotimes_{i=1}^{\mathcal{D}} \varphi_i$ we obtain

$$\langle \phi, \mathcal{X}_i^0 \phi \rangle = \frac{\langle \phi, \phi \rangle}{(\varphi_i, \varphi_i)_i} (\varphi_i, X_i^0 \varphi_i)_i > 0. \tag{3.2.12}$$

From the analogous formula

$$\langle \phi, \mathcal{X}_i^\alpha \phi \rangle = \frac{\langle \phi, \phi \rangle}{(\varphi_i, \varphi_i)_i} (\varphi_i, X_i^\alpha \varphi_i)_i \tag{3.2.13}$$

it follows that the rows of the matrix $\|\langle \phi, \mathcal{X}_i^\alpha \phi \rangle\|^{-1}$ have the same signs as the rows of the matrix $\|(\varphi_i, X_i^\alpha \varphi_i)_i\|^{-1}$. This proves the statement.

Statement 3.7 (analogue of statement 3.5). *Let operators X_i^0 and X_i^α, $\alpha = 1, \ldots, \mathcal{D}$, $i = 1, \ldots, \mathcal{D}$ be replaced by operators $X_i'^0$ and $X_i'^\alpha$, $\alpha = 1, \ldots, \mathcal{D}$, $i = 1, \ldots, \mathcal{D}$ of the form*

$$X_i'^0 = B_i A_i X_i^0 B_i^{-1}, \quad X_i'^\alpha = B_i A_i X_i^\alpha B_i^{-1} \tag{3.2.14}$$

where A_i and B_i are arbitrary non-singular operators acting in the spaces V_i, $i = 1, \ldots, \mathcal{D}$.

Then the resulting class of one-parameter spectral equations connected with the transformed MPS *equations (3.2.8) is the same as in the untransformed case.*

Proof. Introducing the operators

$$\mathcal{A}_i = I_1 \otimes \ldots \otimes I_{i-1} \otimes A_i \otimes I_{i+1} \otimes \ldots \otimes I_{\mathcal{D}} \qquad (3.2.15a)$$

and

$$\mathcal{B}_i = I_1 \otimes \ldots \otimes I_{i-1} \otimes B_i \otimes I_{i+1} \otimes \ldots \otimes I_{\mathcal{D}}, \qquad (3.2.15b)$$

acting in the space $\mathcal{V} = \mathcal{V}_1 \otimes \ldots \otimes \mathcal{V}_{\mathcal{D}}$, it is easy to verify that they satisfy both conditions 1 and 2 of statement 3.5. This proves the statement.

Thus, we see that in order to construct the family of commuting integrals of motion for a certain stable and completely integrable quantum model, it is sufficient to have the system (3.2.8) of MPS equations satisfying the conditions of statement 3.6. According to statement 3.7, the form of the resulting Schrödinger equations depends on the concrete choice of the system only up to an equivalence transformation of the type (3.2.14).

Because of the importance of the system (3.2.8) it is reasonable to discuss it in more detail.

First of all note that this system can be interpreted as a generalization of ordinary one-parameter spectral equations in the Hilbert space. In order to show this, let us consider some properties of system (3.2.8) and compare them with analogous properties of one-parameter equations.

a. Orthogonality of solutions. Let $\varphi_i^{(n)}$ and $\varphi_i^{(m)}$, $i = 1, \ldots, \mathcal{D}$ be two different solutions of system (3.2.8) with the corresponding sets of spectral parameters $\varepsilon_\alpha^{(n)}$ and $\varepsilon_\alpha^{(m)}$, $\alpha = 1, \ldots, \mathcal{D}$. Then one can write

$$X_i^0 \varphi_i^{(n)} = \left(\sum_{\alpha=1}^{\mathcal{D}} X_i^\alpha \varepsilon_\alpha^{(n)} \right) \varphi_i^{(n)}, \quad \varphi_i^{(n)} \in \mathcal{W}_i, \quad i = 1, \ldots \mathcal{D}$$

$$(3.2.16a)$$

and

$$X_i^0 \varphi_i^{(m)} = \left(\sum_{\alpha=1}^{\mathcal{D}} X_i^\alpha \varepsilon_\alpha^{(m)} \right) \varphi_i^{(m)}, \quad \varphi_i^{(m)} \in \mathcal{W}_i, \quad i = 1, \ldots \mathcal{D}.$$

$$(3.2.16b)$$

Multiplying (3.2.16a) by $\varphi_i^{(m)}$ and (3.2.16b) by $\varphi_i^{(n)}$ (in a scalar sense) we obtain

$$\left(\varphi_i^{(m)}, X_i^0 \varphi_i^{(n)} \right)_i = \sum_{\alpha=1}^{\mathcal{D}} \left(\varphi_i^{(m)}, X_i^\alpha \varphi_i^{(n)} \right)_i \varepsilon_\alpha^{(n)}, \quad i = 1, \ldots, \mathcal{D}$$

$$(3.2.17a)$$

and

$$\left(\varphi_i^{(n)}, X_i^0 \varphi_i^{(m)}\right)_i = \sum_{\alpha=1}^{\mathcal{D}} \left(\varphi_i^{(m)}, X_i^\alpha \varphi_i^{(m)}\right)_i \varepsilon_\alpha^{(m)}, \quad i = 1, \ldots, \mathcal{D}.$$

(3.2.17b)

Due to the hermiticity of X_i^0 and X_i^α in the spaces $\mathcal{W}_i$, the left-hand sides of both these equations coincide. The kernels in the right-hand sides also coincide. Therefore, subtracting (3.2.17a) from (3.2.17b) we get

$$\sum_{\alpha=1}^{\mathcal{D}} (\varphi_i^{(n)}, X_i^\alpha \varphi_i^{(m)})_i (\varepsilon_\alpha^{(m)} - \varepsilon_\alpha^{(n)}) = 0, \quad i = 1, \ldots, \mathcal{D}.$$

(3.2.18)

Let us first assume that the sets of spectral parameters $\varepsilon_\alpha^{(n)}$ and $\varepsilon_\alpha^{(m)}$ do not coincide ($n \neq m$). Then equation (3.2.18) can be satisfied if and only if

$$\det \| \left(\varphi_i^{(n)}, X_i^\alpha \varphi_i^{(m)}\right)_i \|_{i,\alpha=1}^{\mathcal{D}} = 0, \quad n \neq m.$$

(3.2.19a)

Now suppose that the sets of spectral parameters coincide ($m = n$). Remember that, according to condition 5 of statement 3.6, the matrix $\|(\varphi_i^{(n)}, X_i^\alpha \varphi_i^{(n)})_i\|_{i,\alpha=1}^{\mathcal{D}}$ is assumed to be invertible. This means that its determinant must differ from zero. But then we can renormalize the solutions $\varphi_i^{(n)}$ in such a way that

$$\det \| \left(\varphi_i^{(n)}, X_i^\alpha \varphi_i^{(n)}\right)_i \|_{i,\alpha=1}^{\mathcal{D}} = 1.$$

(3.2.19b)

Combining the conditions (3.2.19a) and (3.2.19b) we obtain the generalized orthonormalization condition

$$\det \| \left(\varphi_i^{(n)}, X_i^\alpha \varphi_i^{(m)}\right)_i \|_{i,\alpha=1}^{\mathcal{D}} = \delta_{nm},$$

(3.2.20)

which in the one-parameter case ($\mathcal{D} = 1$) is reduced to the standard orthonormalization condition for solutions of an ordinary spectral equation with a weight operator.

b. Variation principle. Multiplying system (3.2.8) by φ_i we obtain the equalities

$$(\varphi_i, X_i^0 \varphi_i) = \sum_{\alpha=1}^{\mathcal{D}} (\varphi_i, X_i^\alpha \varphi_i) \varepsilon_\alpha, \quad i = 1, \ldots, \mathcal{D},$$

(3.2.21)

which can be viewed as equations for spectral parameters ε_α, $\alpha = 1, \ldots, \mathcal{D}$. Solving system (3.2.21) we obtain

$$\varepsilon_\alpha = \sum_{i=1}^{\mathcal{D}} \|(\varphi_i, X_i^\alpha \varphi_i)\|^{-1}(\varphi_i, X_i^0 \varphi_i), \quad \alpha = 1, \ldots, \mathcal{D}. \qquad (3.2.22)$$

Substitution of (3.2.22) into formula (3.1.60) for the spectrum gives

$$E(\lambda) = \sigma^0(\lambda) + \sum_{i,\alpha=1}^{\mathcal{D}} \sigma^\alpha(\lambda)\|(\varphi_i, X_i^\alpha \varphi_i)_i\|^{-1}(\varphi_i, X_i^0 \varphi_i)_i. \qquad (3.2.23)$$

Let us now disregard the fact that the elements φ_i, $i = 1, \ldots, \mathcal{D}$ in (3.2.23) are solutions of system (3.2.8). Consider the right-hand side of (3.2.23) as a functional in the spaces $\mathcal{W}_1, \ldots, \mathcal{W}_{\mathcal{D}}$.

$$E(\lambda, \phi) \equiv \sigma^0(\lambda) + \sum_{i,\alpha=1}^{\mathcal{D}} \sigma^\alpha(\lambda)\|(\varphi_i, X_i^\alpha \varphi_i)_i\|^{-1}(\varphi_i, X_i^0 \varphi_i)_i,$$

$$\varphi_i \in \mathcal{W}_i, \quad i = 1, \ldots, \mathcal{D}. \qquad (3.2.24)$$

According to statement 3.6, this functional is bounded from below and this gives us the possibility of stating the problem of finding its minimum. For this purpose we can use a variation principle.

Let $\varphi_i^{(0)} \in \mathcal{W}_i$ be the elements minimizing the functional (3.2.24), and let $E^{(0)}(\lambda)$ be its minimum value. Consider small variations

$$\varphi = \varphi_i^{(0)} + \delta\eta_i, \qquad (3.2.25)$$

in which $\delta\eta_i \in \mathcal{W}_i$ and $\|\delta\eta_i\|_i \to 0$. Then, substituting (3.2.25) into (3.2.24) we obtain the variation equation

$$\sum_{i,\beta=1}^{\mathcal{D}} \sigma^\beta(\lambda)\|(\varphi_i^{(0)}, X_i^\beta \varphi_i^{(0)})_i\|^{-1}(\delta\eta_i, X_i^0 \varphi_i^{(0)})_i$$

$$- \sum_{i,k,\alpha,\beta=1}^{\mathcal{D}} \sigma^\beta(\lambda)\|(\varphi_i^{(0)}, X_i^\beta \varphi_i^{(0)})_i\|^{-1}(\delta\eta_i, X_i^\alpha \varphi_i^{(0)})_i$$

$$\times \|(\varphi_k^{(0)}, X_k^\alpha \varphi_k^{(0)})_k\|^{-1}(\varphi_k^{(0)}, X_k^0 \varphi_k^{(0)})_k = 0, \qquad (3.2.26)$$

from which it follows that

$$X_i^0 \varphi_i^{(0)} = \left(\sum_{\alpha=1}^{\mathcal{D}} X_i^\alpha \varepsilon_\alpha^{(0)}\right), \quad i = 0, \ldots, \mathcal{D}, \qquad (3.2.27)$$

where

$$\varepsilon_\alpha^{(0)} = \sum_{k=1}^{\mathcal{D}} \|(\varphi_k^{(0)}, X_k^\alpha \varphi_k^{(0)})_k\|^{-1} (\varphi_k^{(0)}, X_k^{(0)} \varphi_k^{(0)}). \qquad (3.2.28)$$

But then the minimum of the functional (3.2.24) can be represented as

$$E^{(0)}(\lambda) = \sigma^0(\lambda) + \sum_{\alpha=1}^{\mathcal{D}} \sigma^\alpha(\lambda) \varepsilon_\alpha^{(0)}. \qquad (3.2.29)$$

We see that system (3.2.28) coincides with the system of MPS equations (3.2.8), and expression (3.2.29) with expression (3.1.60)! The solution

$$\varepsilon_\alpha = \varepsilon_\alpha^{(0)}, \quad \alpha = 1, \dots, \mathcal{D}; \qquad \varphi_i = \varphi_i^{(0)}, \quad i = 1, \dots, \mathcal{D} \qquad (3.2.30)$$

is an obvious analogue of the ground state for the ordinary one-parameter spectral equation.

Let us now show that the analogues of excited states can also be obtained in the framework of the variation method. For this we assume that the first N states $\varphi_i^{(0)}, \varphi_i^{(1)}, \dots, \varphi_i^{(N-1)}, i = 1, \dots, \mathcal{D}$ are already known. Consider the problem of finding the minimum of the functional (3.2.24) under N additional conditions

$$\det \|(\varphi_i, X_i^\alpha \varphi_i^{(n)})_i\| = 0, \quad n = 0, 1, \dots, N-1, \qquad (3.2.31)$$

which express the orthogonality of the minimizing elements $\varphi_i^{(N)}, i = 1, \dots, \mathcal{D}$ to the elements $\varphi_i^{(0)}, \varphi_i^{(1)}, \dots, \varphi_i^{(N-1)}, i = 1, \dots, \mathcal{D}$ (in a generalized sense). We denote the corresponding minimal value of (3.2.24) by $E^{(N)}(\lambda)$. In this case, the variation equation takes the form

$$\sum_{i,\beta=1}^{\mathcal{D}} \sigma^\beta(\lambda) \|(\varphi_i^{(N)}, X_i^\beta \varphi_i^{(N)})_i\|^{-1} (\delta\eta_i, X_i^0 \varphi_i^{(N)})_i$$

$$- \sum_{i,k,\alpha,\beta=1}^{\mathcal{D}} \sigma^\beta(\lambda) \|(\varphi_i^{(N)}, X_i^\beta \varphi_i^{(N)})_i\|^{-1} (\delta\eta_i, X_i^\alpha \varphi_i^{(N)})_i$$

$$\times \|(\varphi_k^{(N)}, X_k^\alpha \varphi_k^{(N)})_k\|^{-1} (\varphi_k^{(N)}, X_k^0 \varphi_k^{(N)})_k$$

$$= \sum_{n=0}^{N-1} \Lambda_n(\lambda) \det \|(\varphi_i^{(N)}, X_i^\alpha \varphi_i^{(n)})_i\|$$

$$\times \sum_{i,\alpha=1}^{\mathcal{D}} \|(\varphi_i^{(N)}, X_i^\alpha \varphi_i^{(n)})_i\|^{-1} (\delta\eta_i, X_i^\alpha \varphi_i^{(n)})_i, \qquad (3.2.32)$$

where $\Lambda_n(\lambda), n = 0, \ldots, N-1$ are Lagrange multipliers. Taking in (3.2.22)

$$\delta\eta_i = \delta \cdot \varphi_i^{(n)}, \quad n = 0, 1, \ldots, N-1, \tag{3.2.33}$$

we find that

$$\Lambda_n(\lambda) = 0, \quad n = 0, 1, \ldots, N-1. \tag{3.2.34}$$

This leads us to the following system of equations for $\varphi_i^{(N)}$:

$$X_i^0 \varphi_i^{(N)} = \left(\sum_{\alpha=1}^{\mathcal{D}} X_i^\alpha \varepsilon_\alpha^{(N)} \right) \varphi_i^{(N)}, \quad i = 0, \ldots, \mathcal{D}, \tag{3.2.35}$$

in which

$$\varepsilon_\alpha^{(N)} = \sum_{k=1}^{\mathcal{D}} |(\varphi_k^{(N)}, X_k^\alpha \varphi_k^{(N)})_k\|^{-1} (\varphi_k^{(N)}, X_k^0 \varphi_k^{(N)}). \tag{3.2.36}$$

For the conditional minimum of the functional (3.2.24) we obtain

$$E^{(N)}(\lambda) = \sigma^0(\lambda) + \sum_{\alpha=1}^{\mathcal{D}} \sigma^\alpha(\lambda) \varepsilon_\alpha^{(N)}. \tag{3.2.37}$$

Thus, we have again obtained both the system of MPS equations (3.2.8) and correct expressions for the spectral function $E(\lambda)$.

Taking successively $N = 1, 2, 3, \ldots$, we can list all solutions of this system. The values of the corresponding spectral functions $E^{(N)}(\lambda)$ form in this case the non-decreasing sequence

$$E^0(\lambda) \le E^{(1)}(\lambda) \le \ldots \le E^{(N)}(\lambda) \le \ldots . \tag{3.2.38}$$

The variation principle described above is, obviously, a most natural generalization of the well known minimal principle for the ordinary one-parameter spectral equations.

The above-mentioned facts clearly demonstrate that there is no essential difference between ordinary one-parameter spectral equations and systems of MPS equations of the type (3.2.8). The latter can obviously be viewed as rather natural generalizations of the former and can be studied by means of the same mathematical methods. At present we are not ready to discuss the properties of systems (3.2.8) in the most general case (when the operators X_i^0 and X_i^α are chosen arbitrarily and satisfy only the conditions of statement 3.6) since many aspects are not yet clear for us. However, there is one particular case the study of which does not encounter serious difficulties. Below we consider this case in detail and show that, in spite of its comparative simplicity, it is rather rich and has many interesting and important physical and mathematical applications. This case appears when operators X_i^0 and X_i^α, forming the system (3.2.8) and differing, in general, from each other (for different values of i), turn out to coincide if we identify the spaces V_i in which they act. Such a coincidence means that the system (3.2.8) of $\mathcal{D}$ MPS equations becomes equivalent to a single MPS equation and, therefore, its study simplifies considerably.

Summarizing the results listed in statements 3.4–3.7, we can formulate the final theorem establishing the relationship between the single MPS equation and completely integrable quantum models.

Theorem 3.1. *Let V be an infinite-dimensional linear vector space and W_i be certain Hilbert subspaces of V characterized by different scalar products $(\ , \ )_i, \quad i = 1, \ldots, \mathcal{D}$.*

Also let X^0 and $X^\alpha, \alpha = 1, \ldots, \mathcal{D}$ be linear operators in V satisfying the following conditions:

1. Operators $X^\alpha, \alpha = 1, \ldots, \mathcal{D}$ commute with each other and are linearly independent.

2. Operators X^0 and $X^\alpha, \alpha = 1, \ldots, \mathcal{D}$ are hermitian in the spaces W_i with respect to the corresponding scalar products $(\ , \ )_i, \quad i = 1, \ldots, \mathcal{D}$.

3. Operator X^0 is positive definite $(\varphi_i, X^0\varphi_i)_i > 0$ for any $\varphi_i \in W_i, \quad i = 1, \ldots, \mathcal{D}$.

4. The matrix $\|(\varphi_i, X^\alpha\varphi_i)_i\|^{-1}, i, \alpha = 1, \ldots, \mathcal{D}$ contains a non-zero number of rows (or their linear combinations) which are positive definite for any $\varphi_i \in W_i, \quad i = 1, \ldots, \mathcal{D}$.

5. The $\mathcal{D}$-parameter spectral equation

$$X^0\varphi = \left(\sum_{\alpha=1}^{\mathcal{D}} X^\alpha \varepsilon_\alpha \right) \varphi, \quad \varphi \in W_1 \cap \ldots \cap W_\mathcal{D} \qquad (3.2.39)$$

has only a discrete set of solutions.

Denote by $\sigma^0(\lambda)$ and

$$\vec{\sigma}(\lambda) = \{\sigma^\alpha(\lambda)\}_{\alpha=1}^{\mathcal{D}} \qquad (3.2.40)$$

the scalar and vector functions of the single variable λ, *and introduce the scalar, vector and matrix operators* T,

$$\vec{X}^0 = \{\underbrace{I \otimes \ldots \otimes I}_{i-1\ times} \otimes \underbrace{X^0}_{ith\ place} \otimes \overbrace{I \otimes \ldots \otimes I}^{\mathcal{D}-i\ times}\}_{i=1}^{\mathcal{D}}, \quad (3.2.41a)$$

$$\hat{X} = \{\underbrace{I \otimes \ldots \otimes I}_{i-1\ times} \otimes \underbrace{X^\alpha}_{ith\ place} \otimes \overbrace{I \otimes \ldots \otimes I}^{\mathcal{D}-i\ times}\}_{i,\alpha=1}^{\mathcal{D}}, \quad (3.2.41b)$$

acting in the space

$$\mathcal{V} \equiv \underbrace{V \otimes \ldots \otimes V}_{\mathcal{D}\ times}. \quad (3.2.42)$$

Denote also by

$$\mathcal{W}_T = T(\det \hat{X})^{\frac{1}{2}}(W_1 \otimes \ldots W_{\mathcal{D}}) \quad (3.2.43)$$

a Hilbert subspace of the space $\mathcal{V}$ *characterized by the scalar product* $\langle\ ,\ \rangle_T$ *which, for the basis elements* $\psi_T \in \mathcal{W}_T$ *given by formula*

$$\psi_T = T(\det \hat{X})^{\frac{1}{2}}\varphi_1 \otimes \ldots \otimes \varphi_{\mathcal{D}}, \quad \varphi_1 \in W_1, \ldots, \varphi_{\mathcal{D}} \in W_{\mathcal{D}} \quad (3.2.44)$$

we define as

$$\langle\psi_T, \psi_T\rangle_T = \det \|(\varphi_i, X^\alpha \varphi_i)_i\|_{i,\alpha=1}^{\mathcal{D}}. \quad (3.2.45)$$

Then, the operators

$$H_T(\lambda) = \sigma^0(\lambda) + T(\det \hat{X})^{\frac{1}{2}}\{\vec{\sigma}(\lambda) \cdot \hat{X}^{-1}\vec{X}^0\}(\det \hat{X})^{-\frac{1}{2}}T^{-1} \quad (3.2.46)$$

form a commuting family and are hermitian in the space $\mathcal{W}_T$. *Their spectral problem*

$$H_T(\lambda)\psi_T = E(\lambda)\psi_T, \quad \psi_T \in \mathcal{W}_T \quad (3.2.47)$$

has a discrete set of solutions

$$E(\lambda) = \sigma^0(\lambda) + \vec{\sigma}(\lambda) \cdot \vec{\varepsilon}, \quad (3.2.48a)$$

$$\psi_T = T(\det \hat{X})^{\frac{1}{2}}(\underbrace{\varphi \otimes \ldots \otimes \varphi}_{\mathcal{D}\ times}), \quad (3.2.48b)$$

which are expressed in terms of solutions $\vec{\varepsilon} = \{\varepsilon_\alpha\}_{\alpha=1}^{\mathcal{D}}$ and φ of the initial MPS *equation (3.2.39). The spectral function $E(\lambda)$ is bounded from below:*

$$E(\lambda) \geq \sigma^0(\lambda). \tag{3.2.49}$$

The homogenous change of operators X^0 and $X^\alpha, \alpha = 1, \ldots, \mathcal{D}$ by the transformations

$$
\begin{aligned}
X'^0 &= BAX^0B^{-1}, \\
X'^\alpha &= BAX^\alpha B^{-1}, \qquad \alpha = 1, \ldots, \mathcal{D}
\end{aligned}
\tag{3.2.50}
$$

in which A and B are arbitrary non-singular operators in V affects only the form of the operator T, and not the general form of expressions (3.2.46) and (3.2.48) for $H_T(\lambda)$ and ψ_T. (This is the end of the theorem).

Proof. Denote by X_i^0 and $X_i^\alpha, \alpha = 1, \ldots, \mathcal{D}, i = 1, \ldots, \mathcal{D}$ the copies of the operators X^0 and $X^\alpha, \alpha = 1, \ldots, \mathcal{D}$ in the spaces V_i. It is easy to verify that these copies satisfy all conditions of statement 3.6. This gives us the possibility of constructing operators $H_T(\lambda)$ with the properties listed in the present theorem. Note that formula (3.2.45) for the scalar product in the space W_T is a trivial consequence of formulas (3.1.52) and (3.2.3). Note also that the copies A_i and B_i of operators A and B in the spaces V_i satisfy the conditions of statement 3.7. This completes the proof of the theorem.

From this very important theorem it follows that any MPS equation of the form (3.2.39) satisfying conditions 1–5 can be used as a starting point in constructing completely integrable and stable quantum mechanical models with discrete spectra.

3.3   The case of differential equations

A most interesting type of completely integrable quantum model, which can be obtained by means of the method described in the preceding section, arises when the hamiltonians $H_T(\lambda)$ are second-order linear differential operators and describe systems of quantum particles moving in a certain coordinate space. Obviously, to obtain such a form of resulting operators $H_T(\lambda)$ we must start with second-order differential MPS equations of the type (3.2.29). Below we consider in detail the case when the initial MPS equation (3.2.29) is one dimensional.

Let V be the space of functions of a single variable λ. The most general form of second-order linear differential operators X^0 and $X^\alpha, \alpha = 1, \ldots, \mathcal{D}$

acting in this space is

$$X^0 = z^0(\lambda)\frac{\partial^2}{\partial\lambda^2} + y^0(\lambda)\frac{\partial}{\partial\lambda} + x^0(\lambda) \tag{3.3.1a}$$

$$X^\alpha = z^\alpha(\lambda)\frac{\partial^2}{\partial\lambda^2} + y^\alpha(\lambda)\frac{\partial}{\partial\lambda} + x^\alpha(\lambda). \tag{3.3.1b}$$

We define the Hilbert subspaces W_i of the space V as the sets of sufficiently slow functions vanishing at the ends of certain intervals $[\lambda_i^-, \lambda_i^+]$. We assume that these intervals do not intersect, but may have common end points. We define the scalar products in the spaces W_i as

$$\left(\varphi^{(1)}, \varphi^{(2)}\right)_i \equiv \int_{\lambda_i^-}^{\lambda_i^+} \varphi^{(1)}(\lambda)\varphi^{(2)}(\lambda)\, d\lambda. \tag{3.3.2}$$

Let us now try to satisfy the conditions of theorem 3.1. The necessary constraints on operator X^0 guaranteeing its hermitian symmetry with respect to the scalar product (3.2.2) can be written as

$$z^0(\lambda) = -1, \quad y^0(\lambda) = 0. \tag{3.3.3}$$

Such a choice does not lead to any loss of generality, since the most general case when $z^0(\lambda) = z^0 = \text{constant}$ can be reduced to our case $z^0(\lambda) = -1$ by the transformation (3.2.50) with $B = 1$ and $A = -\frac{1}{z_0}$. Furthermore, the free term $x^0(\lambda)$ must be regular within the intervals $[\lambda_i^-, \lambda_i^+]$ and its possible singularities at the endpoints $\lambda_i^\pm$ must be integrable.

The positive definiteness of X^0 in W_i also imposes some constraints on the admissible form of the function $x^0(\lambda)$. For our purposes it is sufficient to assume that it is non-negative in all intervals $[\lambda_i^-, \lambda_i^+]$.

In order to guarantee the commutativity of the operators (3.3.1b) and, simultaneously, their hermiticity, we must take

$$z^\alpha(\lambda) = 0, \quad y^\alpha(\lambda) = 0, \quad \alpha = 1, \ldots, \mathcal{D}, \tag{3.3.4}$$

assuming, as before, that the terms $x^\alpha(\lambda)$ are regular in the intervals $[\lambda_i^-, \lambda_i^+]$ and their possible singularities at the end points $\lambda_i^\pm$ are integrable. Note also that functions $x^\alpha(\lambda)$ must be linearly independent.

For the fourth condition of theorem 3.1 to be satisfied, the matrix $\|x^\alpha(\lambda_i)\|$ must be invertible for any $\lambda_i \in [\lambda_i^-, \lambda_i^+]$. The inverse matrix $\|x^\alpha(\lambda_i)\|^{-1}$ must contain a non-zero number of rows (or their linear combinations) with the elements being positive for any $\lambda_i \in [\lambda_i^-, \lambda_i^+]$.

Substituting the resulting expressions for X^0 and X^α, $\alpha = 1, \ldots, \mathcal{D}$ into the MPS equation (3.2.29), we reduce it to the form

$$\left\{ -\frac{\partial^2}{\partial\lambda^2} + x^0(\lambda) \right\} \varphi(x) = \left\{ \sum_{\alpha=1}^{\mathcal{D}} x^\alpha(\lambda)\varepsilon_\alpha \right\} \varphi(\lambda),$$

$$\varphi(\lambda) \in W_1 \cap \ldots \cap W_\mathcal{D}. \qquad (3.3.5)$$

Our aim is now to use the prescriptions given in theorem 3.1 and to construct the class of Schrödinger-type equations describing a certain completely integrable quantum system.

First of all, note that, according to definition (3.2.42), the space $\mathcal{V}$ is formed by all functions of $\mathcal{D}$ variables $\vec{\lambda} = (\lambda_1, \ldots, \lambda_\mathcal{D})$.

The Hilbert subspace W_T is formed by functions that are regular within the parallelogram

$$\mathcal{P} = \bigotimes_{i=1}^{\mathcal{D}} [\lambda_i^-, \lambda_i^+], \qquad (3.3.6)$$

and vanishing at its boundaries. The scalar product $\langle \, , \, \rangle_T$ for $\psi_T^{(1)} \subset W_T$ and $\psi_T^{(2)} \subset W_T$ can be defined as

$$\left\langle \psi_T^{(1)}, \psi_T^{(2)} \right\rangle_T \equiv \int_\mathcal{P} T^{-2}(\vec{\lambda}) \psi_T^{(1)}(\vec{\lambda}) \psi_T^{(2)}(\vec{\lambda}) \, \mathrm{d}^\mathcal{D}\lambda, \qquad (3.3.7)$$

where T is a certain function of λ:

$$T = T(\vec{\lambda}). \qquad (3.3.8)$$

For the vector and matrix operators $\vec{X}^0$ and $\hat{X}$ acting in the space W_T we have:

$$\vec{X}^0 = \left\{ -\frac{\partial^2}{\partial\lambda_i^2} + x^0(\lambda_i) \right\}_{i=1}^{\mathcal{D}} \qquad (3.3.9a)$$

and

$$\hat{X} = \{ x^\alpha(\lambda_i) \}_{i,\alpha=1}^{\mathcal{D}}. \qquad (3.3.9b)$$

Introducing the notation

$$\Delta(\vec{\lambda}) \equiv \det \|x^\alpha(\lambda_i)\|_{i,\alpha=1}^{\mathcal{D}}, \qquad (3.3.10)$$

we can rewrite the expression for the inverse operator (3.3.9b) as

$$\hat{X}^{-1} = \Delta^{-1}(\vec{\lambda}) \left\{ x_\alpha^i(\vec{\lambda}) \right\}_{i,\alpha=1}^{\mathcal{D}}, \tag{3.3.11}$$

where $x_\alpha^i(\vec{\lambda})$ is the cofactor of the element $x^\alpha(\lambda_i)$ in the matrix $\hat{X}$. It is easy to see that the functions $x_\alpha^i(\lambda)$, $\alpha = 1, \ldots, \mathcal{D}$ do not depend on λ_i.

Now we are ready to write down the explicit expression for the operators $H_T(\lambda)$:

$$H_T(\lambda) = \sigma^0(\lambda)$$

$$+ T(\vec{\lambda}) \left\{ \Delta^{-\frac{1}{2}}(\vec{\lambda}) \sum_{i,\alpha=1}^{\mathcal{D}} [\sigma^\alpha(\lambda) x_\alpha^i(\vec{\lambda})] \left[-\frac{\partial^2}{\partial \lambda_i^2} + x^0(\lambda_i) \right] \Delta^{-\frac{1}{2}}(\vec{\lambda}) \right\} T^{-1}(\vec{\lambda}),$$

$$\tag{3.3.12}$$

which are hermitian in the space $\mathcal{W}_T$ and commute with each other for different values of λ. We see that they are $\mathcal{D}$-dimensional second-order differential operators. If the matrix (3.3.9b) satisfies the fourth condition of theorem 3.1, it is always possible to choose the functions $\sigma^\alpha(\lambda)$ in such a way that the relations

$$\sum_{\alpha=1}^{\mathcal{D}} \sigma^\alpha(\lambda) x_\alpha^i(\vec{\lambda}) \geq 0, \quad i = 1, \ldots, \mathcal{D} \tag{3.3.13}$$

hold for any $\vec{\lambda} \in \mathcal{P}$. In this case (3.3.12) are elliptic operators. According to the positivity of $x^0(\lambda_i)$, the second term in (3.3.12) is positive definite in $\mathcal{W}_T$ and, therefore, the spectra of the operators $H_T(\lambda)$ are bounded from below by the function $\sigma^0(\lambda)$. All this means that $H_T(\lambda)$ can be interpreted as hamiltonians of a certain completely integrable and stable quantum mechanical system.

A most interesting case of these hamiltonians arises when the operator T entering into formula (3.3.12) takes the form

$$T = T(\mu, \vec{\lambda}) = \left\{ -\Delta^{\mathcal{D}}(\vec{\lambda}) \prod_{i=1}^{\mathcal{D}} \left(\sum_{\alpha=1}^{\mathcal{D}} \sigma^\alpha(\mu) x_\alpha^i(\vec{\lambda}) \right) \right\}^{-\frac{1}{4}}, \tag{3.3.14}$$

in which μ is a certain additional parameter. At first sight, this case is not more remarkable than the general one. However, if λ is equal to μ the situation changes.

Indeed, let $\lambda = \mu$. Then, introducing the matrix function

$$g_{ik}(\mu, \vec{\lambda}) = \begin{cases} \Delta^{-1}(\vec{\lambda}) \sum_{\alpha=1}^{D} \sigma^{\alpha}(\mu) x_{\alpha}^{i}(\vec{\lambda}), & k = i, \\ 0, & k \neq i, \end{cases} \qquad (3.3.15)$$

and using the notations

$$\begin{aligned} g(\mu, \vec{\lambda}) &= \det\|g_{ik}(\mu\vec{\lambda})\|_{i,k=1}^{D} \\ &= \Delta^{-D}(\vec{\lambda}) \prod_{i=1}^{D} \left(\sum_{\alpha=1}^{D} \sigma^{\alpha}(\mu) x_{\alpha}^{i}(\vec{\lambda}) \right) \end{aligned} \qquad (3.3.16)$$

and

$$\begin{aligned} V(\mu, \vec{\lambda}) &= \sigma^{0}(\mu) + \sum_{i=1}^{D} g_{ii}(\mu, \vec{\lambda}) x^{0}(\lambda_i) \\ &+ \sum_{i=1}^{D} \Delta^{-1}(\vec{\lambda}) \frac{\partial}{\partial \lambda_i} \left\{ g_{ii}(\mu, \vec{\lambda}) \Delta(\vec{\lambda}) \right\} \\ &+ \sum_{i=1}^{D} g_{ii}(\mu, \vec{\lambda}) \left\{ g(\mu, \vec{\lambda}) \Delta^2(\vec{\lambda}) \right\}^{\frac{1}{4}} \frac{\partial^2}{\partial \lambda_i^2} \left\{ g(\mu, \vec{\lambda}) \Delta^2(\vec{\lambda}) \right\}^{-\frac{1}{4}}, \end{aligned}$$
$$(3.3.17)$$

we can rewrite the spectral equations for the operators $H_T(\mu)$ in the form

$$\left\{ -\sqrt{g(\mu, \vec{\lambda})} \sum_{i,k=1}^{D} \frac{\partial}{\partial \lambda_i} \left[\frac{g_{ik}(\mu, \vec{\lambda})}{\sqrt{g(\mu, \vec{\lambda})}} \frac{\partial}{\partial \lambda_k} \right] + V(\mu, \vec{\lambda}) \right\} \psi(\mu, \vec{\lambda})$$
$$= E(\mu) \psi(\mu, \vec{\lambda}). \quad (3.3.18)$$

The solutions of these equations can be represented as

$$E(\mu) = \sigma^{0}(\mu) + \sum_{\alpha=1}^{D} \sigma^{\alpha}(\mu) \varepsilon_{\alpha} \qquad (3.3.19a)$$

and

$$\psi(\mu, \vec{\lambda}) = \left[g(\mu, \vec{\lambda}) \Delta^2(\vec{\lambda}) \right]^{\frac{1}{4}} \prod_{i=1}^{D} \varphi(\lambda_i), \qquad (3.3.19b)$$

where ε_α, $\alpha = 1, \ldots, \mathcal{D}$ and $\varphi(\lambda)$ are solutions of the MPS equation (3.3.5). The eigenvalues (3.3.19b) satisfy the following normalization conditions:

$$\int \frac{\psi^2(\mu, \vec{\lambda})}{\sqrt{g(\mu, \vec{\lambda})}} \, \mathrm{d}^{\mathcal{D}} \lambda < \infty. \tag{3.3.20}$$

It is not difficult to see that the expression

$$\Delta_g \equiv \sqrt{g(\mu, \vec{\lambda})} \sum_{i,k=1}^{\mathcal{D}} \frac{\partial}{\partial \lambda_i} \left(\frac{g_{ik}(\mu, \vec{\lambda})}{\sqrt{g(\mu, \vec{\lambda})}} \frac{\partial}{\partial \lambda_k} \right) \tag{3.3.21}$$

entering into equation (3.3.18) is none other than the ordinary Laplace–Beltrami operator for the (curved) space specified by the covariant metric tensor $g_{ik}(\mu, \vec{\lambda})$. (Remember that the inverse, contravariant metric tensor $g^{ik}(\mu, \vec{\lambda})$ determines the form of the interval in curved space.)

We have obtained a very interesting result: starting with the single one-dimensional differential MPS equation (3.3.5) we have obtained the class of Schrödinger equations (3.3.18) describing a certain completely integrable and stable quantum mechanical model on a $\mathcal{D}$-dimensional, in general, curved manifold!

We see that the procedure of going over from the MPS equation (3.3.5) to the Schrödinger equations (3.3.18) essentially solves the inverse problem of separation of variables in the most general form. Here (3.3.5) can be viewed as a one-dimensional equation arising as a result of separation of variables in the $\mathcal{D}$-dimensional Schrödinger equation (3.3.18). The spectral parameters ε_α, $\alpha = 1, \ldots, \mathcal{D}$ play the role of separation constants, while the operators

$$H_\alpha \equiv T(\vec{\lambda}) \Delta^{-\frac{1}{2}}(\vec{\lambda}) \sum_{i=1}^{\mathcal{D}} x_\alpha^i(\vec{\lambda}) \left[-\frac{\partial^2}{\partial \lambda_i^2} + x^0(\lambda_i) \right] \Delta^{-\frac{1}{2}}(\vec{\lambda}) T^{-1}(\vec{\lambda}) \tag{3.3.22}$$

(whose eigenvalues they are) can be considered as symmetry operators, i.e., operators commuting with the hamiltonians $H_T(\lambda)$.

Let us now consider transformations conserving the general form of MPS equation (3.3.5) and elucidate how they influence the form of the resulting Schrödinger equations (3.3.18). These transformations can be described by the formulas

$$x^0(\lambda) \to \widetilde{x}^0(\widetilde{\lambda}) = \frac{1}{b^4(\lambda)} \left\{ x^0(\lambda) + \left[\frac{b'(\lambda)}{b(\lambda)} \right]' - \left[\frac{b'(\lambda)}{b(\lambda)} \right]^2 \right\}, \tag{3.3.23a}$$

$$x^\alpha(\lambda) \to \tilde{x}^\alpha(\tilde{\lambda}) = \frac{1}{b^4(\lambda)} x^\alpha(\lambda), \quad \alpha = 1, \ldots, \mathcal{D}, \tag{3.3.23b}$$

and

$$\lambda \to \tilde{\lambda} = \int b^2(\lambda)\, \mathrm{d}\lambda, \tag{3.3.24}$$

provided that

$$\varphi(\lambda) \to \tilde{\varphi}(\tilde{\lambda}) = b(\lambda)\varphi(\lambda). \tag{3.3.25}$$

It is not difficult to verify that the transformed MPS equation (3.3.5) satisfies the same conditions as the untransformed one, and, therefore, application of the inverse procedure of separation of variables to it gives us again a certain completely integrable $\mathcal{D}$-dimensional quantum model. Surprising though it is, the Schrödinger equations for this model turn out to coincide with equations (3.3.18) describing the untransformed case. To elucidate the reason for such a coincidence, note that formulas (3.3.23) and (3.3.24) describe two different transformations, one of which is the ordinary equivalence transformation (3.2.50) with $B = b(\lambda)$ and $A = b^{-4}(\lambda)$, while the other is a change of the variable λ. We know from theorem 3.1 that transformation (3.2.50) does not change the form of the resulting Schrödinger equations. The change of variable λ (which implies an analogous change of variable λ_i), being a particular case of general coordinate transformations, can only affect the form of the metric tensor $g_{ik}(\mu, \vec{\lambda})$, and not the potential $V(\mu, \vec{\lambda})$ and corresponding solutions $\psi(\mu, \vec{\lambda})$.

As noted in the preceding sections, we shall be interested mainly in the case when the quantum mechanical models obtained by the inverse method of separation of variables have discrete spectra. It is clear that for this to happen, the spectrum of the initial MPS equation (3.3.5) must also be discrete. In order to convince ourselves that the existence of such MPS equations is possible, let us consider the case when the intervals $[\lambda_i^-, \lambda_i^+]$, $i = 1, \ldots, \mathcal{D}$ describing the structure of the spaces $\mathcal{W}_i$, $i = 1, \ldots, \mathcal{D}$ have common end points, as is shown in figure 3.1. Now note that equation

Figure 3.1. The disposition of intervals $[\lambda_i^-, \lambda_i^+]$, $i = 1, \ldots, \mathcal{D}$ on the real λ-axis.

(3.3.5) is a second-order linear differential equation and, therefore, for any

values of spectral parameters $\varepsilon_1, \ldots, \varepsilon_{\mathcal{D}}$ it has two linearly independent solutions which we denote by $\varphi_1(\lambda; \varepsilon_1, \ldots, \varepsilon_{\mathcal{D}})$ and $\varphi_2(\lambda; \varepsilon_1, \ldots, \varepsilon_{\mathcal{D}})$. The general solution of (3.3.5) can be written in the form

$$\begin{aligned}
\varphi(\lambda; \varepsilon_1, \ldots, \varepsilon_{\mathcal{D}}, \theta) &= \cos\theta \varphi_1(\lambda; \varepsilon_1, \ldots, \varepsilon_{\mathcal{D}}, \theta) \\
&+ \sin\theta \varphi_2(\lambda; \varepsilon_1, \ldots, \varepsilon_{\mathcal{D}}, \theta),
\end{aligned} \tag{3.3.26}$$

where θ is an additional arbitrary parameter. Remember that functions belonging to the spaces W_i must vanish at the ends of the intervals $[\lambda_i^-, \lambda_i^+]$, $i = 1, \ldots, \mathcal{D}$. However, the functions (3.3.26) belong simultaneously to all Hilbert spaces $W_1, \ldots, W_{\mathcal{D}}$ (to their intersection $W_1 \cup \ldots \cup W_{\mathcal{D}}$), and, therefore, they must vanish at all $\mathcal{D}+1$ points $\lambda_1^0, \ldots, \lambda_{\mathcal{D}+1}^0$, where $\lambda_i^0 = \lambda_i^-$, for $i = 1, \ldots, \mathcal{D}$, and $\lambda_{\mathcal{D}+1}^0 = \lambda_{\mathcal{D}}^+$. Writing this condition in the form

$$\varphi(\lambda_i^0; \varepsilon_1, \ldots, \varepsilon_{\mathcal{D}}, \theta) = 0, \quad i = 1, \ldots, \mathcal{D}+1, \tag{3.3.27}$$

we obtain a system of $\mathcal{D} + 1$ numerical equations for $\mathcal{D} + 1$ unknown quantities $\varepsilon_1, \ldots, \varepsilon_{\mathcal{D}}$ and θ. Evidently, the set of solutions of this system is discrete. This means that the MPS equation (3.3.5) also has a discrete spectrum. Of course, this reasoning has only a theoretical meaning, since, in general, the explicit expression for the function (3.3.26) is not known. However, there are many special cases when the construction of explicit solutions of MPS equations (3.3.5) becomes possible. These cases, which will be discussed in detail in the next sections of this chapter, are especially interesting for us since they lead to completely integrable quantum models of the type (3.3.18) with discrete and exactly calculable spectra.

3.4 Algebraically solvable multi-parameter spectral equations

In this section we discuss the problem of constructing one-dimensional differential MPS equations which can be solved exactly by means of algebraic methods. For the sake of convenience we formulate the basic ideas in a most general and abstract form. This gives us the possibility of simplifying much of the reasoning and avoiding various particularities which might appear on considering the specific cases of concrete MPS equations.

a. Generalized MPS equations. Denote by ε the vectors belonging to a $\mathcal{D}$-dimensional vector space $\mathcal{E}$. Denote by $X(\varepsilon)$ the function on $\mathcal{E}$ with values being linear operators in a certain infinite-dimensional vector space V. Denote also by Φ a certain subset of V which does not necessarily form a linear space.

Consider the equation

$$X(\varepsilon)\varphi = 0, \quad \varphi \in \Phi, \tag{3.4.1}$$

which we call the "generalized" MPS equation, stressing the fact that the function $X(\varepsilon)$ may be arbitrary (not necessarily linear). The vector ε plays in (3.4.1) the role of a spectral parameter. We shall call the set of all vectors ε for which (3.4.1) has non-zero solutions in Φ the spectrum of equation (3.4.1).

Below we shall distinguish between two types of linear transformation in the space V conserving the form of equation (3.4.1). They are left multiplications $X'(\varepsilon) = A(\varepsilon)X(\varepsilon)$ and ordinary homogeneous transformations $X'(\varepsilon) = B(\varepsilon)X(\varepsilon)B^{-1}(\varepsilon)$ which can be realized by non-singular operators $A(\varepsilon)$ and $B(\varepsilon)$ depending, generally, on ε. The transformed equation (3.4.1) can be written in the form

$$X'(\varepsilon)\varphi' = 0, \quad \varphi' \in \Phi', \tag{3.4.2}$$

where

$$X'(\varepsilon) = B(\varepsilon)A(\varepsilon)X(\varepsilon)B^{-1}(\varepsilon), \tag{3.4.3}$$
$$\Phi' = B(\mathcal{E})\Phi, \tag{3.4.4}$$

and

$$\varphi' = B(\varepsilon)\varphi. \tag{3.4.5}$$

Two operators $X(\varepsilon)$ and $X'(\varepsilon)$ which are related as in formula (3.4.3) will be called equivalent.

Introducing a basis in the space V and taking

$$\varphi = \{\varphi_k\}_{k=0,...,\infty}, \quad X(\varepsilon) = \{X_m^k(\varepsilon)\}_{m=1,...,\infty}^{k=1,...,\infty}, \tag{3.4.6}$$

we can rewrite (3.4.1) in the form

$$\sum_{k=0}^{\infty} X_m^k(\varepsilon)\varphi_k = 0, \quad m = 0, 1, \ldots, \infty; \quad \{\varphi_k\}_{k=0}^{\infty} \in \Phi. \tag{3.4.7}$$

This is an infinite-dimensional matrix version of the generalized MPS equation (3.4.1).

b. Finite-dimensional matrix MPS **equations.** As noted in preceding sections, in our approach a central role is played by the MPS equations with discrete spectra. To demonstrate how such equations appear, it is sensible to consider first the simplest particular case, when vectors φ and matrices $X(\varepsilon)$ are finite dimensional,

$$\varphi = \{\varphi_k\}_{k=0,...,K}, \quad X(\varepsilon) = \{X_m^k(\varepsilon)\}_{m=0,...,M}^{k=0,...,K}, \tag{3.4.8}$$

and $\Phi = \Phi_M$ is a linear space. Then (3.4.7) takes the form of a finite-dimensional rectangular matrix equation

$$\sum_{m=0}^{M} X_k^m(\varepsilon_1,\ldots,\varepsilon_D)\varphi_m = 0, \quad k = 0,1,\ldots,K. \qquad (3.4.9)$$

Due to the homogeneity of (3.4.9), we can impose some normalization conditions on φ, for example

$$\sum_{m=0}^{M} \varphi_m^2 = 1. \qquad (3.4.10)$$

Then the number of unknown quantities entering into system (3.4.9) becomes $M + D$. Consider three cases:

(i) $K+1 > M+D$. The number of equations (3.4.9) exceeds the number of unknown quantities. The spectrum is empty.

(ii) $K + 1 < M + D$. The number of equations (3.4.9) is less than the number of unknown quantities. The spectrum is continuous.

(iii) $K + 1 = M + D$. The number of equations and unknown quantities coincide, and, therefore, the spectrum is discrete.

The last case is especially interesting for us. We see that in order to obtain a finite-dimensional D-parameter spectral equation with a discrete spectrum, the operator $X(\varepsilon)$ must have the form of a rectangular matrix of size $(M + D) \times (M + 1)$ where M can be chosen arbitrarily. In particular, when $D = 1$, we come to the square $(M+1) \times (M+1)$ matrices which lead us to the standard one-parameter spectral matrix equations.

Using (3.4.9) it is not diffcult to obtain the equation immediately for the spectrum. To this end consider D square matrices:

$$X_n(\varepsilon_1,\ldots,\varepsilon_D) \equiv \|X_{n+k}^m(\varepsilon_1,\ldots,\varepsilon_D)\|_{k,m=1}^{M}, \quad n = 0,\ldots,D-1,$$

$$(3.4.11)$$

and rewrite the system (3.4.9) in the form:

$$X_n(\varepsilon_1,\ldots,\varepsilon_D)\varphi = 0, \quad \varphi \in \Phi_M, \quad n = 0,\ldots,D-1. \qquad (3.4.12)$$

We know that these (ordinary) linear matrix equations admit non-zero solutions if and only if

$$\det X_n(\varepsilon_1,\ldots,\varepsilon_D) = 0, \quad n = 0,\ldots,D-1. \qquad (3.4.13)$$

Thus, we have obtained the system of D secular equations from which the admissible values of D spectral parameters $\varepsilon_1,\ldots,\varepsilon_D$ can be found. Note that in the one-parameter case $(\mathcal{D} = 1)$ system (3.4.13) is reduced to a single secular equation.

c. Infinite-dimensional matrix MPS **equations.** Much more interesting cases appear when the space Φ in which we seek solutions of equations (3.4.1) is infinite dimensional. The construction of MPS equations with discrete spectra remains possible in this case too, although it is difficult to speak about "rectangular" matrices of size $(\infty + \mathcal{D}) \times (\infty + 1)$: the notion of rectangularity in the infinite-dimensional case is not defined.

Nevertheless, we can consider a special class of infinite-dimensional matrices for which the use of the term "rectangularity" is more or less justified. These matrices are distinguished by the fact that they act as finite-dimensional rectangular matrices in the finite-dimensional subspaces of the space Φ. Such a property guarantees discreteness of the spectra of corresponding MPS equations and gives the possibility of algebraizing the problem of finding their solutions.

First of all, let us construct an appropriate infinite-dimensional space Φ. We denote by Φ_M the linear span of the first $M + 1$ basis elements of V. Thus, Φ_M is an $(M + 1)$-dimensional space. Let us define Φ as a space, any element of which is an element of a certain space Φ_M (with a finite M). This means that Φ consists of vectors having a finite number of components only. This number may be arbitrarily large and, therefore, the space Φ is infinite dimensional. Note however, that Φ does not necessarily coincide with the initial space V since the latter may contain vectors with an infinite number of components.

Now, let us assume that the operator $X(\varepsilon)$, having, in general, the form of an infinite-dimensional matrix, admits for any ε and $M = 0, 1, 2, \ldots$ the block-decomposition shown in figure 3.2. in which $X_M(\varepsilon)$ are finite rectangular matrices of size $(M + \mathcal{D}) \times (M + 1)$, and $X'_M(\varepsilon)$ and $X''_M(\varepsilon)$ are infinite blocks, the concrete form of which is non-essential for us. Using this block structure we come to the following chain of embeddings:

$$X(\varepsilon)\Phi_M \subset \Phi_{M+\mathcal{D}-1}, \quad M = 0, 1, 2, \ldots, \tag{3.4.14}$$

from which it follows that the infinite-dimensional matrix $X(\varepsilon)$ acts in any subspace Φ_M of the space Φ as a finite-dimensional rectangular matrix $X_M(\varepsilon)$ of size $(M + \mathcal{D}) \times (M + 1)$. In turn, this means that the infinite-dimensional spectral problem for the operator $X(\varepsilon)$

$$X(\varepsilon)\varphi = 0, \quad \varphi \in \Phi \tag{3.4.15}$$

breaks up into an infinite number of finite-dimensional spectral problems

$$X_M(\varepsilon)\varphi = 0, \quad \varphi \in \Phi_M, \tag{3.4.16}$$

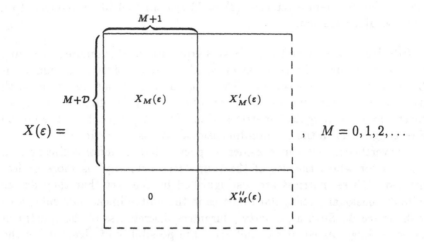

Figure 3.2. The block structure of the matrix $X(\varepsilon)$.

each of which has a discrete spectrum. This gives us reason to assert that the initial equation (3.4.15) also has a discrete spectrum in Φ.

Note that for the block decomposition of $X(\varepsilon)$ depicted in figure 3.2 to be possible for any $M = 0, 1, 2, \ldots$, it is necessary that

$$X_k^m(\varepsilon) = 0, \quad \text{for any } k - m \geq \mathcal{D}. \tag{3.4.17}$$

Matrices of such a type can be depicted as shown in figure 3.3. They have $\mathcal{D}-1$ lower diagonals only. In particular, when $\mathcal{D} = 1$ these matrices become upper triangular and, thus, the spaces Φ_M become invariant subspaces of Φ:

$$X(\varepsilon)\Phi_M \subset \Phi_M, \quad M = 0, 1, 2, \ldots. \tag{3.4.18}$$

d. Algebraically solvable MPS equations. It is natural to call the MPS equation algebraically solvable if all its solutions can be obtained by means of a finite algebraic procedure. Any finite-dimensional MPS matrix equation is algebraically solvable according to this definition. The infinite-dimensional MPS equations discussed in the preceding subsection are also algebraically solvable, since they allow reduction to sets of finite-dimensional spectral problems. Note also that starting with the known algebraically solvable MPS equation and applying to it various explicit equivalence transformations of the type (3.4.3) we obtain a new class of MPS equations all of which will be algebraically solvable by construction.

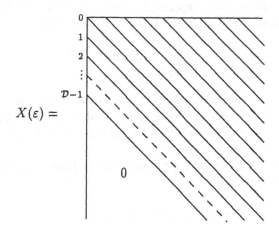

Figure 3.3. The diagonal structure of the matrix $X(\varepsilon)$.

e. Construction of algebraically solvable MPS equations. Consider the Heisenberg algebra, whose elements, a, a^+ and I, satisfy the following commutation relations:

$$[a, a^+] = I; \quad [a, I] = [a^+, I] = 0. \tag{3.4.19}$$

It is known that this algebra admits an infinite-dimensional representation with lowest weight. In this case I acts as the unit operator, and the elements a^+ and a take the meaning of creation and annihilation operators, respectively. The lowest-weight vector $|0\rangle$ is defined as the vector annihilated by the operator a:

$$a|0\rangle = 0. \tag{3.4.20}$$

The corresponding representation space is formed by the vectors

$$|n\rangle = (a^+)^n |0\rangle, \tag{3.4.21}$$

playing the role of the basis vectors in the representation space. From the known relations

$$a^+|n\rangle = |n+1\rangle, \quad a|n\rangle = n|n-1\rangle, \tag{3.4.22}$$

it follows that

$$(a+)^k a^m |n\rangle \sim \left\{ \begin{array}{ll} |k+n-m\rangle, & k+n-m \geq 0, \\ 0, & k+n-m < 0. \end{array} \right\} \tag{3.4.23}$$

This formula allows us to express the operators $X(\varepsilon)$ satisfying condition (3.4.17) in terms of the generators a, a^+. Indeed, taking

$$X(\varepsilon) = \sum_{k<\mathcal{D}+m} X_k^m(\varepsilon)(a^+)^k a^m, \qquad (3.4.24)$$

and recalling that the spaces Φ_M are defined as linear spans of the vectors $|0\rangle, |1\rangle, \ldots, |M\rangle$, it is not difficult to see that condition (3.4.17) is really satisfied. But this means that the MPS equation with the generator $X(\varepsilon)$ is algebraically solvable.

Now, remember that the generators of the Heisenberg algebra allow a differential realization

$$a^+ = \lambda, \quad a = \frac{\partial}{\partial\lambda}, \quad I = 1. \qquad (3.4.25)$$

In this case the role of the lowest-weight vector is played by the constant function

$$|0\rangle = 1 \qquad (3.4.26)$$

and the basis elements $|n\rangle$ have the form

$$|n\rangle = \lambda^n, \quad n = 0, 1, 2, \ldots. \qquad (3.4.27)$$

Therefore, the space Φ takes the form of the space of all polynomials in λ and Φ_M becomes the space of the polynomials of a given order M.

The operators $X(\varepsilon)$ acting in Φ can be rewritten as

$$X(\varepsilon) = \sum_{k<\mathcal{D}+m} X_k^m(\varepsilon)\lambda^k \left(\frac{\partial}{\partial\lambda}\right)^m \qquad (3.4.28)$$

and, thus, we come to the differential MPS equations having algebraic solutions within the class of polynomials.

Of course, the case when operators (3.4.28) are second-order differential operators is most interesting for us. This case is realized when the coefficients $X_k^m(\varepsilon)$ entering into formula (3.4.28) vanish if m exceeds two. Then we obtain

$$X(\varepsilon) = P_{\mathcal{D}+1}(\lambda;\varepsilon)\frac{\partial^2}{\partial\lambda^2} + P_{\mathcal{D}}(\lambda;\varepsilon)\frac{\partial}{\partial\lambda} + P_{\mathcal{D}-1}(\lambda;\varepsilon), \qquad (3.4.29)$$

where $P_n(\varepsilon,\lambda)$ are arbitrarily fixed polynomials of orders $n = \mathcal{D}-1, \mathcal{D}, \mathcal{D}+1$, with the coefficients being given functions of $\mathcal{D}$ spectral parameters $\varepsilon_1, \ldots, \varepsilon_{\mathcal{D}}$.

The corresponding second-order differential MPS equation has the form

$$\left\{ P_{\mathcal{D}+1}(\lambda;\varepsilon)\frac{\partial^2}{\partial\lambda^2} + P_{\mathcal{D}}(\lambda;\varepsilon)\frac{\partial}{\partial\lambda} + P_{\mathcal{D}-1}(\lambda;\varepsilon) \right\} \varphi(\lambda) = 0, \qquad (3.4.30)$$

and for any $M = 0, 1, 2, \ldots$ admits exact solutions belonging to the class of Mth-order polynomials

$$\varphi(\lambda) = \sum_{n=0}^{M} \varphi_n \lambda^n. \qquad (3.4.31)$$

According to the remark given in the previous subsection, the algebraically solvable MPS equations (3.4.30) can be used as a starting point for constructing more complex algebraically solvable MPS equations. For this it is sufficient to apply to (3.4.29) some kind of equivalence transformation (3.4.3). Consider, for example, transformation (3.4.3) in which

$$A(\varepsilon) = -\frac{1}{P_{\mathcal{D}+1}(\lambda;\varepsilon)}, \qquad (3.4.32a)$$

$$B(\varepsilon) = \exp\left\{ \frac{1}{2} \int \frac{P_{\mathcal{D}}(\varepsilon,\lambda)}{P_{\mathcal{D}+1}(\varepsilon,\lambda)} \, \mathrm{d}\lambda \right\}. \qquad (3.4.32b)$$

Then, for the transformed operator $X(\varepsilon)$ we obtain:

$$X(\varepsilon) = -\frac{\partial^2}{\partial\lambda^2} + \frac{1}{2}\left[\frac{P_{\mathcal{D}}(\lambda;\varepsilon)}{P_{\mathcal{D}+1}(\lambda;\varepsilon)} \right]' + \frac{1}{4}\left[\frac{P_{\mathcal{D}}(\lambda;\varepsilon)}{P_{\mathcal{D}+1}(\lambda;\varepsilon)} \right]^2 - \frac{P_{\mathcal{D}-1}(\lambda;\varepsilon)}{P_{\mathcal{D}+1}(\lambda;\varepsilon)}. \qquad (3.4.33)$$

Note that, instead of the three polynomials $P_{\mathcal{D}+1}(\lambda;\varepsilon), P_{\mathcal{D}}(\lambda;\varepsilon)$ and $P_{\mathcal{D}-1}(\lambda;\varepsilon)$, it is convenient to introduce two polynomials $P_{\mathcal{D}+1}(\lambda;\varepsilon)$ and

$$P_{2\mathcal{D}}(\lambda;\varepsilon) = \frac{1}{4}\left\{ 2P'_{\mathcal{D}}(\lambda;\varepsilon)P_{\mathcal{D}+1}(\lambda;\varepsilon) - 2P_{\mathcal{D}}(\lambda;\varepsilon)P'_{\mathcal{D}+1}(\lambda;\varepsilon) \right.$$

$$\left. + P^2_{\mathcal{D}}(\lambda;\varepsilon) - 4P_{\mathcal{D}-1}(\lambda;\varepsilon)P_{\mathcal{D}+1}(\lambda;\varepsilon) \right\} \qquad (3.4.34)$$

of orders $\mathcal{D} + 1$ and $2\mathcal{D}$, respectively. If these polynomials are given, the form of the remaining polynomials $P_{\mathcal{D}}(\lambda;\varepsilon)$ and $P_{\mathcal{D}-1}(\lambda;\varepsilon)$ can easily be recovered from equation (3.4.34). In the new notation operator $X(\varepsilon)$ takes the form

$$X(\varepsilon) = -\frac{\partial^2}{\partial\lambda^2} + \frac{P_{2\mathcal{D}}(\lambda;\varepsilon)}{P^2_{\mathcal{D}+1}(\lambda;\varepsilon)} \qquad (3.4.35)$$

and this gives us the following new MPS equation:

$$\left\{ -\frac{\partial^2}{\partial\lambda^2} + \frac{P_{2\mathcal{D}}(\lambda;\varepsilon)}{P_{\mathcal{D}+1}^2(\lambda;\varepsilon)} \right\} \varphi(\lambda) = 0. \tag{3.4.36}$$

As follows from our derivation, equation (3.4.36) is again algebraically solvable, and for any given $M = 0, 1, 2, \ldots$ has solutions belonging to the class of functions

$$\varphi(\lambda) = \exp\left\{ \frac{1}{2} \int \frac{P_{\mathcal{D}}(\lambda;\varepsilon)}{P_{\mathcal{D}+1}(\lambda;\varepsilon)} \, d\lambda \right\} \sum_{n=0}^{M} \varphi_n \lambda^n. \tag{3.4.37}$$

We emphasize that the coefficients in the polynomial $P_{\mathcal{D}+1}(\lambda;\varepsilon)$ are certain given functions of ε, the coefficients in the polynomial $P_{\mathcal{D}}(\lambda;\varepsilon)$ can be obtained from equation (3.4.34) and $\varepsilon_1, \ldots, \varepsilon_{\mathcal{D}}$ and $\varphi_0, \ldots, \varphi_M$ are unknown numbers.

f. Algebraically solvable MPS equations with linear dependence on spectral parameters. Consider the case when the operator $X(\varepsilon)$ entering into the MPS equation (3.4.1) depends on the parameter ε linearly. The most general form of such an operator is

$$X(\varepsilon) = X^0 - \sum_{\alpha=1}^{\mathcal{D}} X^\alpha \varepsilon_\alpha, \tag{3.4.38}$$

where X^0 and $X^\alpha, \alpha = 1, \ldots, \mathcal{D}$ are certain given operators in V. Substituting (3.4.38) into (3.4.1), we obtain the MPS equation

$$X^0 \varphi = \left(\sum_{\alpha=1}^{\mathcal{D}} X^\alpha \varepsilon_\alpha \right) \varphi, \quad \varphi \in \Phi, \tag{3.4.39}$$

the form of which almost coincides with (3.2.39). The only difference is that the set Φ in which solutions of (3.4.39) are being sought does not necessarily coincide with the space $W = W_1 \cap \ldots \cap W_{\mathcal{D}}$ appearing in equation (3.2.39). Here Φ is chosen in such a way as to guarantee only an algebraic solvability of the equation (3.4.39), and not its physical sensibility.

The simplest way to construct algebraically solvable linearized MPS equations is to construct a class (as wide as possible) of algebraically solvable generalized MPS equations, and then extract the cases when the dependence of the operators $X(\varepsilon)$ on the spectral parameter ε becomes linear. As a starting point we can use the classes of algebraically solvable second-order differential MPS equations (3.4.30) and (3.4.36) constructed in the preceding subsection.

Let us start with equation (3.4.30). It is evident that its linearized version arises when the coefficients $P_n(\lambda; \varepsilon), n = \mathcal{D} - 1, \mathcal{D}, \mathcal{D} + 1$ depend on ε linearly:

$$P_n(\lambda; \varepsilon) = P_n^0(\lambda) - \sum_{\alpha=1}^{\mathcal{D}} P_n^\alpha(\lambda)\varepsilon_\alpha, \quad n = \mathcal{D} - 1, \mathcal{D}, \mathcal{D} + 1. \quad (3.4.40)$$

Here $P_n^0(\lambda)$ and $P_n^\alpha(\lambda), \alpha = 1, \ldots, \mathcal{D}$ are certain fixed polynomials. Substitution of (3.4.40) into equation (3.4.30) gives:

$$\left[P_{\mathcal{D}+1}^0(\lambda)\frac{\partial^2}{\partial\lambda^2} + P_{\mathcal{D}}^0(\lambda)\frac{\partial}{\partial\lambda} + P_{\mathcal{D}-1}^0(\lambda) \right] \varphi(\lambda)$$

$$= \left\{ \sum_{\alpha=1}^{\mathcal{D}} \left[P_{\mathcal{D}+1}^\alpha(\lambda)\frac{\partial^2}{\partial\lambda^2} + P_{\mathcal{D}}^\alpha(\lambda)\frac{\partial}{\partial\lambda} + P_{\mathcal{D}-1}^\alpha(\lambda) \right] \varepsilon_\alpha \right\} \varphi(\lambda). \quad (3.4.41)$$

This equation is algebraically solvable by construction and its solutions belong to the class of polynomials

$$\varphi(\lambda) = Q_M(\lambda), \quad M = 0, 1, 2, \ldots. \quad (3.4.42)$$

Now consider another type of MPS equation described by formula (3.4.36). It is easy to see that for (3.4.36) to be a linear MPS equation, the denominator of the free term must be independent of ε, while the numerator must depend on ε linearly:

$$P_{\mathcal{D}+1}(\lambda; \varepsilon) = P_{\mathcal{D}+1}(\lambda), \quad (3.4.43a)$$

$$P_{2\mathcal{D}}(\lambda; \varepsilon) = P_{2\mathcal{D}}^0(\lambda) - \sum_{\alpha=1}^{\mathcal{D}} P_{2\mathcal{D}}^\alpha(\lambda)\varepsilon_\alpha. \quad (3.4.43b)$$

Substituting (3.4.43) into (3.4.36), we obtain the needed MPS equation

$$\left\{ -\frac{\partial^2}{\partial\lambda^2} + \frac{P_{2\mathcal{D}}^0(\lambda)}{[P_{\mathcal{D}+1}(\lambda)]^2} \right\} \varphi(\lambda) = \left\{ \sum_{\alpha=1}^{\mathcal{D}} \frac{P_{2\mathcal{D}}^\alpha(\lambda)}{[P_{\mathcal{D}+1}(\lambda)]^2}\varepsilon_\alpha \right\} \varphi(\lambda), \quad (3.4.44)$$

which is also algebraically solvable and has solutions of the form

$$\varphi(\lambda) = \exp\left\{ \int \frac{R_{\mathcal{D}}(\lambda)}{P_{\mathcal{D}+1}(\lambda)} \, d\lambda \right\} Q_M(\lambda), \quad M = 0, 1, 2, \ldots, \quad (3.4.45)$$

where $R_{\mathcal{D}}(\lambda)$ and $Q_M(\lambda)$ are unknown polynomials of orders $\mathcal{D}$ and M, respectively.

3.5   An analytic method of constructing algebraically solvable multi-parameter spectral equations

In this section we discuss another (analytic) method of constructing one-dimensional second-order differential MPS equations with linear dependence on spectral parameters ε_α. We find this method more attractive than the algebraic one discussed in the previous section, since it leads immediately to the most general form of such equations. Besides, this method is direct, since it does not require preliminary construction of MPS equations with an arbitrary dependence on spectral parameters. Finally, it gives the possibility of obtaining rather compact explicit expressions for both the MPS equations and their solutions, which is especially important for their analysis and classification.

The idea of this method is very simple. In order to formulate it, let us first consider an arbitrary differential MPS equation of the form

$$\left\{-\frac{\partial^2}{\partial\lambda^2} + x^0(\lambda)\right\}\varphi(\lambda) = \left\{\sum_{\alpha=1}^{D} x^\alpha(\lambda)\varepsilon_\alpha\right\}\varphi(\lambda), \quad \varphi(\lambda) \in \Phi, \quad (3.5.1)$$

and note that it can be interpreted as a particular case of the more general equation for two functions $\varphi(\lambda)$ and $\omega(\lambda)$

$$\left\{-\frac{\partial^2}{\partial\lambda^2} + \omega(\lambda)\right\}\varphi(\lambda) = 0, \quad \omega(\lambda) \in \Omega, \quad \varphi(\lambda) \in \Phi, \quad (3.5.2)$$

provided that Ω is a finite-dimensional functional space with basis $x^0(\lambda), x^\alpha(\lambda), \alpha = 1, \ldots, D$. In fact, any function $\omega(\lambda)$ satisfying equation (3.5.2) is a linear combination of $D + 1$ functions $x^0(\lambda), x^\alpha(\lambda), \alpha = 1, \ldots, D$,

$$\omega(\lambda) = x^0(\lambda)\varepsilon_0 - \sum_{\alpha=1}^{D} x^\alpha(\lambda)\varepsilon_\alpha, \quad (3.5.3)$$

and, therefore, any particular solution of (3.5.2) with $\varepsilon_0 = 1$ is automatically a solution of equation (3.5.1).

Now note that equation (3.5.2) admits one more generalization. In order to construct it, note that all dependence of the function $\omega(\lambda)$ on the sort of solution is concentrated in the coefficients $\varepsilon_0, \varepsilon_\alpha, \alpha = 1, \ldots, D$ of its expansion (3.5.3). The basis functions $x^0(\lambda)$ and $x^\alpha(\lambda), \alpha = 1, \ldots, D$ are assumed to be fixed and, therefore, the location of all singular points of the function $\omega(\lambda)$ is also fixed for all solutions. Thus, we can treat (3.5.2) as a particular case of the more general equation

$$\left\{-\frac{\partial^2}{\partial\lambda^2} + \omega(\lambda)\right\}\varphi(\lambda) = 0, \quad (3.5.4)$$

supplemented by the following additional condition: the location of singularities of $\omega(\lambda)$ does not depend on the sort of solution.

Now let us formulate the method of constructing algebraically solvable MPS equations of the type (3.5.4). This method is naturally divided into two stages. First, we construct the most general solution of equation (3.5.4) and then find the conditions when the functions $\omega(\lambda)$ satisfying this equation can be treated as elements of a certain $(\mathcal{D}+1)$-dimensional functional space and represented in the form (3.5.3) with $\varepsilon_0 = 1$. In this case we obtain the most general expressions for both the one-dimensional second-order differential MPS equation (3.5.1) and its solutions.

In order to realize this program let us first study the behaviour of the second unknown function $\varphi(\lambda)$ in the vicinity of singular points. Introducing the logarithmic derivative of $\varphi(\lambda)$

$$y(\lambda) = \frac{\varphi'(\lambda)}{\varphi(\lambda)}, \qquad (3.5.5)$$

we can rewrite equation (3.5.4) in the Riccati form:

$$y'(\lambda) + y^2(\lambda) = \omega(\lambda). \qquad (3.5.6)$$

We shall distinguish between two different types of singularity of function $y(\lambda)$. Singularities of the first type are those for which the corresponding function $\omega(\lambda)$ is regular. All other singularities are of the second type. It is quite obvious that the former can be located more or less arbitrarily, while the location of the latter is fixed and cannot depend on the sort of solution.

Assume that the function $y(\lambda)$ has at the point $\lambda = \xi$ the first type of singularity. This means that the corresponding function $\omega(\lambda)$ must behave at this point as

$$\omega(\lambda) = \omega_0 + (\lambda - \xi)\omega_1 + (\lambda - \xi)^2\omega_2 + \ldots, \quad \lambda \to \xi. \qquad (3.5.7)$$

Substituting expansion (3.5.7) into equation (3.5.6) and solving this equation we find

$$y(\lambda) = \frac{1}{\lambda - \xi} + \frac{\omega_0}{3}(\lambda - \xi) + \ldots, \quad \lambda \to \xi. \qquad (3.5.8)$$

From (3.5.8) it follows that the role of the first-type singularities of $y(\lambda)$ can be played by the simple poles with unit residues only. Therefore, the most general form of functions $y(\lambda)$ having singularities of both types is

$$y(\lambda) = F(\lambda) + \sum_{i=1}^{M} \frac{1}{\lambda - \xi_i}, \quad M = 0, 1, 2, \ldots. \qquad (3.5.9)$$

Here $F(\lambda)$ are arbitrary functions with a fixed location of singular points, and $\xi_1, \ldots, \xi_M$ are the coordinates of simple poles, the location of which can, in principle, depend on the sort of solution. The number M of these poles can be chosen arbitrarily.

In order to find the admissible values of the numbers $\xi_1, \ldots, \xi_M$, we substitute (3.5.9) into (3.5.6). The resulting expression for $\omega(\lambda)$ takes the form

$$
\begin{aligned}
\omega(\lambda) &= \left[F(\lambda) + \sum_{i=1}^{M} \frac{1}{\lambda - \xi_i} \right]' + \left[F(\lambda) + \sum_{i=1}^{M} \frac{1}{\lambda - \xi_i} \right]^2 \\
&= F'(\lambda) + F^2(\lambda) + 2 \sum_{i=1}^{M} \frac{F(\lambda)}{\lambda - \xi_i} \\
&\quad + \sum_{i \neq k}^{M} \frac{1}{(\lambda - \xi_i)(\lambda - \xi_k)} \\
&= F'(\lambda) + F^2(\lambda) + 2 \sum_{i=1}^{M} \frac{F(\lambda) - F(\xi_i)}{\lambda - \xi_i} \\
&\quad + 2 \sum_{i=1}^{M} \frac{1}{\lambda - \xi_i} \left\{ \sum_{k=1}^{M}{}' \frac{1}{\xi_i - \xi_k} + F(\xi_i) \right\}. \quad (3.5.10)
\end{aligned}
$$

We see that the first two terms in (3.5.10) are singular at the same points as the function $F(\lambda)$. These are second-type singularities, the location of which is assumed to be fixed. The third term in (3.5.10) is obviously regular at the points $\xi_1, \ldots, \xi_M$. These singularities have the form of simple poles with the residues

$$
r_i = \sum_{k=1}^{M}{}' \frac{1}{\xi_i - \xi_k} + F(\xi_i), \quad i = 1, \ldots, M. \quad (3.5.11)
$$

According to the condition of regularity of $\omega(\lambda)$ at these points, we must take

$$
r_i = 0, \quad i = 1, \ldots, M. \quad (3.5.12)
$$

This leads us to a system of numerical equations for $\xi_1, \ldots, \xi_M$ and a rather compact expression for the function $\omega(\lambda)$. Note that the corresponding form of the second function $\varphi(\lambda)$ can be obtained from

$$
\varphi(\lambda) = \exp\left\{ \int y(\lambda)\, d\lambda \right\}. \quad (3.5.13)
$$

Collecting these results we come to the following important statement.

Statement 3.8. *The most general solution of equation (3.5.4) is*

$$\omega(\lambda) = F'(\lambda) + F^2(\lambda) + 2 \sum_{i=1}^{M} \frac{F(\lambda) - F(\xi_i)}{\lambda - \xi_i}, \qquad (3.5.14)$$

$$\varphi(\lambda) = \exp\left\{ \int F(\lambda) \, d\lambda \right\} \prod_{i=1}^{M} (\lambda - \xi_i), \qquad (3.5.15)$$

where $F(\lambda)$ is an arbitrary function with fixed positions of singular points, M is an arbitrary non-negative integer, and the numbers $\xi_1, \ldots, \xi_M$ satisfy the system of M numerical equations

$$\sum_{k=1}^{M}{}' \frac{1}{\xi_i - \xi_k} + F(\xi_i) = 0, \quad i = 1, \ldots, M. \qquad (3.5.16)$$

This statement completes the first stage of our programme. Indeed, the most general solution of the equation (3.5.4) is constructed. Now we must find the conditions under which $\omega(\lambda)$, having the form (3.5.14), belongs to a finite-dimensional functional space. We show that this is always possible when $F(\lambda)$ is a rational function.

First of all, remember that the unique singularities of rational functions are the poles which may be located at finite points or at infinity. Any rational function has only a finite number of poles in the complex plane.

Let $\vec{a} = (a_1, \ldots, a_K)$ be the points in which the given rational function $r(\lambda)$ has poles of orders $\vec{n} = (n_1, \ldots, n_K)$, respectively. Assume also that $r(\lambda)$ behaves at infinity as λ^n. This means that there it has the pole of order $n + 2$ (if $n \geq -1$). Any function of such sort can be represented as a fraction

$$r(\lambda) = \frac{Q_{\mathcal{D}}(\lambda)}{(\lambda - a_1)^{n_1} \ldots (\lambda - a_K)^{n_K}}, \qquad (3.5.17)$$

in which $Q_{\mathcal{D}}(\lambda)$ is a polynomial of order

$$\mathcal{D} = \sum_{\alpha=1}^{K} n_\alpha + n. \qquad (3.5.18)$$

Denote by $R_n\left(\frac{\vec{a}}{\vec{n}}\right) = R_n\left(\begin{smallmatrix}a_1,\ldots,a_K\\n_1,\ldots,n_K\end{smallmatrix}\right)$ the space of all linear combinations of rational functions having the form (3.5.17). From (3.5.17) and (3.5.18) it follows that

$$\dim R_n\left(\frac{\vec{a}}{\vec{n}}\right) = \mathcal{D} + 1 = \sum_{\alpha=1}^{K} a_\alpha + n + 1. \qquad (3.5.19)$$

Now, let us formulate two important statements.

Statement 3.9. *Let $F(\lambda) \in R_n\left(\frac{\vec{a}}{\vec{n}}\right)$. Then*

$$\sum_{\alpha=1}^{M} \frac{F(\lambda) - F(\xi_i)}{\lambda - \xi_i} \in R_{n-1}\left(\frac{\vec{a}}{\vec{n}}\right), \qquad (3.5.20)$$

if the numbers ξ_i satisfy the system (3.5.16).

Proof. First of all note that the function in (3.5.20) is regular at the points $\lambda = \xi_i$, $i = 1, \ldots, M$, and therefore in any finite part of the complex λ-plane it has the same singularities as the function $F(\lambda)$.

If $\lambda \to \infty$, one can write

$$\sum_{i=1}^{M} \frac{F(\lambda) - F(\xi_i)}{\lambda - \xi_i} = M F(\lambda)\left\{\frac{1}{\lambda} + O\left(\frac{1}{\lambda^2}\right)\right\}$$
$$- \left\{\left[\sum_{i=1}^{M} F(\xi_i)\right]\frac{1}{\lambda} + O\left(\frac{1}{\lambda^2}\right)\right\}. \qquad (3.5.21)$$

Using the equality

$$\sum_{i=1}^{M} F(\xi_i) = 0 \qquad (3.5.22)$$

which follows from the system (3.5.16), we can see that, if $F(\lambda) \sim \lambda^n$, then

$$\sum_{i=1}^{M} \frac{F(\lambda) - F(\xi_i)}{\lambda - \xi_i} \sim \lambda^{n-1}, \qquad (3.5.23)$$

for any $n \geq -1$ and $\lambda \to \infty$. This proves the statement.

Statement 3.10. *Let $F(\lambda) \in R_n\left(\frac{\vec{a}}{\vec{n}}\right)$. Then*

$$\omega(\lambda) = F'(\lambda) + F^2(\lambda) + 2\sum_{i=1}^{M} \frac{F(\lambda) - F(\xi_i)}{\lambda - \xi_i} \in R_{2n}\left(\frac{\vec{a}}{\vec{n}}\right), \qquad (3.5.24)$$

if the numbers ξ_i satisfy the system (3.5.16).

Proof. The assertion of this statement follows immediately from the definition of the spaces $R_n\left(\frac{\vec{a}}{\vec{n}}\right)$ and from formula (3.5.20).

Thus, we have shown that all the functions $\omega(\lambda)$ satisfying equation (3.5.4) belong to the space $R_{2n}\left(\frac{\vec{a}}{\vec{n}}\right)$, which is finite dimensional. Its dimension is given by the formula

$$\dim R_{2n}\left(\frac{\vec{a}}{\vec{n}}\right) = 2\mathcal{D} + 1, \qquad (3.5.25)$$

where $\mathcal{D}$ is defined by (3.5.18).

The last step is to reduce this equation to the form (3.5.1). This can be done as follows. Denote by $r^\beta(\lambda)$, $\beta = 0, 1, \ldots, \mathcal{D}$ the basis in the space $R_n\left(\frac{\vec{a}}{\vec{n}}\right)$. Then, for any function $F(\lambda)$ belonging to this space, one can write the expansion

$$F(\lambda) = \sum_{\beta=0}^{\mathcal{D}} F_\beta r^\beta(\lambda). \qquad (3.5.26)$$

Denoting by $r^\beta(\lambda)$, $\beta = 0, 1, \ldots, 2\mathcal{D}$ the basis of the space $R_{2n}\left(\frac{\vec{a}}{2\vec{n}}\right)$ we can write an analogous expansion for $\omega(\lambda)$

$$\omega(\lambda) = \sum_{\beta=0}^{2\mathcal{D}} \omega_\beta(\vec{F}, \vec{\xi}) r^\beta(\lambda), \qquad (3.5.27)$$

where $\omega_\beta(\vec{F}, \vec{\xi})$, $\beta = 0, 1, \ldots, 2\mathcal{D}$ are the coefficients depending on both the $(\mathcal{D} + 1)$-dimensional vector $\vec{F} = (F_0, \ldots, F_\mathcal{D})$ and the M-dimensional vector $\vec{\xi} = (\xi_1, \ldots \xi_M)$.

Now let $x^0(\lambda)$ and $x^\alpha(\lambda)$, $\alpha = 1, \ldots, \mathcal{D}$ be arbitrarily fixed functions belonging to the space $R_{2n}\left(\frac{\vec{a}}{2\vec{n}}\right)$. Then we represent them in the form:

$$x^0(\lambda) = \sum_{\beta=0}^{2\mathcal{D}} x_\beta^0 r^\beta(\lambda), \qquad (3.5.28a)$$

$$x^\alpha(\lambda) = \sum_{\beta=0}^{2\mathcal{D}} x_\beta^\alpha r^\beta(\lambda), \quad \alpha = 1, \ldots, \mathcal{D}, \qquad (3.5.28b)$$

where x_β^0 and x_β^α, $\alpha = 1, \ldots, \mathcal{D}$, $\beta = 1, \ldots, \mathcal{D}$ are certain given coefficients.

Consider the equation

$$\omega(\lambda) = x^0(\lambda) - \sum_{\alpha=1}^{\mathcal{D}} x^\alpha(\lambda)\varepsilon_\alpha, \qquad (3.5.29)$$

in which ε_α, $\alpha = 1, \ldots, \mathcal{D}$ are certain unknown numbers. Substituting expansions (3.5.27) and (3.5.28) into (3.5.29), we obtain the system of numerical equations

$$\omega_\beta(\vec{F}, \vec{\xi}) = x_\beta^0 - \sum_{\alpha=1}^{\mathcal{D}} x_\beta^\alpha \varepsilon_\alpha, \qquad \beta = 1, \ldots, 2\mathcal{D}+1, \qquad (3.5.30)$$

while the substitution of (3.5.26) into (3.5.15) gives

$$\sum_{k=1}^{M}{}' \frac{1}{\xi_i - \xi_k} + \sum_{\beta=0}^{\mathcal{D}} F_\beta r^\beta(\lambda) = 0, \quad i = 1, \ldots, M. \qquad (3.5.31)$$

We see that equations (3.5.30) and (3.5.31) together form the system of $M + 2\mathcal{D}+1$ equations for $M+2\mathcal{D}+1$ unknown quantities $\xi_1, \ldots, \xi_M$, $F_0, \ldots, F_{\mathcal{D}}$ and $\varepsilon_1, \ldots, \varepsilon_{\mathcal{D}}$. Evidently, for any set of fixed numbers x_β^0 and x_β^α this system has algebraic solutions. This gives us the possibility of asserting that equation (3.5.4) takes the form of an algebraically solvable MPS equation. Summarizing these results we can formulate the following basic theorem.

Theorem 3.2. *Let $x^0(\lambda)$ and $x^\alpha(\lambda)$, $\alpha = 1, \ldots, \mathcal{D}$ be the arbitrarily fixed rational functions belonging to the space $R_{2n}\binom{\vec{a}}{\vec{n}}$. Then the MPS equation*

$$\left\{ -\frac{\partial^2}{\partial \lambda^2} + x^0(\lambda) \right\} \varphi(\lambda) = \left\{ \sum_{\alpha=1}^{\mathcal{D}} x^\alpha(\lambda)\varepsilon_\alpha \right\} \varphi(\lambda) \qquad (3.5.32)$$

is algebraically solvable. Its solutions have the form

$$\varphi(\lambda) = \prod_{i=1}^{M} (\lambda - \xi_i) \exp\left\{ \int F(\lambda)\, d\lambda \right\}, \qquad (3.5.33)$$

where M is an arbitrary non-negative integer, $\xi_1, \ldots, \xi_M$ are certain unknown parameters and $F(\lambda)$ is an unknown function belonging to the space $R_n\binom{\vec{a}}{\vec{n}}$. A concrete form of function $F(\lambda)$, the values of $\xi_1, \ldots, \xi_M$ and the corresponding admissible values of the spectral parameters $\varepsilon_1, \ldots, \varepsilon_{\mathcal{D}}$ can be obtained from the following system of algebraic equations:

$$\sum_{k=1}^{M}{}' \frac{1}{\xi_i - \xi_k} + F(\xi_i) = 0, \quad i = 1, \ldots, M, \qquad (3.5.34)$$

$$F'(\lambda) + F^2(\lambda) + 2\sum_{i=1}^{M} \frac{F(\lambda) - F(\xi_i)}{\lambda - \xi_i} = x^0(\lambda) - \sum_{\alpha=1}^{D} x^\alpha(\lambda)\varepsilon_\alpha. \qquad (3.5.35)$$

(This is the end of the theorem.)

This theorem completes the construction of the most general classes of algebraically solvable one-dimensional second-order differential MPS equations with a linear dependence on spectral parameters.

Now let us try to elucidate how many solutions the obtained MPS equations have. Quite obviously, this number is equal to the number of solutions of the system of algebraic equations (3.5.34) and (3.5.35).

First of all, consider the first equation (3.5.34) and treat it as a system of M numerical equations for M unknown numbers $\xi_1, \ldots, \xi_M$, assuming that the function $F(\lambda)$ determining the form of this system is given. Since the function belongs to the space $R_n\left(\frac{d}{n}\right)$, it can be represented as the fraction

$$F(\lambda) = \frac{Q_D(\lambda)}{Q_{D-n}(\lambda)}, \qquad (3.5.36)$$

in which $Q_D(\lambda)$ and $Q_{D-n}(\lambda)$ are certain given polynomials of degrees D and $D - n$, respectively.

Substituting (3.5.36) into (3.5.34) we obtain

$$\sum_{k=1}^{M}{}' \frac{1}{\xi_i - \xi_k} + \frac{Q_D(\xi_i)}{Q_{D-n}(\xi_i)} = 0, \quad i = 1, \ldots, M. \qquad (3.5.37)$$

Multiplying (3.5.37) by the denominator of the second term gives:

$$Q_{D-n}(\xi_i) \sum_{k=1}^{M}{}' \frac{1}{\xi_i - \xi_k} + Q_D(\xi_i) = 0, \quad i = 1, \ldots, M. \qquad (3.5.38)$$

Note that this system is equivalent to the system of M equations

$$\sum_{i,k=1}^{M}{}' \frac{Q_{D-n}(\xi_i)}{\xi_i - \xi_k} + \sum_{k=1}^{M} Q_D(\xi_i)\xi_i^m = 0, \quad m = 0, \ldots, M - 1, \qquad (3.5.39)$$

which can be obtained from (3.5.38) by multiplying it by ξ_i^m, $m = 0, \ldots, M - 1$ and summing over $i = 1, \ldots, M$.

The first term in (3.5.39) is a polynomial in ξ_i of degree $D - n + m - 1$. This follows from the evident identity

$$\sum_{i,k=1}^{M}{}' \frac{Q_{D-n}(\xi_i)}{\xi_i - \xi_k} = \frac{1}{2} \sum_{i,k=1}^{M}{}' \frac{Q_{D-n}(\xi_i)\xi_i^m - Q_{D-n}(\xi_k)\xi_k^m}{\xi_i - \xi_k}. \qquad (3.5.40)$$

At the same time, the second term in (3.5.39) is a polynomial of degree $\mathcal{D}+m$. Since $n \geq 1$ by assumption, the mth equation (3.5.39) is an algebraic equation of degree $m + \mathcal{D}$.

Thus, we have M different algebraic equations (3.5.39) of degrees $\mathcal{D}, \mathcal{D} + 1, \ldots, \mathcal{D} + M - 1$, and therefore this system has, generally, $\mathcal{D} \times (\mathcal{D} - 1) \ldots (\mathcal{D} + M - 1) = \frac{(M+\mathcal{D}-1)!}{(\mathcal{D}-1)!}$ solutions. However, many of these solutions are equivalent. In fact, the numbers $\xi_1, \ldots, \xi_M$ enter into the second system (3.5.35) symmetrically. This means that all the solutions of the first system (3.5.34), which can be transformed into each other by means of a permutation of the numbers $\xi_1, \ldots, \xi_M$, must be considered as equivalent solutions. Thus, in order to obtain the number of non-equivalent solutions of the system (3.5.34), we must divide the total number of solutions, $\frac{(M+\mathcal{D}-1)!}{(\mathcal{D}-1)!}$, by the total number of all permutations, $M!$. This results in the following statement.

Statement 3.11. *Let $F(\lambda)$ be a given rational function, belonging to the space $R_n\left(\frac{\vec{a}}{\vec{n}}\right)$ of dimension $\mathcal{D} + 1$. Then the total number of non-equivalent solutions of the system (3.5.34) is $\frac{(M+\mathcal{D}-1)!}{M!(\mathcal{D}-1)!}$.*

Now, let us consider the second equation (3.5.35), and interpret it as a system of $2\mathcal{D} + 1$ algebraic equations for $2\mathcal{D} + 1$ numbers $\varepsilon_1, \ldots, \varepsilon_{\mathcal{D}}$ and $F_0, \ldots, F_{\mathcal{D}}$, assuming that the parameters $\xi_1, \ldots, \xi_M$ are fixed. Remember that the right-hand side of (3.5.35) as well as the two first terms in the left-hand side belong to the space $R_n\left(\frac{\vec{a}}{\vec{n}}\right)$, while the third term in the left-hand side belongs to the subspace $R_{n-1}\left(\frac{\vec{a}}{\vec{n}}\right)$. Denoting the projection operator from $R_{2n}\left(\frac{\vec{a}}{2\vec{n}}\right)$ to $R_{n-1}\left(\frac{\vec{a}}{\vec{n}}\right)$ by $\mathcal{P}$, we can rewrite (3.5.35) in the form of two independent equations:

$$\mathcal{P}\left\{F'(\lambda) + F^2(\lambda) + 2\sum_{i=1}^{M}\frac{F(\lambda) - F(\xi_i)}{\lambda - \xi_i}\right\}$$

$$= \mathcal{P}\left\{x^0(\lambda) - \sum_{\alpha=1}^{\mathcal{D}} x^\alpha(\lambda)\varepsilon_\alpha\right\} \tag{3.5.41a}$$

and

$$(1 - \mathcal{P})\left\{F'(\lambda) + F^2(\lambda)\right\}$$

$$= (1 - \mathcal{P})\left\{x^0(\lambda) - \sum_{\alpha=1}^{\mathcal{D}} x^\alpha(\lambda)\varepsilon_\alpha\right\}. \tag{3.5.41b}$$

Due to the fact that $\dim R_{n-1}\left(\frac{\vec{a}}{\vec{n}}\right) = \mathcal{D}$ and $\dim R_{2n}\left(\frac{\vec{a}}{2\vec{n}}\right) - \dim R_{n-1}\left(\frac{\vec{a}}{\vec{n}}\right) = \mathcal{D} + 1$, equations (3.5.41a) and (3.5.41b) are equivalent to systems of $\mathcal{D}$ and

$\mathcal{D}+1$ algebraic equations. Considering (3.5.41a) as a system of $\mathcal{D}$ algebraic equations for $\mathcal{D}$ unknown numbers $\varepsilon_1, \ldots, \varepsilon_{\mathcal{D}}$, we can express these numbers via the parameters $F_0, \ldots, F_{\mathcal{D}}$. Substituting the obtained expressions for $\varepsilon_1, \ldots, \varepsilon_{\mathcal{D}}$ into (3.5.41b), we obtain a system of $\mathcal{D}+1$ equations for $\mathcal{D}+1$ unknown parameters $F_0, \ldots, F_{\mathcal{D}}$ determining the form of the function $F(\lambda)$. Assume now that the total number of singularities of functions $x^0(\lambda)$ and $x^\alpha(\lambda)$, $\alpha = 1, \ldots, \mathcal{D}$ is $K+1$. Obviously, the function $F(\lambda)$ is singular at the same points as the functions $x^0(\lambda)$ and $x^\alpha(\lambda)$, $\alpha = 1, \ldots, \mathcal{D}$. Denoting the coefficients of these singularities by F_k, $k = 0, \ldots, \mathcal{D}$ and equating the leading (most singular) terms in both sides of (3.5.41b), we get $K+1$ separate quadratic equations for $F_0, \ldots, F_K$. Each of these equations has two solutions, which gives us 2^{K+1} different sets of coefficients F_k, $k = 1, \ldots, K$. As soon as one such set is chosen, the other coefficients F_k, $k = K+1, \ldots, \mathcal{D}$ of the function $F(\lambda)$ can be determined uniquely by equating the non-leading terms in equation (3.5.41b). This means that the total number of solutions of this equation is 2^{K+1}. Summarizing, we come to the following statement.

Statement 3.12. *Let the total number of singularities of rational functions belonging to the space $R_{2n}\left(\frac{d}{2n}\right)$ be $K+1$. Then for any given $\xi_1, \ldots, \xi_M$ equation (3.5.35) has 2^{K+1} different solutions for $F(\lambda) \in R_n\left(\frac{d}{n}\right)$ and $\varepsilon_1, \ldots, \varepsilon_{\mathcal{D}}$.*

Thus, on the one hand, we have shown that equation (3.5.34) has $\frac{(M+\mathcal{D}-1)!}{M!(\mathcal{D}-1)!}$ solutions for $\xi_1, \ldots, \xi_M$ if the function $F(\lambda)$ is fixed. On the other hand, we see that equation (3.5.35) has 2^{K+1} solutions for $F(\lambda)$ if the numbers $\xi_1, \ldots, \xi_M$ are fixed. Combining these two assertions, we can conclude that, in general case, the system of equations (3.5.34) and (3.5.35) must have $2^{K+1} \frac{(M+\mathcal{D}-1)!}{M!(\mathcal{D}-1)!}$ different solutions for any given M. Of course, this result is not rigorous from the mathematical point of view. However, it is very reliable and, at least at the present time, we do not know any counter-example. This gives us a reason to conjecture the following theorem.

Theorem 3.3. *Let the functions $x^0(\lambda)$ and $x^\alpha(\lambda)$, $\alpha = 1, \ldots, \mathcal{D}$, entering into the* MPS *equation (3.5.32), belong to a $(2\mathcal{D}+1)$-dimensional space $R_{2n}\left(\frac{d}{2n}\right)$ of functions having $K+1$ poles at the points $a_1, \ldots, a_K$ and at infinity. Then for any $M = 0, 1, 2, \ldots$ the number of algebraic solutions of this equation, given by formulas (3.5.33)–(3.5.35), is equal to*

$$N_M = 2^{K+1} \frac{(M+\mathcal{D}-1)!}{M!(\mathcal{D}-1)!}, \quad M = 0, 1, 2, \ldots. \qquad (3.5.42)$$

The total number of such solutions is infinite, and thus, (3.5.32) is an MPS *equation with an infinite and discrete spectrum.*

Let us now discuss the classification problem for algebraically solvable MPS equations of the type (3.5.32). From theorems 3.2 and 3.3 it follows that in order to solve this problem it is sufficient to classify the spaces of rational functions $R_{2n}\left(\genfrac{}{}{0pt}{}{\vec{a}}{2\vec{n}}\right)$. For this we can use a graphical method. Before formulating the essence of this method, let us first give an alternative definition of the spaces $R_{2n}\left(\genfrac{}{}{0pt}{}{\vec{a}}{2\vec{n}}\right)$ in terms of so-called "double poles". We shall say that a rational function has at the point a the double pole of order n if it behaves near this point as $\frac{1}{(\lambda-a)^{2n}}$ when a is finite, and as λ^{2n-4} when a is infinite. In this language $R_{2n}\left(\genfrac{}{}{0pt}{}{\vec{a}}{2\vec{n}}\right)$ can be interpreted as the space of all rational functions having at the points $a_1,\ldots,a_K$ and $a_0=\infty$ double poles of orders $n_1,\ldots,n_K$ and $n_0 = n+2$, respectively. Depicting the points a_α by n_α small concentric circles we come to a diagram describing the space $R_{2n}\left(\genfrac{}{}{0pt}{}{\vec{a}}{2\vec{n}}\right)$. The technical difficulties connected with depicting the infinite point a_0 can easily be overcome if we map the complex λ-plane onto the sphere of a finite radius (by means of a stereographic projection) and then look at this sphere from the point $-i\infty$. Then we see a disc with the boundary being the image of the real λ-axis with identified plus and minus infinity, and with the interior formed by identified complex conjugate points. In order to distinguish between the points with positive and negative imaginary parts, we depict the former by the upper half-circles, and the latter by the lower half-circles. We shall also depict the complex conjugate pairs as well as the real points by whole circles.

Thus, we come to a disc-like diagram consisting of several small circles (or half-circles) describing the position and orders of double poles determining the structure of the space $R_{2n}\left(\genfrac{}{}{0pt}{}{\vec{a}}{2\vec{n}}\right)$. We shall call such diagrams simple. When listing simple diagrams, one can list the types of

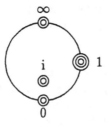

Figure 3.4. An example of a simple diagram.

spaces $R_{2n}\left(\genfrac{}{}{0pt}{}{\vec{a}}{2\vec{n}}\right)$ and, consequently, the corresponding classes of algebraically solvable MPS equations (3.5.32).

For example, the simple diagram depicted in figure 3.4 describes the space $R_{2n}\binom{\vec{a}}{2\vec{n}}$ which consists of rational functions having double poles of orders 1, 3, 2, 2 and 2 at the points 0, 1, i, −i and ∞. Thus, in this case, $\vec{a} = (0, 1, i, -i)$, $2\vec{n} = (2, 6, 4, 4)$, $2n = 0$ and $\mathcal{D} = 9$. The class of corresponding MPS equations (3.5.32) is determined by the functions $x^0(\lambda)$ and $x^\alpha(\lambda), \alpha = 1, \ldots, 9$ belonging to this space.

We call two simple diagrams equivalent if they describe the rational functions with equal numbers of double poles and identical sets of orders. For example, the diagrams depicted in figure 3.5 are equivalent since both describe the rational functions having double poles of orders 1, 2, 2 and 3.

Figure 3.5. An example of equivalent simple diagrams.

The MPS equations connected with two equivalent (simple) diagrams can be obtained from each other by means of a certain continuous deformation. Note that in an important particular case these deformations are reduced to the ordinary equivalence transformations described by formulas (3.3.21) and (3.3.22). This case appears when the function $b(\lambda)$ entering into the above-mentioned formulas takes the form of an inverse linear function. In this case these formulas can be rewritten as

$$\lambda = \frac{\alpha\tilde{\lambda} + \beta}{\gamma\tilde{\lambda} + \delta} \tag{3.5.43}$$

and

$$\tilde{x}^0(\tilde{\lambda}) = \frac{(\alpha\delta - \beta\gamma)^2}{(\gamma\tilde{\lambda} + \delta)^4} x^0\left(\frac{\alpha\tilde{\lambda} + \beta}{\gamma\tilde{\lambda} + \delta}\right), \tag{3.5.44a}$$

$$\tilde{x}^\alpha(\tilde{\lambda}) = \frac{(\alpha\delta - \beta\gamma)^2}{(\gamma\tilde{\lambda} + \delta)^4} x^\alpha\left(\frac{\alpha\tilde{\lambda} + \beta}{\gamma\tilde{\lambda} + \delta}\right), \quad \alpha = 1, \ldots, \mathcal{D}. \tag{3.5.44b}$$

It is not difficult to understand that formulas (3.5.43) and (3.5.44) describe the most general transformations conserving the rationality of functions $x^0(\lambda)$ and $x^\alpha(\lambda)$ and the orders of their double poles. Therefore, they conserve not only the form of the MPS equation

$$\left\{ -\frac{\partial^2}{\partial\tilde{\lambda}^2} + \tilde{x}^0(\tilde{\lambda}) \right\} \tilde{\varphi}(\tilde{\lambda}) = \left\{ \sum_{\alpha=1}^{D} \tilde{x}^\alpha(\tilde{\lambda})\varepsilon_\alpha \right\} \tilde{\varphi}(\tilde{\lambda}), \qquad (3.5.45)$$

but also the general form of its solutions

$$\tilde{\varphi}(\tilde{\lambda}) = \exp\left\{ \int \tilde{F}(\tilde{\lambda})\,d\tilde{\lambda} \right\} \prod_{i=1}^{M} (\tilde{\lambda} - \tilde{\xi}_i) \qquad (3.5.46)$$

complemented by spectral conditions

$$\sum_{k=1}^{M}{}' \frac{1}{\tilde{\xi}_i - \tilde{\xi}_k} + \tilde{F}(\tilde{\xi}_i) = 0, \quad i = 1, \ldots, M \qquad (3.5.47)$$

and

$$\tilde{F}'(\tilde{\lambda}) + \tilde{F}^2(\tilde{\lambda}) + 2 \sum_{i=1}^{M} \frac{\tilde{F}(\tilde{\lambda}) - \tilde{F}(\tilde{\xi}_i)}{\tilde{\lambda} - \tilde{\xi}_i} = \tilde{x}^0(\tilde{\lambda}) - \sum_{\alpha=1}^{D} \tilde{x}^\alpha(\tilde{\lambda})\tilde{\varepsilon}_\alpha. \qquad (3.5.48)$$

Here we have taken

$$\xi_i = \frac{\alpha\tilde{\xi}_i + \beta}{\gamma\tilde{\xi}_i + \delta} \qquad (3.5.49)$$

and also

$$\tilde{F}(\tilde{\lambda}) = \frac{(1-M)\gamma}{\gamma\tilde{\lambda} + \delta} + \frac{\alpha\delta - \beta\gamma}{(\gamma\tilde{\lambda} + \delta)^2} F\left(\frac{\alpha\tilde{\lambda} + \beta}{\gamma\tilde{\lambda} + \delta} \right). \qquad (3.5.50)$$

We see that transformation properties of functions $F(\lambda)$ depend on a concrete choice of values of the non-negative integer parameter M. This is a trivial consequence of the fact that both the parameter M and functions $F(\lambda)$ determine the form of solutions of equation (3.5.1), and this form, obviously, is not invariant under transformations (3.5.43)–(3.5.44).

In conclusion, note that linear-fractional transformations (3.5.43)–(3.5.44) are determined by three independent parameters and describe certain representation of the $SL(2)$ group. This makes us able to assert that the most general equivalence transformations for algebraically solvable differential MPS equations discussed in this section form the $SL(2)$ group, provided that the type of the corresponding simple diagram is fixed.

3.6   Reduction to exactly solvable models

In the preceding section we have constructed the class of one-dimensional second-order differential MPS equations (3.5.32) having infinite and discrete sets of algebraic solutions (3.5.33). However, we have not discussed the question of whether or not these equations are physically meaningful. In this section we will attempt to remedy this deficiency and list the cases when a physical interpretation of constructed MPS equations becomes possible.

First of all, let us rewrite formulas (3.5.32)–(3.5.35) in a more convenient form. Consider for example the simplest solution of equation (3.5.32), arising when $M = 0$:

$$\varphi_0(\lambda) = \exp\left\{\int F_0(\lambda)\,d\lambda\right\}. \qquad (3.6.1)$$

Here $F_0(\lambda)$ is a rational function, satisfying the system (3.5.34)–(3.5.35), which, due to the absence of the numbers ξ_i, is reduced to the single Riccati equation

$$F_0'(\lambda) + F_0^2(\lambda) = x^0(\lambda) - \sum_{\alpha=1}^{\mathcal{D}} x^\alpha(\lambda)\varepsilon_{0\alpha}. \qquad (3.6.2)$$

This equation can always be interpreted as a system of $2\mathcal{D} + 1$ algebraic equations for $\mathcal{D}$ unknown spectral parameters $\varepsilon_{0\alpha}$, $\alpha = 1, \ldots, \mathcal{D}$ and $\mathcal{D} + 1$ unknown parameters characterizing the function $F_0(\lambda)$.

Now note that the function $x^0(\lambda)$, entering into equation (3.6.2) and belonging by assumption to a $(2\mathcal{D} + 1)$-dimensional space of rational functions, is also characterized by $2\mathcal{D} + 1$ numerical parameters which are assumed to be fixed for a given MPS equation. The fact that the number of these parameters and the number of equations forming the system (3.6.2) coincide allows us to invert this system and interpret it as an equation for $x^0(\lambda)$. In this case, instead of fixing the function $x^0(\lambda)$ and solving (3.6.2) for $F_0(\lambda)$ and $\varepsilon_{0\alpha}$, we can fix the function $F_0(\lambda)$ and the numbers $\varepsilon_{0\alpha}$ and then recover from (3.6.2) the function $x^0(\lambda)$ for which $F_0(\lambda)$ and $\varepsilon_{0\alpha}$ are solutions. In other words, instead of $2\mathcal{D} + 1$ independent external parameters characterizing the function $x^0(\lambda)$, we can introduce $2\mathcal{D} + 1$ new independent parameters, characterizing the simplest solution of equation (3.5.32). Substituting $x^0(\lambda)$ obtained from (3.6.2) into (3.5.32), we come to the following form of this equation:

$$\left\{-\frac{\partial}{\partial\lambda^2} + F_0'(\lambda) + F_0^2(\lambda) - \sum_{\alpha=1}^{\mathcal{D}} x^\alpha(\lambda)(\varepsilon_\alpha - \varepsilon_{0\alpha})\right\}\varphi(\lambda) = 0, \qquad (3.6.3a)$$

or, equivalently,

$$\left\{ F_0(\lambda) + \frac{\partial}{\partial \lambda} \right\} \left\{ F_0(\lambda) - \frac{\partial}{\partial \lambda} \right\} \varphi(\lambda)$$

$$= \left\{ \sum_{\alpha=1}^{\mathcal{D}} x^\alpha(\lambda)(\varepsilon_\alpha - \varepsilon_{0\alpha}) \right\} \varphi(\lambda). \tag{3.6.3b}$$

According to (3.6.1), the simplest solution of (3.6.3) is

$$\varphi_0(\lambda) = \exp \left\{ \int F_0(\lambda) \, d\lambda \right\}, \tag{3.6.4a}$$

$$\varepsilon_\alpha = \varepsilon_{0\alpha}, \quad \alpha = 1, \ldots, \mathcal{D}, \tag{3.6.4b}$$

and other solutions must be sought in the form

$$\varphi(\lambda) = \prod_{i=1}^{M} (\lambda - \xi_i) \exp \left\{ \int F(\lambda) \, d\lambda \right\}, \tag{3.6.5a}$$

$$\varepsilon_\alpha = \varepsilon_{0\alpha} + \epsilon_\alpha, \quad \alpha = 1, \ldots, \mathcal{D}, \tag{3.6.5b}$$

where $M = 1, 2, 3, \ldots$. The numbers ϵ_α, $\alpha = 1, \ldots, \mathcal{D}$ and ξ_i, $i = 1, \ldots, M$ and functions $F(\lambda)$ entering into (3.6.5) can be found from the system

$$\sum_{k=1}^{M} \frac{1}{\xi_i - \xi_k} + F(\xi_i) = 0, \quad i = 1, \ldots, M, \tag{3.6.6}$$

$$F'(\lambda) + F^2(\lambda) + 2 \sum_{i=1}^{M} \frac{F(\lambda) - F(\xi_i)}{\lambda - \xi_i}$$

$$+ \sum_{\alpha=1}^{\mathcal{D}} x^\alpha(\lambda)\epsilon_\alpha = F_0'(\lambda) + F_0^2(\lambda), \tag{3.6.7}$$

if the functions $F_0(\lambda)$ and $x^\alpha(\lambda)$, $\alpha = 1, \ldots, \mathcal{D}$ are given.

Thus, we have reduced the algebraically solvable MPS equation (3.5.32) to the form (3.6.3) when at least one of its solutions, namely (3.6.4), is explicitly known. In fact, it plays the role of an external parameter in this equation and can be chosen arbitrarily.

We call the $\mathcal{D}$-parameter spectral equation (3.6.3) physically sensible if one can choose $\mathcal{D}$ different Hilbert spaces $W_1, \ldots, W_{\mathcal{D}}$ in such a way that all conditions of theorem 3.1 for both the equations (3.6.3) and its simplest solution (3.6.4) are satisfied.

Applying to the physically sensible MPS equation the inverse method of separation of variables described in sections 3.2 and 3.3, we obtain a class of completely integrable and stable quantum mechanical models having at least one bound state for which the spectral problem can be solved exactly.

To classify such models, it is sufficient to solve the classification problem for the physically sensible MPS equations of the type (3.6.3). In turn, the problem is reduced to constructing the systems of Hilbert spaces $W = 1, \ldots, W_D$ satisfying the conditions of theorem 3.1.

A very important condition, which must be satisfied, is the condition of the hermiticity of operators

$$X^0 \;=\; \left\{ F_0(\lambda) + \frac{\partial}{\partial \lambda} \right\} \left\{ F_0(\lambda) - \frac{\partial}{\partial \lambda} \right\} \tag{3.6.8}$$

and

$$X^\alpha \;=\; x^\alpha(\lambda), \quad \alpha = 1, \ldots, \mathcal{D} \tag{3.6.9}$$

in all spaces $W_1, \ldots, W_D$. To guarantee their hermiticity it is necessary to require that the functions $F_0(\lambda)$ and $x^\alpha(\lambda)$, $\alpha = 1, \ldots, \mathcal{D}$ are real. For this to happen, the points a_α, $\alpha = 1, \ldots, K$, in which they have poles, must lie on the real λ-axis or form complex conjugate pairs. The real points a_α divide the real axis into a set of intervals which we shall refer to as fundamental intervals and which can be finite, semi-finite or infinite. The number of fundamental intervals is $K' + 1$ where K' is the number of real points a_α.

Let us now assume that $K' + 1 \geq \mathcal{D}$ and consider the set of $\mathcal{D}$ arbitrarily chosen fundamental intervals $[\lambda_i^-, \lambda_i^+]$, $i = 1, \ldots, \mathcal{D}$, which we call the physical intervals.

We know that the functions $F_0(\lambda)$ and $x^\alpha(\lambda)$, $\alpha = 1, \ldots, \mathcal{D}$ are regular within the physical intervals, and singular at their ends $\lambda_i^\pm = a_1, \ldots, a_{K'}, \infty$. Their behaviour near the points $\lambda_i^\pm$ is seen from the following formulas

$$x^\alpha(\lambda) \;\sim\; \begin{cases} x^\alpha_{k,n_k}(\lambda - a_k)^{-2n_k}, & \lambda \to a_k, \quad k = 1, \ldots, K' \\[2mm] x^\alpha_n \lambda^{2n}, & \lambda \to \infty, \end{cases} \tag{3.6.10}$$

and

$$F_0(\lambda) \;\sim\; \begin{cases} F_{0k,n_k}(\lambda - a_k)^{-n_k}, & \lambda \to a_k, \quad k = 1, \ldots, K' \\[2mm] F_{0,n}\lambda^n, & \lambda \to \infty. \end{cases} \tag{3.6.11}$$

Now let us denote by

$$W_i \equiv W_i \left[\lambda_i^-, \lambda_i^+ \right], \quad i = 1, \ldots, \mathcal{D} \tag{3.6.12}$$

the spaces of functions being regular within the intervals $[\lambda_i^-, \lambda_i^+]$ and vanishing at their ends more rapidly than $(\lambda - a_k)^{n_k - \frac{1}{2}}$ if $\lambda_i^\pm = a_k$ and more rapidly than $\lambda^{-n - \frac{1}{2}}$ if $\lambda_i^\pm = \infty$. Introducing the scalar products

$$(\varphi_1, \varphi_2)_i = \int\limits_{\lambda_i^-}^{\lambda_i^+} \varphi_1(\lambda)\varphi_2(\lambda)\, \mathrm{d}\lambda, \quad i = 1, \ldots, \mathcal{D} \qquad (3.6.13)$$

in W_i, one can easily see that for any $\varphi_1, \varphi_2 \in W_i$ the integrals

$$(\varphi_1, X^0 \varphi_2)_i = \int\limits_{\lambda_i^-}^{\lambda_i^+} \varphi_1(\lambda) \left\{ F_0(\lambda) + \frac{\partial}{\partial \lambda} \right\} \left\{ F_0(\lambda) - \frac{\partial}{\partial \lambda} \right\} \varphi_2(\lambda)\, \mathrm{d}\lambda$$

$$(3.6.14)$$

and

$$(\varphi_1, X^\alpha \varphi_2)_i = \int\limits_{\lambda_i^-}^{\lambda_i^+} \varphi_1(\lambda) x^\alpha(\lambda) \varphi_2(\lambda)\, \mathrm{d}\lambda, \quad \alpha = 1, \ldots, \mathcal{D} \qquad (3.6.15)$$

exist and, therefore, the operators X^0 and X^α are hermitian in W_i. But this means that we can identify the spaces $W_1, \ldots, W_{\mathcal{D}}$ with the Hilbert spaces appearing in equation (3.3.5). In this case condition 2 of theorem 3.1 will be satisfied automatically.

The other conditions of this theorem can also be satisfied without difficulty. Indeed, the positive definiteness of the operator X^0 follows immediately from its representation (3.6.8). The linear independence of $\mathcal{D}$ functions $x^\alpha(\lambda)$ can always be guaranteed since the space $R_{2n}\binom{\vec{a}}{2\vec{n}}$, to which they belong, is $(2\mathcal{D} + 1)$ dimensional. Besides, we can always choose the functions $x^\alpha(\lambda)$ in such a way as to gurantee the invertibility of the matrix $\|x^\alpha(\lambda_i)\|_{i, \alpha = 1, \ldots, \mathcal{D}}$ with $\lambda_i \in [\lambda_i^-, \lambda_i^+]$ and ensure the existence of positive-definite rows (or their linear combinations) in the inverse matrix $\|x^\alpha(\lambda_i)\|_{i, \alpha = 1, \ldots, \mathcal{D}}^{-1}$.

To satisfy the last condition of theorem 3.1 we must elucidate when the function (3.6.4) satisfying equation (3.6.3) belongs simultaneously to all the spaces $W_1, \ldots, W_{\mathcal{D}}$. To this end it is sufficient to look at the behaviour of this function near the points $\lambda_i^\pm$.

Using formulas (3.6.11) and (3.6.4) we obtain

$$\varphi_0(\lambda) \sim \begin{cases} \exp\left\{ \frac{F_{0k, n_k}}{n_k - 1} (\lambda - a_k)^{1 - n_k} \right\}, & n_k > 1, \ \lambda \to \alpha_k, \\[2ex] (\lambda - a_k)^{F_{0k, n_k}}, & n_k = 1, \ \lambda \to \alpha_k, \end{cases} \qquad (3.6.16a)$$

for finite end points, and

$$\varphi_0(\lambda) \sim \begin{cases} \exp\left\{\frac{F_{0,n}}{n+1}\lambda^{n+1}\right\}, & n > -1,\ \lambda \to \infty, \\ \\ \lambda^{F_{0,n}}, & n = -1,\ \lambda \to \infty, \end{cases} \qquad (3.6.16b)$$

for infinite end points. From these asymptotic formulas it follows that for the function $\varphi_0(\lambda)$ to belong to the spaces W_i, $i = 1,\ldots,\mathcal{D}$, the numbers F_{0k,n_k} and $F_{0,n}$ determining its behaviour near the points $\lambda_i^{\pm}$, $i = 1,\ldots,\mathcal{D}$ must satisfy the following constraints:

$$\begin{cases} F_{0k,n_k} > 0, & \text{if } \lambda_i^- = a_k \text{ and } n_k > 1 \\ F_{0k,n_k} > \frac{1}{2}, & \text{if } \lambda_i^- = a_k \text{ and } n_k = 1 \end{cases} \qquad (3.6.17a)$$

$$\begin{cases} F_{0k,n_k} < 0, & \text{if } \lambda_i^+ = a_k \text{ and } n_k > 1 \text{ is even} \\ F_{0k,n_k} > 0, & \text{if } \lambda_i^+ = a_k \text{ and } n_k > 1 \text{ is odd} \\ F_{0k,n_k} > \frac{1}{2}, & \text{if } \lambda_i^+ = a_k \text{ and } n_k = 1 \end{cases} \qquad (3.6.17b)$$

$$\begin{cases} F_{0,n} > 0, & \text{if } \lambda_i^- = -\infty \text{ and } n > -1 \text{ is even} \\ F_{0,n} < 0, & \text{if } \lambda_i^- = -\infty \text{ and } n > -1 \text{ is odd} \\ F_{0,n} < \frac{1}{2}, & \text{if } \lambda_i^- = -\infty \text{ and } n = -1 \end{cases} \qquad (3.6.17c)$$

$$\begin{cases} F_{0,n} < 0, & \text{if } \lambda_i^+ = +\infty \text{ and } n > -1 \\ F_{0,n} < \frac{1}{2}, & \text{if } \lambda_i^+ = +\infty \text{ and } n = -1. \end{cases} \qquad (3.6.17d)$$

We call a system of physical intervals $[\lambda_i^-, \lambda_i^+]$, $i = 1,\ldots,\mathcal{D}$, for which all the conditions (3.6.17) can be simultaneously satisfied, an admissible system.

To list all such systems let us at first consider the case of two neighbouring intervals $[\lambda_i^-, \lambda_i^+]$ and $[\lambda_{i+1}^-, \lambda_{i+1}^+]$ having a common finite end point a_k. We see that a_k is the right end point ($a_k = \lambda_i^+$) for the left interval and, simultaneously, it is the left end point ($a_k = \lambda_{i+1}^-$) for the right interval. Using formulas (3.6.17a) and (3.6.17b) it is not difficult to understand that for even values of n_k these formulas contradict each other,

$$F_{0k,n_k} > 0, \qquad F_{0k,n_k} < 0, \qquad (3.6.18)$$

which leads us to the following exclusion principle.

(i) Two admissible physical intervals cannot have a common finite end point a_k if the corresponding number n_k is even.

Analogously, we can consider the case of two semi-infinite physical intervals (or a single infinite physical interval) having end points at plus or minus infinity. Using formulas (3.6.17c) and (3.6.17d), it is easy to see that for even values of N they also contradict each other:

$$F_{0,n} > 0, \qquad F_{0,n} < 0. \qquad (3.6.19)$$

This gives us two new exclusion principles:

(ii) The existence of two admissible semi-infinite physical intervals is impossible when the number n is even.

(iii) The existence of a single admissible infinite interval is impossible if the number n is even.

Taking into account these exclusion principles, we can list all admissible systems of physical intervals. Quite obviously, any such system will describe a certain class of algebraically solvable MPS equations (3.6.3) and their simplest solutions (3.6.4) satisfying all conditions of theorem 3.1.

The above description makes us able to solve the classification problem for physically sensible MPS equations. For this purpose it is convenient to use the graphical method described in the preceeding section. We consider here an extended version of this method. Namely, starting with "simple diagrams" determining the "double-pole structure" of functions $x^\alpha(\lambda)$, $\alpha = 1, \ldots, \mathcal{D}$ we supplement them by admissible systems of physical intervals, which we depict by thick lines between the real double poles. We call such extended diagrams physical diagrams.

We define two physical diagrams as equivalent if they can be obtained from each other by means of reflections or arbitrary continuous deformations conserving the relative desposition of real double poles and physical intervals. Note that, according to this definition, all the diagrams connected by the linear-fractional transfomations of the (real) $SL(2)$ group are equivalent. Note also that the $SL(2)$ transformations do not change conditions (3.6.17).

Below we write down all admissible non-equivalent physical diagrams obtained by taking into account the exclusion principles 1, 2 and 3. We start with the simplest particular cases when the number of spectral parameters is 1 or 2.

$\mathcal{D} = 1$. In this case the function $x^\alpha(\lambda)$, $\alpha = 1$ has double poles of total order three. The number of stable intervals is unity. This gives us four non-equivalent diagrams depicted in figure 3.6.

$\mathcal{D} = 2$. In this case the function $x^\alpha(\lambda)$, $\alpha = 1, 2$ must have double poles of total order four. The number of physical intervals is two. Using the exclusion principles, we come again to four non-equivalent diagrams depicted in figure 3.7.

$\mathcal{D} \geq 3$. It is not difficult to see that in the general case the function $x^\alpha(\lambda)$, $\alpha = 1, \ldots, \mathcal{D}$ will have double poles of total order $\mathcal{D} + 2$. The number of physical intervals becomes $\mathcal{D}$. Therefore, we shall have again only four

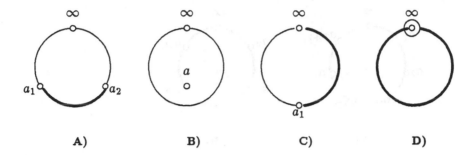

Figure 3.6. Non-equivalent physical diagrams for the one-dimensional case.

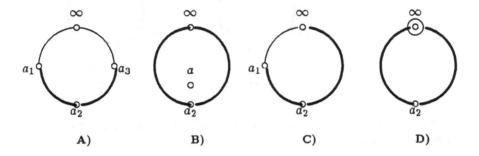

Figure 3.7. Non-equivalent physical diagrams for the two-dimensional case.

admissible types of non-equivalent physical diagram for any given $\mathcal{D}$ (see figure 3.8).

The diagrams depicted in figure 3.8 describe the following systems of physical intervals:

$$
\begin{array}{llll}
\text{A)} & \lambda_i \in [a_i, a_{i+1}], & i = 1, \ldots, \mathcal{D}; & \text{(3.6.20a)} \\
\text{B)} & \lambda_1 \in [-\infty, a_2]; & & \\
& \lambda_i \in [a_i, a_{i+1}], & i = 2, \ldots, \mathcal{D}; & \text{(3.6.20b)} \\
\text{C)} & \lambda_i \in [a_i, a_{i+1}], & i = 1, \ldots, \mathcal{D} - 1; & \\
& \lambda_{\mathcal{D}} \in [a_{\mathcal{D}}, \infty]; & & \text{(3.6.20c)} \\
\text{D)} & \lambda_1 \in [-\infty, a_2]; & & \\
& \lambda_i \in [a_i, a_{i+1}], & i = 2, \ldots, \mathcal{D} - 1; & \\
& \lambda_{\mathcal{D}} \in [a_{\mathcal{D}}, \infty]. & & \text{(3.6.20d)}
\end{array}
$$

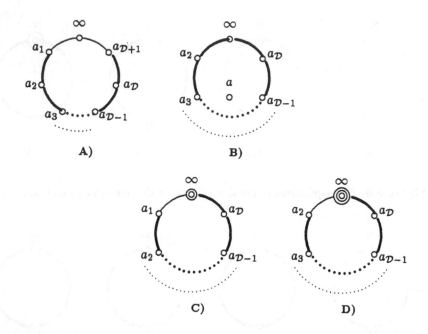

Figure 3.8.   Non-equivalent physical diagrams for the general (multi-dimensional) case.

The functions $x^\alpha(\lambda)$, $\alpha = 1, \ldots, \mathcal{D}$ corresponding to these diagrams have the form

$$\text{A)} \quad x^\alpha(\lambda) = \frac{Q_{2\mathcal{D}}^\alpha(\lambda)}{\prod_{k=1}^{\mathcal{D}+1}(\lambda - a_k)^2}, \tag{3.6.21a}$$

$$\text{B)} \quad x^\alpha(\lambda) = \frac{Q_{2\mathcal{D}}^\alpha(\lambda)}{\prod_{k=2}^{\mathcal{D}}(\lambda - a_k)^2 \cdot (\lambda - a)^2 (\lambda - a^*)^2}, \tag{3.6.21b}$$

$$\text{C)} \quad x^\alpha(\lambda) = \frac{Q_{2\mathcal{D}}^\alpha(\lambda)}{\prod_{k=1}^{\mathcal{D}}(\lambda - a_k)^2}, \tag{3.6.21c}$$

$$\text{D)} \quad x^\alpha(\lambda) = \frac{Q_{2\mathcal{D}}^\alpha(\lambda)}{\prod_{k=2}^{\mathcal{D}}(\lambda - a_k)^2}, \tag{3.6.21d}$$

where $Q_{2\mathcal{D}}^\alpha(\lambda)$, $\alpha = 1, \ldots, \mathcal{D}$ are arbitrary linearly independent polynomials of order $2\mathcal{D}$ chosen such that they guarantee the invertibility of the matrix $\|Q_{2\mathcal{D}}^\alpha(\lambda_i)\|_{i,\alpha=1,\ldots,\mathcal{D}}$ and the existence of positive-definite rows (or their linear combinations) in the inverse matrix $\|Q_{2\mathcal{D}}^\alpha(\lambda_i)\|_{i,\alpha=1,\ldots,\mathcal{D}}^{-1}$ for any λ_i belonging to the physical intervals (3.6.20). The corresponding functions

$F_0(\lambda)$ are described by the formulas

$$\text{A)} \quad F_0(\lambda) \;=\; \sum_{k=1}^{\mathcal{D}+1} \frac{F_{0k}}{\lambda - a_k}, \tag{3.6.22a}$$

$$\text{B)} \quad F_0(\lambda) \;=\; \sum_{k=2}^{\mathcal{D}} \frac{F_{0k}}{\lambda - a_k} + \frac{F_0}{\lambda - a} + \frac{F_0^*}{\lambda - a^*}, \tag{3.6.22b}$$

$$\text{C)} \quad F_0(\lambda) \;=\; \sum_{k=1}^{\mathcal{D}} \frac{F_{0k}}{\lambda - a_k} + F_{0,\mathcal{D}+1}, \tag{3.6.22c}$$

$$\text{D)} \quad F_0(\lambda) \;=\; \sum_{k=2}^{\mathcal{D}} \frac{F_{0k}}{\lambda - a_k} + F_{01} + \lambda F_{0,\mathcal{D}+1}, \tag{3.6.22d}$$

which must be supplemented by the conditions

$$\text{A)} \quad F_{0k} > \frac{1}{2}, \quad \alpha = 1,\ldots,\mathcal{D}+1, \tag{3.6.23a}$$

$$\text{B)} \quad F_{0k} > \frac{1}{2}, \quad \alpha = 2,\ldots,\mathcal{D}; \quad F_0 < \frac{1}{2}, \tag{3.6.23b}$$

$$\text{C)} \quad F_{0k} > \frac{1}{2}, \quad \alpha = 1,\ldots,\mathcal{D}; \quad F_{0,\mathcal{D}+1} < 0, \tag{3.6.23c}$$

$$\text{D)} \quad F_{0k} > \frac{1}{2}, \quad \alpha = 2,\ldots,\mathcal{D}; \quad F_{0,\mathcal{D}+1} < 0, \tag{3.6.23d}$$

in which

$$F_0 \equiv \sum_{k=2}^{\mathcal{D}} F_{0k} + F_0 + F_0^*. \tag{3.6.24}$$

The diagrams given in figure 3.8 and formulas (3.6.20)–(3.6.23) give us the final solution of the classification problem for physically sensible MPS equations of the type (3.6.3). Remember that all these equations can be considered as starting points in constructing completely integrable and stable quantum mechanical models having at least one explicit solution of the spectral problem. Thus, we have given a classification of such models.

Now let us discuss the spectral properties of the resulting completely integrable models in more detail. First of all note that the explicit solution in question, whose normalizability is guaranteed by conditions (3.6.23), is simply the ground state solution. This follows immediately from the fact that the function (3.6.1) from which it is constructed has no zeros within the physical intervals forming the domain in $\mathcal{D}$-dimensional space (a parallelogram) in which the resulting Schrödinger problem is formulated.

In order to elucidate whether the models under consideration have other explicit and normalizable solutions, it is necessary to return to the initial MPS equations and study their spectral properties in the space $W = W_1 \cap \ldots \cap W_D$.

We denote by S_W the set of all solutions of equation (3.6.3) in W, and by S_Φ the set of all its exact (algebraic) solutions described by formulas (3.6.4)–(3.6.7). Consider three possible cases:

1. The set S_W has no intersection with S_Φ (see figure 3.9).

2. The set S_W has partial intersection with S_Φ (see figure 3.10).

3. The set S_W belongs to the set S_Φ (see figure 3.11).

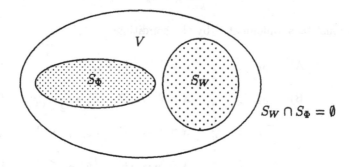

$$S_W \cap S_\Phi = \emptyset$$

Figure 3.9. Exactly non-solvable equations.

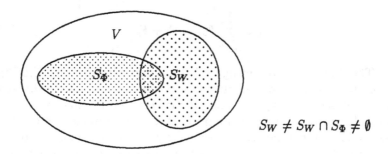

$$S_W \neq S_W \cap S_\Phi \neq \emptyset$$

Figure 3.10. Quasi-exactly solvable equations.

In the first case the physical solutions of equation (3.6.3) are unknown. We shall call such MPS equations exactly non-solvable (in W). In the second case we know a certain number of physical solutions, but not all solutions. We call such MPS equations quasi-exactly solvable (in W). In the third

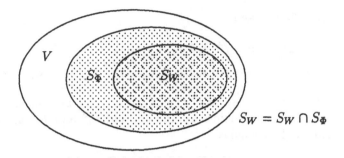

Figure 3.11. Exactly solvable equations.

case all physical solutions are known. We call such MPS equations exactly solvable (in W).

We can omit consideration of the first case, since we know that equation (3.6.3) has at least one solution in W, so that the set S_W has non-empty intersection with S_Φ. Thus, our aim is to ascertain which of the remaining two cases is realized.

To answer this question, let us try to construct the most general form of functions belonging to the space W and satisfying equation (3.6.3). This can be done as follows.

First of all, note that (3.6.3) is a second-order linear differential equation and, therefore, the most general form of its solutions is

$$
\begin{aligned}
\varphi(\lambda) \quad \sim \quad & (\cos\theta)\varphi_1(\lambda;\varepsilon_1,\dots,\varepsilon_\mathcal{D}) \\
+ \quad & (\sin\theta)\varphi_2(\lambda;\varepsilon_1,\dots,\varepsilon_\mathcal{D}),
\end{aligned}
\tag{3.6.25}
$$

where φ_1 and φ_2 are two linearly independent partial solutions and θ is an arbitrary parameter.

Let us now consider a certain concrete physically sensible MPS equation, assuming for definiteness that it is specified by the diagram A (see figure 3.8).

Equation (3.6.3) has in this case $\mathcal{D} + 2$ singularities located at the points $a_1,\dots,a_{\mathcal{D}+1}$ and ∞. Therefore, the solution (3.6.25) must also be singular at the same points. The character of these singularities can easily be derived from equation (3.6.3). Using formulas (3.6.21) and (3.6.22) we obtain

$$
\begin{aligned}
\varphi(\lambda) \quad = \quad & f_k^+(\lambda;\theta;\varepsilon_1,\dots,\varepsilon_\mathcal{D})(\lambda - a_k)^{F_k^+} \\
+ \quad & f_k^-(\lambda;\theta;\varepsilon_1,\dots,\varepsilon_\mathcal{D})(\lambda - a_k)^{F_k^-}, \quad \lambda \to \alpha_k,
\end{aligned}
\tag{3.6.26}
$$

for $k = 1, \ldots, \mathcal{D} + 1$, and

$$
\begin{aligned}
\varphi(\lambda) &= f^+(\lambda; \theta; \varepsilon_1, \ldots, \varepsilon_{\mathcal{D}}) \lambda^{F^+} \\
&+ f^-(\lambda; \theta; \varepsilon_1, \ldots, \varepsilon_{\mathcal{D}}) \lambda^{F^-}, \quad \lambda \to \infty, \quad (3.6.27)
\end{aligned}
$$

where $f_k^{\pm}(\lambda; \theta, \varepsilon_1, \ldots, \varepsilon_{\mathcal{D}})$ and $f^{\pm}(\lambda; \theta, \varepsilon_1, \ldots, \varepsilon_{\mathcal{D}})$ are certain functions that are regular at the points a_k and ∞, and $F_k^{\pm}$ and $F^{\pm}$ are constants determined by the formulas

$$
F_k^{\pm} = \frac{1}{2} \pm \sqrt{(F_{0k} - \tfrac{1}{2})^2 - \sum_{\alpha=1}^{\mathcal{D}} x_k^{\alpha}(\varepsilon_{\alpha} - \varepsilon_{0\alpha})} \qquad (3.6.28)
$$

and

$$
F^{\pm} = \frac{1}{2} \pm \sqrt{(\sum_{k=1}^{\mathcal{D}+1} F_{0k} - \tfrac{1}{2})^2 - \sum_{\alpha=1}^{\mathcal{D}} x^{\alpha}(\varepsilon_{\alpha} - \varepsilon_{0\alpha})}. \qquad (3.6.29)
$$

Here we have denoted by x_k^{α} and x^{α} the "double-pole residues" of functions $x^{\alpha}(\lambda)$ at the points a_k and ∞:

$$
x_k^{\alpha} = \lim_{\lambda \to a_k} (\lambda - a_k)^2 x^{\alpha}(\lambda), \quad x^{\alpha} = \lim_{\lambda \to \infty} \lambda^2 x^{\alpha}(\lambda). \qquad (3.6.30)
$$

From expressions (3.6.26) and (3.6.28), and conditions (3.6.23), it follows that for the functions $\varphi(\lambda)$ to be elements of the space W, all the terms in (3.6.26) proportional to the leading singularities $(\lambda - a_K)^{F_k^-}$ must vanish. This gives us a system of $\mathcal{D} + 1$ numerical equations for $\mathcal{D} + 1$ unknown quantities θ and $\varepsilon_1, \ldots, \varepsilon_{\mathcal{D}}$:

$$
F_k^-(a_k; \theta, \varepsilon_1, \ldots, \varepsilon_{\mathcal{D}}) = 0, \quad k = 1, \ldots, \mathcal{D} + 1. \qquad (3.6.31)
$$

Solving this system and substituting the obtained values of θ and $\varepsilon_1, \ldots, \varepsilon_{\mathcal{D}}$ into (3.6.25), we obtain a set of functions $\varphi(\lambda)$ satisfying the equation (3.6.3) and belonging to the space W by construction. Due to the absence of the leading singularities $(\lambda - a_k)^{F_k^-}$ in $\varphi(\lambda)$, the most general form of these solutions is

$$
\varphi(\lambda) = \prod_{k=1}^{\mathcal{D}+1} (\lambda - a_k)^{F_k^+} f(\lambda), \qquad (3.6.32)
$$

where $f(\lambda)$ is a certain function behaving at infinity as a power function and being regular at all other points. But this means that $f(\lambda)$ is a polynomial.

Representing $f(\lambda)$ in the form

$$f(\lambda) = \prod_{i=1}^{M}(\lambda - \xi_i) \qquad (3.6.33)$$

and using the identity

$$\prod_{k=1}^{\mathcal{D}+1}(\lambda - a_k)^{F_k^+} = \exp\left\{\int\left[\sum_{k=1}^{\mathcal{D}+1}\frac{F_k^+}{\lambda - a_k}\right]\,d\lambda\right\} \qquad (3.6.34)$$

we can rewrite (3.6.32) as

$$\varphi(\lambda) = \exp\left\{\int F(\lambda)\,d\lambda\right\}\prod_{i=1}^{M}(\lambda - \xi_i) \qquad (3.6.35)$$

where $F(\lambda)$ is a certain rational function having simple poles at the points $a_1,\ldots,a_{\mathcal{D}+1}$ and ∞. But this is simply the correct *ansatz* (3.6.5) for the equations (3.6.3) described by the diagram A! Other MPS equations described by the diagrams B, C and D can be considered analogously.

We have obtained a remarkable result: functions $\varphi(\lambda)$ belonging to the space W and satisfying the physically sensible MPS equation have the form (3.6.35) and therefore, can be found algebraically. But this means that the set S_W defined above belongs to the set S_Φ and thus, according to our definition, we deal with exactly solvable MPS equations. Hence, we can formulate the following important theorem.

Theorem 3.4. *All physically sensible MPS equations (3.6.3) described by diagrams listed in figure 3.8 and formulas (3.6.20)–(3.6.23) are exactly solvable in the corresponding spaces $W = W_1 \cap \ldots \cap W_{\mathcal{D}}$. Therefore, all completely integrable and stable quantum mechanical models obtained from (3.6.3), by means of the inverse method of separation of variables, are also exactly solvable in the standard sense of this word.*

This is the main result of this chapter, completing the procedure of constructing and classifying exactly solvable models of quantum mechanics obtained by means of the inverse procedure of separation of variables.

In conclusion of this section, we note that the fact that the models constructed in such a way are exactly solvable does not necessarily mean that the number of their exact solutions is infinite. In fact, the term "exact solvability" only means that all bound states in the model (i.e. states described by the normalizable wavefunctions) can be found exactly by means of an algebraic procedure. However, we know that the number of

such states may be finite if the potential describing the model has the form
of a well of finite depth. In terms of the initial MPS equation (3.6.3) this
means that the number of its physically sensible solutions (3.6.5) is finite.

Note that conditions for solutions (3.6.5) to be physically sensible (to
belong to the space W) can be expressed in terms of functions $F(\lambda)$. The
latter evidently have the same functional structure as the functions $F^{(0)}(\lambda)$
and thus can be sought in the form

$$\text{A)} \quad F(\lambda) \;=\; \sum_{k=1}^{\mathcal{D}+1} \frac{F_k}{\lambda - a_k}, \tag{3.6.36a}$$

$$\text{B)} \quad F(\lambda) \;=\; \sum_{k=2}^{\mathcal{D}} \frac{F_k}{\lambda - a_k} + \frac{F}{\lambda - a} + \frac{F^*}{\lambda - a^*}, \tag{3.6.36b}$$

$$\text{C)} \quad F(\lambda) \;=\; \sum_{k=1}^{\mathcal{D}} \frac{F_k}{\lambda - a_k} + F_{\mathcal{D}+1}, \tag{3.6.36c}$$

$$\text{D)} \quad F(\lambda) \;=\; \sum_{k=2}^{\mathcal{D}} \frac{F_k}{\lambda - a_k} + F_1 + \lambda F_{\mathcal{D}+1}. \tag{3.6.36d}$$

Substituting these formulas into (3.6.5) and recalling the definition of the
spaces $W_1, \ldots, W_{\mathcal{D}}$, we find the needed conditions

$$\text{A)} \qquad F_k > \frac{1}{2}, \quad \alpha = 1, \ldots, \mathcal{D}+1, \tag{3.6.37a}$$

$$\text{B)} \qquad F_k > \frac{1}{2}, \quad \alpha = 2, \ldots, \mathcal{D}; \quad F < \frac{1}{2}, \tag{3.6.37b}$$

$$\text{C)} \qquad F_k > \frac{1}{2}, \quad \alpha = 1, \ldots, \mathcal{D}; \quad F_{\mathcal{D}+1} < 0, \tag{3.6.37c}$$

$$\text{D)} \qquad F_k > \frac{1}{2}, \quad \alpha = 2, \ldots, \mathcal{D}; \quad F_{\mathcal{D}+1} < 0, \tag{3.6.37d}$$

with

$$F \equiv \sum_{k=2}^{\mathcal{D}} F_k + F + F^* + M, \tag{3.6.38}$$

which allow us to formulate the following theorem.

Theorem 3.5. *The total number of bound states in exactly solvable models
connected with the physically sensible MPS equation (3.6.3) is equal to
the total number of solutions of the system (3.6.6)–(3.6.7) satisfying the
conditions (3.6.37).*

Unfortunately, in the general case we do not know any more explicit criterion which would allow us to compute this number before solving the spectral equations (3.6.6)–(3.6.7). However, there exists one special case when such a criterion can be given.

To show how this case arises, let us consider in detail the system (3.6.6)–(3.6.7).

First of all, let us look at equation (3.6.7). Remember that the functions $x^\alpha(\lambda)$, $\alpha = 1, \ldots, \mathcal{D}$ entering are elements of a $(2\mathcal{D} + 1)$-dimensional space of rational functions $R_{2n}\left(\frac{\vec{a}}{2\vec{n}}\right)$. Therefore, equation (3.6.7) is equivalent to a system of $2\mathcal{D} + 1$ algebraic equations. In general, this system is rather complicated and its analysis is far from being trivial. However, the situation changes if the functions $x^\alpha(\lambda)$, $\alpha = 1, \ldots, \mathcal{D}$ belong to a $\mathcal{D}$-dimensional subspace $R_{n-1}\left(\frac{\vec{a}}{\vec{n}}\right)$ of the space $R_{2n}\left(\frac{\vec{a}}{2\vec{n}}\right)$.

Indeed, in this case the last (fourth) term in the left-hand side of (3.6.7) is an element of the space $R_{n-1}\left(\frac{\vec{a}}{\vec{n}}\right)$. This enables us to rewrite equation (3.6.7) in the form

$$F'(\lambda) + F'^2(\lambda) - F_0'(\lambda) - F_0^2(\lambda) \in R_{n-1}\left(\frac{\vec{a}}{\vec{n}}\right). \qquad (3.6.39)$$

It is absolutely obvious that this condition can be satisfied only if the functions $F(\lambda)$ and $F_0(\lambda)$ coincide:

$$F(\lambda) = F_0(\lambda). \qquad (3.6.40)$$

This means that we have partially solved system (3.6.6)–(3.6.7), since a solution for $F(\lambda)$ is already known.

Substitution of (3.6.40) into the remaining equations of this system gives

$$\sum_{k=1}^{M}{}' \frac{1}{\xi_i - \xi_k} + F_0(\xi_i) = 0, \quad i = 1, \ldots, M \qquad (3.6.41)$$

and

$$2 \sum_{i=1}^{M} \frac{F_0(\lambda) - F_0(\xi_i)}{\lambda - \xi_i} = \sum_{\alpha=1}^{\mathcal{D}} x^\alpha(\lambda)\epsilon_\alpha. \qquad (3.6.42)$$

The second equation (3.6.42) for ϵ_α, $\alpha = 1, \ldots, \mathcal{D}$ can easily be solved if we fix $\mathcal{D}$ arbitrary numbers λ_β, $\beta = 1, \ldots, \mathcal{D}$ and substitute them into (3.6.42) in place of λ. Then, we obtain a system of $\mathcal{D}$ algebraic equations

$$\sum_{\alpha=1}^{\mathcal{D}} x^\alpha(\lambda_\beta)\epsilon_\alpha = 2 \sum_{i=1}^{M} \frac{F_0(\lambda_\beta) - F_0(\xi_i)}{\lambda_\beta - \xi_i}, \quad \beta = 1, \ldots, \mathcal{D}, \qquad (3.6.43)$$

for $\mathcal{D}$ unknown quantities ϵ_α, which has the following simple solution:

$$\epsilon_\alpha = 2 \sum_{\beta=1}^{\mathcal{D}} \|x^\alpha(\lambda_\beta)\|^{-1} \sum_{i=1}^{M} \frac{F_0(\lambda_\beta) - F_0(\xi_i)}{\lambda_\beta - \xi_i}, \quad \alpha = 1, \ldots, \mathcal{D}. \tag{3.6.44}$$

Thus, we see that, in order to construct the solution of a $\mathcal{D}$-parameter spectral equation (3.6.3) when the functions $x^\alpha(\lambda)$, $\alpha = 1, \ldots, \mathcal{D}$ forming this equation belong to the space $R_{n-1}\left(\frac{\partial}{2\pi}\right)$, it is sufficient to solve only one non-trivial system of algebraic equations (3.7.3) for the numbers $\xi_1, \ldots, \xi_M$, and then, substituting the obtained values of $\xi_1, \ldots, \xi_M$ into the explicit expressions (3.6.44), find the corresponding values of spectral parameters $\varepsilon_1, \ldots, \varepsilon_{\mathcal{D}}$. According to statement 3.11, equation (3.6.3) has in this case $\frac{(M+\mathcal{D}-1)!}{M!(\mathcal{D}-1)!}$ solutions of the form (3.6.5) for any given M.

In order to find the total number of solutions of equation (3.6.3) satisfying conditions (3.6.37), we must replace the coefficients of the functions $F(\lambda)$ entering into (3.6.37) by the coefficients of the function $F_0(\lambda)$. Then conditions (3.6.37a), (3.6.37c) and (3.6.37d) take the form of the analogous conditions (3.6.23a), (3.6.23c) and (3.6.23d), which are assumed to be satisfied. However, this means that equation (3.6.3) has an infinite number of physically sensible solutions and, therefore, the exactly solvable and completely integrable quantum models associated with it have also infinite and discrete spectra. One can say that they describe the quantum motion in potential wells of an infinite depth. Another situation arises if we make such a replacement in condition (3.6.37b). It is not difficult to verify that after this it takes the form

$$F_{0k} > \frac{1}{2}, \quad k = 1, \ldots, \mathcal{D}; \quad \sum_{k=2}^{\mathcal{D}} F_{0k} + F_0 + F_0^* + M < \frac{1}{2}. \tag{3.6.45}$$

We see that (3.6.45) does not coincide with the analogous condition (3.6.23b): the number M enters explicitly into (3.6.45), and it is quite obvious that for sufficiently large M the condition (3.6.45) will be violated. This means that in this case equation (3.6.3) has a finite number of physically sensible solutions only, and thus, the quantum mechanical models associated with it have finite spectra. This is a typical situation for wells of finite depth. The number of bound states in these wells can easily be obtained from (3.6.45) and the results of statement 3.11. It is

$$N_{\text{tot}} = \sum_{M=0}^{M_{\text{max}}} \frac{(M + \mathcal{D} - 1)!}{M!(\mathcal{D} - 1)!} \tag{3.6.46}$$

where

$$M_{\mathrm{max}} = \text{integer part of } \left(\tfrac{1}{2} - \sum_{k=2}^{\mathcal{D}} F_{0k} - F_0 - F_0^* \right). \qquad (3.6.47)$$

Concrete examples of exactly solvable models with finite and infinite spectra will be considered in detail in sections 3.7 and 3.8.

3.7   The one-dimensional case. Classification

In the preceding section we have described the classes of $\mathcal{D}$-parameter algebraically solvable spectral equations which, after applying the inverse method of separation of variables, can be reduced to $\mathcal{D}$-dimensional exactly solvable models of quantum mechanics.

The simplest case of this reduction procedure is realized when $\mathcal{D} = 1$. Then the initial MPS equations take the form of ordinary one-parameter spectral equations and the resulting exactly solvable models become one dimensional.

Below we consider this case in detail and show that in spite of its comparative simplicity it is rather interesting and leads to wide classes of one-particle problems with exactly calculable spectra.

We start with equation (3.6.3) which, for $\mathcal{D} = 1$, takes an especially simple form:

$$\left\{ -\frac{\partial^2}{\partial \lambda^2} + F_0'(\lambda) + F_0^2(\lambda) \right\} \varphi(\lambda) = (E - E_0)\rho(\lambda)\varphi(\lambda). \qquad (3.7.1)$$

Here $\rho(\lambda)$ and $F_0(\lambda)$ are given rational functions belonging to the spaces $R_{2n}\left(\frac{d}{2\pi}\right)$ and $R_n\left(\frac{d}{\pi}\right)$ of dimensions three and two, respectively, and E_0 is a given number. According to formulas (3.6.4)–(3.6.7), solutions of this equation can be represented as

$$\varphi(\lambda) \;=\; \exp\left\{ \int F(\lambda)\, \mathrm{d}\lambda \right\} \prod_{i=1}^{M} (\lambda - \xi_i), \qquad (3.7.2a)$$

$$E \;=\; E_0 + \epsilon, \qquad (3.7.2b)$$

where the numbers ξ_i, $i = 1, \ldots, M$ and ϵ, as well as the functions $F(\lambda)$ belonging to the space $R_n\left(\frac{d}{\pi}\right)$, can be found from the following system of equations:

$$\sum_{k=1}^{M}{}' \frac{1}{\xi_i - \xi_k} + F_0(\xi_i) = 0, \quad i = 1, \ldots, M, \qquad (3.7.3)$$

$$F'(\lambda) + F^2(\lambda) + 2 \sum_{i=1}^{M} \frac{F(\lambda) - F(\xi)}{\lambda - \xi_i} + \epsilon\rho(\lambda) = F'_0(\lambda) + F_0^2(\lambda). \qquad (3.7.4)$$

We define the Hilbert space W, in which the solutions of equation (3.7.1) must be sought, as the space of functions regular in a certain physical interval (i.e. the interval between the neighbouring singular points of function $\rho(\lambda)$), and vanishing at its ends in such a way that the integrability of functions $\varphi^2(\lambda)$ with the weight $\rho(\lambda)$ is guaranteed.

The function $\rho(\lambda)$ plays the role of the weight function for the spectral equation (3.7.1). Therefore, it must be positive in the physical interval in which the spectral problem is formulated. As to function $F_0(\lambda)$, also entering into (3.7.1), it must satisfy the appropriate conditions of the type (3.6.23). Then reduction of (3.7.1) to Schrödinger form becomes possible.

Introducing a new variable x and a new function $\psi(x)$ by the formulas

$$x = x(\lambda), \qquad (3.7.5a)$$
$$\psi(x) = g(\lambda)\varphi(\lambda), \qquad (3.7.5b)$$

and substituting them into (3.7.1) we obtain

$$-[x'(\lambda)]^2 \frac{\partial^2 \psi(x)}{\partial x^2} - \left[x''(\lambda) - 2\frac{g'(\lambda)}{g(\lambda)} x'(\lambda) \right] \frac{\partial \psi(x)}{\partial x}$$

$$+ \left\{ \left[\frac{g'(\lambda)}{g(\lambda)} \right]' - \left[\frac{g'(\lambda)}{g(\lambda)} \right]^2 + F'_0(\lambda) + F_0^2(\lambda) + \rho(\lambda)E_0 \right\} \psi(x)$$

$$= \rho(\lambda)E\psi(x). \qquad (3.7.6)$$

Requiring that

$$x''(\lambda) - 2\frac{g'(\lambda)}{g(\lambda)} x'(\lambda) = 0 \qquad (3.7.7a)$$

and

$$[x'(\lambda)]^2 = \rho(\lambda) \qquad (3.7.7b)$$

and solving the system (3.7.1) we obtain instead of (3.7.6) the Schrödinger equation

$$\left\{ -\frac{\partial^2}{\partial x^2} + V(x) \right\} \psi(x) = E\psi(x) \qquad (3.7.8)$$

with potential

$$V(x) = E_0 + \frac{1}{\rho(\lambda)} \left\{ \frac{1}{4} \left[\frac{\rho'(\lambda)}{\rho(\lambda)} \right]' - \frac{1}{16} \left[\frac{\rho'(\lambda)}{\rho(\lambda)} \right]^2 \right.$$

$$+F_0'(\lambda) + F_0^2(\lambda) \bigg\} \bigg|_{\lambda=\lambda(x)} \qquad (3.7.9)$$

and solutions

$$\psi(x) = \sqrt[4]{\rho(\lambda)} \exp\left\{ \int F(\lambda)\, d\lambda \right\} \prod_{i=1}^{M} (\lambda - \xi_i) \bigg|_{\lambda=\lambda(x)}, \qquad (3.7.10a)$$

$$E = E_0 + \epsilon, \qquad (3.7.10b)$$

in which $\lambda = \lambda(x)$ is a function determined from the equation

$$x = \int_{\lambda_0}^{\lambda(x)} \sqrt{\rho(\lambda)}\, d\lambda. \qquad (3.7.11)$$

Here λ_0 is a number belonging to a chosen physical interval.

The concrete form of the Schrödinger equation (3.7.8) depends on the concrete choice of the functions $\rho(\lambda)$ and $F_0(\lambda)$ and the physical interval. According to the results of the preceding section, there are four non-equivalent possibilities for such a choice:

$$\text{A)} \quad \rho(\lambda) = \frac{\rho_0(\lambda - \rho_1)(\lambda - \rho_2)}{(\lambda - a_1)^2(\lambda - a_2)^2},$$

$$F_0(\lambda) = \frac{F_{01}}{\lambda - a_1} + \frac{F_{02}}{\lambda - a_2}, \qquad \lambda \in [a_1, a_2]; \qquad (3.7.12a)$$

$$\text{B)} \quad \rho(\lambda) = \frac{\rho_0(\lambda - \rho_1)(\lambda - \rho_2)}{(\lambda - a)^2(\lambda - a^*)^2},$$

$$F_0(\lambda) = \frac{F_0}{\lambda - a} + \frac{F_0^*}{\lambda - a^*}, \qquad \lambda \in [-\infty, +\infty]; \qquad (3.7.12b)$$

$$\text{C)} \quad \rho(\lambda) = \frac{\rho_0(\lambda - \rho_1)(\lambda - \rho_2)}{(\lambda - a)^2},$$

$$F_0(\lambda) = \frac{F_{01}}{\lambda - a} + F_{02}, \qquad \lambda \in [a, \infty]; \qquad (3.7.12c)$$

$$\text{D)} \quad \rho(\lambda) = \rho_0(\lambda - \rho_1)(\lambda - \rho_2),$$

$$F_0(\lambda) = F_{01} + \lambda F_{02}, \qquad \lambda \in [-\infty, +\infty]. \qquad (3.7.12d)$$

All these possibilities can be described by diagrams listed in figure 3.6. In order to make these diagrams more informative, we can depict on them the points ρ_1 and ρ_2 which, obviously, must lie outside the chosen physical interval (due to the condition of positivity of function $\rho(\lambda)$ in it). Depicting these points by small crosses × or by double crosses # (in case of their coincidence) we can list all possible cases of their positions. Of course, we must take into account the fact that these points can merge with other singular points of the function $\rho(\lambda)$ or move out to infinity. Then we arrive at four series of non-equivalent extended physical diagrams depicted in figures 3.12, 3.13, 3.14 and 3.15, each of which describes a certain exactly solvable model of quantum mechanics.

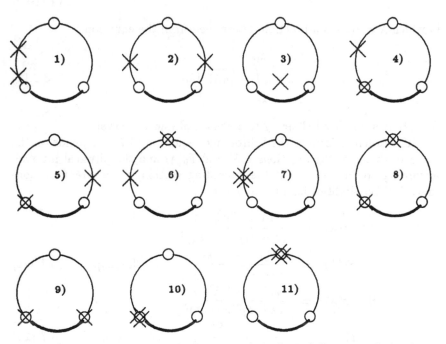

Figure 3.12. The non-equivalent extended physical diagrams. Series A.

Figure 3.13. The non-equivalent extended physical diagrams. Series B.

As an example demonstrating this correspondence, let us consider the first diagram in figure 3.12 and obtain the concrete model associated with

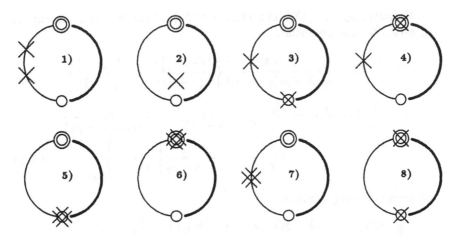

Figure 3.14. The non-equivalent extended physical diagrams. Series C.

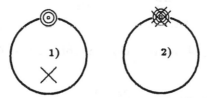

Figure 3.15. The non-equivalent extended physical diagrams. Series D.

it.

Using $SL(2)$ transformations which do not influence the form of the resulting model, we can assume that the double poles depicted on this diagram are located at $+1$ and -1. In this case the model is completely characterized by the number E_0 and two functions

$$F_0(\lambda) = \frac{F_{0-}}{\lambda + 1} + \frac{F_{0+}}{\lambda - 1}, \tag{3.7.13a}$$

$$\rho(\lambda) = \frac{\rho_0(\lambda - \rho_1)(\lambda - \rho_2)}{(\lambda - 1)^2(\lambda + 1)^2}, \tag{3.7.13b}$$

in which $F_{0\pm}$ and ρ_i are external (real) parameters satisfying the following constraints:

$$F_{0-} > \frac{1}{2}, \quad F_{0+} > \frac{1}{2}, \quad \rho_0 > 0, \quad \rho_1 < \rho_2 < -1. \tag{3.7.14}$$

The role of the physical interval is played by the interval

$$\lambda \in [-1, 1]. \tag{3.7.15}$$

Substituting (3.7.13) into (3.7.9) and (3.7.10) we find an explicit expression for the potential

$$V(x) = E_0 + \frac{(\lambda^2 - 1)^2}{\rho_0(\lambda - \rho_1)(\lambda - \rho_2)} \left\{ \frac{(F_{0-} - \frac{1}{2})^2}{(\lambda + 1)^2} + \frac{(F_{0+} - \frac{1}{2})^2}{(\lambda - 1)^2} \right.$$

$$+ \frac{2F_{0+}F_{0-} - \frac{1}{2}}{\lambda - 1} - \frac{\frac{5}{16}}{(\lambda - \rho_1)^2} - \frac{\frac{5}{16}}{(\lambda - \rho_2)^2} - \frac{\frac{1}{8}}{(\lambda - \rho_1)(\lambda - \rho_2)}$$

$$\left. + \frac{1}{2} \left[\frac{1}{\lambda - \rho_1} + \frac{1}{\lambda - \rho_2} \right] \frac{\lambda}{\lambda^2 - 1} \right\} \Bigg|_{\lambda = \lambda(x)} \qquad (3.7.16)$$

and corresponding solutions

$$\psi(x) = (\lambda - \rho_1)^{\frac{1}{4}}(\lambda - \rho_2)^{\frac{1}{4}}(\lambda + 1)^{F_- - \frac{1}{2}}(\lambda - 1)^{F_+ - \frac{1}{2}}$$

$$\times \prod_{i=1}^{M}(\lambda - \xi_i) \Bigg|_{\lambda = \lambda(x)}, \qquad (3.7.17a)$$

$$E = E_0 + \epsilon. \qquad (3.7.17b)$$

Evaluating the integral in (3.7.11) we obtain the following implicit expression for $\lambda(x)$:

$$x = -\sqrt{\rho_0}\operatorname{arcch}\frac{2\lambda - \rho_2 - \rho_1}{\rho_2 - \rho_1}$$

$$-\sqrt{\rho_0(\rho_1 + 1)(\rho_2 + 1)}\operatorname{arcth}\sqrt{\frac{(\rho_2 + 1)(\lambda - \rho_1)}{(\rho_1 + 1)(\lambda - \rho_2)}}$$

$$+\sqrt{\rho_0(\rho_1 - 1)(\rho_2 - 1)}\operatorname{arcth}\sqrt{\frac{(\rho_1 - 1)(\lambda - \rho_2)}{(\rho_2 - 1)(\lambda - \rho_1)}}. \qquad (3.7.18)$$

From (3.7.18) it follows that the Schrödinger problem is formulated on the whole x-axis:

$$x \in [-\infty, +\infty] \qquad (3.7.19)$$

and

$$V(x) \rightarrow E_0 + \begin{cases} \frac{(2F_{0-} - 1)^2}{\rho_0(\rho_1 + 1)(\rho_2 + 1)}, & \text{if } x \rightarrow -\infty, \\ \frac{(2F_{0+} - 1)^2}{\rho_0(\rho_1 - 1)(\rho_2 - 1)}, & \text{if } x \rightarrow +\infty. \end{cases} \qquad (3.7.20)$$

This means that the potential (3.7.16) has the form of a well of finite depth, so that the quantum system corresponding to this potential may only have a finite number of bound states.

In order to find these states, let us consider equations (3.7.3)–(3.7.4) for $\xi_1, \ldots, \xi_M$, ϵ and

$$F(\lambda) = \frac{F_-}{\lambda + 1} + F_+ \lambda - 1, \tag{3.7.21}$$

and supplement them by the constraints

$$F_- > \frac{1}{2}, \quad F_+ > \frac{1}{2}, \tag{3.7.22}$$

guaranteeing the normalizability of wavefunctions (3.7.17). It is not difficult to show that the admissible values of numbers $F_\pm$ and ϵ can be obtained from the second equation (3.7.4). In order to reduce it to a more convenient form, let us first simplify the third term in the left-hand side of (3.7.4). Using (3.7.21) we obtain

$$\sum_{i=1}^{M} \frac{F(\lambda) - F(\xi_i)}{\lambda - \xi_i}$$

$$= -\frac{1}{\lambda + 1} \sum_{i=1}^{M} \frac{F_-}{\xi_i + 1} - \frac{1}{\lambda - 1} \sum_{i=1}^{M} \frac{F_+}{\xi_i - 1}$$

$$= -\frac{\lambda}{\lambda^2 - 1} \sum_{i=1}^{M} \left(\frac{F_-}{\xi_i + 1} + \frac{F_+}{\xi_i - 1} \right) + \frac{1}{\lambda^2 - 1} \sum_{i=1}^{M} \left(\frac{F_-}{\xi_i + 1} + \frac{F_+}{\xi_i - 1} \right). \tag{3.7.23}$$

Taking into account the first equation (3.7.3) we obtain

$$\sum_{i=1}^{M} \left(\frac{F_-}{\xi_i + 1} + \frac{F_+}{\xi_i - 1} \right) = -\sum_{i,k=1}^{M}{}' \frac{1}{\xi_i - \xi_k} = 0, \tag{3.7.24}$$

and

$$\sum_{i=1}^{M} \left(\frac{F_-}{\xi_i + 1} + \frac{F_+}{\xi_i - 1} \right) = \sum_{i=1}^{M} (F_- + F_+) + \sum_{i,k=1}^{M}{}' \frac{\xi_i}{\xi_i - \xi_k}$$

$$= M(F_- + F_+) + \frac{M(M-1)}{2}, \tag{3.7.25}$$

which gives

$$2 \sum_{i=1}^{M} \frac{F(\lambda) - F(\xi_i)}{\lambda - \xi_i} = \frac{M(M-1) + 2M(F_- + F_+)}{\lambda^2 - 1}. \tag{3.7.26}$$

Substituting expressions (3.7.21), (3.7.26) and (3.7.13) into (3.7.4) and taking in the equations obtained $\lambda \to \pm 1$ and $\lambda \to \infty$, we find three algebraic equations for the numbers $F_\pm$ and ϵ,

$$\left[F_\pm - \frac{1}{2}\right]^2 - \left[F_{0\pm} - \frac{1}{2}\right]^2 + \frac{\rho_0(\rho_1 \pm 1)(\rho_2 \pm 1)}{4}\epsilon = 0, \qquad (3.7.27)$$

$$\left[F_- + F_+ + M - \frac{1}{2}\right]^2 - \left[F_{0-} + F_{0+} - \frac{1}{2}\right]^2 + \rho_0\epsilon = 0. \qquad (3.7.28)$$

Solving these equations for $F_\pm - \frac{1}{2}$ and $F_- + F_+ + M - \frac{1}{2}$ and chosing the true branches of solutions by means of (3.7.22) we obtain

$$F_\pm - \frac{1}{2} = \sqrt{\left[F_{0\pm} - \frac{1}{2}\right]^2 - \frac{\rho_0(\rho_1 \pm 1)(\rho_2 \pm 1)}{4}\epsilon}, \qquad (3.7.29)$$

$$(F_- - \frac{1}{2}) + (F_+ - \frac{1}{2}) + (M + \frac{1}{2})$$
$$= \sqrt{\left[F_{0-} + F_{0+} - \frac{1}{2}\right]^2 - \rho_0\epsilon}. \qquad (3.7.30)$$

Substitution of (3.7.29) into (3.7.30) gives us immediately the equation for the spectrum $E = E_0 + \epsilon$:

$$\sqrt{\left[F_{0-} - \frac{1}{2}\right]^2 - \frac{\rho_0(\rho_1 + 1)(\rho_2 + 1)}{4}\epsilon}$$
$$+ \sqrt{\left[F_{0+} - \frac{1}{2}\right]^2 - \frac{\rho_0(\rho_1 - 1)(\rho_2 - 1)}{4}\epsilon}$$
$$+ M + \frac{1}{2} = \sqrt{\left[F_- + F_+ - \frac{1}{2}\right]^2 - \rho_0\epsilon}. \qquad (3.7.31)$$

Thus, we see that the solution of the spectral problem for the model (3.7.16) includes three stages. First, we solve equation (3.7.31) and find the spectrum E. Then, we restore the admissible values of parameters $F_\pm$ by formulas (3.7.29), and substitute them into the system

$$\sum_{k=1}^{M}{}' \frac{1}{\xi_i - \xi_k} + \frac{F_-}{\xi_i + 1} + \frac{F_+}{\xi_i - 1} = 0, \quad i = 1, \dots, M. \qquad (3.7.32)$$

Finally, we solve this system and construct the wavefunctions $\psi(\lambda)$.

From (3.7.31) it follows that

$$E \leq E_0 + \min\left\{\frac{(2F_{0+} - 1)^2}{\rho_0(\rho_1 - 1)(\rho_2 - 1)},\right.$$

$$\left.\frac{(2F_{0-} - 1)^2}{\rho_0(\rho_1 + 1)(\rho_2 + 1)}, \frac{(2F_{0+} + 2F_{0-} - 1)^2}{4\rho_0}\right\} \tag{3.7.33}$$

which gives an upper bound for the spectrum in model (3.7.16). The number M of the highest level can be obtained from the other inequality

$$M \leq F_{0-} + F_{0+} - 1 \tag{3.7.34}$$

which also follows from (3.7.31). These formulas complete the construction of exactly solvable models associated with diagram 1 in figure 3.12.

Other models described by the diagrams depicted in figures 3.12, 3.13, 3.14 and 3.15 can be considered analogously. For more details see section 3.8.

3.8 Reference list of elementary exactly solvable models

In this section we consider one-dimensional exactly solvable models associated with the diagrams depicted in figures 3.12, 3.13, 3.14 and 3.15. Since the list of these diagrams is rather long, it is reasonable to restrict ourselves by discussing only those cases in which the resulting models and their solutions can be expressed in terms of elementary functions. Below we give (without explicit derivation) a complete list of all such models and their solutions.

Series A. First of all, let us consider series A, for which

$$F_0(\lambda) = \frac{F_0^-}{\lambda - 1} + \frac{F_0^+}{\lambda + 1}, \quad F_0^\pm > \frac{1}{2}, \quad \lambda \in [-1, +1]. \tag{3.8.1}$$

Model 1. The extended physical diagram is depicted in figure 3.16. Weight function:

$$\rho(\lambda) = \frac{\rho_0(\lambda + \frac{n+1}{n-1})^2}{(\lambda^2 - 1)^2}, \quad \rho_0 > 0, \quad n = 2, 3, 4. \tag{3.8.2}$$

Potential:

$$V(x) = E_0 + \frac{[\lambda^2(x) - 1]^2}{\rho_0[\lambda(x) + \frac{n+1}{n-1}]^2}\left\{\frac{[F_0^- - \frac{1}{2}]^2}{[\lambda(x) + 1]^2} + \frac{[F_0^+ - \frac{1}{2}]^2}{[\lambda(x) - 1]^2}\right.$$

$$\left. + \frac{2F_0^- F_0^+ - \frac{1}{2}}{\lambda^2(x) - 1} - \frac{\frac{3}{4}}{[\lambda(x) + \frac{n+1}{n-1}]^2} + \frac{\lambda(x)}{[\lambda^2(x) - 1][\lambda(x) + \frac{n+1}{n-1}]}\right\}. \tag{3.8.3}$$

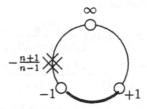

Figure 3.16. Diagram A7.

Solutions:

$$\psi(x) = \prod_{i=1}^{M} [\lambda(x) - \xi_i] \left[\lambda(x) + \frac{n+1}{n-1} \right]^{\frac{1}{2}}$$

$$\times \ [\lambda(x) - 1]^{\sqrt{[F_0^- - \frac{1}{2}]^2 - \frac{\rho_0}{(n-1)^2}[E - E_0]}}$$

$$\times \ [\lambda(x) + 1]^{\sqrt{[F_0^+ - \frac{1}{2}]^2 - \frac{\rho_0}{(n-1)^2}[E - E_0]}}. \qquad (3.8.4)$$

Function $\lambda(x)$ satisfies the equation

$$[1 - \lambda(x)]^n = \exp\left(-\frac{n-1}{\sqrt{\rho_0}} x\right) [1 + \lambda(x)] \qquad (3.8.5)$$

which for $n = 2, 3, 4$ can be solved in radicals.
Spectral equations:

$$\sqrt{\left[F_0^- - \frac{1}{2}\right]^2 - \frac{\rho_0}{(n-1)^2}[E - E_0]}$$

$$+ \ \sqrt{\left[F_0^+ - \frac{1}{2}\right]^2 - \frac{\rho_0 n^2}{(n-1)^2}[E - E_0] + M + \frac{1}{2}}$$

$$= \ \sqrt{\left[F_0^- + F_0^+ - \frac{1}{2}\right]^2 - \rho_0[E - E_0]}, \qquad (3.8.6)$$

$$\sum_{k=1}^{M}{}' \frac{1}{\xi_i - \xi_k} \ + \ \frac{\frac{1}{2} + \sqrt{\left[F_0^- - \frac{1}{2}\right]^2 - \frac{\rho_0}{(n-1)^2}[E - E_0]}}{\xi_i + 1}$$

$$+ \ \frac{\frac{1}{2} + \sqrt{\left[F_0^+ - \frac{1}{2}\right]^2 - \frac{\rho_0 n^2}{(n-1)^2}[E - E_0]}}{\xi_i - 1} = 0,$$

$$i = 1, \ldots, M. \qquad (3.8.7)$$

Spectral inequalities:

$$E \leq E_0 + \min\left\{\frac{(n-1)^2}{n^{1\pm 1}\rho_0}\left[F_0^\pm - \frac{1}{2}\right]^2\right\}, \qquad (3.8.8)$$

$$M \leq F_0^- + F_0^+ - 1. \qquad (3.8.9)$$

Physical interval:

$$x \in [-\infty, +\infty]. \qquad (3.8.10)$$

Asymptotic properties:

$$V(\pm\infty) = E_0 + \frac{(n-1)^2}{n^{1\pm 1}\rho_0}\left[F_0^\pm - \frac{1}{2}\right]^2. \qquad (3.8.11)$$

The model describes a potential well of finite depth.

Model 2. The extended physical diagram is depicted in figure 3.17.

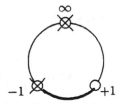

Figure 3.17. Diagram A8.

Weight function:

$$\rho(\lambda) = \frac{\rho_0}{(\lambda^2 - 1)(\lambda + 1)}, \qquad \rho_0 > 0. \qquad (3.8.12)$$

Potential:

$$V(x) = E_0 + \frac{2}{\rho_0\,\mathrm{ch}^2\frac{x}{\sqrt{2\rho_0}}}\left\{\frac{\left[F_0^- - \frac{3}{4}\right]\left[F_0^+ - \frac{1}{4}\right]}{\mathrm{sh}^2\frac{x}{\sqrt{2\rho_0}}}\right.$$

$$\left. + \left[F_0^+ - \frac{1}{2}\right]^2\mathrm{sh}^2\frac{x}{\sqrt{2\rho_0}} - \left[2F_0^- F_0^+ - \frac{1}{4}\right]\right\}. \qquad (3.8.13)$$

Solutions:

$$\psi(x) = \frac{\left[\text{th}^2 \frac{x}{\sqrt{2\rho_0}}\right]^{F_0^- - \frac{1}{4}}}{\left[\text{ch}^2 \frac{x}{\sqrt{2\rho_0}}\right]^{F_0^+ - \frac{1}{2} - M}} \prod_{i=1}^{M} \left[\text{th}^2 \frac{x}{\sqrt{2\rho_0}} - \frac{1 + \xi_i}{2}\right],$$

(3.8.14)

$$E = E_0 + \frac{2M}{\rho_0} \left[2F_0^+ - 1 - M\right].$$

(3.8.15)

Spectral equations:

$$\sum_{k=1}^{M}{}' \frac{1}{\xi_i - \xi_k} + \frac{F_0^-}{\xi_i + 1} + \frac{F_0^+ - M}{\xi_i - 1} = 0, \qquad i = 1, \dots, M.$$

(3.8.16)

Spectral inequalities:

$$E \leq E_0 + \frac{2\left[F_0^+ - \frac{1}{2}\right]^2}{\rho_0},$$

(3.8.17)

$$M \leq F_0^+ - \frac{1}{2}.$$

(3.8.18)

Physical interval:

$$x \in [0, \infty].$$

(3.8.19)

Asymptotic properties:

$$V(-\infty) = +\infty, \qquad V(+\infty) = \frac{2}{\rho_0} \left[F_0^+ - \frac{1}{2}\right]^2.$$

(3.8.20)

This model describes a potential well of finite depth which is known as the hyperbolic Pöschel–Teller potential well.

Model 3. The extended physical diagram is depicted in figure 3.18. Weight function:

$$\rho(\lambda) = \frac{\rho_0}{1 - \lambda^2}, \qquad \rho_0 > 0.$$

(3.8.21)

Potential:

$$V(x) = E_0 + \frac{1}{\rho_0} \left\{ \left[F_0^- - \frac{1}{4}\right] \left[F_0^- - \frac{3}{4}\right] \frac{1 - \sin \frac{x}{\sqrt{\rho_0}}}{1 + \sin \frac{x}{\sqrt{\rho_0}}} \right.$$

$$\left. + \left[F_0^+ - \frac{1}{4}\right] \left[F_0^- - \frac{3}{4}\right] \frac{1 + \sin \frac{x}{\sqrt{\rho_0}}}{1 - \sin \frac{x}{\sqrt{\rho_0}}} - \left[2F_0^- F_0^+ - \frac{1}{8}\right] \right\}.$$

(3.8.22)

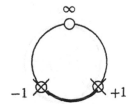

Figure 3.18. Diagram A9.

Solutions:

$$\psi(x) = \left(1 - \sin\frac{x}{\sqrt{\rho_0}}\right)^{F_0^- - \frac{1}{4}} \left(1 + \sin\frac{x}{\sqrt{\rho_0}}\right)^{F_0^+ - \frac{1}{4}}$$

$$\times \prod_{i=1}^{M}\left[\sin\frac{x}{\sqrt{\rho_0}} - \xi_i\right], \tag{3.8.23}$$

$$E = E_0 + \frac{M}{\rho_0}\left[F_0^- + F_0^+ + M - 1\right]. \tag{3.8.24}$$

Spectral equations:

$$\sum_{k=1}^{M}{}' \frac{1}{\xi_i - \xi_k} + \frac{F_0^-}{\xi_i + 1} + \frac{F_0^+}{\xi_i - 1} = 0, \quad i = 1, \ldots, M. \tag{3.8.25}$$

Spectral inequalities:

$$E \leq \infty, \quad M \leq \infty. \tag{3.8.26}$$

Physical interval:

$$x \in \left[-\sqrt{\rho_0}\frac{\pi}{2}, +\sqrt{\rho_0}\frac{\pi}{2}\right]. \tag{3.8.27}$$

Asymptotic properties:

$$V\left(\pm\sqrt{\rho_0}\frac{\pi}{2}\right) = \infty. \tag{3.8.28}$$

This model describes a potential well of infinite depth, which is known as the trigonometric Pöschel–Teller potential well.

Model 4. The extended physical diagram is depicted in figure 3.19.

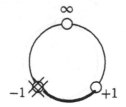

Figure 3.19. Diagram A10.

Weight function:

$$\rho(\lambda) = \frac{\rho_0}{(\lambda - 1)^2}, \quad \rho_0 > 0. \qquad (3.8.29)$$

Potential:

$$V(x) = E_0 + \frac{1}{\rho_0}\left\{ \frac{F_0^- \left[F_0^- - 1\right]}{\left[\exp\left(\frac{x}{\sqrt{\rho_0}}\right) - 1\right]^2} \right.$$

$$+ \left. \frac{2F_0^- F_0^+}{\exp\left(\frac{x}{\sqrt{\rho_0}}\right) - 1} + \left[F_0^+ - \tfrac{1}{2}\right]^2 \right\}. \qquad (3.8.30)$$

Solutions:

$$\psi(x) = \left[1 - \exp\left(-\frac{x}{\sqrt{\rho_0}}\right)\right]^{F_0^-} \left[\exp\left(-\frac{x}{\sqrt{\rho_0}}\right)\right]^{\sqrt{\left[F_0^+ - \frac{1}{2}\right]^2 - \rho_0[E - E_0]}}$$

$$\times \prod_{i=1}^{M}\left\{1 - \exp\left(-\frac{x}{\sqrt{\rho_0}}\right) - \frac{1 + \xi_i}{2}\right\}. \qquad (3.8.31)$$

Spectral equations:

$$F_0^- + M + \sqrt{\left[F_0^+ - \tfrac{1}{2}\right]^2 - \rho_0\left[E - E_0\right]}$$

$$= \sqrt{\left[F_0^- + F_0^+ - \tfrac{1}{2}\right]^2 - \rho_0\left[E - E_0\right]}, \qquad (3.8.32)$$

$$\sum_{k=1}^{M}{}' \frac{1}{\xi_i - \xi_k} + \frac{F_0^-}{\xi_i + 1} + \frac{\tfrac{1}{2} + \sqrt{\left[F_0^+ - \tfrac{1}{2}\right]^2 - \rho_0\left[E - E_0\right]}}{\xi_i - 1} = 0,$$

$$i = 1, \ldots, M. \qquad (3.8.33)$$

Spectral inequalities:

$$E \;\le\; E_0 + \frac{\left[F_0^+ - \frac{1}{2}\right]^2}{\rho_0}, \tag{3.8.34}$$

$$M \;\le\; F_0^+ - \tfrac{1}{2}. \tag{3.8.35}$$

Physical interval:

$$x \in [0, \infty]. \tag{3.8.36}$$

Asymptotic properties:

$$V(0) = \infty, \qquad V(\infty) \;=\; \tfrac{1}{\rho_0}\left[F_0^+ - \tfrac{1}{2}\right]. \tag{3.8.37}$$

This model describes a potential well of finite depth. It is known as the Eckart potential.

Model 5. The extended physical diagram is depicted in figure 3.20.

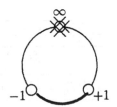

Figure 3.20. Diagram A11.

Weight function:

$$\rho(\lambda) = \frac{\rho_0}{(\lambda^2 - 1)^2}, \qquad \rho_0 > 0. \tag{3.8.38}$$

Potential:

$$
\begin{aligned}
V(x) \;=\; E_0 &+ \frac{4}{\rho_0}\left\{ \left[F_0^- - \tfrac{1}{2}\right]^2 \frac{\exp\left(\frac{4x}{\sqrt{\rho_0}}\right)}{\left[\exp\left(\frac{2x}{\sqrt{\rho_0}}\right)+1\right]^2}\right. \\
&+ \left[F_0^+ - \tfrac{1}{2}\right]^2 \frac{1}{\left[\exp\left(\frac{2x}{\sqrt{\rho_0}}\right)+1\right]^2} \\
&- \left. \left[2F_0^- F_0^+ - \tfrac{1}{2}\right] \frac{\exp\left(\frac{2x}{\sqrt{\rho_0}}\right)}{\left[\exp\left(\frac{2x}{\sqrt{\rho_0}}\right)+1\right]^2}\right\}. \tag{3.8.39}
\end{aligned}
$$

Solutions:

$$\psi(x) \;=\; \left[\dfrac{1}{\exp\left(\frac{2x}{\sqrt{\rho_0}}\right)+1}\right]^{F_0^- - \frac{1}{2}} \left[\dfrac{\exp\left(\frac{2x}{\sqrt{\rho_0}}\right)}{\exp\left(\frac{2x}{\sqrt{\rho_0}}\right)+1}\right]^{F_0^+ - \frac{1}{2}}$$

$$\times \; \prod_{i=1}^{M}\left[\dfrac{1}{\exp\left(\frac{2x}{\sqrt{\rho_0}}\right)+1} - \dfrac{1+\xi_i}{2}\right]. \qquad (3.8.40)$$

Spectral equations:

$$\sqrt{\left[F_0^- - \tfrac{1}{2}\right]^2 - \dfrac{\rho_0}{4}\left[E - E_0\right]}$$

$$+ \; \sqrt{\left[F_0^+ - \tfrac{1}{2}\right]^2 - \dfrac{\rho_0}{4}\left[E - E_0\right]} + M$$

$$= \; F_0^- + F_0^+ - 1, \qquad (3.8.41)$$

$$\sum_{k=1}^{M}{}' \dfrac{1}{\xi_i - \xi_k} \; + \; \dfrac{\tfrac{1}{2}\sqrt{\left[F_0^- - \tfrac{1}{2}\right]^2 - \frac{\rho_0}{4}\left[E - E_0\right]}}{\xi_i + 1}$$

$$+ \; \dfrac{\tfrac{1}{2}\sqrt{\left[F_0^+ - \tfrac{1}{2}\right]^2 - \frac{\rho_0}{4}\left[E - E_0\right]}}{\xi_i - 1} = 0,$$

$$i = 1,\ldots,M. \qquad (3.8.42)$$

Spectral inequalities:

$$E \;\le\; E_0 + \min\left\{\dfrac{4}{\rho_0}\left[F_0^- - \tfrac{1}{2}\right]^2, \dfrac{4}{\rho_0}\left[F_0^+ - \tfrac{1}{2}\right]^2\right\}, \qquad (3.8.43)$$

$$M \;\le\; F_0^- + F_0^+ - 1. \qquad (3.8.44)$$

Physical interval:

$$x \in [-\infty, +\infty]. \qquad (3.8.45)$$

Asymptotic properties:

$$V(\pm\infty) = \dfrac{4}{\rho_0}\left[F_0^\mp - \tfrac{1}{2}\right]^2. \qquad (3.8.46)$$

This model describes a potential well of finite depth. It also is known as the Eckart potential.

Series B. Now let us consider series B, for which

$$F_0(\lambda) = \frac{F_0}{\lambda - i} + \frac{F_0^*}{\lambda + i},$$

$$F_0 + F_0^* + M < \tfrac{1}{2}, \quad \lambda \in [-\infty, +\infty]. \qquad (3.8.47)$$

Model 6. The extended physical diagram is depicted in figure 3.21.

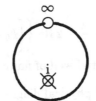

Figure 3.21. Diagram B2.

Weight function:

$$\rho(\lambda) = \frac{\rho_0}{\lambda^2 + 1}, \quad \rho_0 > 0. \qquad (3.8.48)$$

Potential:

$$\begin{aligned}
V(x) &= E_0 + \frac{2\|F_0\|^2 - \tfrac{1}{8}}{\rho_0} \\
&+ \frac{2\operatorname{Re}\left\{\left[F_0 - \tfrac{1}{4}\right]\left[F_0 - \tfrac{3}{4}\right]\right\}\left[\operatorname{sh}^2 \frac{x}{\sqrt{\rho_0}} - 1\right]}{\rho_0 \operatorname{ch}^2 \frac{x}{\sqrt{\rho_0}}} \\
&- \frac{4\operatorname{Im}\left\{\left[F_0 - \tfrac{1}{4}\right]\left[F_0 - \tfrac{3}{4}\right]\right\}\operatorname{sh} \frac{x}{\sqrt{\rho_0}}}{\rho_0 \operatorname{ch}^2 \frac{x}{\sqrt{\rho_0}}}.
\end{aligned} \qquad (3.8.49)$$

Solutions:

$$\psi(x) = \left|\left[\operatorname{sh}\frac{x}{\sqrt{\rho_0}} + i\right]^{F_0 - \tfrac{1}{4}}\right|^2 \prod_{i=1}^{M}\left[\operatorname{sh}\frac{x}{\sqrt{\rho_0}} - \xi_i\right], \qquad (3.8.50)$$

$$E = E_0 - \frac{M}{\rho_0}(4\operatorname{Re} F_0 - 1 - M). \qquad (3.8.51)$$

Spectral equations:

$$\sum_{k=1}^{M}{}' \frac{1}{\xi_j - \xi_k} + \frac{F_0}{\xi_j - i} + \frac{F_0^*}{\xi_j + i} = 0, \quad j = 1, \dots, M. \qquad (3.8.52)$$

Spectral inequalities:

$$E \leq E_0 + \frac{\left[2\,\mathrm{Re}\,F_0 - \frac{1}{2}\right]^2}{\rho_0}, \qquad (3.8.53)$$

$$M \leq 2\,\mathrm{Re}\,F_0. \qquad (3.8.54)$$

Physical interval:

$$x \in [-\infty, +\infty]. \qquad (3.8.55)$$

Asymptotic properties:

$$V(\pm\infty) = \frac{1}{\rho_0}\left\{2|F_0|^2 - \frac{1}{8} + 2\,\mathrm{Re}\left\{\left[F_0 - \frac{1}{4}\right]\left[F_0 - \frac{3}{4}\right]\right\}\right\}. \qquad (3.8.56)$$

This model describes a potential well of finite depth.

Model 7. The extended physical diagram is depicted in figure 3.22.

Figure 3.22. Diagram B3.

Weight function:

$$\rho(\lambda) = \frac{\rho_0}{(\lambda^2 + 1)^2}, \qquad \rho_0 > 0. \qquad (3.8.57)$$

Potential:

$$V(x) = E_0 + \frac{2|F_0|^2 - \frac{1}{2}}{\rho_0}\frac{1}{\cos^2\frac{x}{\sqrt{\rho_0}}} + \frac{2}{\rho_0}\,\mathrm{Re}\left[F_0 - \frac{1}{2}\right]^2$$

$$\times \left(\mathrm{tg}^2\frac{x}{\sqrt{\rho_0}} - 1\right) - \frac{4}{\rho_0}\,\mathrm{Im}\left[F_0 - \frac{1}{2}\right]^2\,\mathrm{tg}\frac{x}{\sqrt{\rho_0}}. \qquad (3.8.58)$$

Solutions:

$$\psi(x) = \left\|\left[\mathrm{tg}\frac{x}{\sqrt{\rho_0}} - \mathrm{i}\right]^{\sqrt{[F_0 - \frac{1}{2}]^2 + \frac{\rho_0}{4}[E - E_0]}}\right\|^2$$

$$\times \prod_{i=1}^{M}\left[\mathrm{tg}\frac{x}{\sqrt{\rho_0}} - \xi_i\right]. \qquad (3.8.59)$$

Spectral equations:

$$2\operatorname{Re}\sqrt{\left[F_0 - \frac{1}{2}\right]^2 + \frac{\rho_0}{4}[E - E_0]} + M = 2\operatorname{Re} F_0 - 1, \qquad (3.8.60)$$

$$\sum_{k=1}^{M}{}' \frac{1}{\xi_j - \xi_k} \; + \; \frac{\frac{1}{2} + \sqrt{\left[F_0 - \frac{1}{2}\right]^2 + \frac{\rho_0}{4}[E - E_0]}}{\xi_j - i}$$

$$+ \; \frac{\frac{1}{2} + \sqrt{\left[F_0^* - \frac{1}{2}\right]^2 + \frac{\rho_0}{4}[E - E_0]}}{\xi_j + i} = 0,$$

$$j = 1, \ldots, M. \qquad (3.8.61)$$

Spectral inequalities:

$$E \leq \infty, \quad M \leq \infty. \qquad (3.8.62)$$

Physical interval:

$$x \in \left[-\sqrt{\rho_0}\frac{\pi}{2}, +\sqrt{\rho_0}\frac{\pi}{2}\right]. \qquad (3.8.63)$$

Asymptotic properties:

$$V\left(\pm\sqrt{\rho_0}\frac{\pi}{2}\right) = \infty. \qquad (3.8.64)$$

This model describes a potential well of infinite depth.

Series C. For series C we have

$$F_0 = \frac{F_{01}}{\lambda} + F_{02}, \quad F_{01} > \frac{1}{2}, \quad F_{02} < 0, \quad \lambda \in [0, \infty]. \qquad (3.8.65)$$

Model 8. The extended physical diagram is depicted in figure 3.23. Weight function:

$$\rho(\lambda) = \rho_0, \quad \rho_0 > 0. \qquad (3.8.66)$$

Potential:

$$V(x) = E_0 + \frac{F_{01}\left[F_{01} - 1\right]}{x^2} + \frac{2F_{01}F_{02}}{\sqrt{\rho_0}\,x} + \frac{[F_{02}]^2}{\rho_0}. \qquad (3.8.67)$$

Figure 3.23. Diagram C5.

Solutions:

$$\psi(x) = x^{F_{01}} \exp\left(\frac{F_{01}F_{02}}{M + F_{0i}}\frac{x}{\sqrt{\rho_0}}\right) \prod_{i=1}^{M}\left[\frac{x}{\sqrt{\rho_0}} - \xi_i\right], \qquad (3.8.68)$$

$$E = E_0 + \frac{[F_{02}]^2}{\rho_0}\left\{1 - \left[\frac{F_{01}}{M + F_{01}}\right]^2\right\}. \qquad (3.8.69)$$

Spectral equations:

$$\sum_{k=1}^{M}{}' \frac{1}{\xi_i - \xi_k} + \frac{F_{01}}{\xi_i} + \frac{F_{01}F_{02}}{M + F_{01}} = 0, \qquad i = 1, \ldots, M. \qquad (3.8.70)$$

Spectral inequalities:

$$E \leq E_0 + \frac{[F_{02}]^2}{\rho_0}, \qquad (3.8.71)$$

$$M \leq \infty. \qquad (3.8.72)$$

Physical interval:

$$x \in [0, \infty]. \qquad (3.8.73)$$

Asymptotic properties:

$$V(0) = \infty, \quad V(\infty) = \frac{[F_{02}]^2}{\rho_0}. \qquad (3.8.74)$$

This is the well known Kratzer potential.

Model 9. The extended physical diagram is depicted in figure 3.24.

Figure 3.24. Diagram C6.

Weight function:

$$\rho(\lambda) = \frac{\rho_0}{\lambda^2}, \quad \rho_0 > 0. \tag{3.8.75}$$

Potential:

$$V(x) = E_0 + \frac{\left[F_{01} - \frac{1}{2}\right]^2}{\rho_0} + \frac{2F_{01}F_{02}}{\rho_0} \exp\left(\frac{x}{\sqrt{\rho_0}}\right) + \frac{\left[F_{02}\right]^2}{\rho_0} \exp\left(\frac{2x}{\sqrt{\rho_0}}\right). \tag{3.8.76}$$

Solutions:

$$\psi(x) = \left[\exp\left(\frac{x}{\sqrt{\rho_0}}\right)\right]^{F_{01} - \frac{1}{2}} \exp\left[F_{02} \exp\left(\frac{x}{\sqrt{\rho_0}}\right)\right]$$

$$\times \prod_{i=1}^{M}\left[\exp\left(\frac{x}{\sqrt{\rho_0}}\right) - \xi_i\right], \tag{3.8.77}$$

$$E = E_0 - \frac{M}{\rho_0}\left[2F_{01} - 1 - M\right]. \tag{3.8.78}$$

Spectral equations:

$$\sum_{k=1}^{M}{}' \frac{1}{\xi_i - \xi_k} + \frac{F_{01} - M}{\xi_i} + F_{02} = 0, \quad i = 1, \ldots, M. \tag{3.8.79}$$

Spectral inequalities:

$$E \leq E_0 + \frac{\left[F_{01} - \frac{1}{2}\right]^2}{\rho_0}, \tag{3.8.80}$$

$$M \leq F_{01} - \frac{1}{2}. \tag{3.8.81}$$

Physical interval:

$$x \in [-\infty, +\infty]. \qquad (3.8.82)$$

Asymptotic properties:

$$V(-\infty) = E_0 + \frac{\left[F_{01} - \frac{1}{2}\right]^2}{\rho_0}, \qquad V(\infty) = \infty. \qquad (3.8.83)$$

This is the well known Morse potential.

Model 10. The extended physical diagram is depicted in figure 3.25.

Figure 3.25. Diagram C8.

Weight function:

$$\rho(\lambda) = \frac{\rho_0}{\lambda}, \qquad \rho_0 > 0. \qquad (3.8.84)$$

Potential:

$$V(x) = E_0 + \frac{4\left[F_{01} - \frac{1}{4}\right]\left[F_{01} - \frac{3}{4}\right]}{x^2}$$
$$+ \frac{2F_{01}F_{02}}{\rho_0} + \frac{\left[F_{02}\right]^2}{4\rho_0} x^2. \qquad (3.8.85)$$

Solutions:

$$\psi(x) = (x^2)^{F_{01} - \frac{1}{4}} \exp\left(F_{02}\frac{x^2}{4\rho_0}\right) \prod_{i=1}^{M}\left(\frac{x^2}{4\rho_0} - \xi_i\right), \qquad (3.8.86)$$

$$E = E_0 + \frac{2F_{02}}{\rho_0} M. \qquad (3.8.87)$$

Spectral inequalities:

$$E \le \infty, \quad M \le \infty. \tag{3.8.88}$$

Physical interval:

$$x \in [0, \infty]. \tag{3.8.89}$$

Asymptotic properties:

$$V(0) = \infty, \quad V(\infty) = \infty. \tag{3.8.90}$$

This is a harmonic oscillator with a centrifugal barrier.

Series D. The last series D is characterized by

$$F_0(\lambda) = F_{01} + F_{02}\lambda, \quad F_{02} < 0, \quad \lambda \in [-\infty, +\infty]. \tag{3.8.91}$$

Model 11. The extended physical diagram is depicted in figure 3.26. Weight function:

Figure 3.26. Diagram D2.

$$\rho(\lambda) = \rho_0, \quad \rho_0 > 0. \tag{3.8.92}$$

Potential:

$$V(x) = E_0 + \frac{[F_{01}]^2 + F_{02}}{\rho_0} + \frac{2F_{01}F_{02}}{\rho_0^{3/2}} x + \frac{[F_{02}]^2}{\rho_0^2} x^2. \tag{3.8.93}$$

Solutions:

$$\psi(x) = \exp\left(F_{01} \frac{x}{\sqrt{\rho_0}} + F_{02} \frac{x^2}{2\rho_0} \right) \prod_{i=1}^{M} \left[\frac{x}{\sqrt{\rho_0}} - \xi_i \right], \tag{3.8.94}$$

$$E = E_0 - \frac{2M}{\rho_0} F_{02}. \tag{3.8.95}$$

Spectral equations:

$$\sum_{k=1}^{M}{}' \frac{1}{\xi_i - \xi_k} + F_{01} + F_{02}\xi_i = 0, \quad i = 1, \ldots, M. \tag{3.8.96}$$

Spectral inequalities:

$$E \leq \infty, \quad M \leq \infty. \tag{3.8.97}$$

Physical interval:

$$x \in [-\infty, +\infty]. \tag{3.8.98}$$

Asymptotic properties:

$$V(\pm\infty) = \infty. \tag{3.8.99}$$

This is the model of a simple harmonic oscillator.

3.9   The multi-dimensional case. Classification

In this section we discuss multi-dimensional exactly solvable models connected with MPS equations (3.6.3). We know that the metric tensor, characterizing the space in which these models are formulated, is expressed via the functions $x^\alpha(\lambda)$ and $\sigma^\alpha(\mu)$, $\alpha = 1, \ldots, \mathcal{D}$ (see formula (3.3.15)). This expression, which is rather complicated in the general case, is considerably simplified if the functions $x^\alpha(\lambda)$, $\alpha = 1, \ldots, \mathcal{D}$ take the form

$$x^\alpha(\lambda) = x(\lambda)(\lambda - c)^{\alpha-1}, \quad \alpha = 1, \ldots, \mathcal{D}, \tag{3.9.1}$$

where $x(\lambda)$ is a certain rational function, the form of which will be determined below, and c is a parameter. Then the matrix $\|x^\alpha(\lambda_i)\|_{i,\alpha=1,\ldots,\mathcal{D}}$ becomes explicitly invertible and we find the following simple expressions for non-zero (diagonal) components of the metric tensor (3.3.15):

$$g_{ii}(\mu, \vec{\lambda}) = \left\{ \sum_{\alpha=1}^{\mathcal{D}} (-1)^\alpha \sigma^\alpha(\mu) \sum_{\substack{i_1 < \cdots < i_{\mathcal{D}-\alpha} \\ i_1, \ldots, i_{\mathcal{D}-\alpha} \neq i}} (\lambda_{i_1} - c) \ldots (\lambda_{i_{\mathcal{D}-\alpha}} - c) \right\}$$

$$\times \left\{ x(\lambda_i) \prod_{k=1}^{\mathcal{D}}{}' (\lambda_i - \lambda_k) \right\}^{-1}. \tag{3.9.2}$$

For the Schrödinger equation to be an elliptic equation, all these components must be positive. Since (3.9.2) has the form of a fraction, it is sufficient to require that both the numerator and denominator are positive for any $i = 1, \ldots, \mathcal{D}$.

First of all, note that positive definiteness of the numerator can be ensured by choosing the signs of functions $\sigma^\alpha(\lambda)$, $\alpha = 1, \ldots, \mathcal{D}$. Indeed, consider three possible cases:

   a) the number c lies to the left of any physical interval,
   b) the number c lies to the right of any physical interval, and
   c) the number c is contained in the system of physical intervals.

In the first case all factors $\lambda_i - c$ are positive and therefore the positive definiteness of the numerator can be obtained by taking

$$\text{sign}\, \sigma^\alpha(\mu) = (-1)^\alpha. \tag{3.9.3a}$$

In the second case all factors $\lambda_i - c$ are negative, and therefore, the corresponding condition becomes

$$\text{sign}\, \sigma^\alpha(\mu) = (-1)^\mathcal{D}. \tag{3.9.3b}$$

In the third case, the sign of the product $(\lambda_{i_1} - c) \ldots (\lambda_{i_{\mathcal{D}-\alpha}} - c)$ is indefinite and therefore the only possibility to make the numerator positive is to take

$$\sigma^\alpha(\mu) = 0, \quad \alpha =, 1 \ldots, \mathcal{D} - 1; \quad \text{sign}\, \sigma^\mathcal{D}(\mu) = (-1)^\mathcal{D}. \tag{3.9.3c}$$

Now consider the positivity condition for the denominator in the fraction (3.9.2). Quite obviously, for it to be positive the function $x(\lambda)$ must have the form

$$x(\lambda) = \frac{\rho(\lambda)}{\prod_{k=2}^{\mathcal{D}}(\lambda - a_k)}, \tag{3.9.4}$$

where $a_2 < a_3 < \ldots < a_{\mathcal{D}-1} < a_{\mathcal{D}}$ are the common end points of $\mathcal{D}$ physical intervals, and $\rho(\lambda)$ is a certain positive function. Its form can easily be recovered if we remember that the functions $x^\alpha(\lambda)$ are elements of $(2\mathcal{D} + 1)$-dimensional rational spaces $R_{2n}\left(\frac{\vec{a}}{2\vec{n}}\right)$ described by diagrams A, B, C and D (see figure 3.8 in section 3.6). Then we obtain the following expressions for $\rho(\lambda)$:

$$\text{A)} \quad \rho(\lambda) = \frac{\rho_0(\lambda - \rho_1)(\lambda - \rho_2)}{(\lambda - a_1)^2(\lambda - a_{\mathcal{D}+1})^2}, \tag{3.9.5a}$$

$$\text{B)} \quad \rho(\lambda) = \frac{\rho_0(\lambda - \rho_1)(\lambda - \rho_2)}{(\lambda - a)^2(\lambda - a^*)^2}, \tag{3.9.5b}$$

$$\text{C)} \quad \rho(\lambda) = \frac{\rho_0(\lambda - \rho_1)(\lambda - \rho_2)}{(\lambda - a_1)^2}, \tag{3.9.5c}$$

$$\text{D)} \quad \rho(\lambda) = \rho_0(\lambda - \rho_1)(\lambda - \rho_2). \tag{3.9.5d}$$

For $\rho(\lambda)$ to be positive, the points ρ_1 and ρ_2 must either lie outside the system of physical intervals or coincide with one of the internal points a_k: $\rho_1 = \rho_2 = a_k$; $2 \leq k \leq \mathcal{D}$. Depicting these points by crosses we obtain the series of extended diagrams expressing the positivity conditions for the denominators in a most compact and visual form. These series are depicted in figures 3.27, 3.28, 3.29 and 3.30.

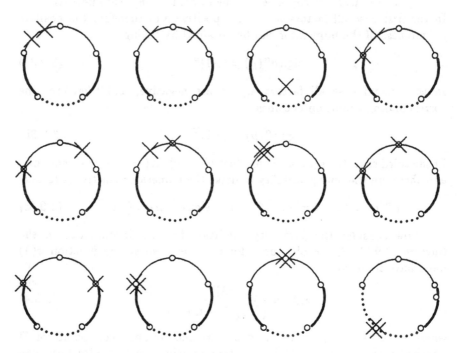

Figure 3.27. Extended physical diagrams for multi-dimensional exactly solvable models. Series A.

Note that the diagrams of series A are compatible with conditions (3.9.3a), (3.9.3b) and (3.9.3c), the diagrams of series C with conditions (3.9.3a) and (3.9.3c), and the diagrams of series B and D with condition (3.9.3c).

It is not difficult to see that the list of these diagrams completely solves the classification problem for multi-dimensional exactly solvable models associated with MPS equations (3.6.3). Indeed, let us assume that a diagram belonging to this list is given. This means that function $\rho(\lambda)$ is known. Using formulas (3.9.4), (3.9.1), (3.9.2) and (3.3.10), we can easily restore the form of functions $\Delta(\vec{\lambda})$, x^α and $g_{ik}(\mu, \vec{\lambda})$ entering into

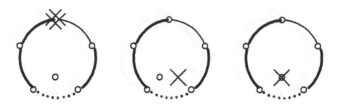

Figure 3.28. Extended physical diagrams for multi-dimensional exactly solvable models. Series B.

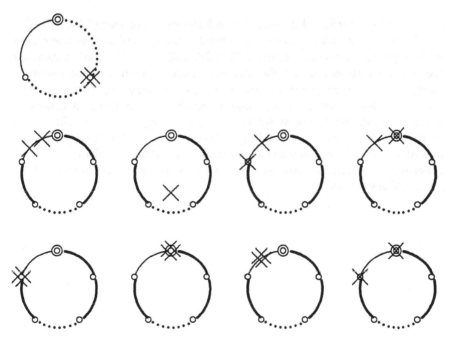

Figure 3.29. Extended physical diagrams for multi-dimensional exactly solvable models. Series C.

the expression (3.3.17) for the potential $V(\mu, \vec{\lambda})$, and determining the form of the corresponding Schrödinger equation (3.3.18) and the curvature of the space in which it is formulated. The remaining function $x^0(\lambda)$, also entering into the expression (3.3.17), can be found from equation (3.6.2) after fixing the function $F_0(\lambda)$, whose general structure is completely determined by the extended physical diagram (see formulas (3.6.5) and conditions (3.6.6)). This completes the construction procedure. The solutions of the obtained

Figure 3.30. Extended physical diagrams for multi-dimensional exactly solvable models. Series D.

model can be determined by formulas (3.3.19) and equations (3.6.4)–(3.6.7).

We shall not consider here any concrete examples of models associated with diagrams depicted in figures 3.27, 3.28, 3.29 and 3.30. Their potentials are rather cumbersome and the fact that most of them describe quantum motion in a curved space makes their analysis very non-trivial. True, there is at least one exception relating to the second (most degenerate) diagram belonging to the series D (see figure 3.30). It is not difficult to verify that the space corresponding to this diagram is flat. Computing the potential (3.3.17), we can see that it describes the class of multi-dimensional spherically asymmetric harmonic oscillators with centrifugal barriers along the coordinate axes.

Chapter 4

Classification of quasi-exactly solvable models with separable variables

4.1 Preliminary comments

According to the inverse method of separation of variables, the problem of constructing $\mathcal{D}$-dimensional exactly or quasi-exactly solvable models of quantum mechanics is reduced to the problem of constructing exactly or quasi-exactly solvable one-dimensional $\mathcal{D}$-parameter spectral equations of the form

$$\left\{-\frac{\partial^2}{\partial\lambda^2} + x^0(\lambda)\right\}\varphi(\lambda) = \left\{\sum_{\alpha=1}^{\mathcal{D}} x^\alpha(\lambda)\varepsilon_\alpha\right\}\varphi(\lambda). \qquad (4.1.1)$$

The cases when equations (4.1.1) become exactly solvable were discussed in detail in section 3.6. Omitting details, one can say that they are realized when the functions $x^0(\lambda)$ and $x^\alpha(\lambda)$, $\alpha = 1, \ldots, \mathcal{D}$ forming equation (4.1.1) belong to $(2\mathcal{D}+1)$-dimensional spaces of rational functions and satisfy all conditions of theorem 3.1. In this chapter we show that for equations (4.1.1) to be quasi-exactly solvable, the spaces to which the functions $x^0(\lambda)$ and $x^\alpha(\lambda)$, $\alpha = 1, \ldots, \mathcal{D}$ belong, must be $(2\mathcal{D}+3)$-dimensional. We describe a regular method of constructing such equations and demonstrate that they actually lead to wide classes of quasi-exactly solvable problems of quantum mechanics, both one dimensional and multi-dimensional.

Before formulating the basic idea of this method, let us first reduce equation (4.1.1) to a more convenient form. For sake of generality, assume that the weight functions $x^\alpha(\lambda)$, $\alpha = 1, \ldots, \mathcal{D}$ are elements of a $[2(\mathcal{D}+K)+1]$-dimensional space $R_{2n}\binom{\vec{n}}{2\vec{d}}$ and denote by $F_0(\lambda)$ an arbitrary function belonging to the corresponding $[(D + K) + 1]$-dimensional space $R_n\binom{\vec{n}}{\vec{d}}$. Denote also by ξ_{0i}, $i = 1, \ldots, M_0$ the numbers satisfying the system of equations

$$\sum_{k=1}^{M_0} \frac{1}{\xi_{0i} - \xi_{0k}} + F_0(\xi_{0i}) = 0, \quad i = 1, \ldots, M_0. \qquad (4.1.2)$$

It is not difficult to see that the function

$$x^0(\lambda) = F_0'(\lambda) + F_0^2(\lambda) + 2 \sum_{i=1}^{M_0} \frac{F_0(\lambda) - F_0(\xi_{0i})}{\lambda - \xi_{0i}} - \sum_{\alpha=1}^{D} x^\alpha(\lambda)\varepsilon_{0\alpha}, \quad (4.1.3)$$

in which $\varepsilon_{0\alpha}$, $\alpha = 1, \ldots, D$ are arbitrarily chosen real numbers, belongs to the space $R_{2n}\binom{\vec{n}}{2\vec{d}}$. Substituting (4.1.3) into equation (4.1.1) we can rewrite it as

$$\left\{ -\frac{\partial^2}{\partial \lambda^2} + F_0'(\lambda) + F_0^2(\lambda) + 2 \sum_{i=1}^{M_0} \frac{F_0(\lambda) - F_0(\xi_{0i})}{\lambda - \xi_{0i}} \right\} \varphi(\lambda)$$

$$= \left\{ \sum_{\alpha=1}^{D} x^\alpha(\lambda)(\varepsilon_\alpha - \varepsilon_{0\alpha}) \right\} \varphi(\lambda), \qquad (4.1.4a)$$

or, after using conditions (4.1.2), as

$$\left[F_0(\lambda) + \sum_{i=1}^{M_0} \frac{1}{\lambda - \xi_{0i}} + \frac{\partial}{\partial \lambda} \right] \left[F_0(\lambda) + \sum_{i=1}^{M_0} \frac{1}{\lambda - \xi_{0i}} - \frac{\partial}{\partial \lambda} \right] \varphi(\lambda)$$

$$= \left\{ \sum_{\alpha=1}^{D} x^\alpha(\lambda)(\varepsilon_\alpha - \varepsilon_{0\alpha}) \right\} \varphi(\lambda). \qquad (4.1.4b)$$

The form (4.1.5) of equation (4.1.1) allows us to write down its explicit solution

$$\varphi(\lambda) = \exp\left[\int F_0(\lambda)\, d\lambda \right] \prod_{i=1}^{M_0} (\lambda - \xi_{0i}) \qquad (4.1.5a)$$

$$\varepsilon_\alpha = \varepsilon_{0\alpha}, \quad \alpha = 1, \ldots, D. \qquad (4.1.5b)$$

According to the results of the preceding chapter, other solutions of (4.1.4) must be sought in an analogous form:

$$\varphi(\lambda) = \exp\left[\int F(\lambda)\, d\lambda \right] \prod_{i=1}^{M} (\lambda - \xi_i), \qquad (4.1.6a)$$

$$\varepsilon_\alpha = \varepsilon_{0\alpha} + \epsilon_\alpha, \quad \alpha = 1, \ldots, D. \qquad (4.1.6b)$$

Substitution of (4.1.6) into (4.1.4) leads to the system of equations for $F(\lambda)$, ξ_i, $i = 1, \ldots, M$ and ϵ_α, $\alpha = 1, \ldots, \mathcal{D}$

$$\sum_{\alpha=1}^{\mathcal{D}} x^\alpha(\lambda)\epsilon_\alpha = \left\{ F_0'(\lambda) + F_0^2(\lambda) + 2\sum_{i=1}^{M_0} \frac{F_0(\lambda) - F_0(\xi_{0i})}{\lambda - \xi_{0i}} \right\}$$

$$- \left\{ F'(\lambda) + F^2(\lambda) + 2\sum_{i=1}^{M} \frac{F(\lambda) - F(\xi_i)}{\lambda - \xi_i} \right\}, \qquad (4.1.7)$$

$$\sideset{}{'}\sum_{k=1}^{M} \frac{1}{\xi_i - \xi_k} + F(\xi_i) = 0, \quad i = 1, \ldots, M, \qquad (4.1.8)$$

which will be studied in detail in subsequent sections in this chapter.

Thus, we see that for any $F_0(\lambda) \in R_n\left(\frac{d}{n}\right)$ $(\dim R_n\left(\frac{d}{n}\right) = \mathcal{D} + K + 1)$ and $x^\alpha(\lambda) \in R_{2n}\left(\frac{d}{2n}\right)$, $\alpha = 1, \ldots, \mathcal{D}$ $(\dim R_{2n}\left(\frac{d}{2n}\right) = 2(\mathcal{D}+K)+1)$ equation (4.1.4) has at least one explicit solution described by formula (4.1.6). The functions $F_0(\lambda)$ and $x^\alpha(\lambda)$, $\alpha = 1, \ldots, \mathcal{D}$ can always be chosen in such a way that all conditions of theorem 3.1 are satisfied and both equation (4.1.4) and its solutions (4.1.6) are physically sensible. This means that exact or quasi-exact solvability of equation (4.1.4) is guaranteed for any $K \geq 0$.

The case when $K = 0$ was discussed in detail in the preceding chapter. We demonstrated that in this case equation (4.1.4) can be interpreted only as an exactly solvable equation, since the condition for its solution $\varphi(\lambda)$ to belong to the space $W_1 \cap \ldots \cap W_{\mathcal{D}}$ automatically determines the functional structure (4.1.6a) of $\varphi(\lambda)$.

Now let us consider the case when $K = 1$. Suppose for definiteness that all double poles of the functions $x^\alpha(\lambda)$ are simple and real. Then their number is $\mathcal{D} + 3$ and they all lie on the circle forming the simple diagram depicted in figure 4.1.

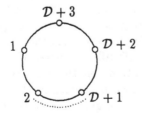

Figure 4.1. The diagram describing $\mathcal{D} + 3$ simple and real double poles.

In order to construct the spaces $W_1, \ldots, W_{\mathcal{D}}$ we need $\mathcal{D}$ physical intervals which must be located between the $\mathcal{D} + 3$ singular points on this diagram. There exist two essentially different possibilities. The first possibility is realized when the system of physical intervals is connected (see figure 4.2), while the second one corresponds to the systems consisting

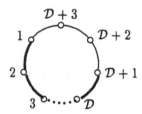

Figure 4.2. The diagram describing a connected system of physical intervals.

of two disconnected parts (see figure 4.3).

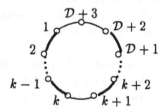

Figure 4.3. The diagram describing a disconnected system of physical intervals.

At first sight there is one more possibility, when the system of physical intervals consists of three disconnected parts. However, diagrams of such a sort contradict the normalization conditions (3.6.20) and cannot describe any physically sensible model.

Let us first consider the diagram in figure 4.2. We see that the number of singular points belonging to the system of physical intervals is here $\mathcal{D} + 2$. Therefore, in order to construct the solutions of equation (4.1.1) belonging to the system of $\mathcal{D}$ Hilbert spaces $W_1, \ldots, W_{\mathcal{D}}$, we must impose $\mathcal{D} + 2$ normalization conditions on the general solution of this equation, depending on $\mathcal{D} + 1$ unknown parameters: $\mathcal{D}$ spectral parameters ε_α and one mixing parameter θ. In the general case the system of such conditions is overdetermined and has no solutions. However, it is always possible to choose the functions $x^0(\lambda)$ and $x^\alpha(\lambda)$, $\alpha = 1, \ldots, \mathcal{D}$ in such a way that they make this system degenerate and solvable. In this case, we obtain the MPS equation (4.1.4) having a non-zero number of solutions in the space $W_1 \cap \ldots \cap W_{\mathcal{D}}$. The functional structure (4.1.6a) of these solutions can easily be recovered if we repeat the reasonings of section 3.6, and note that there is only one singular point lying beyond the system of physical intervals. But this means that the problem of solving equation (4.1.4) in $W_1 \cap \ldots \cap W_{\mathcal{D}}$ is purely algebraic, so that this equation is exactly solvable.

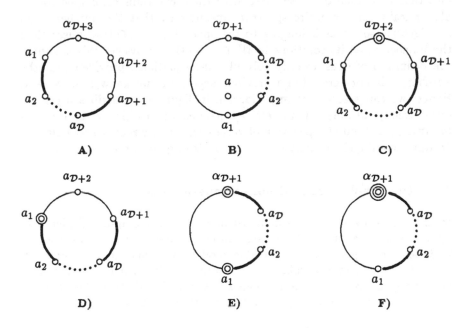

Figure 4.4. Non-equivalent diagrams for quasi-exactly solvable models.

Let us now consider the diagram in figure 4.3. The situation for this diagram differs drastically from the situation described above. Indeed, the number of double poles belonging to the system of physical intervals is now $\mathcal{D} + 1$, and, therefore, the number of normalization conditions and the number of unknown parameters characterizing the general solution of equation (4.1.4) coincide. The system is well defined and has a discrete set of solutions determining the spectrum of equation (4.1.4) in $W_1 \cap \ldots \cap W_{\mathcal{D}}$. However the functional structure of these solutions is now unknown. Now, we cannot restore it by the help of the reasonings of section 3.6, since the number of singular points lying outside the system of physical intervals is two. This means that physical solutions of equations (4.1.1) cannot be characterized in this case by a finite number of parameters and, therefore, the problem of finding these solutions is not an algebraic problem. A simple analysis shows that the same is true for equations (4.1.4) associated with other non-degenerate diagrams, in which the simple double poles lying outside the physical intervals are complex conjugated. Moreover, this is also true for the equations connected with degenerate diagrams arising when the simple double poles merge or go off to infinity. For any $\mathcal{D}$ the number of such (non-equivalent) diagrams is six. They are depicted in figure 4.4. The fact that MPS equations associated with these diagrams cannot be solved algebraically for the entire spectrum means only that the number N of its algebraic solutions belonging to the space $W_1 \cap \ldots \cap W_{\mathcal{D}}$ is less than the number of all its solutions in $W_1 \cap \ldots \cap W_{\mathcal{D}}$. However this is simply the definition of quasi-exactly solvable MPS equations, provided that the number N is non-zero. From previous arguments we know that $N \geq 1$. Hence, we can assert that any diagram in figure 4.4 describes a certain quasi-exactly solvable $\mathcal{D}$-parameter spectral equation which, after applying the inverse method of separation of variables, is easily reduced to a class of $\mathcal{D}$-dimensional quasi-exactly solvable models of quantum mechanics.

4.2   The one-dimensional non-degenerate case

In the preceding section we demonstrated that any diagram depicted in figure 4.4 and containing $\mathcal{D}$ physical intervals describes a certain class of $\mathcal{D}$-dimensional quasi-exactly solvable models of quantum mechanics. If $\mathcal{D} = 1$, these diagrams take an especially simple form (see figure 4.5). and the quasi-exactly solvable models associated with them become one dimensional. The Schrödinger equation for these models

$$\left\{ -\frac{\partial^2}{\partial x^2} + V(x) \right\} \psi(x) = E\psi(x) \tag{4.2.1}$$

coincides essentially with equation (4.1.4) which, after introducing more convenient notations $\rho(\lambda) \equiv x^1(\lambda)$, $\varepsilon_0 = \varepsilon_{01}$ and $\epsilon = \varepsilon_1 - \varepsilon_{10}$, can be rewritten as

$$\left[F_0(\lambda) + \sum_{i=1}^{M_0} \frac{1}{\lambda - \xi_{0i}} + \frac{\partial}{\partial \lambda} \right] \left[F_0(\lambda) + \sum_{i=1}^{M_0} \frac{1}{\lambda - \xi_{0i}} - \frac{\partial}{\partial \lambda} \right] \varphi(\lambda)$$
$$= \epsilon \rho(\lambda) \varphi(\lambda). \tag{4.2.2}$$

The only difference is that (4.2.1) is formulated in terms of other independent variables x and unknown functions $\psi(x)$. The connection

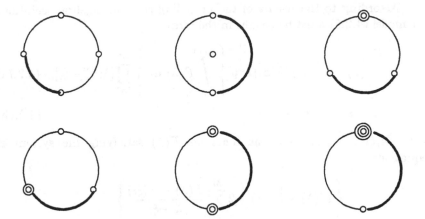

Figure 4.5. Non-equivalent diagrams for one-dimensional quasi-exactly solvable models.

between (4.2.1) and (4.2.2) is given by

$$x = \int \sqrt{\rho(\lambda)}\, d\lambda \qquad (4.2.3)$$

and

$$\psi(x) = \sqrt[4]{\rho(\lambda)}\varphi(\lambda). \qquad (4.2.4)$$

Denoting by $\lambda(x)$ the inverse function (4.2.3) one can write down the potential

$$V(x) = \varepsilon_0 + \frac{1}{\rho(\lambda(x))}\left\{ \frac{1}{4}\left[\frac{\rho'(\lambda(x))}{\rho(\lambda(x))}\right]' - \frac{1}{16}\left[\frac{\rho'(\lambda(x))}{\rho(\lambda(x))}\right]^2 \right.$$

$$\left. +F_0'(\lambda(x)) + F_0^2(\lambda(x)) + 2\sum_{i=1}^{M_0}\frac{F_0(\lambda(x)) - F_0(\xi_{0i})}{\lambda(x) - \xi_{0i}} \right\}, \qquad (4.2.5)$$

which, as we see, is completely characterized by the functions $\rho(\lambda)$ and $F_0(\lambda)$, the numbers M_0 and ε_0 and the parameters ξ_{0i}, $i = 1,\ldots, M_0$ satisfying the system of constraints

$$\sum_{k=1}^{M_0}{}'\frac{1}{\xi_{0i} - \xi_{0k}} + F_0(\xi_{0i}) = 0, \quad i = 1,\ldots, M_0. \qquad (4.2.6)$$

According to the results of the preceding section, algebraic solutions for model (4.2.5) must be sought in the form

$$\psi(x) = \sqrt[4]{\rho(\lambda(x))} \exp \left\{ \int^{\lambda(x)} F(\lambda)\, d\lambda \right\} \prod_{i=1}^{M} [\lambda(x) - \xi_i], \quad (4.2.7a)$$

$$e = \varepsilon_0 + \epsilon, \quad (4.2.7b)$$

with $M, \xi_1, \ldots, \xi_M, \epsilon$ and the functions $F(\lambda)$ satisfying the system of equations

$$\left\{ F'(\lambda) + F^2(\lambda) + 2 \sum_{i=1}^{M} \frac{F(\lambda) - F(\xi_i)}{\lambda - \xi_i} \right\}$$

$$- \left\{ F_0'(\lambda) + F_0^2(\lambda) + 2 \sum_{i=1}^{M_0} \frac{F_0(\lambda) - F_0(\xi_{0i})}{\lambda - \xi_{0i}} \right\} = -\epsilon \rho(\lambda) \quad (4.2.8)$$

and

$$\sum_{i=1}^{M} \frac{1}{\xi_i - \xi_k} + F(\xi_i) = 0, \quad i = 1, \ldots, M. \quad (4.2.9)$$

Now note that the functions $\rho(\lambda)$ and $F_0(\lambda)$ are fixed only up to conformal transformations

$$\rho(\lambda) \to \tilde{\rho}(\lambda) = \frac{(\alpha\delta - \beta\gamma)^2}{(\gamma\lambda + \delta)^4} \rho\left(\frac{\alpha\lambda + \beta}{\gamma\lambda + \delta} \right), \quad (4.2.10a)$$

$$F_0(\lambda) \to \tilde{F}_0(\lambda) = \frac{(1 - M_0)\gamma}{(\gamma\lambda + \delta)^2} F_0\left(\frac{\alpha\lambda + \beta}{\gamma\lambda + \delta} \right) \quad (4.2.10b)$$

which, according to the results of section 3.5, do not influence the final form of the potential (4.2.5) and corresponding solutions (4.2.7). This means that by using these transformations we can simplify the specific form of functions $\rho(\lambda)$ and $F_0(\lambda)$. In particular, it is always possible to ensure the following asymptotic behaviour of these functions at infinity:

$$\rho(\lambda) \approx \frac{\rho_0}{\lambda^4}, \quad \lambda \to \infty, \quad (4.2.11)$$

$$F_0(\lambda) \approx \frac{1 - M_0}{\lambda}, \quad \lambda \to \infty. \quad (4.2.12)$$

We know that for equation (4.2.1) to be quasi-exactly solvable, the function $\rho(\lambda)$ must have double poles of total order four. The condition (4.2.11) means that all the points at which $\rho(\lambda)$ has double poles are finite.

In the most general case, when all the double poles are simple, $\rho(\lambda)$ takes the form

$$\rho(\lambda) = \rho_0 \prod_{\alpha=1}^{4} \frac{(\lambda - \rho_\alpha)}{(\lambda - a_\alpha)^2} \qquad (4.2.13)$$

and the corresponding function $F_0(\lambda)$ can be represented as

$$F_0(\lambda) = \sum_{\alpha=1}^{4} \frac{\beta_{0\alpha} + \frac{1}{2}}{\lambda - a_\alpha}, \qquad (4.2.14)$$

where $\beta_{0\alpha}, \alpha = 1, \ldots, 4$ are certain given numbers satisfying the constraints

$$\sum_{\alpha=1}^{4} \beta_{0\alpha} = -(M_0 + 1). \qquad (4.2.15)$$

The unknown functions $F(\lambda)$ must have an analogous form

$$F(\lambda) = \sum_{\alpha=1}^{4} \frac{\beta_\alpha + \frac{1}{2}}{\lambda - a_\alpha}, \qquad (4.2.16)$$

where $\beta_\alpha, \alpha = 1, \ldots, 4$ are unknown numbers satisfying also a constraint of the type (4.2.15):

$$\sum_{\alpha=1}^{4} \beta_\alpha = -(M + 1). \qquad (4.2.17)$$

In this non-degenerate case the integral (4.2.3) can be evaluated explicitly. Indeed, using (4.2.13) and rewriting (4.2.3) in the form

$$
\begin{aligned}
x \;=\; & \sqrt{\rho_0} \int \left(\prod_{\alpha=1}^{4} \frac{\lambda - \rho_\alpha}{\lambda - a_\alpha} \right) \\
& \times \frac{d\lambda}{\sqrt{(\lambda - \rho_1)(\lambda - \rho_2)(\lambda - \rho_3)(\lambda - \rho_4)}} \\
=\; & \sqrt{\rho_0} \int \frac{d\lambda}{\sqrt{(\lambda - \rho_1)(\lambda - \rho_2)(\lambda - \rho_3)(\lambda - \rho_4)}} \\
+\; & \sqrt{\rho_0} \sum_{\alpha=1}^{4} \frac{\displaystyle\prod_{\beta=1}^{4} (a_\alpha - \rho_\beta)}{\displaystyle\prod_{\beta=1}^{4}{}^{\prime} (a_\alpha - a_\beta)} \\
& \times \int \frac{d\lambda}{(\lambda - a_\alpha)\sqrt{(\lambda - \rho_1)(\lambda - \rho_2)(\lambda - \rho_3)(\lambda - \rho_4)}}, \qquad (4.2.18)
\end{aligned}
$$

we see that the first term in (4.2.18) is an elliptic integral of the first type, while the four remaining terms are expressed in terms of elliptic integrals of the third type.

Unfortunately, in the general case, the function $x(\lambda)$ defined by formula (4.2.18) is explicitly non-invertible. This means that we cannot write a general expression for the function $\lambda(x)$ in a more or less closed form. However, there are some exceptions when an explicit inversion of the function $x(\lambda)$ becomes possible. All these cases will be discussed in detail in the next sections.

Now let us discuss the spectral equations (4.2.8), (4.2.9). In order to reduce them to a more convenient form we use another, absolutely equivalent representation of the weight function $\rho(\lambda)$:

$$\rho(\lambda) = \frac{1}{\displaystyle\prod_{\alpha=1}^{4}(\lambda - a_\alpha)} \left\{ \sum_{\alpha=1}^{4} \frac{\displaystyle\prod_{\beta=1}^{4}{}'(a_\alpha - a_\beta)}{\lambda - a_\alpha} R_\alpha + R_0 \right\} \qquad (4.2.19)$$

in which

$$R_0 = \rho_0, \qquad R_\alpha = \frac{\rho_0 \displaystyle\prod_{\beta=1}^{4}(a_\alpha - \rho_\beta)}{\displaystyle\prod_{\beta=1}^{4}{}'(a_\alpha - a_\beta)^2}, \qquad \alpha = 1,\dots,4. \qquad (4.2.20)$$

Substituting expressions (4.2.14), (4.2.16) and (4.2.19) into the system (4.2.8), (4.2.9) we obtain the following constraints for the parameters β_α, $\alpha = 1,\dots,4, M$, ξ_i, $i = 1,\dots,M$ and ϵ:

$$\beta_\alpha = \sigma_\alpha\sqrt{\beta_{0\alpha}^2 - R_\alpha\epsilon}, \qquad \sigma_\alpha = \pm 1, \qquad \alpha = 1,\dots,4, \qquad (4.2.21)$$

$$\sum_{\alpha=1}^{4} \sigma_\alpha\sqrt{\beta_{0\alpha}^2 - R_\alpha\epsilon} = -(M+1), \qquad (4.2.22)$$

$$\left\{ \left[\sum_{\alpha=1}^{4} a_\alpha^2 \left(\tfrac{1}{2} + \sigma_\alpha\sqrt{\beta_{0\alpha}^2 - R_\alpha\epsilon} \right) + \sum_{i=1}^{M} \xi_i^2 \right] \right.$$
$$\left. - \left[\sum_{\alpha=1}^{4} a_\alpha \left(\tfrac{1}{2} + \sigma_\alpha\sqrt{\beta_{0\alpha}^2 - R_\alpha\epsilon} \right) + \sum_{i=1}^{M} \xi_i \right]^2 \right\}$$

$$- \left\{ \left[\sum_{\alpha=1}^{4} a_\alpha^2 \left(\tfrac{1}{2} + \beta_{0\alpha} \right) + \sum_{i=1}^{M_0} \xi_{0i}^2 \right] \right.$$

$$\left. - \left[\sum_{\alpha=1}^{4} a_\alpha \left(\tfrac{1}{2} + \beta_{0\alpha} \right) + \sum_{i=1}^{M_0} \xi_{0i} \right]^2 \right\} = R_0 \epsilon, \qquad (4.2.23)$$

$$\sum_{k=1}^{M} {}' \frac{1}{\xi_i - \xi_k} + \sum_{\alpha=1}^{4} \frac{\tfrac{1}{2} + \sigma_\alpha \sqrt{\beta_{0\alpha}^2 - R_\alpha \epsilon}}{\xi_i - a_\alpha} = 0, \quad i = 1, \dots, M. \qquad (4.2.24)$$

The signs of the roots $\sqrt{\beta_{0\alpha}^2 - R_\alpha \epsilon}$ are chosen such that

$$\mathrm{Re} \sqrt{\beta_{0\alpha}^2 - R_\alpha \epsilon} > 0. \qquad (4.2.25)$$

The constraints (4.2.22)–(4.2.24) determine the system of $M + 2$ equations for $M + 1$ unknown quantities ϵ and $\xi_1, \dots, \xi_M$. This system is overdetermined and in general has no solutions, except the trivial one

$$\beta_\alpha = \beta_{0\alpha}, \quad \alpha = 1, \dots, 4; \qquad M = M_0;$$
$$\xi_i = \xi_{0i}, \quad i = 1, \dots, M_0; \qquad \epsilon = 0. \qquad (4.2.26)$$

However, there are some special cases when the first equation (4.2.22) of this system trivializes (becomes an identity). In this case it can be excluded from consideration, and we arrive at a system in which the numbers of equations and unknown quantities coincide. The number of solutions of such systems may exceed one and we obtain quasi-exactly solvable models of a certain finite order.

The cases when trivialization of equation (4.2.22) becomes possible arise when:

(i)   All roots in (4.2.22) do not depend on ϵ.

(ii) Two roots in (4.2.22) depend on ϵ but cancel, while the remaining two roots do not depend on ϵ.

(iii) Three roots in (4.2.22) depend on ϵ but cancel, while the fourth remaining root does not depend on ϵ.

In the next three sections we will discuss these three cases in detail and describe the sets of one-dimensional quasi-exactly solvable models associated with them.

4.3   The non-degenerate case. The first type

Let us assume that all four roots in (4.2.22) do not depend on ϵ. This is possible if and only if

$$R_1 = R_2 = R_3 = R_4 = 0, \quad R_0 = S, \tag{4.3.1}$$

where the sign of S is chosen from the condition of positivity of the weight function $\rho(\lambda)$. Then equation (4.2.22) is reduced to the form

$$\sigma_1\sqrt{\beta_{01}^2} + \sigma_2\sqrt{\beta_{02}^2} + \sigma_3\sqrt{\beta_{03}^2} + \sigma_4\sqrt{\beta_{04}^2} = -(M+1), \tag{4.3.2}$$

and can be interpreted as an additional constraint for the free parameters $\beta_{0\alpha}$, $\alpha = 1, \ldots, 4$ determining the form of potential (4.2.5). It can also be interpreted as an equation for the signs σ_α, $\alpha = 1, \ldots, 4$ and M, if the parameters $\beta_{0\alpha}$, $\alpha = 1, \ldots, 4$ and M_0 satisfying condition (4.2.15) are given.

In any event, equation (4.3.2) is absolutely independent of the remaining equations (4.2.23) and (4.2.24) which, after imposing the constraints (4.3.1), take an especially simple form:

$$
\begin{aligned}
\epsilon =\ & \frac{1}{S}\left\{ \left[\sum_{\alpha=1}^{4} a_\alpha^2 \left(\frac{1}{2} + \sigma_\alpha\sqrt{\beta_{0\alpha}^2} \right) + \sum_{i=1}^{M} \xi_i^2 \right] \right. \\
& - \left. \left[\sum_{\alpha=1}^{4} a_\alpha \left(\frac{1}{2} + \sigma_\alpha\sqrt{\beta_{0\alpha}^2} \right) + \sum_{i=1}^{M} \xi_i \right]^2 \right\} \\
& - \frac{1}{S}\left\{ \left[\sum_{\alpha=1}^{4} a_\alpha^2 \left(\frac{1}{2} + \beta_{0\alpha} \right) + \sum_{i=1}^{M_0} \xi_{0i}^2 \right] \right. \\
& - \left. \left[\sum_{\alpha=1}^{4} a_\alpha \left(\frac{1}{2} + \beta_{0\alpha} \right) + \sum_{i=1}^{M_0} \xi_{0i} \right]^2 \right\},
\end{aligned}
\tag{4.3.3}
$$

$$\sum_{k=1}^{M}{}' \frac{1}{\xi_i - \xi_k} + \sum_{\alpha=1}^{4} \frac{\frac{1}{2} + \sigma_\alpha\sqrt{\beta_{0\alpha}^2}}{\xi_i - a_\alpha} = 0, \quad i = 1, \ldots, M. \tag{4.3.4}$$

Here only the system (4.3.4) is non-trivial. Solving it and substituting the obtained values of ξ_i, $i = 1, \ldots, M$ into (4.3.3) we find the admissible values of ϵ.

According to (4.3.1), the weight function $\rho(\lambda)$ is in this case

$$\rho(\lambda) = \frac{S}{(\lambda - a_1)(\lambda - a_2)(\lambda - a_3)(l - a_4)}. \tag{4.3.5}$$

Substituting (4.3.5) into (4.2.5) and (4.2.7) and taking for definiteness

$$\varepsilon_0 = \frac{1}{S}\left\{\left[\sum_{\alpha=1}^{4}a_\alpha^2\left(\frac{1}{2}+\beta_{0\alpha}\right)+\sum_{i=1}^{M_0}\xi_{0i}^2\right]\right.$$
$$\left. - \left[\sum_{\alpha=1}^{4}a_\alpha\left(\frac{1}{2}+\beta_{0\alpha}\right)+\sum_{i=1}^{M_0}\xi_{0i}\right]^2\right\}, \tag{4.3.6}$$

we obtain the following simple expressions for the potential

$$V(x) = V_0 + \frac{1}{S}\sum_{\alpha=1}^{4}\left[\beta_{0\alpha}^2-\frac{1}{16}\right]\prod_{\beta=1}^{4}{}'(a_\alpha-a_\beta)\frac{1}{\lambda(x)-a_\alpha} \tag{4.3.7}$$

and the corresponding solutions

$$\psi(x) = \prod_{\alpha=1}^{4}[\lambda(x)-a_\alpha]^{\sigma_\alpha\sqrt{\beta_{0\alpha}^2}+\frac{1}{4}}\prod_{i=1}^{M}[\lambda(x)-\xi_i], \tag{4.3.8a}$$

$$E = \frac{1}{S}\left\{\left[\sum_{\alpha=1}^{4}a_\alpha^2\left(\frac{1}{2}+\sigma_\alpha\sqrt{\beta_{0\alpha}^2}\right)+\sum_{i=1}^{M}\xi_i^2\right]\right.$$
$$\left. - \left[\sum_{\alpha=1}^{4}a_\alpha\left(\frac{1}{2}+\sigma_\alpha\sqrt{\beta_{0\alpha}^2}\right)+\sum_{i=1}^{M}\xi_i\right]^2\right\}. \tag{4.3.8b}$$

Here

$$V_0 = \frac{1}{16S}\left[\left(\sum_{\alpha=1}^{4}a_\alpha\right)^2-4\sum_{\alpha=1}^{4}a_\alpha^2\right], \tag{4.3.9}$$

the numbers ξ_i satisfy system (4.3.4) and the signs σ_α and M can be found from (4.3.2) provided that the parameters $\beta_{0\alpha}$, $\alpha = 1,\ldots,4$ are given.

In order to obtain a concrete form of functions $V(x)$ and $\psi(x)$ in the x-representation, we must choose the numbers a_α and the physical interval, establish the stability conditions and evaluate the integral

$$x = \int\frac{\sqrt{S}\,d\lambda}{\sqrt{(\lambda-a_1)\ldots(\lambda-a_4)}} \tag{4.3.10}$$

determining the function $\lambda(x)$.

It is not difficult to see that there are two different (non-equivalent) possibilities which lead to two different classes of quasi-exactly solvable models. Below we consider these possibilities separately.

1.   The first possibility. This is realized when all four points a_α, $\alpha = 1, \ldots, 4$ are real. Taking into account the conformal covariance of the functions $\rho(\lambda)$ and $F_0(\lambda)$ we can take (without loss of generality)

$$a_1 = -1, \quad a_2 = +1, \quad a_3 = -\alpha, \quad a_4 = +\alpha, \qquad (4.3.11)$$

where

$$\alpha > 1. \qquad (4.3.12)$$

The role of physical intervals can be played in this case by the intervals $[-1, +1]$, $[1, \alpha]$, $[\alpha, -\alpha]$, $[-\alpha, -1]$. Owing to the same conformal covariance, all these intervals are absolutely equivalent. For definiteness we choose the first interval:

$$\lambda \in [-1, 1]. \qquad (4.3.13)$$

In this case the weight function $\rho(\lambda)$ takes the form

$$\rho(\lambda) = \frac{S}{(1 - \lambda^2)(\alpha^2 - \lambda^2)}. \qquad (4.3.14)$$

Obviously, it is positive definite if

$$S > 0. \qquad (4.3.15)$$

This case can be described by the diagram in figure 4.6 from which it is

Figure 4.6. The diagram for the weight function (4.3.14).

clear that the zeros of function $\rho(\lambda)$ coincide with the double poles at the points a_α.

Now integral (4.3.10) takes the form

$$\begin{aligned}
x &= \sqrt{S} \int_0^{\lambda(x)} \frac{d\lambda}{\sqrt{(1 - \lambda^2)(\alpha^2 - \lambda^2)}} \\
&= \frac{\sqrt{S}}{\alpha} F\left(\arcsin\lambda(x), \frac{1}{\alpha^2}\right),
\end{aligned} \qquad (4.3.16)$$

where $F(\varphi, m)$ is an elliptic integral of the first type with amplitude φ and modulus m. The function $x(\lambda)$ defined by (4.3.16) is invertible. We can write

$$\lambda(x) = \text{sn}(\omega x, m), \quad \omega = \frac{\alpha}{\sqrt{S}}, \quad m = \frac{1}{\alpha^2}. \tag{4.3.17}$$

The normalization conditions for the chosen interval (4.3.13) have the form

$$\sigma_1 = +1, \quad \sigma_2 = +1. \tag{4.3.18}$$

Substituting formulas (4.3.1), (4.3.17) and (4.3.18) into (4.3.7) and (4.3.8), and taking into account the fact that the reality of the α_α implies the reality of the $\beta_{0\alpha}$:

$$\sqrt{\beta_{0\alpha}^2} = |\beta_{0\alpha}|, \tag{4.3.19}$$

we obtain the final expression for the potential

$$
\begin{aligned}
V(x) \;=\; & -\omega^2 \frac{1+m}{2} + 2\omega^2(1-m)\left\{ \frac{\beta_{01}^2 - \frac{1}{16}}{1 + \text{sn}(\omega x, m)} + \frac{\beta_{02}^2 - \frac{1}{16}}{1 - \text{sn}(\omega x, m)} \right. \\
& \left. - \frac{\beta_{03}^2 - \frac{1}{16}}{1 + \sqrt{m}\,\text{sn}(\omega x, m)} - \frac{\beta_{04}^2 - \frac{1}{16}}{1 - \sqrt{m}\,\text{sn}(\omega x, m)} \right\}
\end{aligned}
\tag{4.3.20}
$$

and the corresponding solutions

$$
\begin{aligned}
\psi(x) \;=\; & [1 + \text{sn}(\omega x, m)]^{|\beta_{01}|} [1 - \text{sn}(\omega x, m)]^{|\beta_{02}|} \\
& \times\; [1 + \sqrt{m}\,\text{sn}(\omega x, m)]^{\sigma_3|\beta_{03}|} [1 - \sqrt{m}\,\text{sn}(\omega x, m)]^{\sigma_4|\beta_{04}|} \\
& \times\; \prod_{i=1}^{M} [\text{sn}(\omega x, m) - \xi_i],
\end{aligned}
\tag{4.3.21a}
$$

$$
\begin{aligned}
E \;=\; & \omega^2 m \left\{ 1 + \frac{1}{m} + \left[|\beta_{01}| + |\beta_{02}| + \frac{\sigma_3|\beta_{03}| + \sigma_4|\beta_{04}|}{m} + \sum_{i=1}^{M} \xi_i^2 \right] \right. \\
& \left. - \left[|\beta_{02}| - |\beta_{01}| + \frac{\sigma_4|\beta_{04}| - \sigma_3|\beta_{03}|}{\sqrt{m}} + \sum_{i=1}^{M} \xi_i \right]^2 \right\},
\end{aligned}
\tag{4.3.21b}
$$

where $\xi_1, \ldots, \xi_M$ are numbers satisfying the equation

$$
\begin{aligned}
\sum_{k=1}^{M} \frac{1}{\xi_i - \xi_k} + \frac{|\beta_{01}| + \frac{1}{2}}{\xi + 1} + \frac{|\beta_{02}| + \frac{1}{2}}{\xi - 1} \\
+ \frac{\sigma_3|\beta_{03}| + \frac{1}{2}}{\xi + m^{-\frac{1}{2}}} + \frac{\sigma_4|\beta_{04}| + \frac{1}{2}}{\xi - m^{-\frac{1}{2}}} = 0, \quad i = 1, \ldots, M
\end{aligned}
\tag{4.3.22}
$$

and the integers $\sigma_3 = \pm 1$, $\sigma_4 = \pm 1$ and M can be found from the equation

$$|\beta_{01}| + |\beta_{02}| + \sigma_3|\beta_{03}| + \sigma_4|\beta_{01}| = -(M+1), \qquad (4.3.23)$$

provided that the numbers $\beta_{0\alpha}$ satisfy the condition

$$\beta_{01} + \beta_{02} + \beta_{03} + \beta_{04} = -(M_0 + 1). \qquad (4.3.24)$$

The potential (4.3.20) is defined on the interval

$$x \in \left[-\frac{K(m)}{\omega}, +\frac{K(m)}{\omega} \right] \qquad (4.3.25)$$

where $K(m)$ is a complete elliptic integral. We see that it is singular at the points $x = \pm\frac{K(m)}{\omega}$, and, thus, has the form of a potential well of infinite depth (see figure 4.7). This means that the spectrum of the model (4.3.20) is

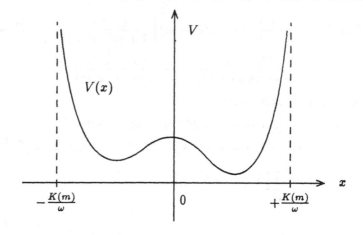

Figure 4.7. The form of the potential (4.3.20).

infinite and discrete. At the same time, the number of algebraic solutions in the model (4.3.20) is finite. It is equal to the number of solutions of system (4.3.22) which, according to statement 3.11, is $M+1$. Thus, we see that model (4.3.20) is a quasi-exactly solvable model of order $M+1$.

Strictly speaking the above is true only if equations (4.3.23) and (4.3.24) are satisfied. In order to elucidate when this is possible, assume for definiteness that

$$|\beta_{03}| < |\beta_{04}|. \qquad (4.3.26)$$

Then we have two possibilities

$$\sigma_3 = -1, \quad \sigma_4 = -1 \tag{4.3.27a}$$

and

$$\sigma_3 = +1, \quad \sigma_4 = -1. \tag{4.3.27b}$$

We shall distinguish between the following three cases:

(i) The expression $|\beta_{01}| + |\beta_{02}| - |\beta_{03}| - |\beta_{04}|$ is positive. Then the model (4.3.20) has no algebraic solutions.

(ii) The expression $|\beta_{01}| + |\beta_{02}| - |\beta_{03}| - |\beta_{04}|$ is negative and integer while the expression $|\beta_{01}| + |\beta_{02}| + |\beta_{03}| - |\beta_{04}|$ is either negative and non-integer, or positive. Then (4.3.20) is a quasi-exactly solvable model of order

$$M + 1 = |\beta_{03}| + |\beta_{04}| - |\beta_{01}| - |\beta_{02}|. \tag{4.3.28}$$

(iii) Both the expressions $|\beta_{01}| + |\beta_{02}| \pm |\beta_{03}| + |\beta_{04}|$ are negative and integer. Then the potential (4.3.20) describes two quasi-exactly solvable models of orders

$$M_1 + 1 = -|\beta_{03}| + |\beta_{04}| - |\beta_{01}| - |\beta_{02}| \tag{4.3.29a}$$

and

$$M_2 + 1 = |\beta_{03}| + |\beta_{04}| - |\beta_{01}| - |\beta_{02}|. \tag{4.3.29b}$$

This means that the infinite-dimensional hamiltonian matrix admits two different block decompositions depicted in figure 4.8. In other words, we know two different invariant subspaces for the hamiltonian of the model (4.3.20). Solutions belonging to these subspaces are described by formulas (4.3.21) with $\sigma_3 = +1$, $\sigma_4 = -1$ and $M = M_1$ of $\sigma_3 = -1$, $\sigma_4 = -1$ and $M = M_2$. These solutions describe the levels with the numbers $0, 1, \ldots, M_1$ in the first case, and the levels with the numbers $0, 1, \ldots, M_2$ in the second case. But since the spectra of one-dimensional quantum models are always non-degenerate, the first $M_1 + 1$ solutions of both these series must coincide. This is possible only if the large block in figure 4.8 has itself a block structure depicted in figure 4.9. In this case the energy levels corresponding to the numbers $0, 1, \ldots, M_1$ and $M_1 + 1, \ldots, M_2$ are described by two absolutely disconnected analytic functions and form two disconnected Riemann surfaces consisting of $M_1 + 1$ and $M_2 - M_1$ sheets, respectively. The energy levels belonging to these two sheets cannot be obtained from each other by analytic continuation.

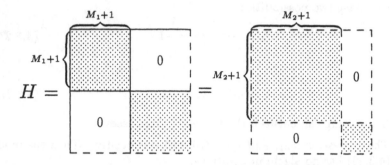

Figure 4.8. Two different block decompositions for the hamiltonian in the model (4.3.20).

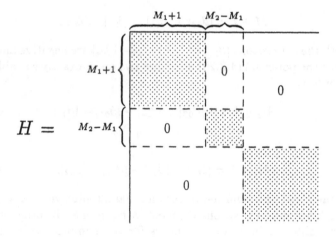

Figure 4.9. The block structure of the hamiltonian in the model (4.3.20).

This concludes the discussion of the case when the points a_α, $\alpha = 1, \ldots, 4$, in which the function $\rho(\lambda)$ has its poles, are real.

2. The second possibility. This is realized when two points, for example, a_1 and a_2, are real, while the remaining points a_3 and a_4 form a complex conjugate pair.

Due to conformal covariance we can take

$$a_1 = -1, \quad a_2 = +1, \quad a_3 = -i\alpha, \quad a_4 = +i\alpha \qquad (4.3.30)$$

with

$$\alpha > 0. \tag{4.3.31}$$

Now we have only two possibilities of choosing the physical interval $[-1, +1]$ and $[+1, -1]$, which, obviously, are equivalent. We choose the first interval

$$\lambda \in [-1, +1]. \tag{4.3.32}$$

In this case the weight function can be written as

$$\rho(\lambda) = \frac{-S}{(1 - \lambda^2)(\lambda^2 + \alpha^2)}. \tag{4.3.33}$$

This is positive definite in (4.3.32) when

$$S < 0. \tag{4.3.34}$$

The corresponding physical diagram is depicted in figure 4.10. In this case

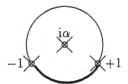

Figure 4.10. The diagram for the weight function (4.3.33).

integral (4.3.10) takes the form

$$
\begin{aligned}
x &= \sqrt{|S|} \int_0^\lambda \frac{d\lambda}{\sqrt{(1 - \lambda^2)(\lambda^2 + \alpha^2)}} \\
&= \frac{\sqrt{|S|}}{\sqrt{1 + \alpha^2}} F\left[\arcsin\left(\lambda\sqrt{\frac{1 + \alpha^2}{\lambda^2 + \alpha^2}} \right), \frac{1}{1 + \alpha^2} \right]. \tag{4.3.35}
\end{aligned}
$$

Inverting (4.3.35) we obtain

$$\lambda = \sqrt{1 - m}\, \mathrm{sd}(\omega x, m), \quad \omega = \frac{\sqrt{1 + \alpha^2}}{\sqrt{|S|}}, \quad m = \frac{1}{1 + \alpha^2}, \tag{4.3.36}$$

where sd is the elliptic function defined as

$$\mathrm{sd} \equiv \frac{\mathrm{sn}}{\mathrm{dn}} \equiv \mathrm{sn} \cdot \mathrm{nd}. \tag{4.3.37}$$

The stability condition for this diagram is again

$$\sigma_1 = +1, \quad \sigma_2 = +1. \tag{4.3.38}$$

The choice (4.3.30) implies that β_{01} and β_{02} are real while β_{03} and β_{04} are complex conjugated. Thus, the signs σ_3 and σ_4 must be negative:

$$\sigma_3 = -1, \quad \sigma_4 = -1. \tag{4.3.39}$$

Substituting formulas (4.3.30), (4.3.36), (4.3.38) and (4.3.39) into expressions (4.3.7) and (4.3.8), we obtain

$$
\begin{aligned}
V(x) \;=\; 2\omega^2 \Bigg\{ &\frac{\beta_{01}^2 - \frac{1}{16}}{1 + \sqrt{1-m}\,\mathrm{sd}(\omega x, m)} + \frac{\beta_{02}^2 - \frac{1}{16}}{1 - \sqrt{1-m}\,\mathrm{sd}(\omega x, m)} \\
&+ 2\frac{(\mathrm{Im}\,\beta_{03}^2)\sqrt{m}\,\mathrm{sd}(\omega x, m) - (\mathrm{Re}\,\beta_{03}^2 - \frac{1}{16})}{\mathrm{nd}^2(\omega x, m)} \Bigg\} + \omega^2 \frac{2m-1}{2}
\end{aligned}
\tag{4.3.40}
$$

and

$$
\begin{aligned}
\psi(x) \;=\; & [1 + \sqrt{1-m}\,\mathrm{sd}(\omega x, m)]^{|\beta_{01}| + \frac{1}{4}} \\
\times\; & [1 - \sqrt{1-m}\,\mathrm{sd}(\omega x, m)]^{|\beta_{02}| + \frac{1}{4}} \\
\times\; & [\mathrm{nd}^2(\omega x, m)]^{-\,\mathrm{Re}\,\sqrt{\beta_{03}^2} + \frac{1}{4}} \\
\times\; & \exp\left\{ -2\,\mathrm{Im}\,\sqrt{\beta_{03}^2}\,\mathrm{arctg}\,\left[\sqrt{m}\,\mathrm{sd}(\omega x, m) \right] \right\} \\
\times\; & \prod_{i=1}^{M} \left[\sqrt{1-m}\,\mathrm{sd}(\omega x, m) - \xi_i \right],
\end{aligned}
\tag{4.3.41a}
$$

$$
\begin{aligned}
E \;=\; & \frac{3}{2}\omega^2(1-2m) \\
& - \omega^2 m \Bigg\{ \left[|\beta_{01}| + |\beta_{02}| + 2\frac{1-m}{m}\,\mathrm{Re}\,\sqrt{\beta_{03}^2} + \sum_{i=1}^{M} \xi_i^2 \right] \\
& - \left[|\beta_{02}| - |\beta_{01}| - 2\frac{1-m}{m}\,\mathrm{Im}\,\sqrt{\beta_{03}^2} + \sum_{i=1}^{M} \xi_i \right]^2 \Bigg\}, \tag{4.3.41b}
\end{aligned}
$$

where the numbers $\xi_1, \ldots, \xi_M$ satisfy the equations

$$
\sum_{k=1}^{M}{}' \frac{1}{\xi_i - \xi_k} + \frac{|\beta_{01}| + \frac{1}{2}}{\xi_i + 1} + \frac{|\beta_{02}| + \frac{1}{2}}{\xi_i - 1} + 2\,\mathrm{Re}\,\frac{-\sqrt{\beta_{03}^2} + \frac{1}{2}}{\xi_i + i\sqrt{\frac{1-m}{m}}} = 0,
$$

$$
i = 1, \ldots, M, \tag{4.3.42}
$$

and M can be found from the condition

$$|\beta_{01}| + |\beta_{02}| - 2\operatorname{Re}\sqrt{\beta_{03}^2} = -(M+1), \qquad (4.3.43)$$

provided that

$$\beta_{01} + \beta_{02} + 2\operatorname{Re}\beta_{03} = -(M_0+1). \qquad (4.3.44)$$

The potential is defined on the interval

$$x \in \left[-\frac{K(m)}{\omega}, +\frac{K(m)}{\omega}\right]. \qquad (4.3.45)$$

It is singular at the points $x = \pm\frac{K(m)}{\omega}$ and has the form depicted in figure 4.7. Thus, its spectrum is again infinite and discrete. If conditions (4.3.43) and (4.3.44) are satisfied, potential (4.3.40) describes a quasi-exactly solvable model of order $M+1$.

4.4 The non-degenerate case. The second type

As noted in section 4.2, the second case of trivialization of equation (4.2.22) is realized when two roots in (4.2.22), for example, the first and second roots, depend on ϵ explicitly but cancel, while the third and fourth roots do not depend on ϵ.

This is possible when

$$R_0 = S, \quad R_1 = R_2 = R, \quad R_3 = R_4 = 0, \qquad (4.4.1)$$

where R and S are certain numbers, the signs of which are chosen from the condition of positivity of the weight function $\rho(\lambda)$. The additional conditions of cancellation of these roots have the form

$$\beta_{01} = \pi_1\beta_0, \quad \beta_{02} = \pi_2\beta_0, \qquad (4.4.2)$$

$$\sigma_1 = \sigma, \quad \sigma_2 = -\sigma, \qquad (4.4.3)$$

where β_0 is a certain number and π_1, π_2 and σ take the values ± 1. If all these conditions are satisfied, equation (4.2.22) takes the form

$$\sigma_3\sqrt{\beta_{03}^2} + \sigma_4\sqrt{\beta_{04}^2} = -(M+1). \qquad (4.4.4)$$

Taking into account the conformal covariance and letting

$$a_1 = a, \quad a_2 = -a, \qquad (4.4.5)$$

we can rewrite the remaining equations (4.2.23), (4.2.24) as

$$S\epsilon = \left\{ \left[a^2 + a_3^2 \left(\frac{1}{2} + \sigma_3 \sqrt{\beta_{03}^2} \right) + a_4^2 \left(\frac{1}{2} + \sigma_4 \sqrt{\beta_{04}^2} \right) + \sum_{i=1}^{M} \xi_i^2 \right] \right.$$

$$- \left. \left[2a\sigma\sqrt{\beta_0^2 - R\epsilon} + a_3 \left(\frac{1}{2} + \sigma_3 \sqrt{\beta_{03}^2} \right) + a_4 \left(\frac{1}{2} + \sigma_4 \sqrt{\beta_{04}^2} \right) + \sum_{i=1}^{M} \xi_i \right]^2 \right\}$$

$$- \left\{ \left[a^2 + a^2 \beta_0 (\pi_1 + \pi_2) + a_3^2 \left(\frac{1}{2} + \beta_{03} \right) + a_4^2 \left(\frac{1}{2} + \beta_{04} \right) + \sum_{i=1}^{M_0} \xi_{0i}^2 \right] \right.$$

$$- \left. \left[a\beta_0 (\pi_1 - \pi_2) + a_3 \left(\frac{1}{2} + \beta_{03} \right) + a_4 \left(\frac{1}{2} + \beta_{04} \right) + \sum_{i=1}^{M_0} \xi_{0i} \right]^2 \right\} \qquad (4.4.6)$$

and

$$\sum_{k=1}^{M}{}' \frac{1}{\xi_i - \xi_k} + \frac{\frac{1}{2} + \sigma\sqrt{\beta_0^2 - R\epsilon}}{\xi_i - a} + \frac{\frac{1}{2} - \sigma\sqrt{\beta_0^2 - R\epsilon}}{\xi_i + a}$$

$$+ \frac{\frac{1}{2} + \sigma_3 \sqrt{\beta_{03}^2}}{\xi_i - a_3} + \frac{\frac{1}{2} + \sigma_4 \sqrt{\beta_{04}^2}}{\xi_i - a_4} = 0, \quad i = 1, \dots, M. \qquad (4.4.7)$$

From this system of $M + 1$ equations it is possible to find $M + 1$ unknown quantities ϵ and $\xi_1, \dots, \xi_M$.

The most general form of weight functions satisfying conditions (4.4.1) is

$$\rho(\lambda) = R \left(\frac{1}{\lambda - a} - \frac{1}{\lambda + a} \right)^2 + \frac{S - 4Ra^2}{(\lambda - a)(\lambda + a)(\lambda - a_3)(\lambda - a_4)}. \qquad (4.4.8)$$

It is convenient to rewrite (4.4.8) in the form

$$\rho(\lambda) = \frac{S}{(\lambda - a)^2 (\lambda + a)^2} \frac{(\lambda - \rho_1)(\lambda - \rho_2)}{(\lambda - a_3)(\lambda - a_4)}, \qquad (4.4.9)$$

where ρ_1 and ρ_2 are the roots of the quadratic equation

$$(\lambda - a_3)(\lambda - a_4) + \left(\frac{S}{4a^2 R} - 1 \right) (\lambda - a_1)(\lambda - a_2) = 0. \qquad (4.4.10)$$

In this case the potential $V(x)$ takes the form:

$$V(x) = \varepsilon_0 + \frac{1}{S} \frac{(\lambda - a)(\lambda + a)}{(\lambda - \rho_1)(\lambda - \rho_2)}$$

$$\times \left\{ \left[\frac{1/2}{(\lambda - a)^2} + \frac{1/2}{(\lambda + a)^2} + \frac{1/4}{(\lambda - a_3)^2} \right. \right.$$

$$+ \frac{1/4}{(\lambda - a_4)^2} - \frac{1/4}{(\lambda - \rho_1)^2} - \frac{1/4}{(\lambda - \rho_2)^2}$$

$$\left. - \left(\frac{1/2}{\lambda - a} + \frac{1/2}{\lambda + a} + \frac{1/4}{\lambda - a_3} + \frac{1/4}{\lambda - a_4} - \frac{1/4}{\lambda - \rho_1} - \frac{1/4}{\lambda - \rho_2} \right)^2 \right]$$

$$\times \ (\lambda - a)(\lambda + a)(\lambda - a_3)(\lambda - a_4)$$

$$+ \left(\beta_0^2 - \frac{1}{4} \right) \left[\frac{2a(a - a_3)(a - a_4)}{\lambda - a} - \frac{2a(a - a_3)(a - a_4)}{\lambda + a} \right]$$

$$+ \left(\beta_{03}^2 - \frac{1}{4} \right) \frac{(a_3 - a)(a_3 + a)(a_3 - a_4)}{\lambda - a_3}$$

$$+ \left(\beta_{04}^2 - \frac{1}{4} \right) \frac{(a_4 - a)(a_4 + a)(a_4 - a_3)}{\lambda - a_4}$$

$$- \left[a^2 + a^2(\pi_1 + \pi_2)\beta_0 + a_3^2 \left(\frac{1}{2} + \beta_{03} \right) + a_4^2 \left(\frac{1}{2} + \beta_{04} \right) + \sum_{i=1}^{M_0} \xi_{0i}^2 \right]$$

$$+ \left[a(\pi_1 - \pi_2)\beta_0 + a_3 \left(\frac{1}{2} + \beta_{03} \right) + a_4 \left(\frac{1}{2} + \beta_{04} \right) + \sum_{i=1}^{M_0} \xi_{0i} \right]^2 \right\} \Bigg|_{\lambda = \lambda(x)} \quad (4.4.11)$$

and for the corresponding solutions we have

$$\psi(x) \ \sim \ (\lambda - \rho_1)^{\frac{1}{4}} (\lambda - \rho_1)^{\frac{1}{4}} \left(\frac{\lambda - a}{\lambda + a} \right)^{\sigma \sqrt{\beta_0^2 - R\epsilon}}$$

$$\times \ (\lambda - a_3)^{\sigma_3 \sqrt{\beta_{03}^2} + \frac{1}{4}} (\lambda - a_4)^{\sigma_4 \sqrt{\beta_{04}^2} + \frac{1}{4}}$$

$$\times \ \prod_{i=1}^{M} (\lambda - \xi_i) \Bigg|_{\lambda = \lambda(x)} , \quad (4.4.12a)$$

$$E \ = \ \varepsilon_0 + \epsilon, \quad (4.4.12b)$$

where the numbers ϵ and ξ_i satisfy equations (4.4.6) and (4.4.7), and σ_3, σ_4 and M can be found from condition (4.4.4).

As in the previous case, in order to obtain the concrete form of $V(x)$ and $\psi(x)$ it is necessary to choose the numbers a, a_3 and a_4, the physical interval and establish the stability conditions. Besides, we must evaluate integral (4.2.3) determining the form of the functions $\lambda(x)$.

Let us consider again two non-equivalent possibilities leading to different types of quasi-exactly solvable model.

1. The first possibility. This is realized when all points a, a_3 and a_4 are real. We can take (without loss of generality)

$$a = -1, \quad a_3 = 0, \quad a_4 = -\alpha, \qquad (4.4.13)$$

where

$$\alpha > 0. \qquad (4.4.14)$$

At first sight, we have four possibilities in choosing the physical interval. However, some of these possibilities are forbidden. For example, we cannot take $\lambda \in [1, -1]$ or $\lambda \in [-\alpha, 0]$. Indeed, the first choice implies the stability condition $\sigma_1 = \sigma_2 = 1$ which contradicts the requirement (4.4.3). The second choice leads to another stability condition $\sigma_3 = \sigma_4 = 1$ which also cannot be satisfied since the left-hand side of (4.4.4) must be negative. The remaining two possibilities are equivalent to the following choice of the physical interval

$$\lambda \in [0, 1], \qquad (4.4.15)$$

provided that $0 < \alpha < 1$ or $1 < \alpha < \infty$. These cases can be described by the diagrams depicted in figure 4.11 and can be considered simultaneously.

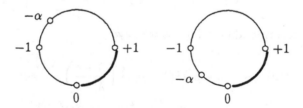

Figure 4.11. Two non-equivalent diagrams corresponding to the weight function (4.4.19).

The stability condition for this diagram is

$$\sigma_2 = +1, \quad \sigma_3 = +1. \qquad (4.4.16)$$

Using (4.4.3) and the condition of negativity of the left-hand side of (4.4.4) we obtain

$$\sigma_1 = -1, \quad \sigma_4 = -1. \qquad (4.4.17)$$

The weight function $\rho(\lambda)$ is described by the formula

$$\rho(\lambda) = \frac{4R}{(\lambda^2 - 1)^2} + \frac{S - 4R}{(\lambda^2 - 1)\lambda(\lambda + \alpha)}. \qquad (4.4.18)$$

This function is positive in the physical interval (4.4.15) when

$$R > 0, \quad S \le 4R. \tag{4.4.19}$$

Let us now list the extended diagrams associated with the function (4.4.18) and describing different types of quasi-exactly solvable model.

In the case when $0 < \alpha < 1$ we have six non-equivalent diagrams depicted in figures 4.12–4.17. Here we have used the notation $R_\pm = \frac{1}{2}(1 \pm \sqrt{1 - \alpha^2})$. In the case when $1 < \alpha < \infty$ we have only two non-equivalent diagrams depicted in figures 4.18 and 4.19.

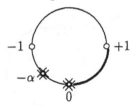

Figure 4.12. Diagram 1. $0 < \alpha < 1$, $S = 4R$.

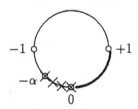

Figure 4.13. Diagram 2. $0 < \alpha < 1$, $4R_+ < S < 4R$.

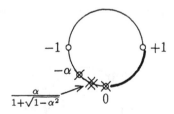

Figure 4.14. Diagram 3. $0 < \alpha < 1$, $S = 4R_+$.

The integrals (4.2.3) for diagrams 2, 4, 6 and 8 are expressed via elliptic integrals of the first and third type. The integrals for diagrams 1, 3, 5 and 7 are simpler and can be expressed in terms of inverse hyperbolic functions.

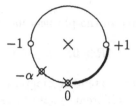

Figure 4.15. Diagram 4. $0 < \alpha < 1$, $4R_- < S < 4R_+$.

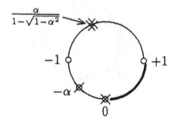

Figure 4.16. Diagram 5. $0 < \alpha < 1$, $S = 4R_-$.

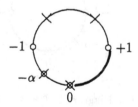

Figure 4.17. Diagram 6. $0 < \alpha < 1$, $S < 2R_-$.

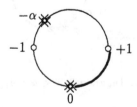

Figure 4.18. Diagram 7. $1 < \alpha < \infty$, $S = 4R$.

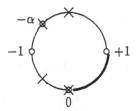

Figure 4.19. Diagram 8. $1 < \alpha < \infty$, $S < 4R$.

The integrals for diagrams 1 and 7 have the simplest form. They describe rather interesting quasi-exactly solvable models which will be discussed below.

Using the weight function

$$\rho(\lambda) = \frac{4R}{(1 - \lambda^2)^2} \qquad (4.4.20)$$

and evaluating the integral (4.2.3)

$$x = 2\sqrt{R} \int\limits_0^\lambda \frac{d\lambda}{1 - \lambda^2} = 2\sqrt{R}\,\text{arcth}\,\lambda, \qquad (4.4.21)$$

we obtain

$$\lambda(x) = \text{th}\,\omega x, \quad \omega = \frac{1}{2\sqrt{R}}. \qquad (4.4.22)$$

Substituting (4.4.22) into (4.4.11) and taking for definiteness

$$\varepsilon_0 = -\omega^2, \qquad (4.4.23)$$

we obtain the potential

$$
\begin{aligned}
V(x) \; = \; & \frac{\omega^2}{\text{sh}\,\omega x \;\; \text{ch}\,\omega x \;\; (\text{th}\,\omega x + \alpha)} \\
\times \; & \left\{ \left(\beta_0^2 - \frac{1}{4} \right) \left[\frac{2(1 - \alpha)}{1 + \text{th}\,\omega x} + \frac{2(1 + \alpha)}{1 - \text{th}\,\omega x} \right] \right. \\
+ \; & \left(\beta_{03}^2 - \frac{1}{4} \right) \frac{\alpha}{\text{th}\,\omega x} + \left(\beta_{04}^2 - \frac{1}{4} \right) \frac{\alpha(\alpha^2 - 1)}{\alpha + \text{th}\,\omega x} \\
+ \; & \left. \left[(\pi_1 + \pi_2)\beta_0 + \alpha^2 \beta_{04} + 1 + \frac{\alpha^2}{2} + \sum_{i=1}^{M_0} \xi_{0i}^2 \right] \right.
\end{aligned}
$$

$$- \left[(\pi_2 - \pi_1)\beta_0 - \alpha\beta_{04} - \frac{\alpha}{2} + \sum_{i=1}^{M_0} \xi_{0i} \right]^2 \Bigg\}, \qquad (4.4.24)$$

in which the numbers ξ_{0i}, $i = 1, \ldots, M$ satisfy the system of equations

$$\sum_{k=1}^{M_0} {}' \frac{1}{\xi_{0i} - \xi_{0k}} + \frac{\pi_1\beta_0 + \frac{1}{2}}{\xi_{0i} + 1} + \frac{\pi_2\beta_0 + \frac{1}{2}}{\xi_{0i} - 1}$$

$$+ \frac{\beta_{03} + \frac{1}{2}}{\xi_{0i}} + \frac{\beta_{04} + \frac{1}{2}}{\xi_{0i} + \alpha} = 0, \quad i = 1, \ldots, M_0. \qquad (4.4.25)$$

Substitution of (4.4.22) into (4.4.12) gives us the solution of the corresponding Schrödinger equation:

$$\psi(x) = \left[\frac{1 - \operatorname{th}\omega x}{1 + \operatorname{th}\omega x} \right]^{\sqrt{\beta_0^2 - R\epsilon}} (\operatorname{th}\omega x)^{|\beta_{03}| + \frac{1}{2}}$$

$$\times \ (\operatorname{th}\omega x + \alpha)^{-|\beta_{04}| + \frac{1}{2}} \prod_{i=1}^{M} (\operatorname{th}\omega x - \xi_i), \qquad (4.4.26a)$$

$$E = -\omega^2 + \epsilon. \qquad (4.4.26b)$$

The values of parameters ϵ and $\xi_1, \ldots, \xi_M$ can be found from the system:

$$\omega^2 \epsilon = \left\{ \left[1 + \frac{\alpha^2}{2} |\beta_{04}| + \sum_{i=1}^{M} \xi_i^2 \right] \right.$$

$$- \left[2\sqrt{\beta_0^2 - R\epsilon} + \alpha|\beta_{04}| - \frac{\alpha}{2} + \sum_{i=1}^{M} \xi_i \right]^2 \right\}$$

$$- \left\{ \left[(\pi_1 + \pi_2)\beta_0 + \alpha^2\beta_{04} + 1 + \frac{\alpha^2}{2} + \sum_{i=1}^{M_0} \xi_{0i}^2 \right] \right.$$

$$- \left[(\pi_2 - \pi_1)\beta_0 - \alpha\beta_{04} - \frac{\alpha}{2} + \sum_{i=1}^{M_0} \xi_{0i} \right] \right\}, \qquad (4.4.27)$$

$$\sum_{k=1}^{M} {}' \frac{1}{\xi_i - \xi_k} + \frac{\frac{1}{2} - \sqrt{\beta_0^2 - R\epsilon}}{\xi_i + 1} + \frac{\frac{1}{2} + \sqrt{\beta_0^2 - R\epsilon}}{\xi_i - 1}$$

$$+ \frac{|\beta_{03}| + \frac{1}{2}}{\xi_i} + \frac{-|\beta_{04}| + \frac{1}{2}}{\xi_i + \alpha} = 0, \quad i = 1, \ldots, M. \qquad (4.4.28)$$

Here the number M is determined by the formula

$$M = |\beta_{04}| - |\beta_{03}| - 1, \tag{4.4.29}$$

which means that the difference $|\beta_{04}| - |\beta_{03}|$ must be a natural number. Note also that the numbers $\beta_{0\alpha}$ and M_0 entering into the formulas (4.4.24)–(4.4.29) must satisfy the condition

$$(\pi_1 + \pi_2)\beta_0 + \beta_{03} + \beta_{04} = -(M_0 + 1). \tag{4.4.30}$$

If both the conditions (4.4.29) and (4.4.30) are satisfied, the Schrödinger equation for the potential (4.4.24) has $M + 1$ algebraic solutions and, thus, describes a quasi-exactly solvable model of order $M + 1$.

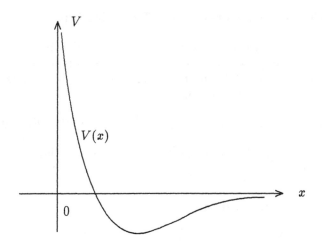

Figure 4.20. The form of the potential (4.4.24).

This model is defined on the semi-infinite interval

$$x \in [0, \infty]. \tag{4.4.31}$$

From the explicit expression (4.4.24) it follows that the potential $V(x)$ is singular at $x = 0$ and vanishes if $x \to \infty$. Thus, it has the form of a potential well of finite depth (see figure 4.20).

2. The second possibility. Let us now discuss the case when two of the points a_α are complex conjugated. Since both the potential (4.4.11) and the solutions (4.4.12) must be real, it is sufficient to examine two cases: 1) when a is imaginary and a_3 and a_4 are real and 2) when a is real and a_3 and a_4 are complex conjugated.

In the first case the real points a_3 and a_4 play the role of the ends of the physical interval. Therefore, the normalization condition implies that $\sigma_3 = \sigma_4 = 1$. However, this contradicts the condition of negativity of the left-hand side of (4.4.4).

In the second case the physical interval must have its ends at the points a and $-a$. Therefore the normalization condition results in $\sigma_1 = \sigma_2 = 1$. However, this contradicts the condition of cancellation of roots in (4.2.22).

Thus, we can conclude that the second possibility of trivialization of equation (4.2.22) cannot be realized.

4.5   The non-degenerate case. The third type

In this section we discuss the last possibility of trivializing equation (4.2.22). For this possibility to be realized the first three roots in (4.2.22) (depending on ϵ) must cancel, while the fourth root must be independent of ϵ. The necessary conditions for this are

$$R_0 = S, \quad R_1 = A_1^2 R, \quad R_2 = A_2^2 R, \quad R_3 = R, \quad R_4 = 0, \qquad (4.5.1)$$

where A_1 and A_2 are certain positive numbers such that

$$A_1 + A_2 = 1. \qquad (4.5.2)$$

Simultaneously, we must take

$$\beta_{01} = A_1 \pi_1 \beta_0, \quad \beta_{02} = A_2 \pi_2 \beta_0, \quad \beta_{03} = \beta_0, \qquad (4.5.3)$$

and

$$\sigma_1 = \sigma_2 = \sigma, \quad \sigma_3 = -\sigma. \qquad (4.5.4)$$

Then the equation (4.2.22) becomes

$$\sigma_4 \sqrt{\beta_{04}^2} = -(M + 1), \qquad (4.5.5)$$

which gives us immediately

$$\sigma_4 = -1. \qquad (4.5.6)$$

Taking for definiteness

$$a_1 = a, \quad a_2 = -a, \quad a_3 = 0 \qquad (4.5.7)$$

(and remembering that such a choice does not lead to loss of generality), we obtain for the remaining equations (4.2.23) and (4.2.24):

$$
\left\{\left[a^2\left(1+\sigma\sqrt{\beta_0^2-R\epsilon}\right)-a_4^2\left(M+\tfrac{1}{2}\right)+\sum_{i=1}^{M}\xi_i^2\right]\right.
$$

$$
-\left[a(A_1-A_2)\sigma\sqrt{\beta_0^2-R\epsilon}-a_4\left(M+\tfrac{1}{2}\right)+\sum_{i=1}^{M}\xi_i\right]^2\right\}
$$

$$
-\left\{\left[a^2(1+\beta_{01}+\beta_{02})+a_4^2\left(\beta_{04}+\tfrac{1}{2}\right)+\sum_{i=1}^{M_0}\xi_{0i}^2\right]\right.
$$

$$
\left.-\left[a(\beta_{01}-\beta_{02})+a_4\left(\beta_{04}+\tfrac{1}{2}\right)+\sum_{i=1}^{M_0}\xi_{0i}\right]^2\right\}=S\epsilon, \qquad (4.5.8)
$$

$$
\sum_{k=1}^{M}{}' \frac{1}{\xi_i-\xi_k} + \frac{\tfrac{1}{2}+A_1\sigma\sqrt{\beta_0^2-R\epsilon}}{\xi_i-a}
$$

$$
+ \frac{\tfrac{1}{2}+A_2\sigma\sqrt{\beta_0^2-R\epsilon}}{\xi_i+a} + \frac{\tfrac{1}{2}-\sigma\sqrt{\beta_0^2-R\epsilon}}{\xi_i}
$$

$$
- \frac{M-\tfrac{1}{2}}{\xi_i-a_4}=0, \qquad i=1,\ldots,M. \qquad (4.5.9)
$$

The most general form of the weight function satisfying conditions (4.5.1) is

$$
\rho(\lambda) = R\left(\frac{A_1}{\lambda-a}+\frac{A_2}{\lambda+a}-\frac{1}{\lambda}\right)^2
$$

$$
+ \frac{S-Ra^2(A_1-A_2)^2}{(\lambda-a)(\lambda+a)\lambda(\lambda-a_4)}. \qquad (4.5.10)
$$

It is convenient to rewrite it in the form

$$
\rho(\lambda) = \frac{S}{(\lambda-a)^2(\lambda+a)^2\lambda^2}\frac{(\lambda-\rho_1)(\lambda-\rho_2)(\lambda-\rho_3)}{(\lambda-a_4)}, \qquad (4.5.11)
$$

where ρ_1,ρ_2 and ρ_3 are zeros of $\rho(\lambda)$, which can be found from the cubic equation

$$
\left(\lambda+\frac{a}{A_1-A_2}\right)^2(\lambda-a_4)+\left[\frac{S}{Ra^2(A_1-A_2)^2}-1\right](\lambda^2-a^2)\lambda=0.
$$

$$
(4.5.12)
$$

The potential $V(x)$ takes the form

$$
\begin{aligned}
V(x) \;=\; & \varepsilon_0 + \frac{1}{S}\frac{(\lambda-a)(\lambda+a)\lambda}{(\lambda-\rho_1)(\lambda-\rho_2)(\lambda-\rho_3)} \\
& \times \left\{\left[\frac{1/2}{(\lambda-a)^2}+\frac{1/2}{(\lambda+a)^2}+\frac{1/2}{\lambda^2}+\frac{1/4}{(\lambda-a_4)^2}-\frac{1/4}{(\lambda-\rho_1)^2}\right.\right. \\
& \quad -\frac{1/4}{(\lambda-\rho_2)^2}-\frac{1/4}{(\lambda-\rho_3)^2}-\left(\frac{1/2}{\lambda-a}+\frac{1/2}{\lambda+a}+\frac{1/2}{\lambda}\right. \\
& \quad \left.\left.+\frac{1/4}{\lambda-a_4}-\frac{1/4}{\lambda-\rho_1}-\frac{1/4}{\lambda-\rho_2}-\frac{1/4}{\lambda-\rho_3}\right)^2\right] \\
& \times (\lambda-a)(\lambda+a)\lambda(\lambda-a_4) \\
& +\left(A_1^2\beta_0^2-\tfrac14\right)\frac{2a^2(a-a_4)}{\lambda-a}-\left(A_2^2\beta_0^2-\tfrac14\right)\frac{2a^2(a-a_4)}{\lambda+a} \\
& +\left(\beta_0^2-\tfrac14\right)\frac{a^2a_4}{\lambda}+\left(\beta_{04}^2-\tfrac14\right)\frac{(a_4-a)(a_4+a)a_4}{\lambda-a_4} \\
& -\left[a^2+(\pi_1A_1+\pi_2A_2)a^2\beta_0+a_4^2\left(\beta_{04}+\tfrac12\right)+\sum_{i=1}^{M_0}\xi_{0i}^2\right] \\
& +\left[a(\pi_1A_1-\pi_2A_2)\beta_0+a_4\left(\beta_{04}+\tfrac12\right)+\sum_{i=1}^{M_0}\xi_{0i}\right]^2\Bigg\}\Bigg|_{\lambda=\lambda(x)},
\end{aligned}
$$

(4.5.13)

and for the corresponding solutions we have

$$
\begin{aligned}
\psi(x) \;=\; & (\lambda-\rho_1)^{\frac14}(\lambda-\rho_2)^{\frac14}(\lambda-\rho_3)^{\frac14} \\
& \times \left[\frac{(\lambda-a)^{A_1}(\lambda+a)^{A_2}}{\lambda}\right]^{\sigma\sqrt{\beta_0^2-R\epsilon}} \\
& \times (\lambda-a_4)^{-M-\frac34}\prod_{i=1}^{M}(\lambda-\xi_i)\Bigg|_{\lambda=\lambda(x)}
\end{aligned}
$$

(4.5.14a)

and

$$
E \;=\; \varepsilon_0+\epsilon,
$$

(4.5.14b)

where ϵ and ξ_i satisfy equations (4.5.8) and (4.5.9), and the sign σ can be found from the normalization condition. There are several ways to make the models described by formulas (4.5.13), (4.5.14) physically sensible.

1. The first possibility. The parameters a and a_4 are real. We can take:

$$a_1 = -1, \quad a_2 = +1, \quad a_3 = 0, \quad a_4 = -\alpha, \tag{4.5.15}$$

with

$$\alpha > 0. \tag{4.5.16}$$

As in the previous case, the physical interval can be chosen in many ways. However, the unique way which does not contradict the normalization conditions is

$$\lambda \in [1, -1]. \tag{4.5.17}$$

The corresponding diagram is depicted in figure 4.21 and the normalization

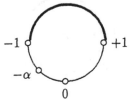

Figure 4.21. The diagram describing the weight function (4.5.19).

condition for this diagram is

$$\sigma_1 = \sigma_2 = -\sigma_3 = \sigma = 1. \tag{4.5.18}$$

For the function $\rho(\lambda)$ we have

$$\rho(\lambda) = R\left(\frac{A_1}{\lambda+1} + \frac{A_2}{\lambda-1} - \frac{1}{\lambda}\right)^2 + \frac{S - R(A_1 - A_2)^2}{(\lambda^2 - 1)\lambda(\lambda + \alpha)}. \tag{4.5.19}$$

This is positive in (4.5.17) if

$$S > R(A_1 - A_2)^2, \quad R > 0. \tag{4.5.20}$$

Let us now list the extended diagrams associated with the function (4.5.19) and describing different types of quasi-exactly solvable model.

Denote by S_0 a positive root of the system of two equations

$$(\lambda_0 - \gamma)^2 (\lambda_0 + \alpha) = \left[1 - \frac{S_0\gamma^2}{R}\right](\lambda_0^2 - 1)\lambda_0, \tag{4.5.21a}$$

$$(\lambda_0 - \gamma)(3\lambda_0 + 2\alpha - \gamma) = \left[1 - \frac{S_0\gamma^2}{R}\right](3\lambda_0^2 - 1) \tag{4.5.21b}$$

in which $\gamma = (A_1 - A_2)^{-1}$. (It is not difficult to verify that this system may have only one positive root.) Then we have three inequivalent cases:

1. $R(A_1 - A_2)^2 < S < S_0$. Then

$$\rho(\lambda) = \frac{S}{(\lambda^2 - 1)^2 \lambda^2} \cdot \frac{(\lambda - \rho_1)(\lambda - \rho)(\lambda - \rho^*)}{\lambda + \alpha}, \tag{4.5.22}$$

where ρ_1, ρ and ρ^* are real and complex conjugated roots of the first (cubic) equation (4.5.21a). We have

$$-1 < \rho_1 < -\alpha. \tag{4.5.23}$$

The diagram describing this case is depicted in figure 4.22.

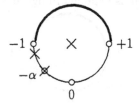

Figure 4.22. Diagram 1. $\alpha > 0$, $R(A_1 - A_2)^2 < S < S_0$.

2. $S = S_0$. Then

$$\rho(\lambda) = \frac{S_0}{(\lambda^2 - 1)^2 \lambda^2} \frac{(\lambda - \rho_1)(\lambda - \rho)^2}{(\lambda + \alpha)}, \tag{4.5.24}$$

where ρ_1 and ρ are single and double roots of (4.5.21). We have the inequalities

$$-1 < \rho_1 < -\alpha, \quad 0 < \rho < 1, \tag{4.5.25}$$

and the diagram describing this case is depicted in figure 4.23.

3. $S > S_0$. Then

$$\rho(\lambda) = \frac{S}{(\lambda^2 - 1)^2 \lambda^2} \frac{(\lambda - \rho_1)(\lambda - \rho_2)(\lambda - \rho_3)}{\lambda + \alpha}, \tag{4.5.26}$$

where ρ_1, ρ_2 and ρ_3 are three different real roots of (4.5.21), such that

$$-1 < \rho_1 < -\alpha, \quad 0 < \rho_2 \neq \rho_3 < 1. \tag{4.5.27}$$

The corresponding diagram has the form shown in figure 4.24.

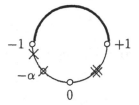

Figure 4.23. Diagram 2. $\alpha > 0$, $S = S_0$.

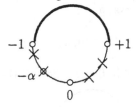

Figure 4.24. Diagram 3. $\alpha > 0$, $S > S_0$.

Note that integrals (4.2.3) for diagrams 1 and 3 are expressed in terms of elliptic integrals, while the integral for diagram 2 is expressed in terms of inverse hyperbolic functions. Unfortunately, the resulting solutions are too cumbersome and therefore we shall not write down explicit expressions for them. We only note that the potentials corresponding to all these cases are defined on the whole x-axis

$$x \in [-\infty, +\infty]. \tag{4.5.28}$$

They are regular at infinity and tend to constants if x tends to $\pm\infty$. Therefore, they describe potential wells of finite depth (see figure 4.25).

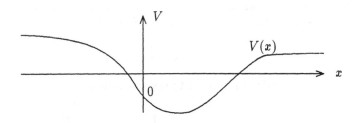

Figure 4.25. The form of the potentials associated with diagrams 1, 2 and 3.

2. The second possibility. This can be realized when the point a is complex while the points a_3 and a_4 are real. Without loss of generality, we can take

$$a_1 = -i, \quad a_2 = +i, \quad a_3 = 0, \quad a_4 = \alpha, \tag{4.5.29}$$

with

$$\alpha < 0. \tag{4.5.30}$$

Choosing the physical interval as

$$\lambda \in [0, \alpha], \tag{4.5.31}$$

we get the diagram depicted in figure 4.26 and the normalization conditions

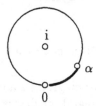

Figure 4.26. The diagram describing the weight function (4.5.35).

$$\sigma_3 = \sigma_4 = 1. \tag{4.5.32}$$

From (4.5.4) it also follows that

$$\sigma_1 = \sigma_2 = -1. \tag{4.5.33}$$

Note that the condition of cancellation of the roots with the indices 1, 2 and 3 implies that the parameters β_0 and R are real and

$$A_1 = A_2 = \tfrac{1}{2}. \tag{4.5.34}$$

In this case the weight function $\rho(\lambda)$ takes the form

$$\rho(\lambda) = R \left(\frac{1}{2(\lambda + i)} + \frac{1}{2(\lambda - i)} - \frac{1}{\lambda} \right)^2 + \frac{S}{(\lambda^2 + 1)\lambda(\lambda - \alpha)}. \tag{4.5.35}$$

The condition of its positivity in (4.5.31) is

$$R > 0, \quad S \leq 0. \tag{4.5.36}$$

In order to classify the types of quasi-exactly solvable model associated with function (4.5.35), let us introduce the number S_0 defined as a negative solution of the system

$$(\lambda_0^2 + 1)\lambda_0 + \frac{R}{S_0}(\lambda_0 - \alpha) = 0, \qquad (4.5.37a)$$

$$3\lambda_0^2 + 1 + \frac{R}{S_0} = 0. \qquad (4.5.37b)$$

Then we arrive at the diagrams given as figures 4.27–4.30. The integrals

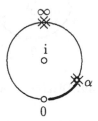

Figure 4.27. Diagram 4. $\alpha < 0$, $S = 0$.

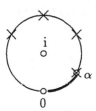

Figure 4.28. Diagram 5. $\alpha < 0$, $S_0 < S < 0$.

for diagrams 5, 6 and 7 are expressed in terms of inverse elliptic and hyperbolic functions. The integral for diagram 4 has the simplest form. Indeed, substituting the corresponding function

$$\rho(\lambda) = \frac{R}{(\lambda^2 + 1)^2 \lambda^2} \qquad (4.5.38)$$

into (4.2.3) we obtain

$$x = -\sqrt{R} \int \frac{d\lambda}{\lambda(\lambda^2 + 1)} = \frac{\sqrt{R}}{2} \ln\left(1 + \frac{1}{\lambda^2}\right). \qquad (4.5.39)$$

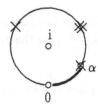

Figure 4.29. Diagram 6. $\alpha < 0$, $S = S_0$.

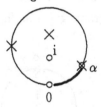

Figure 4.30. Diagram 7. $\alpha < 0$, $S < S_0$.

This gives us the following explicit expression for the function $\lambda(x)$:

$$\lambda(x) = \frac{1}{\sqrt{\exp \omega x - 1}}, \qquad \omega = \frac{2}{\sqrt{R}}. \tag{4.5.40}$$

Substituting (4.5.40) and (4.5.29) into (4.5.13) and (4.5.14) and taking for definiteness

$$\varepsilon_0 = -\frac{1}{R} \tag{4.5.41}$$

and

$$\pi_1 = \pi_2 = \pi, \tag{4.5.42}$$

we obtain the potential

$$
\begin{aligned}
V(x) &= -\frac{3}{16}\omega^2 \left(\frac{\exp \omega x}{\exp \omega x - 1}\right)^2 + \frac{1}{4}\frac{\omega^2 \exp \omega x}{(\exp \omega x - 1)\left[1 - \alpha\sqrt{\exp \omega x - 1}\right]} \\
&\times \left\{ (\beta_0^2 - 1)\frac{\exp \omega x - 1}{\exp \omega x}\left(\frac{\alpha}{\sqrt{\exp \omega x - 1}} + 1\right) - (\beta_0^2 - \tfrac{1}{4}) \right. \\
&\times \left. \alpha\sqrt{\exp \omega x - 1} + (\beta_{04}^2 - \tfrac{1}{4})\,\alpha(a^2 + 1)\frac{\sqrt{\exp \omega x - 1}}{1 - \alpha\sqrt{\exp \omega x - 1}} \right. \\
&\left. + \; 1 + \pi\beta_0 - \alpha^2\left(\beta_{04} + \tfrac{1}{2}\right) - \sum_{i=1}^{M_0} \xi_{0i}^2 \right.
\end{aligned}
$$

$$+ \left[\alpha \left(\beta_{04} + \tfrac{1}{2}\right) + \sum_{i=1}^{M_0} \xi_{0i}\right]^2 \Bigg\} \qquad (4.5.43)$$

and the corresponding solutions

$$\psi(x) = (\exp \, \omega x)^{-\frac{1}{2}\sqrt{\beta_0^2 - R\epsilon}} \left[\frac{1}{\sqrt{\exp \, \omega x - 1}} - \alpha\right]^{-M - \frac{3}{4}}$$

$$\times \prod_{i=1}^{M} \left[\frac{1}{\sqrt{\exp \, \omega x - 1}} - \xi_i\right], \qquad (4.5.44a)$$

$$E = -\frac{\omega^2}{4} + \epsilon. \qquad (4.5.44b)$$

The numbers ϵ and $\xi_1, \ldots, \xi_M$ can be found from the system of equations

$$\left\{\sqrt{\beta_0^2 - R\epsilon} - \alpha^2 \left(M + \tfrac{1}{2}\right) + \sum_{i=1}^{M} \xi_i^2 - \left[-\alpha \left(M + \tfrac{1}{2}\right) + \sum_{i=1}^{M} \xi_i\right]^2\right\}$$

$$= \left\{-\pi \beta_0 + \alpha^2 \left(\beta_{04} + \tfrac{1}{2}\right) + \sum_{i=1}^{M_0} \xi_{0i}^2 - \left[\alpha \left(\beta_{04} + \tfrac{1}{2}\right) + \sum_{i=1}^{M_0} \xi_{0i}\right]^2\right\},$$

$$(4.5.45)$$

$$\sum_{k=1}^{M}{}' \frac{1}{\xi_i - \xi_k} + \frac{1 - \sqrt{\beta_0^2 - R\epsilon}}{2(\xi_i + i)} + \frac{1 - \sqrt{\beta_0^2 - R\epsilon}}{2(\xi_i - i)}$$

$$+ \frac{\tfrac{1}{2} + \sqrt{\beta_0^2 - R\epsilon}}{\xi_i} - \frac{M + \tfrac{1}{2}}{\xi_i - \alpha} = 0, \quad i = 1, \ldots, M \qquad (4.5.46)$$

which have $M + 1$ different solutions provided that parameters β_0 and β_{04} satisfy the following additional conditions

$$|\beta_{04}| = M + 1, \qquad (4.5.47)$$
$$(\pi + 1)\beta_0 + \beta_{04} = (M_0 + 1). \qquad (4.5.48)$$

This gives us a quasi-exactly solvable model of order $M + 1$ defined on the semi-infinite interval

$$x \in \left[\frac{1}{\omega} \ln \left(1 + \frac{1}{\alpha^2}\right), \infty\right]. \qquad (4.5.49)$$

The potential of this model is singular at the point $x_0 = \frac{1}{\omega} \ln(1 + \frac{1}{\alpha^2})$, and tends to a constant if $x \to \infty$. Therefore, we have a singular potential well of finite depth (see figure 4.31).

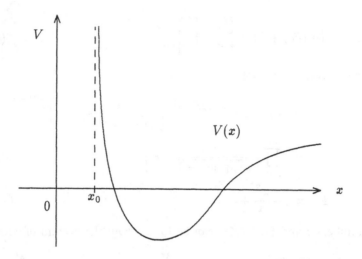

Figure 4.31. The form of the potential (4.5.43).

4.6   The one-dimensional simplest degenerate case

In the preceding three sections we discussed quasi-exactly solvable models connected with a non-degenerate weight function $\rho(\lambda)$ having the form (4.2.13). This function had only simple double poles at the points a_α, $\alpha = 1,\dots,4$. Let us now consider the degenerate cases which appear when the double poles of the function $\rho(\lambda)$ merge, and show that they also lead to wide classes of one-dimensional quasi-exactly solvable models.

The simplest degenerate case arises when any two double poles, for example, a_3 and a_4, merge. Due to conformal covariance we can assume (without loss of generality) that they merge at infinity. In order to obtain the resulting (degenerate) form of the weight function $\rho(\lambda)$ it is sufficient to replace ρ_0 by $\rho_0 a_3^2 a_4^2$ in (4.2.13) and then take the limit $a_3 \to \infty, a_4 \to \infty$. This gives

$$\rho(\lambda) = \rho_0 \frac{(\lambda - \rho_1)(\lambda - \rho_2)(\lambda - \rho_3)(\lambda - \rho_4)}{(\lambda - a_1)^2 (\lambda - a_2)^2}, \qquad (4.6.1)$$

or, equivalently,

$$
\begin{aligned}
\rho(\lambda) \;=\; & R_1 \left[\frac{1}{(\lambda - a_1)^2} - \frac{1}{(\lambda - a_1)(\lambda - a_2)} \right] \\
+ \; & R_2 \left[\frac{1}{(\lambda - a_2)^2} - \frac{1}{(\lambda - a_2)(\lambda - a_1)} \right] + R_3
\end{aligned}
$$

$$+ \ 2R_4 \frac{\lambda - a_1 - a_2}{(\lambda - a_1)(\lambda - a_2)} + R_5 \frac{1}{(\lambda - a_1)(\lambda - a_2)}, \qquad (4.6.2)$$

where $R_1, \ldots, R_5$ are certain given parameters.

To obtain the degenerate form of the corresponding function $F_0(\lambda)$ (defined in the non-degenerate case by (4.2.14)), we must replace the numbers β_{03} and β_{04} by $-\frac{a_3}{2}(\beta_{03} - \beta_{04})$ and β_{04} by $-\frac{a_3}{2}(\beta_{03} + \beta_{04})$ and then take the same limit $a_3 \to \infty$, $a_4 \to \infty$. Then we obtain

$$F_0(\lambda) = \frac{\beta_{01} + \frac{1}{2}}{\lambda - a_1} + \frac{\beta_{02} + \frac{1}{2}}{\lambda - a_2} + \beta_{03}. \qquad (4.6.3)$$

Substitution of (4.6.1) into (4.2.5) gives us the potentials of the corresponding quasi-exactly solvable models. Solutions of these models must be sought in the form (4.2.7). Here the role of the function $F_M(\lambda)$ is played by

$$F_M(\lambda) = \frac{\beta_1 + \frac{1}{2}}{\lambda - a_1} + \frac{\beta_2 + \frac{1}{2}}{\lambda - a_2} + \beta_3, \qquad (4.6.4)$$

where β_1, β_2 and β_3 are certain unknown numbers.

The spectral equations for $\beta_1, \beta_2, \beta_3$ and also for ϵ and $\xi_1, \ldots, \xi_M$ can be obtained after substituting formulas (4.6.2)–(4.6.4) into the general equations (4.2.8) and (4.2.9). This gives

$$\beta_\alpha = \sigma_\alpha \sqrt{\beta_{0\alpha}^2 - R_\alpha \epsilon}, \quad \alpha = 1, 2, 3, \qquad (4.6.5)$$

$$\frac{\beta_{03}(\beta_{01} + \beta_{02} + M_0 + 1) - R_4 \epsilon}{\sigma_3 \sqrt{\beta_{03}^2 - R_3 \epsilon}}$$

$$- \ \sigma_1 \sqrt{\beta_{01}^2 - R_1 \epsilon} - \sigma_2 \sqrt{\beta_{02}^2 - R_2 \epsilon} = M + 1, \qquad (4.6.6)$$

$$\left(\sigma_1 \sqrt{\beta_{01}^2 - R_1 \epsilon} + \sigma_2 \sqrt{\beta_{02}^2 - R_2 \epsilon} + M + 1 \right)$$

$$\times \left(\sigma_1 \sqrt{\beta_{01}^2 - R_1 \epsilon} + \sigma_2 \sqrt{\beta_{02}^2 - R_2 \epsilon} + M \right)$$

$$+ \ 2\sigma_3 \sqrt{\beta_{03} - R_3 \epsilon} \left(\sum_{\alpha=1}^{2} a_\alpha \sigma_\alpha \sqrt{\beta_{0\alpha}^2 - R_\alpha \epsilon} + \frac{a_1 + a_2}{2} + \sum_{i=1}^{M} \xi_i \right)$$

$$= \ (\beta_{01} + \beta_{02} + M_0 + 1)(\beta_{01} + \beta_{02} + M_0) + 2\beta_{03}$$

$$\times \left(\beta_{01} a_1 + \beta_{02} a_2 + \frac{a_1 + a_2}{2} + \sum_{i=1}^{M_0} \xi_{0i} \right) - R_5 \epsilon, \qquad (4.6.7)$$

$$\sum_{k=1}^{M} \frac{1}{\xi_i - \xi_k} + \sum_{\alpha=1}^{2} \frac{\sigma_\alpha \sqrt{\beta_{0\alpha}^2 - R_\alpha \epsilon} + \frac{1}{2}}{\xi_i - a_\alpha} + \sigma_3 \sqrt{\beta_{03}^2 - R_3 \epsilon} = 0,$$

$$i = 1, \dots, M. \quad (4.6.8)$$

As in the non-degenerate case, we have the system (4.6.6)–(4.6.8) of $M + 2$ equations for $M + 1$ quantities ϵ and $\xi_1, \dots, \xi_M$. This system is overdetermined and, in the general case, does not have solutions. The coincidence of the number of equations and the number of unknown quantities becomes possible only if one of these equations is trivialized (becomes an identity). The equation (4.6.6) admits a most natural trivialization. It can be trivialized when

(i)   all roots in (4.6.6) and the numerator of the first term do not depend on ϵ,

(ii) the roots with the indices 1 and 2 depend on ϵ but cancel, while the first term in (4.6.6) does not depend on ϵ, or

(iii) the first term in (4.6.6) cancels with one of the roots 1 or 2, while the remaining root does not depend on ϵ.

These cases will be discussed in the next three sections.

Concluding this short section, we note that the points a_1 and a_2 must be real. They cannot be made complex conjugated as in the non-degenerate case. Indeed if these points were complex, the physical interval would coincide with the whole real λ-axis. But this is impossible according to the exclusion principle 3 given in section 3.6.

The reality of the points a_1 and a_2 enables us to take (without loss of generality)

$$a_1 = -1, \qquad a_2 = +1. \quad (4.6.9)$$

In this case the numbers β_{01}, β_{02} and β_{03} must also be real.

4.7   The simplest degenerate case. The first type

Let us assume that all roots in (4.6.6) as well as the numerator of the first term do not depend on ϵ. This is possible when

$$R_1 = R_2 = R_3 = R_4 = 0, \quad R_5 = S. \quad (4.7.1)$$

Then equation (4.6.6) is reduced to the form

$$\frac{\beta_{03}(\beta_{01} + \beta_{02} + M_0 + 1)}{\sigma_3 |\beta_{03}|} - \sigma_1 |\beta_{01}| - \sigma_2 |\beta_{02}| = M + 1 \quad (4.7.2)$$

and can be interpreted as an equation for the integers $\sigma_1, \sigma_2, \sigma_3 = \pm 1$ and $M = 0, 1, 2, \ldots$, provided that the numbers β_{01}, β_{02} and β_{03} are given. In this case the remaining equations (4.6.7) and (4.6.8) become

$$
\begin{aligned}
&\left\{ (\sigma_1|\beta_{01}| + \sigma_2|\beta_{02}| + M + 1)(\sigma_1|\beta_{01}| + \sigma_2|\beta_{02}| + M) \right. \\
&\qquad \left. + 2\sigma_3|\beta_{03}| \left(\sigma_2|\beta_{02}| - \sigma_1|\beta_{01}| + \sum_{i=1}^{M} \xi_i \right) \right\} \\
&- \left\{ (\beta_{01} + \beta_{02} + M_0 + 1)(\beta_{01} + \beta_{02} + M_0) \right. \\
&\qquad \left. + 2\beta_{03} \left(\beta_{02} - \beta_{01} + \sum_{i=1}^{M_0} \xi_{0i} \right) \right\} = -S\epsilon,
\end{aligned}
\tag{4.7.3}
$$

$$
\sum_{k=1}^{M} \frac{1}{\xi_i - \xi_k} + \frac{\sigma_1|\beta_{01}| + \frac{1}{2}}{\xi_i + 1} + \frac{\sigma_2|\beta_{02}| + \frac{1}{2}}{\xi_i - 1}
$$
$$
+ \ \sigma_3|\beta_{03}| = 0, \quad i = 1, \ldots, M.
\tag{4.7.4}
$$

Here only the system (4.7.4) is non-trivial. Solving it and substituting the result into (4.7.3) we can obtain an explicit expression for the spectral parameter ϵ.

The choice (4.7.1) implies that

$$
\rho(\lambda) = \frac{S}{\lambda^2 - 1}
\tag{4.7.5}
$$

and

$$
F_0(\lambda) = \frac{\beta_{01} + \frac{1}{2}}{\lambda + 1} + \frac{\beta_{02} + \frac{1}{2}}{\lambda - 1} + \beta_{03},
\tag{4.7.6a}
$$

$$
F(\lambda) = \frac{\beta_1 + \frac{1}{2}}{\lambda + 1} + \frac{\beta_2 + \frac{1}{2}}{\lambda - 1} + \beta_3.
\tag{4.7.6b}
$$

Substituting these expressions into (4.2.5) and (4.2.7) and taking for definiteness

$$
\begin{aligned}
\varepsilon_0 = -\frac{1}{S} \left\{ \frac{1}{4} + (\beta_{01} + \beta_{02} + M_0 + 1)(\beta_{01} + \beta_{02} + M_0) \right. \\
\left. + 2\beta_{03} \left(\beta_{02} - \beta_{01} + \sum_{i=1}^{M_0} \xi_{0i} \right) \right\},
\end{aligned}
\tag{4.7.7}
$$

we obtain the following expressions for the potential

$$V(x) = -\frac{2}{S}\left(\beta_{01}^2 - \tfrac{1}{16}\right)\frac{1}{1+\lambda(x)} - \frac{2}{S}\left(\beta_{02}^2 - \tfrac{1}{16}\right)\frac{1}{1-\lambda(x)}$$

$$+\frac{2}{S}\beta_{03}(\beta_{01} + \beta_{02} + M_0 + 1)\lambda(x)$$

$$+\frac{1}{S}\beta_{03}^2\left[\lambda^2(x) - 1\right], \tag{4.7.8}$$

and corresponding solutions

$$\psi(x) = [\lambda(x) + 1]^{\sigma_1|\beta_{01}|+\frac{1}{4}}[\lambda(x) - 1]^{\sigma_2|\beta_{02}|+\frac{1}{4}}$$

$$\times \quad \exp[\sigma_3|\beta_{03}|\lambda(x)]\prod_{i=1}^{M}[\lambda(x) - \xi_i], \tag{4.7.9a}$$

$$E = -\frac{1}{S}\left\{\frac{1}{4} + (\sigma_1|\beta_{01}| + \sigma_2|\beta_{02}| + M_0 + 1)\right.$$

$$\times \quad (\sigma_1|\beta_{01}| + \sigma_2|\beta_{02}| + M_0)$$

$$+ \quad 2\sigma_3|\beta_{03}|\left(\sigma_2|\beta_{02}| - \sigma_1|\beta_{01}| + \sum_{i=1}^{M}\xi_i\right)\right\}, \tag{4.7.9b}$$

in which ξ_i, $i = 1,\ldots,M$ satisfy the system (4.7.4), while the signs $\sigma_1, \sigma_2, \sigma_3$ and M satisfy the constraint (4.7.2).

Now it is not difficult to obtain the concrete form of the potential $V(x)$ and solutions $\psi(x)$ in the x-representation. For this we must choose the physical interval, establish normalization conditions and construct the function $\lambda(x)$ evaluating the integral (4.2.3).

Note that we have only two possibilities in choosing the physical interval. The first possibility is

$$\lambda \in [-1, +1]. \tag{4.7.10}$$

This case is described by the extended physical diagram depicted in figure 4.32 and the positivity condition for the weight function (4.7.5) is

$$S < 0. \tag{4.7.11}$$

Evaluating integral (4.2.3) we get

$$x = \sqrt{|S|}\int_0^{\lambda(x)}\frac{d\lambda}{\sqrt{1-\lambda^2}} = \arcsin\lambda(x) \tag{4.7.12}$$

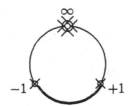

Figure 4.32. Diagram 1. $S < 0$.

and, consequently,

$$\lambda(x) = \sin\omega x, \quad \omega = \frac{1}{\sqrt{|S|}}. \tag{4.7.13}$$

The normalization condition implies that

$$\sigma_1 = +1, \quad \sigma_2 = +1. \tag{4.7.14}$$

Then, substituting (4.7.13) and (4.7.14) into (4.7.8) and (4.7.9), we arrive at the following final expressions for the potential

$$
\begin{aligned}
V(x) = {} & 2\omega^2 \left[\frac{\beta_{01}^2 - \frac{1}{16}}{1 + \sin\omega x} + \frac{\beta_{02}^2 - \frac{1}{16}}{1 - \sin\omega x} \right. \\
& \left. - \beta_{03}(\beta_{01} + \beta_{02} + M_0 + 1)\sin\omega x \right] + \omega^2 \beta_{03}^2 \cos^2\omega x
\end{aligned}
\tag{4.7.15}
$$

and solutions

$$
\begin{aligned}
\psi(x) = {} & (1 + \sin\omega x)^{|\beta_{01}| + \frac{1}{4}}(1 - \sin\omega x)^{|\beta_{02}| + \frac{1}{4}} \\
& \times \ \exp(\sigma_3 |\beta_{03}| \sin\omega x) \prod_{i=1}^{M} (\sin\omega x - \xi_i),
\end{aligned}
\tag{4.7.16a}
$$

$$
\begin{aligned}
E = {} & \omega^2 \left\{ \left(|\beta_{01}| + |\beta_{02}| + M + \tfrac{1}{2} \right)^2 \right. \\
& \left. + \ 2\sigma_3 |\beta_{03}| \left(|\beta_{02}| - |\beta_{01}| + \sum_{i=1}^{M} \xi_i \right) \right\},
\end{aligned}
\tag{4.7.16b}
$$

in which the numbers ξ_i, $i = 1, \ldots, M$ satisfy the system of equations

$$\sum_{k=1}^{M}{}' \frac{1}{\xi_i - \xi_k} + \frac{|\beta_{01}| + \frac{1}{2}}{\xi_i + 1} \frac{|\beta_{02}| + \frac{1}{2}}{\xi_i - 1} + \sigma_3 |\beta_{03}| = 0,$$

$$i = 1, \dots, M, \qquad (4.7.17)$$

and the integers σ_3 and M can be determined from the equation

$$\sigma_3 \frac{\beta_{03}}{|\beta_{03}|} (\beta_{01} + \beta_{02} + M_0 + 1) - |\beta_{01}| - |\beta_{02}| = M + 1. \qquad (4.7.18)$$

If equation (4.7.18) is satisfied, we obtain a quasi-exactly solvable model of order $M + 1$, defined on the interval

$$x \in \left[-\frac{\pi}{2\omega}, +\frac{\pi}{2\omega} \right]. \qquad (4.7.19)$$

The potential (4.7.15) is singular at the points $\pm\frac{\pi}{2\omega}$ and, thus, describes a potential well of infinite depth.

The second possibility is realized when

$$\lambda \in [1, \infty]. \qquad (4.7.20)$$

In this case the extended physical diagram takes the form depicted in figure 4.33 and the corresponding normalization condition is

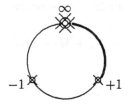

Figure 4.33. Diagram 2. $S > 0$.

$$S > 0. \qquad (4.7.21)$$

Now we have

$$x = \sqrt{S} \int_1^\lambda \frac{d\lambda}{\sqrt{\lambda^2 - 1}} = \sqrt{S} \operatorname{arcch} \lambda \qquad (4.7.22)$$

and, thus,

$$\lambda = \operatorname{ch} \omega x, \quad \omega = \frac{1}{\sqrt{S}}. \qquad (4.7.23)$$

Besides, we have the normalization conditions

$$\sigma_2 = +1, \quad \sigma_3 = -1. \tag{4.7.24}$$

In this case the potential becomes

$$
V(x) = 2\omega^2 \left[\frac{\beta_{01}^2 - \frac{1}{16}}{\operatorname{ch}\omega x + 1} + \frac{\beta_{02}^2 - \frac{1}{16}}{\operatorname{ch}\omega x - 1} \right.
$$
$$
+ \left. \beta_{03}(\beta_{01} + \beta_{02} + M_0 + 1)\operatorname{ch}\omega x \right] + \omega^2 \beta_{03}^2 \operatorname{sh}^2\omega x. \tag{4.7.25}
$$

The corresponding solutions take the form

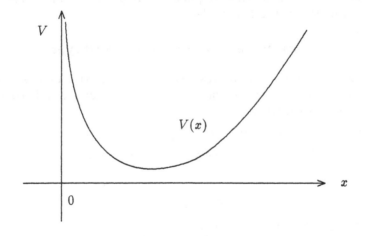

Figure 4.34. The form of the potential (4.7.25).

$$
\psi(x) = [\operatorname{ch}\omega x + 1]^{\sigma_1|\beta_{01}|+\frac{1}{4}}[\operatorname{ch}\omega x - 1]^{|\beta_{02}|+\frac{1}{4}}
$$
$$
\times \quad \exp(-|\beta_{03}|\operatorname{ch}\omega x)\prod_{i=1}^{M}[\operatorname{ch}\omega x - \xi_i], \tag{4.7.26a}
$$
$$
E = -\omega^2 \left\{ (\sigma_1|\beta_{01}| + |\beta_{02}| + M + \tfrac{1}{2})^2 \right.
$$
$$
- \left. 2|\beta_{03}|\left(|\beta_{02}| - \sigma_1|\beta_{01}| + \sum_{i=1}^{M}\xi_i\right) \right\} \tag{4.7.26b}
$$

where the numbers ξ_i, $i = 1, \ldots, M$ now satisfy the system of equations

$$\sum_{k=1}^{M}{}' \frac{1}{\xi_i - \xi_k} + \frac{\sigma_1|\beta_{01}| + \frac{1}{2}}{\xi_i + 1} + \frac{|\beta_{02}| + \frac{1}{2}}{\xi_i - 1} - |\beta_{03}| = 0,$$

$$i = 1, \ldots, M, \qquad (4.7.27)$$

and the integers σ_1 and M are found from the condition

$$-\frac{\beta_{03}}{|\beta_{03}|}(\beta_{01} + \beta_{02} + M_0 + 1) - \sigma_1|\beta_{01}| - |\beta_{02}| = M + 1. \qquad (4.7.28)$$

Thus, we have obtained a new quasi-exactly solvable model of order $M + 1$, defined on the positive half-axis

$$x \in [0, \infty]. \qquad (4.7.29)$$

The potential of this model is singular at the points 0 and ∞ and, thus, has the form depicted in figure 4.34.

4.8   The simplest degenerate case. The second type

In this section we discuss the case when the roots with indices 1 and 2 depend on ϵ but cancel, while the fraction in (4.6.6) does not depend on ϵ. This case is realized when

$$R_1 = R_2 = R, \quad R_3 = R_4 = 0, \quad R_5 = S, \qquad (4.8.1)$$

$$\beta_{01} = \pi_1\beta_0, \quad \beta_{02} = \pi_2\beta_0 \qquad (4.8.2)$$

and

$$\sigma_1 = \sigma, \quad \sigma_2 = -\sigma, \qquad (4.8.3)$$

where π_1, π_2 and σ take the values ± 1.

In this case the equation (4.6.6) takes the form

$$\sigma_3\frac{\beta_{03}}{|\beta_{03}|}(\beta_{01} + \beta_{02} + M_0 + 1) = M + 1, \qquad (4.8.4)$$

and for the other equations we have

$$M(M + 1) + 2\sigma_3|\beta_{03}| \sum_{i=1}^{M} \xi_i$$

$$- \left\{ [(\pi_1 + \pi_2)\beta_0 + M_0 + 1]\,[(\pi_1 + \pi_2)\beta_0 + M_0] \right.$$

$$\left. + 2\beta_{03} \left[(\pi_2 - \pi_1)\beta_0 + \sum_{i=1}^{M_0} \xi_{0i} \right] \right\} = -S\epsilon, \qquad (4.8.5)$$

$$\sum_{k=1}^{M}{}' \frac{1}{\xi_i - \xi_k} + \frac{\frac{1}{2} + \sigma\sqrt{\beta_0^2 - R\epsilon}}{\xi_i + 1}$$

$$+ \frac{\frac{1}{2} - \sigma\sqrt{\beta_0^2 - R\epsilon}}{\xi_i - 1} + \sigma_3|\beta_{03}| = 0, \quad i = 1, \dots, M. \quad (4.8.6)$$

The weight function $\rho(\lambda)$ is in this case

$$\rho(\lambda) = R\left(\frac{1}{\lambda - 1} - \frac{1}{\lambda + 1}\right)^2 + \frac{S}{\lambda^2 - 1} \qquad (4.8.7)$$

or, equivalently,

$$\rho(\lambda) = S\frac{\lambda^2 + \left(\frac{4R}{S} - 1\right)}{(\lambda^2 - 1)^2}. \qquad (4.8.8)$$

This gives us the following expression for the potential:

$$
\begin{aligned}
V(x) \;=\; & \varepsilon_0 + \frac{(\lambda^2 - 1)^2}{S\lambda^2 + 4R - S} \\
& \times \left\{ \frac{\lambda^2 + 1}{(\lambda^2 - 1)^2} - \frac{1}{2}\frac{\lambda^2 - \frac{4R}{S} + 1}{\left(\lambda^2 + \frac{4R}{S} - 1\right)^2} \right. \\
& \quad - \lambda^2 \left.\left(\frac{1}{\lambda^2 - 1} - \frac{1}{2}\frac{1}{\lambda^2 + \frac{4R}{S} - 1} \right)^2 \right\} \\
& + \frac{\lambda^2 - 1}{S\lambda^2 + 4R - S}\left\{ \left(\beta_0^2 - \frac{1}{4}\right)\frac{4}{\lambda^2 - 1} \right. \\
& + \left[(\pi_1 + \pi_2)\beta_0 + M_0 + 1\right]\left[(\pi_1 + \pi_2)\beta_0 + M_0\right] \\
& + 2\beta_{03}\left[(\pi_2 - \pi_1)\beta_0 + \sum_{i=1}^{M_0} \xi_{0i} \right] \\
& + 2\beta_{03}\left[(\pi_1 + \pi_2)\beta_0 + M_0 + 1\right]\lambda \\
& + \left.\beta_{03}^2(\lambda^2 - 1)\right\}\Bigg|_{\lambda = \lambda(x)} \qquad (4.8.9)
\end{aligned}
$$

and the corresponding solutions

$$
\begin{aligned}
\psi(x) \;=\; & \sqrt[4]{S\lambda^2 + 4R - S}\left(\frac{\lambda + 1}{\lambda - 1}\right)^{\sigma\sqrt{\beta_0^2 - R\epsilon}} \\
& \times \exp(\sigma_3|\beta_{03}|\lambda)\prod_{i=1}^{M}(\lambda - \xi_i)\Bigg|_{\lambda = \lambda(x)}, \qquad (4.8.10a) \\
E \;=\; & \varepsilon_0 + \epsilon, \qquad (4.8.10b)
\end{aligned}
$$

in which the numbers ξ_i, $i = 1, \ldots, M$ and ϵ can be found from equations (4.8.5) and (4.8.6), and the integers σ_3 and M satisfy conditions (4.8.4).

Let us now classify the quasi-exactly solvable models corresponding to the choice (4.8.1).

First of all, note that the physical interval $\lambda \in [-1, +1]$ is forbidden. In fact, such an interval implies the normalization condition $\sigma_1 = +1, \sigma_2 = +1$ which, obviously, contradicts condition (4.8.3). The only possibility is to take

$$\lambda \in [1, \infty]. \tag{4.8.11}$$

In this case the stability condition

$$\sigma_2 = +1, \quad \sigma_3 = -1 \tag{4.8.12}$$

implies that

$$\sigma_1 = -1. \tag{4.8.13}$$

The condition of positivity of the function $\rho(\lambda)$ in (4.8.11) is

$$R > 0, \quad S > 0. \tag{4.8.14}$$

Now it is not difficult to write down all admissible weight functions $\rho(\lambda)$ and corresponding physical diagrams.

1. $S = 0$.

$$\rho(\lambda) = \frac{4R}{(\lambda^2 - 1)^2}. \tag{4.8.15}$$

2. $0 < S < 4R$.

$$\rho(\lambda) = S\frac{\lambda^2 + \rho^2}{(\lambda^2 - 1)^2}, \quad \rho = \sqrt{\frac{4R}{S} - 1}. \tag{4.8.16}$$

3. $S = 4R$.

$$\rho(\lambda) = 4R\frac{\lambda^2}{(\lambda^2 - 1)^2}. \tag{4.8.17}$$

4. $S > 4R$.

$$\rho(\lambda) = S\frac{\lambda^2 - \rho^2}{(\lambda^2 - 1)^2}, \quad \rho = \sqrt{1 - \frac{4R}{S}}. \tag{4.8.18}$$

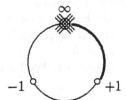

Figure 4.35. Diagram 1. $S = 0$.

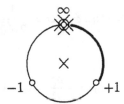

Figure 4.36. Diagram 2. $0 < S < 4R$.

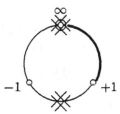

Figure 4.37. Diagram 3. $S = 4R$.

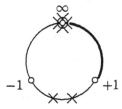

Figure 4.38. Diagram 4. $S > 4R$.

The corresponding extended physical diagrams are depicted in figures 4.35, 4.36, 4.37 and 4.38. Note that the integrals (4.2.3) for these diagrams are expressed in terms of inverse hyperbolic functions. They have the simplest form for diagrams 1 and 3.

Consider, for example, diagram 1. We have

$$x = 2\sqrt{R} \int_{\lambda(x)}^{\infty} \frac{d\lambda}{\lambda^2 - 1} = 2\sqrt{R}\,\mathrm{arcth}\,\lambda(x) \tag{4.8.19}$$

and, therefore,

$$\lambda(x) = \mathrm{cth}\,\omega x, \quad \omega = \frac{1}{2\sqrt{R}}. \tag{4.8.20}$$

Taking

$$\varepsilon_0 = -\frac{\beta_0^2}{R}, \tag{4.8.21}$$

we obtain

$$\begin{aligned}
V(x) &= \omega^2\Bigg\{[(\pi_1 + \pi_2)\beta_0 + M_0 + 1]\,[(\pi_1 + \pi_2)\beta_0 + M_0] \\
&\quad + 2\beta_{03}\left[(\pi_2 - \pi_1)\beta_0 + \sum_{i=1}^{M_0}\xi_{0i}\right]\Bigg\}\frac{1}{\mathrm{sh}^2\,\omega x} \\
&\quad + 2\omega^2\beta_{03}\,[(\pi_1 + \pi_2)\beta_0 + M_0 + 1]\frac{\mathrm{ch}\,\omega x}{\mathrm{sh}^3\,\omega x} \\
&\quad + \omega^2\beta_{03}^2\frac{1}{\mathrm{sh}^4\,\omega x}.
\end{aligned} \tag{4.8.22}$$

The solutions for this potential are:

$$\begin{aligned}
\psi(x) &= \exp\left(-\frac{\omega}{2}\sqrt{\beta_0^2 - R\epsilon}\,x - |\beta_{03}|\,\mathrm{cth}\,\omega x\right) \\
&\quad \times \prod_{i=1}^{M}(\mathrm{cth}\,\omega x - \xi_i), \tag{4.8.23a} \\
E &= -4\omega^2\beta_0^2 + \epsilon. \tag{4.8.23b}
\end{aligned}$$

Here ϵ and ξ_i, $i = 1, \ldots, M$ can be found from the system

$$M(M + 1) - 2|\beta_{03}|\sum_{i=1}^{M}\xi_i$$

$$= \ [(\pi_1 + \pi_2)\beta_0 + M_0 + 1] \, [(\pi_1 + \pi_2)\beta_0 + M_0]$$

$$+ \ 2\beta_{03} \left[(\pi_2 - \pi_1)\beta_0 + \sum_{i=1}^{M_0} \xi_{0i} \right], \tag{4.8.24}$$

$$\sum_{k=1}^{M}{}' \frac{1}{\xi_i - \xi_k} \ + \ \frac{\frac{1}{2} - \sqrt{\beta_0^2 - R\epsilon}}{\xi_i + 1} + \frac{\frac{1}{2} + \sqrt{\beta_0^2 - R\epsilon}}{\xi_i - 1}$$

$$- |\beta_{03}| = 0, \quad i = 1, \dots, M. \tag{4.8.25}$$

Besides, we have the condition

$$- \frac{\beta_{03}}{|\beta_{03}|} (\beta_{01} + \beta_{02} + M_0 + 1) = M + 1 \tag{4.8.26}$$

determining the order of the corresponding quasi-exactly solvable model. This model is defined on the half-axis

$$x \in [0, \infty] \tag{4.8.27}$$

and its potential has the same form as the potential depicted in figure 4.20.

Let us now consider diagram 3. We have

$$x = 2\sqrt{R} \int^{\lambda(x)} \frac{\lambda \, d\lambda}{\lambda^2 - 1} = \sqrt{R} \ln(\lambda^2(x) - 1) \tag{4.8.28}$$

and

$$\lambda(x) = \sqrt{1 + \exp \ \omega x}, \quad \omega = \frac{1}{\sqrt{R}}. \tag{4.8.29}$$

Taking

$$\varepsilon_0 = -\frac{1}{4R}, \tag{4.8.30}$$

we obtain the potential

$$V(x) \ = \ -\frac{3\omega^2}{16} \frac{\exp \ 2\omega x}{(1 + \exp \ \omega x)^2} + \omega^2 \left(\beta_0^2 - \tfrac{1}{4} \right) \frac{1}{1 + \exp \ \omega x}$$

$$+ \ \frac{\omega^2}{4} \Big\{ [(\pi_1 + \pi_2)\beta_0 + M_0 + 1] \, [(\pi_1 + \pi_2)\beta_0 + M_0]$$

$$+ \ 2\beta_{03} \left[(\pi_1 + \pi_2)\beta_0 + \sum_{i=1}^{M_0} \xi_{0i} \right] \Big\} \frac{\exp \ \omega x}{1 + \exp \ \omega x}$$

$$+ \quad \frac{\omega^2}{2}\beta_{03}\left[(\pi_2 - \pi_1)\beta_0 + \sum_{i=1}^{M_0}\xi_{0i}\right]\frac{\exp\,\omega x}{\sqrt{1+\exp\,\omega x}}$$

$$+ \quad \frac{\omega^2}{4}\beta_{03}^2\frac{\exp\,2\omega x}{1+\exp\,\omega x} \tag{4.8.31}$$

and the corresponding solutions

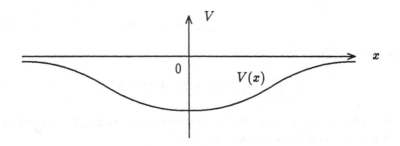

Figure 4.39. The form of the potential (4.8.31).

$$
\begin{aligned}
\psi(x) \quad &= \quad \sqrt[4]{1+\exp\,\omega x}\left(\frac{\sqrt{1+\exp\,\omega x}+1}{\sqrt{1+\exp\,\omega x}-1}\right)^{-\sqrt{\beta_0^2-R\epsilon}} \\
&\times \quad \exp\left(-|\beta_{03}|\sqrt{1+\exp\,\omega x}\right) \\
&\times \quad \prod_{i=1}^{M}\left(\sqrt{1+\exp\,\omega x}-\xi_i\right), \tag{4.8.32a} \\
E \quad &= \quad -\frac{\omega^2}{4}+\epsilon. \tag{4.8.32b}
\end{aligned}
$$

The numbers ϵ and ξ_i, $i=1,\ldots,M$ satisfy the system of equations

$$
\begin{aligned}
M(M+1) &- 2|\beta_{03}|\sum_{i=1}^{M}\xi_i + \frac{4}{\omega^2}\epsilon \\
&= \quad [(\pi_1+\pi_2)\beta_0 + M_0 + 1]\,[(\pi_1+\pi_2)\beta_0 + M_0] \\
&+ \quad 2\beta_{03}\left[(\pi_2-\pi_1)\beta_0 + \sum_{i=1}^{M_0}\xi_{0i}\right] \tag{4.8.33}
\end{aligned}
$$

and

$$\sum_{k=1}^{M}{}'\frac{1}{\xi_i-\xi_k} \quad + \quad \frac{\frac{1}{2}-\sqrt{\beta_0^2-R\epsilon}}{\xi_i+1}+\frac{\frac{1}{2}+\sqrt{\beta_0^2-R\epsilon}}{\xi_i-1}$$

$$-|\beta_{03}| = 0, \quad i = 1, \ldots, M \tag{4.8.34}$$

which must be supplemented by condition (4.8.26) determining the order of the obtained quasi-exactly solvable model. This model is defined on the whole x-axis

$$x \in [-\infty, +\infty] \tag{4.8.35}$$

and its potential has the form depicted in figure 4.39.

4.9 The simplest degenerate case. The third type

Now we consider the last case when a trivialization of equation (4.6.6) is possible. This case is realized when the first term in (4.6.6) depends on ϵ explicitly but cancels with the third term, while the second (remaining) term does not depend on ϵ.

In this case we have

$$R_1 = 0 \tag{4.9.1}$$

and equation (4.6.6) takes the form

$$-\sigma_1 |\beta_{01}| = M + 1, \tag{4.9.2}$$

from which it follows that

$$\sigma_1 = -1 \tag{4.9.3}$$

and

$$|\beta_{01}| = M + 1. \tag{4.9.4}$$

From condition (4.9.3) it also follows that the interval $\lambda \in [-1, +1]$ is forbidden and, therefore, we have only one possibility, to choose the physical interval as

$$\lambda \in [1, \infty]. \tag{4.9.5}$$

Then the stability conditions have the form

$$\sigma_2 = 1, \quad \sigma_3 = -1. \tag{4.9.6}$$

The condition of cancellation can be rewritten as follows:

$$\begin{aligned} - \quad & \beta_{03}(\beta_{01} + \beta_{02} + M_0 + 1) + R_4\epsilon \\ = \quad & \sqrt{\beta_{02}^2 - R_2\epsilon}\sqrt{\beta_{03}^2 - R_3\epsilon}. \end{aligned} \tag{4.9.7}$$

For (4.9.7) to be satisfied for any ϵ, we must take

$$R_2 = A^2 R, \quad R_3 = R, \quad R_4 = -AR, \tag{4.9.8}$$

where A is a positive number. Simultaneously we must require that

$$\beta_{02} = \pi A \beta_0, \quad \beta_{03} = \beta_0 \tag{4.9.9}$$

and

$$-\beta_{03}(\beta_{01} + \beta_{02} + M_0 + 1) = A\beta_0^2 \tag{4.9.10}$$

or, equivalently,

$$\beta_{01} + (\pi + 1)A\beta_0 + M_0 + 1 = 0. \tag{4.9.11}$$

In this case the remaining spectral equations take the form

$$
\begin{aligned}
& A\sqrt{\beta_0^2 - R\epsilon}\left(A\sqrt{\beta_0^2 - R\epsilon} - 1\right) \\
& - 2\sqrt{\beta_0^2 - R\epsilon}\left[M + 1 + A\sqrt{\beta_0^2 - R\epsilon} + \sum_{i=1}^{M}\xi_i\right] \\
& = A\beta_0(A\beta_0 + 1) + 2\beta_0\left(\pi A\beta_0 - \beta_{01} + \sum_{i=1}^{M_0}\xi_{0i}\right) - S\epsilon
\end{aligned}
\tag{4.9.12}
$$

and

$$
\sum_{k=1}^{M}\frac{1}{\xi_i - \xi_k} - \frac{M + \frac{1}{2}}{\xi_i + 1} + \frac{A\sqrt{\beta_0^2 - R\epsilon} + \frac{1}{2}}{\xi_i - 1}
$$
$$
- \sqrt{\beta_0^2 - R\epsilon} = 0, \quad i = 1,\ldots,M. \tag{4.9.13}
$$

The weight function is

$$
\begin{aligned}
\rho(\lambda) &= A^2 R\left[\frac{1}{(\lambda - 1)^2} - \frac{1}{(\lambda - 1)(\lambda + 1)}\right] + R \\
& - 2RA\frac{\lambda}{(\lambda - 1)(\lambda + 1)} + \frac{S}{(\lambda - 1)(\lambda + 1)}
\end{aligned}
\tag{4.9.14}
$$

or, equivalently,

$$\rho(\lambda) = R\left(\frac{A}{\lambda - 1} - 1\right)^2 + \frac{S - A(A - 2)R}{(\lambda - 1)(\lambda + 1)}. \tag{4.9.15}$$

It can also be written in the form

$$\rho(\lambda) = \frac{R(\lambda - \rho_1)(\lambda - \rho_2)(\lambda - \rho_3)}{(\lambda - 1)^2(\lambda + 1)}, \tag{4.9.16}$$

where ρ_1, ρ_2 and ρ_3 are the roots of the cubic equation

$$R\left[A - (\lambda - 1)\right]^2 (\lambda + 1) + [S - A(A - 2)R](\lambda - 1) = 0. \tag{4.9.17}$$

Then we obtain

$$
\begin{aligned}
V(x) \;=\;& \varepsilon_0 + \frac{1}{R}\frac{(\lambda^2 - 1)^2}{\prod\limits_{\alpha=1}^{3}(\lambda - \rho_\alpha)} \\
&\times \left\{ \frac{\frac{1}{2}}{(\lambda + 1)^2} + \frac{\frac{1}{2}}{(\lambda - 1)^2} - \sum_{\alpha=1}^{3}\frac{\frac{1}{4}}{(\lambda - \rho_\alpha)^2} \right. \\
&- \left(\frac{\frac{1}{2}}{\lambda + 1} + \frac{\frac{1}{2}}{\lambda - 1} - \sum_{\alpha=1}^{3}\frac{\frac{1}{4}}{\lambda - \rho_\alpha} \right)^2 \\
&+ \left(M + \tfrac{1}{2} \right)\left(M + \tfrac{3}{2} \right)\frac{2}{(1 + \lambda)^2(1 - \lambda)} \\
&+ \left(A^2\beta_0^2 - \tfrac{1}{4} \right)\frac{2}{(1 - \lambda)^2(1 + \lambda)} + \beta_0^2 + 2A\beta_0^2\frac{\lambda}{(1 - \lambda)(1 + \lambda)} \\
&+ \left[A\beta_0(A\beta_0 + 1) + 2\beta_0\left(\pi A\beta_0 - \beta_{01} + \sum_{i=1}^{M_0}\xi_{0i} \right) \right] \\
&\times \left. \frac{1}{(\lambda + 1)(\lambda - 1)} \right\}\Bigg|_{\lambda = \lambda(x)}, \tag{4.9.18}
\end{aligned}
$$

and also

$$
\begin{aligned}
\psi(x) \;=\;& \prod_{\alpha=1}^{3}(\lambda - \rho_\alpha)^{\frac{1}{4}}(\lambda + 1)^{-M - \frac{3}{4}}(\lambda - 1)^{A\sqrt{\beta_0^2 - R\varepsilon}} \\
&\times \exp\left(-\sqrt{\beta_0^2 - R\varepsilon}\,\lambda \right)\prod_{i=1}^{M}(\lambda - \xi_i)\Bigg|_{\lambda = \lambda(x)}, \tag{4.9.19a}
\end{aligned}
$$

$$E \;=\; \varepsilon_0 + \varepsilon, \tag{4.9.19b}$$

where the numbers ε and $\xi_1, \ldots, \xi_M$ satisfy system (4.9.12) and (4.9.13).

Let us now try to classify the types of quasi-exactly solvable model described by the potential in (4.9.18) and having solutions (4.9.19). For this purpose it is sufficient to classify the weight functions $\rho(\lambda)$ defined by formulas (4.9.14)–(4.9.16).

The condition of positivity of $\rho(\lambda)$ in the interval (4.9.5) is

$$R > 0, \quad S < A(2 - A)R. \tag{4.9.20}$$

When this condition is satisfied, equation (4.9.17) has only one real root belonging to the interval $\lambda \in [-1, +1]$. The other two roots are complex conjugate. Then we have

$$\rho(\lambda) = \frac{R(\lambda - \rho_1)(\lambda - \rho)(\lambda - \rho^*)}{(\lambda - 1)^2(\lambda + 1)}. \tag{4.9.21}$$

The corresponding extended physical diagram describing this case is depicted in figure 4.40. This gives us an unique type of quasi-exactly

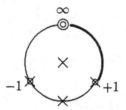

Figure 4.40. The diagram describing the weight function (4.9.22).

solvable model. The potentials of these models are defined on the whole x-axis. They are regular at infinity and, therefore, describe a potential well of finite depth.

The integrals (4.2.3) for this diagram are expressed in terms of inverse elliptic functions. The explicit expressions for the potentials are cumbersome and we shall not present them here.

4.10   The one-dimensional twice-degenerate case

Another type of degeneration arises when the simple double poles of the function (4.6.1) also merge. Without loss of generality we can assume that they merge at zero. As a result, we obtain the twice-degenerate weight function

$$\rho(\lambda) = \rho_0 \frac{(\lambda - \rho_1)(\lambda - \rho_2)(\lambda - \rho_3)(\lambda - \rho_4)}{\lambda^4} \tag{4.10.1}$$

which can also be rewritten in the form

$$\rho(\lambda) = R_0 + \frac{R_1}{\lambda} + \frac{R_2}{\lambda^2} + \frac{R_3}{\lambda^3} + \frac{R_4}{\lambda^4}. \tag{4.10.2}$$

In order to obtain the corresponding degenerate form of the function $F_0(\lambda)$, we must replace in (4.6.3) the numbers β_{01} and β_{02} by $\frac{a_1\beta_{02}-\beta_{01}}{a_1-a_2}$ and $\frac{a_2\beta_{01}-\beta_{02}}{a_2-a_1}$ and take the limit $a_1 \to 0, a_2 \to 0$. The result is

$$F_0(\lambda) = \beta_{00} + \frac{\beta_{01}}{\lambda} + \frac{\beta_{02}}{\lambda^2}. \tag{4.10.3}$$

The function $F_M(\lambda)$ must have an analogous form

$$F_M(\lambda) = \beta_0 + \frac{\beta_1}{\lambda} + \frac{\beta_2}{\lambda^2} \tag{4.10.4}$$

where β_0, β_1 and β_2 are unknowns. If we choose the physical interval as

$$\lambda \in [0, \infty], \tag{4.10.5}$$

then β_2 must be positive and β_0 negative.

Equations for the numbers β and also for ϵ and ξ_i, $i = 1, \ldots, M$ can easily be obtained by substituting (4.10.1), (4.10.3) and (4.10.4) into (4.2.8) and (4.2.9). They have the form:

$$\begin{aligned}
\beta_0 &= -\sqrt{\beta_{00}^2 - R_0\epsilon}, \\
\beta_1 &= 1 + \frac{2\beta_{02}(\beta_{01} - 1) - R_3\epsilon}{2\sqrt{\beta_{02}^2 - R_4\epsilon}}, \\
\beta_2 &= -\sqrt{\beta_{02}^2 - R_4\epsilon},
\end{aligned} \tag{4.10.6}$$

$$-\frac{2\beta_{00}(\beta_{01} + M_0) - R_1\epsilon}{2\sqrt{\beta_{00}^2 - R_0\epsilon}} - \frac{2\beta_{02}(\beta_{01} - 1) - R_3\epsilon}{2\sqrt{\beta_{02}^2 - R_4\epsilon}} = M + 1, \tag{4.10.7}$$

$$\begin{aligned}
&\left(\frac{2\beta_{02}(\beta_{01} - 1) - R_3\epsilon}{2\sqrt{\beta_{02} - R_4\epsilon}} + 1\right) \frac{2\beta_{02}(\beta_{01} - 1) - R_3\epsilon}{2\sqrt{\beta_{02} - R_4\epsilon}} \\
&\quad -2\sqrt{\beta_{02}^2 - R_4\epsilon} \left(\sqrt{\beta_{00}^2 - R_0\epsilon} + \sum_{i=1}^{M} \frac{1}{\xi_i}\right) \\
&\quad -\left[\beta_{01}(\beta_{01} - 1) + 2\beta_{02}\left(\beta_{00} - \sum_{i=1}^{M_0} \frac{1}{\xi_{0i}}\right)\right] = -R_2\epsilon, \tag{4.10.8}
\end{aligned}$$

$$\sideset{}{'}\sum_{k=1}^{M} \frac{1}{\xi_i - \xi_k} - \sqrt{\beta_{00}^2 - R_0\epsilon}$$

$$+ \left[1 + \frac{2\beta_{02}(\beta_{01} - 1) - R_3\epsilon}{2\sqrt{\beta_{02}^2 - R_4\epsilon}}\right]\frac{1}{\xi_i}$$

$$+ \sqrt{\beta_{00}^2 - R_4\epsilon}\frac{1}{\xi_i^2} = 0, \quad i = 1, \ldots, M. \qquad (4.10.9)$$

We have again a system of $M + 2$ equations for $M + 1$ quantities ϵ and $\xi_1, \ldots, \xi_M$. The coincidence of the number of equations and the number of unknown quantities is possible when equation (4.10.7) trivializes. Note that all parameters β_{00}, β_{01} and β_{02} in (4.10.7) are real. Therefore, we come to the following two cases:

(i) both terms in (4.10.7) do not depend on ϵ, or

(ii) one of the terms in (4.10.7) vanishes, while the second one does not depend on ϵ.

In the next two sections we will consider these cases in detail.

4.11   The twice-degenerate case. The first type

Assume that both the terms in (4.10.7) do not depend on ϵ. This is possible if

$$R_0 = R_1 = R_3 = R_4 = 0, \quad R_2 = S. \qquad (4.11.1)$$

Then the equation (4.10.7) is reduced to the form

$$-\frac{\beta_{00}(\beta_{01} + M_0)}{|\beta_{00}|} - \frac{\beta_{02}(\beta_{01} - 1)}{|\beta_{02}|} = M + 1, \qquad (4.11.2)$$

and for the other equations we have

$$\left[\left(\frac{\beta_{02}(\beta_{01} - 1)}{|\beta_{02}|} + 1\right)\frac{\beta_{02}(\beta_{01} - 1)}{|\beta_{02}|}\right.$$

$$\left. - 2|\beta_{00}|\left(\beta_{00} + \sum_{i=1}^{M}\frac{1}{\xi_i}\right)\right]$$

$$- \left[\beta_{01}(\beta_{01} - 1) + 2\beta_{02}\left(\beta_{00} - \sum_{i=1}^{M_0}\frac{1}{\xi_{0i}}\right)\right] = -S\epsilon \qquad (4.11.3)$$

and

$$\sum_{k=1}^{M} \frac{1}{\xi_i - \xi_k} - |\beta_{00}| + \frac{\beta_{02}(\beta_{01} - 1)}{|\beta_{02}|} \frac{1}{\xi_i}$$

$$+ \; |\beta_{02}| \frac{1}{\xi_i^2} = 0, \quad i = 1, \ldots, M. \tag{4.11.4}$$

The weight function

$$\rho(\lambda) = \frac{S}{\lambda^2} \tag{4.11.5}$$

is positive when

$$S > 0. \tag{4.11.6}$$

It can be described by the diagram depicted in figure 4.41.

Figure 4.41. The diagram describing the weight function (4.11.5).

The integral (4.2.3) for (4.11.5) is equal to

$$x = \sqrt{S} \int^{\lambda(x)} \frac{d\lambda}{\lambda} = \sqrt{S} \ln \lambda(x) \tag{4.11.7}$$

and, therefore,

$$\lambda(x) = \exp \omega x, \quad \omega = \frac{1}{\sqrt{S}}. \tag{4.11.8}$$

Taking for definiteness

$$\varepsilon_0 = -\frac{1}{S} \left[(\beta_{01} - \tfrac{1}{2})^2 + 2\beta_{02} \left(\beta_{00} - \sum_{i=1}^{M_0} \frac{1}{\xi_{0i}} \right) \right], \tag{4.11.9}$$

we obtain the potential

$$V(x) = \omega^2 \Big\{ \beta_{02}^2 \exp(-2\omega x) + 2\beta_{02}(\beta_{01} - 1)\exp(-\omega x)$$
$$+ \ 2\beta_{00}(\beta_{01} + M_0)\exp \omega x + \beta_{00}^2 \exp 2\omega x \Big\} \qquad (4.11.10)$$

and the corresponding solutions

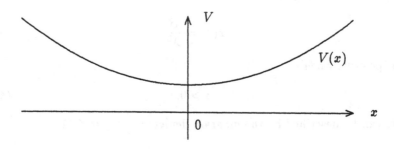

Figure 4.42. The form of the potential (4.11.10).

$$\psi(x) = (\exp \omega x)^{\frac{\beta_{01}}{|\beta_{02}|}(\beta_{01}-1)+\frac{1}{2}}$$
$$\times \ \exp\{-|\beta_{02}|\exp(-\omega x) - |\beta_{00}|\exp \omega x\}, \qquad (4.11.11a)$$

$$E = -\omega^2 \Bigg\{ \left[\frac{\beta_{02}}{|\beta_{02}|}(\beta_{01} - 1) + 1 \right] \frac{\beta_{02}}{|\beta_{02}|}(\beta_{01} - 1)$$
$$- \ 2|\beta_{02}| \left(|\beta_{00}| + \sum_{i=1}^{M} \frac{1}{\xi_i} \right) \Bigg\}, \qquad (4.11.11b)$$

in which the numbers ξ_i, $i = 1, \ldots, M$ satisfy the system of equations
(4.11.4).

Thus, we have obtained a simple quasi-exactly solvable model of order
$M + 1$, the potential of which is defined on the whole x-axis

$$x \in [-\infty, +\infty] \qquad (4.11.12)$$

and has the form depicted in figure 4.42.

4.12   The twice-degenerate case. The second type

Let us now consider the second case of trivialization of equation (4.10.7)
which is realized when the first term in the left-hand side of (4.10.7)

vanishes, while the second one is a constant which does not depend on ϵ. This is possible when

$$R_0 = 0, \quad R_1 = 0, \quad R_2 = S, \quad R_3 = 0, \quad R_4 = R \qquad (4.12.1)$$

and, simultaneously,

$$\beta_{01} = 1. \qquad (4.12.2)$$

Then equation (4.10.7) takes the form

$$-\frac{\beta_{00}}{|\beta_{00}|}(M_0 + 1) = M + 1, \qquad (4.12.3)$$

from which it follows that

$$\beta_{00} = -|\beta_{00}| \qquad (4.12.4)$$

and

$$M = M_0. \qquad (4.12.5)$$

The remaining system becomes

$$2\sqrt{\beta_{02}^2 - R\epsilon}\left(|\beta_{00}| + \sum_{i=1}^{M_0}\frac{1}{\xi_i}\right) - 2\beta_{02}\left(|\beta_{00}| + \sum_{i=1}^{M_0}\frac{1}{\xi_{0i}}\right) = S\epsilon, \quad (4.12.6)$$

$$\sum_{k=1}^{M_0}\frac{1}{\xi_i - \xi_k} - |\beta_{00}| \;+\; \frac{1}{\xi_i} - \frac{\sqrt{\beta_{02}^2 - R\epsilon}}{\xi_i^2} = 0,$$

$$i = 1, \ldots, M. \qquad (4.12.7)$$

The weight function corresponding to this case is

$$\rho(\lambda) = \frac{S}{\lambda^2} + \frac{R}{\lambda^4} \qquad (4.12.8)$$

and, therefore, we have the following potential

$$
\begin{aligned}
V(x) \;=\; & \varepsilon_0 + \frac{1}{4S}\frac{\lambda^4\left(\lambda^2 + 6\frac{R}{S}\right)}{\left(\lambda^2 + \frac{R}{S}\right)^3} \\
& + \frac{1}{4S}\frac{1}{\lambda^2 + \frac{R}{S}}\left\{\beta_{02}^2 - 2\beta_{02}\left(|\beta_{00}| + \sum_{i=1}^{M_0}\frac{1}{\xi_{0i}}\right)\lambda^2\right. \\
& \left. - 2|\beta_{00}|(M_0 + 1)\lambda^3 + \beta_{00}^2\lambda^4\right\}
\end{aligned}
\qquad (4.12.9)
$$

and solutions

$$\psi(x) = \sqrt[4]{\lambda^2 + \frac{R}{S}} \exp\left(-\frac{\sqrt{\beta_{02}^2 - R\epsilon}}{\lambda} - |\beta_{00}|\lambda\right)$$

$$\times \prod_{i=1}^{M_0}(\lambda - \xi_i), \qquad (4.12.10a)$$

$$E = \varepsilon_0 + \epsilon. \qquad (4.12.10b)$$

The function $\rho(\lambda)$ is positive when

$$S \geq 0, \quad R \geq 0. \qquad (4.12.11)$$

Consider two cases.

    1. $S = 0$, $R > 0$. Then

$$\rho(\lambda) = \frac{R}{\lambda^4}. \qquad (4.12.12)$$

    2. $S > 0$, $R > 0$. Then

$$\rho(\lambda) = S\frac{\lambda^2 + \rho^2}{\lambda^4}, \quad \rho = \sqrt{\frac{R}{S}}. \qquad (4.12.13)$$

The corresponding diagrams are depicted in figures 4.43 and 4.44.

Figure 4.43. Diagram 1. $S = 0$, $R > 0$.

    The quasi-exactly solvable model described by diagram 1 and weight function (4.12.12) has the simplest form. Substituting it into (4.2.3) we obtain

$$x = \sqrt{R} \int^{\lambda(x)} \frac{d\lambda}{\lambda^2} = \frac{\sqrt{R}}{\lambda(x)} \qquad (4.12.14)$$

Figure 4.44. Diagram 2. $S > 0$, $R > 0$.

and, hence,

$$\lambda(x) = \frac{1}{\omega x}, \quad \omega = \frac{1}{\sqrt{R}}. \tag{4.12.15}$$

Taking

$$\varepsilon_0 = 0 \tag{4.12.16}$$

we obtain the potential of the model in question

$$
\begin{aligned}
V(x) = \quad & \frac{\omega^2}{4}\left\{ \beta_{02}^2 - 2\beta_{02}\left(|\beta_{00}| + \sum_{i=1}^{M_0} \frac{1}{\xi_{0i}}\right)\frac{1}{\omega^2 x^2}\right. \\
& \left. - 2|\beta_{00}|(M+1)\frac{1}{\omega^3 x^3} + \beta_{00}^2\frac{1}{\omega^4 x^4}\right\}
\end{aligned}
\tag{4.12.17}
$$

and its solutions

$$
\begin{aligned}
\psi(x) = \quad & \exp\left\{-\sqrt{\beta_{02}^2 - R\epsilon}\,\omega x - |\beta_{00}|\frac{1}{\omega x}\right\} \\
& \times \prod_{i=1}^{M_0}\left(\frac{1}{\omega x} - \xi_i\right),
\end{aligned}
\tag{4.12.18a}
$$

$$E = \epsilon, \tag{4.12.18b}$$

where ϵ and ξ_i, $i = 1, \ldots, M_0$ can be found from the equations

$$\sqrt{\beta_{02}^2 - R\epsilon}\left(|\beta_{00}| + \sum_{i=1}^{M_0} \frac{1}{\xi_i}\right) = \beta_{02}\left(|\beta_{00}| + \sum_{i=1}^{M_0} \frac{1}{\xi_{0i}}\right) \tag{4.12.19}$$

and (4.12.7). The order of this quasi-exactly solvable model is $M + 1$. Its potential is defined on the half-axis

$$x \in [0, \infty] \tag{4.12.20}$$

and has the same form as the potential depicted in figure 4.20.

4.13   The one-dimensional most degenerate case

This case arises when three double poles in (4.2.13) merge. We can assume that they merge at infinity and the remaining simple double pole is located at zero. This gives

$$\rho(\lambda) = \rho_0 \frac{(\lambda - \rho_1)(\lambda - \rho_2)(\lambda - \rho_3)(\lambda - \rho_4)}{\lambda^2} \tag{4.13.1}$$

or, equivalently,

$$\rho(\lambda) = \frac{R_{-2}}{\lambda^2} + \frac{R_{-1}}{\lambda} + R_0 + R_1\lambda + R_2\lambda^2. \tag{4.13.2}$$

We also have

$$F_0(\lambda) = \frac{\beta_{01} + \frac{1}{2}}{\lambda} + \beta_{02} + \beta_{03}\lambda. \tag{4.13.3}$$

Taking

$$F(\lambda) = \frac{\beta_1 + \frac{1}{2}}{\lambda} + \beta_2 + \beta_3\lambda \tag{4.13.4}$$

and substituting formulas (4.13.2) and (4.13.3) into the system (4.2.8), (4.2.9) we can rewrite it in the algebraic form

$$\beta_1^2 - \beta_{01}^2 = -R_{-2}\epsilon, \tag{4.13.5}$$

$$(2\beta_1 + 1)\left(\beta_2 - \sum_{i=1}^{M} \frac{1}{\xi_i}\right) - (2\beta_{01} + 1)\left(\beta_{02} - \sum_{i=1}^{M_0} \frac{1}{\xi_{0i}}\right) = -R_{-1}\epsilon, \tag{4.13.6}$$

$$\beta_2^2 + \left(\beta_1 - \tfrac{1}{2}\right)\beta_3 - \beta_{02}^2 - \left(\beta_{01} - \tfrac{1}{2}\right)\beta_{03} = -R_0\epsilon, \tag{4.13.7}$$

$$2\beta_3(\beta_2 + M) - 2\beta_{03}(\beta_{02} + M_0) = -R_1\epsilon, \tag{4.13.8}$$

$$\beta_3^2 - \beta_{03}^2 = -R_2\epsilon, \tag{4.13.9}$$

$$\sum_{k=1}^{M}{}' \frac{1}{\xi_i - \xi_k} + \frac{\beta_1 + \frac{1}{2}}{\xi_i} + \beta_2 + \beta_3\xi_i = 0,$$

$$i = 1, \ldots, M. \tag{4.13.10}$$

For this system to have solutions it is necessary to take

$$R_{-2} = 0, \quad R_{-1} = R, \quad R_0 = R_1 = R_2 = 0 \tag{4.13.11}$$

and, simultaneously,

$$M = M_0. \tag{4.13.12}$$

This gives

$$\beta_1 = \beta_{01}, \quad \beta_2 = \beta_{02}, \quad \beta_3 = \beta_{03}, \tag{4.13.13}$$

and also

$$\epsilon = \frac{2\beta_{01} + 1}{R} \sum_{i=1}^{M_0} \left(\frac{1}{\xi_i} - \frac{1}{\xi_{i0}} \right) \tag{4.13.14}$$

and

$$\sum_{k=1}^{M_0}{}' \frac{1}{\xi_i - \xi_k} + \frac{\beta_{01} + \frac{1}{2}}{\xi_i} + \beta_{02} + \beta_{03}\xi_i = 0,$$
$$i = 1, \ldots, M_0. \tag{4.13.15}$$

In this case the function

$$\rho(\lambda) = \frac{R}{\lambda} \tag{4.13.16}$$

is positive in the interval

$$x \in [0, \infty] \tag{4.13.17}$$

if

$$R > 0, \tag{4.13.18}$$

and is described by the diagram depicted in figure 4.45.

The normalization conditions for this diagram have the form

$$\beta_{01} > 0, \quad \beta_{03} < 0. \tag{4.13.19}$$

Computing the integral (4.2.3)

$$x = \sqrt{R} \int^{\lambda(x)} \frac{d\lambda}{\sqrt{\lambda}} = 2\sqrt{R}\sqrt{\lambda(x)}, \tag{4.13.20}$$

Figure 4.45. The diagram describing the weight function (4.13.16).

we obtain

$$\lambda(x) = \omega x^2, \quad \omega = \frac{1}{4R}. \tag{4.13.21}$$

Taking for definiteness

$$\varepsilon_0 = \frac{1}{R}\sum_{i=1}^{M_0}\frac{1}{\xi_{i0}}, \tag{4.13.22}$$

we obtain the final expression for the potential

$$
\begin{aligned}
V(x) &= (4\beta_{01}^2 - \tfrac{1}{4})x^{-2} \\
&+ 4(2\beta_1 + 1)\omega + 4\left[\beta_{02}^2 + (\beta_{01} - \tfrac{1}{2})\beta_{03}\right]\omega^2 x^2 \\
&+ 8\beta_{03}(\beta_{02} + M_0)\omega^3 x^4 + 4\beta_{03}^2\omega^4 x^6
\end{aligned} \tag{4.13.23}
$$

and the corresponding solutions

$$
\begin{aligned}
\psi(x) &= x^{2\beta_{01}+\frac{1}{2}}\exp\left\{\beta_{03}\frac{\omega^2 x^4}{2} + \beta_{02}\omega x^2\right\} \\
&\times \prod_{i=1}^{M_0}(\omega x^2 - \xi_i),
\end{aligned} \tag{4.13.24a}
$$

$$E = 4(2\beta + 1)\omega\sum_{i=1}^{M_0}\frac{1}{\xi_i}. \tag{4.13.24b}$$

This model (which has already been discussed in chapter 2) completes the classification of one-dimensional quasi-exactly solvable models described by formulas (4.2.5)–(4.2.9) and characterized by the functions $\rho(\lambda)$ having double poles of total order four.

4.14   The multi-dimensional case

In this section we discuss multi-dimensional quasi-exactly solvable models connected with equation (4.1.4). The construction of these models consists of three stages. First, we choose the functions $\sigma^\alpha(\mu)$ and $x^\alpha(\lambda)$ in such a way as to ensure the positive definiteness of the metric $g_{ik}(\mu, \vec{\lambda})$ defined by the formula (3.3.15). Remember that this is a condition for the resulting Schrödinger equation to be an elliptic equation. Second, we satisfy the normalization condition for solutions of this equation by imposing necessary constraints on the function $F_0(\lambda)$. Finally, we find the conditions under which the order of the quasi-exactly solvable model obtained is more than one.

By analogy with the exactly solvable case we can take

$$\operatorname{sign}\sigma^\alpha(\mu) \;=\; (-1)^\alpha, \quad \alpha = 1,\ldots,D, \tag{4.14.1}$$

representing the corresponding weight functions in the form

$$x^\alpha(\lambda) \;=\; \frac{(\lambda - c)^{\alpha-1}}{\prod_{\alpha=2}^{D}(\lambda - a_\alpha)}\rho(\lambda), \quad \alpha = 1,\ldots,D. \tag{4.14.2}$$

Then, for the metric tensor to be positive definite, the number c must satisfy the constraint

$$c < a_1, \tag{4.14.3}$$

and the function $\rho(\lambda)$ must be positive. We stress that the form of this function depends on the sort of physical diagram. So, for the diagrams listed in figure 4.4 we have:

$$\text{A)} \quad \rho(\lambda) \;=\; \frac{\rho_0(\lambda - \rho_1)(\lambda - \rho_2)(\lambda - \rho_3)(\lambda - \rho_4)}{(\lambda - a_1)^2(\lambda - a_{D+1})^2(\lambda - a_{D+2})^2(\lambda - a_{D+3})^2}$$

$$\text{B)} \quad \rho(\lambda) \;=\; \frac{\rho_0(\lambda - \rho_1)(\lambda - \rho_2)(\lambda - \rho_3)(\lambda - \rho_4)}{(\lambda - a_1)^2(\lambda - a_{D+1})^2(\lambda - a)^2(\lambda - a^*)^2}$$

$$\text{C)} \quad \rho(\lambda) \;=\; \frac{\rho_0(\lambda - \rho_1)(\lambda - \rho_2)(\lambda - \rho_3)(\lambda - \rho_4)}{(\lambda - a_1)^2(\lambda - a_{D+1})^2(\lambda - a_{D+2})^4}$$

$$\text{D)} \quad \rho(\lambda) \;=\; \frac{\rho_0(\lambda - \rho_1)(\lambda - \rho_2)(\lambda - \rho_3)(\lambda - \rho_4)}{(\lambda - a_1)^4(\lambda - a_{D+1})^2(\lambda - a_{D+2})^2}$$

$$\text{E)} \quad \rho(\lambda) \;=\; \frac{\rho_0(\lambda - \rho_1)(\lambda - \rho_2)(\lambda - \rho_3)(\lambda - \rho_4)}{(\lambda - a_1)^4(\lambda - a_{D+1})^4}$$

$$\text{F)} \quad \rho(\lambda) \;=\; \frac{\rho_0(\lambda - \rho_1)(\lambda - \rho_2)(\lambda - \rho_3)(\lambda - \rho_4)}{(\lambda - a_1)^2(\lambda - a_{D+1})^6}. \tag{4.14.4}$$

The positivity condition for $\rho(\lambda)$ is ensured if the points $\rho_1, \rho_2, \rho_3, \rho_4$ satisfy one of the following three constraints

(i) $\rho_1, \rho_2, \rho_3, \rho_4 \in (a_1, a_{D+1})$,

(ii) $\rho_1, \rho_2 \notin (a_1, a_{D+1})$; $\rho_3 = \rho_4 = a_\alpha$, $\alpha = 2, \ldots, D$,

(iii) $\rho_1 = \rho_2 = a_\alpha$, $\rho_3 = \rho_4 = a_\beta$, $\alpha, \beta = 2, \ldots, D$.   (4.14.5)

The functions $F_0(\lambda)$ associated with (4.14.4) have the form

A) $F_0(\lambda) = \displaystyle\sum_{\alpha=1}^{D+3} \frac{F_{0\alpha}}{\lambda - a_\alpha}$,

B) $F_0(\lambda) = \displaystyle\sum_{\alpha=1}^{D+1} \frac{F_{0\alpha}}{\lambda - a_\alpha} + \frac{F_0}{\lambda - a} + \frac{F_0^*}{\lambda - a^*}$,

C) $F_0(\lambda) = \displaystyle\sum_{\alpha=1}^{D+2} \frac{F_{0\alpha}}{\lambda - a_\alpha} + \frac{F_{0,D+2}'}{(\lambda - a_{D+2})^2}$,

D) $F_0(\lambda) = \displaystyle\sum_{\alpha=1}^{D+2} \frac{F_{0\alpha}}{\lambda - a_\alpha} + \frac{F_{01}'}{(\lambda - a_1)^2}$,

E) $F_0(\lambda) = \displaystyle\sum_{\alpha=1}^{D+1} \frac{F_{0\alpha}}{\lambda - a_\alpha} + \frac{F_{01}'}{(\lambda - a_1)^2} + \frac{F_{0,D+1}'}{(\lambda - a_{D+1})^2}$,

F) $F_0(\lambda) = \displaystyle\sum_{\alpha=1}^{D+1} \frac{F_{0\alpha}}{\lambda - a_\alpha} + \frac{F_{0,D+1}'}{(\lambda - a_{D+1})^2} + \frac{F_{0,D+1}''}{(\lambda - a_{D+1})^3}$,   (4.14.6)

and their corresponding normalization conditions are

A) $\displaystyle\sum_{\alpha=1}^{D+3} F_{0\alpha} = -(M_0 - 1)$; $F_{0\alpha} > \dfrac{1}{2}$, $\alpha = 1, \ldots, D+1$,

B) $\displaystyle\sum_{\alpha=1}^{D+1} F_{0\alpha} + F_0 + F_0^* = -(M_0 - 1)$; $F_{0\alpha} > \dfrac{1}{2}$, $\alpha = 1, \ldots, D+1$,

C) $\displaystyle\sum_{\alpha=1}^{D+2} F_{0\alpha} = -(M_0 - 1)$; $F_{0\alpha} > \dfrac{1}{2}$, $\alpha = 1, \ldots, D+1$,

D) $\displaystyle\sum_{\alpha=1}^{D+1} F_{0\alpha} = -(M_0 - 1)$; $F_{01}' > 0$, $F_{0\alpha} > \dfrac{1}{2}$, $\alpha = 2, \ldots, D+1$,

E) $\displaystyle\sum_{\alpha=1}^{D+1} F_{0\alpha} = -(M_0 - 1)$; $F_{01}' > 0$, $F_{0,D+1}' < 0$, $\alpha = 1, \ldots, D$,

F) $\displaystyle\sum_{\alpha=1}^{D+1} F_{0\alpha} = -(M_0 - 1); \quad F''_{0,D+1} > 0, \quad \alpha = 1, \ldots, D.$ (4.14.7)

If conditions (4.14.5) and (4.14.7) are satisfied, we can apply the inverse method of separation of variables to equation (4.1.4) and transform it to the class of D-dimensional quasi-exactly solvable Schrödinger equations

$$\left[-\sqrt{g(\mu, \vec{\lambda})} \sum_{i,k=1}^{D} \frac{\partial}{\partial \lambda_i} \left(\frac{g_{ik}(\mu, \vec{\lambda})}{\sqrt{g(\mu, \vec{\lambda})}} \frac{\partial}{\partial \lambda_k} \right) + V(\mu, \vec{\lambda}) \right] \psi(\mu, \vec{\lambda})$$

$$= E(\mu)\psi(\mu, \vec{\lambda}). \quad (4.14.8)$$

The metric tensor $g_{ik}(\mu, \vec{\lambda})$, describing the manifold on which the Schrödinger problem is formulated, is diagonal and is defined by the formulas

$$g_{ii}(\mu, \vec{\lambda}) = \sum_{\alpha=1}^{D} (-1)^{\alpha} \sigma^{\alpha}(\mu) s_i^{D-\alpha}(\vec{\lambda}) \frac{\prod_{\alpha=2}^{D}(\lambda_i - a_\alpha)}{\rho(\lambda_i) \prod_{k=1}^{D}{}'(\lambda_i - \lambda_k)}. \quad (4.14.9)$$

Here we have used the notation

$$s_i^{D-\alpha}(\vec{\lambda}) = \sum_{\substack{i_1 < \cdots < i_{D-\alpha} \\ i_1, \ldots, i_{D-\alpha} \neq i}} (\lambda_{i_1} - c) \cdots (\lambda_{i_{D-\alpha}} - c). \quad (4.14.10)$$

The potential in (4.14.8) has the form

$$V(\mu, \vec{\lambda}) = \sigma^0(\mu) + \sum_{\alpha-1}^{D} \sigma^{\alpha}(\mu)\varepsilon_{0\alpha}$$

$$+ \sum_{i=1}^{D} \sum_{\alpha=1}^{D} (-1)^{\alpha} \sigma^{\alpha}(\mu) s_i^{D-\alpha}(\vec{\lambda}) \frac{\prod_{\alpha=2}^{D}(\lambda_i - a_\alpha)}{\rho(\lambda_i) \prod_{k=1}^{D}{}'(\lambda_i - \lambda_k)}$$

$$\times \left\{ F_0'(\lambda_i) + F_0^2(\lambda_i) + 2 \sum_{j=1}^{M_0} \frac{F_0(\lambda_i) - F_0(\xi_{0j})}{\lambda_i - \xi_{0j}} \right.$$

$$+ \prod_{k=1}^{D} \left[\sum_{\beta=1}^{D} (-1)^{\beta} \sigma^{\beta}(\mu) s_k^{D-\beta}(\vec{\lambda}) \frac{\rho(\lambda_k)}{\prod_{\alpha=2}^{D}(\lambda_k - a_\alpha)} \right]^{\frac{1}{4}}$$

$$\times \frac{\partial^2}{\partial \lambda_i^2} \prod_{k=1}^{D} \left[\sum_{\beta=1}^{D} (-1)^{\beta} \sigma^{\beta}(\mu) s_k^{D-\beta}(\vec{\lambda}) \frac{\rho(\lambda_k)}{\prod_{\alpha=2}^{D}(\lambda_k - a_\alpha)} \right]^{-\frac{1}{4}} \right\},$$

(4.14.11)

where $\xi_{0j}, j = 1, \ldots, M_0$ satisfy equation (4.1.2), and solutions of (4.14.8) related to this potential,

$$
\psi(\mu, \vec{\lambda}) = \prod_{k=1}^{D} \left[\sum_{\beta=1}^{D} (-1)^{\beta} \sigma^{\beta}(\mu) s_k^{D-\beta}(\vec{\lambda}) \frac{\rho(\lambda_k)}{\prod_{\alpha=2}^{D}(\lambda_k - a_\alpha)} \right]^{\frac{1}{4}}
$$

$$
\times \prod_{k=1}^{D} \left[\exp\left(\int F(\lambda_k)\, d\lambda_k \right) \prod_{j=1}^{M} (\lambda_k - \xi_j) \right], \quad (4.14.12a)
$$

$$
E(\mu) = \sigma^0(\mu) + \sum_{\alpha=1}^{D} \sigma^\alpha(\mu)\varepsilon_\alpha, \quad (4.14.12b)
$$

are expressed via the functions $F(\lambda)$, having the same functional structure as $F_0(\lambda)$, and the numbers $M, \xi_1, \ldots, \xi_M, \varepsilon_1, \ldots, \varepsilon_D$, satisfying the system of equations (4.1.7), (4.1.8). The Schrödinger equation (4.14.8) has at least one explicit normalizable solution corresponding to the choice

$$
M = M_0, \quad F(\lambda) = F_0(\lambda), \quad \xi_i = \xi_{0i},
$$
$$
i = 1, \ldots, M_0, \quad \varepsilon_\alpha = \varepsilon_{0\alpha}, \quad \alpha = 1, \ldots, D, \quad (4.14.13)
$$

and, thus, it is actually a quasi-exactly solvable equation.

In order to answer the question of whether or not this equation has other normalizable solutions, let us look at the system (4.1.7), (4.1.8). At first sight, it is not solvable, since, due to the condition

$$
\lim_{\lambda \to \infty} \lambda F(\lambda) = -(M - 1), \quad (4.14.14)
$$

the number of equations entering into it exceeds the number of unknown quantities. However, a closer look reveals that this is not true.

To demonstrate this fact, we restrict ourselves to the simplest (non-degenerate) case described by diagram A in figure 4.4. In this case the function $F(\lambda)$ has the form

$$
F(\lambda) = \sum_{\alpha=1}^{D+3} \frac{F_\alpha}{\lambda - a_\alpha} \quad (4.14.15)
$$

and condition (4.14.14) for it becomes

$$
\sum_{\alpha=1}^{D+3} F_\alpha = -(M - 1). \quad (4.14.16)
$$

For the corresponding wavefunction to be normalizable, the numbers F_α must satisfy the constraints

$$F_\alpha > \frac{1}{2}, \quad \alpha = 1, \ldots, D+1. \tag{4.14.17}$$

Substitution of (4.14.2), (4.14.4), (4.14.6) and (4.14.15) into (4.1.7) gives a system of $2D+3$ algebraic equations, $D+3$ of which can be obtained by equating the "residues" in simple double poles in the similar terms. These $D+3$ equations have the form

$$(F_\alpha - \frac{1}{2})^2 - (F_{0\alpha} - \frac{1}{2})^2 = 0, \quad \alpha = 2, \ldots, D \tag{4.14.18a}$$

and

$$(F_\alpha - \frac{1}{2})^2 - (F_{0\alpha} - \frac{1}{2})^2 = -\sum_{\beta=1}^{D} R_\alpha^\beta (\varepsilon_\beta - \varepsilon_{0\beta}), \quad \alpha \in \Delta, \tag{4.14.18b}$$

where

$$\Delta \equiv \{1, D+1, D+2, D+3\} \tag{4.14.19}$$

and

$$R_\alpha^\beta = \frac{\rho_0(a_\alpha - \rho_1)(a_\alpha - \rho_2)(a_\alpha - \rho_3)(a_\alpha - \rho_4)(a_\alpha - c)^{\beta-1}}{\prod_{\beta=1}^{D+3} {}'(a_\alpha - a_\beta)}, \tag{4.14.20}$$

The other D equations having a more complicated form are, fortunately, not so important to us and we shall not write them down here.

Taking into account the constraints (4.14.17) and solving the system (4.14.18) we obtain

$$F_\alpha = F_{0\alpha}, \quad \alpha = 2, \ldots, D, \tag{4.14.21a}$$

$$F_\alpha = \frac{1}{2} + \sigma_\alpha \sqrt{(F_{0\alpha} - \frac{1}{2})^2 - \sum_{\beta=1}^{D} R_\alpha^\beta (\varepsilon_\beta - \varepsilon_{0\beta})},$$

$$\alpha \in \Delta, \tag{4.14.21b}$$

where $\sigma_\alpha, \alpha \in \Delta$ are the "signs" taking the values ± 1. Substituting (4.14.21) into (4.14.16) we obtain the equation

$$\sum_{\alpha=2}^{D} F_{0\alpha} + \sum_{\alpha \in \Delta} \sigma_\alpha \sqrt{(F_{0\alpha} - \frac{1}{2})^2 - \sum_{\beta=1}^{D} R_\alpha^\beta (\varepsilon_\beta - \varepsilon_{0\beta})} = -(M+1)$$

$$\tag{4.14.22}$$

which plays a central role in our consideration. Indeed, this additional equation, the presence of which makes the system (4.1.7), (4.1.8) non-solvable, is, fortunately, not so strict as other equations of this system and, in some special cases, can be transformed into an identity. Then the number of equations and the number of unknown quantities become equal and this allows the system to have several algebraic solutions.

The cases when this is possible can easily be found from the condition of the cancellation of ε_α-dependent terms in the left-hand side of (4.14.22). As in the one-dimensional case (see section 4.2), these possibilities are realized when

(i)   all the four roots in (4.14.22) do not depend on ε_β,

(ii)   two roots in (4.14.22) depend on ε_β but cancel, while the remaining two roots do not depend on ε_β,

(iii) three roots in (4.14.22) depend on ε_β but cancel, while the fourth root does not depend on ε_β.

The first possibility implies that all the coefficients (4.14.20) are zero:

$$R_\alpha^\beta = 0, \quad \alpha \in \Delta, \quad \beta = 1, \dots, D. \tag{4.14.23}$$

The second possibility occurs when

$$R_{\alpha 1}^\beta = R_{\alpha 2}^\beta = 0, \quad R_{\alpha 3}^\beta = R_{\alpha 4}^\beta = R^\beta, \quad \beta = 1, \dots, D \tag{4.14.24a}$$

and, simultaneously,

$$\left| F_{0\alpha_3} - \frac{1}{2} \right| = \left| F_{0\alpha_4} - \frac{1}{2} \right| \tag{4.14.24b}$$

and

$$\sigma_{\alpha_3} = -\sigma_{\alpha_4}, \tag{4.14.24c}$$

where the indices $\alpha_1, \alpha_2, \alpha_3, \alpha_4 \in \Delta$ differ from each other.

For the third possibility to be realized, we must take

$$R_{\alpha_1}^\beta = A_1^2 R^\beta, \quad R_{\alpha_2}^\beta = A_2^2 R^\beta, \quad R_{\alpha_3}^\beta = R^\beta, \quad R_{\alpha_4}^\beta = 0,$$
$$\beta = 1, \dots, D, \tag{4.14.25a}$$

where A_1 and A_2 are certain positive numbers such that

$$A_1 + A_2 = 1. \tag{4.14.25b}$$

Simultaneously, we must require that

$$\frac{1}{A_1} \left| F_{0\alpha_1} - \frac{1}{2} \right| = \frac{1}{A_2} \left| F_{0\alpha_2} - \frac{1}{2} \right| = \left| F_{0\alpha_3} - \frac{1}{2} \right| \tag{4.14.25c}$$

and

$$\sigma_{\alpha_1} = \sigma_{\alpha_2} = -\sigma_{\alpha_3}. \tag{4.14.25d}$$

Here the indices $\alpha_1, \alpha_2, \alpha_3, \alpha_4 \in \Delta$ also differ from each other.

Imposing the constraints (4.14.23), (4.14.24) or (4.14.25) on the system (4.1.7), (4.1.8) and repeating the reasonings of sections 4.3, 4.4 and 4.5 we can list all solutions satisfying the normalization condition (4.14.17). Using the results of statement 3.11 it is not difficult to show that for any admissible set of integers $\sigma_\alpha, \alpha \in \Delta$ and M entering into equation (4.14.22) the number of such solutions is $\frac{(M+D)!}{M!D!}$ and, therefore, the order of the corresponding quasi-exactly solvable models described by formulas (4.14.8)–(4.14.11) is also $\frac{(M+D)!}{M!D!}$.

The systems associated with other diagrams in figure 4.4 can be studied quite analogously. It can be shown that an appropriate choice of functions $\rho(\lambda)$ and $F_0(\lambda)$ leads to quasi-exactly solvable models of an arbitrary, arbitrarily large order.

Chapter 5

Completely integrable Gaudin models and quasi-exact solvability

5.1 Hidden symmetries in quasi-exactly solvable models

Summarizing the results of chapter 4, we can conclude that for the $\mathcal{D}$-dimensional quantum mechanical model constructed by means of the inverse procedure of separation of variables to be quasi-exactly solvable, the initial $\mathcal{D}$-parameter spectral equation

$$\left\{-\frac{\partial^2}{\partial\lambda^2} + x^o(\lambda)\right\}\varphi(\lambda) = \left\{\sum_{\alpha=1}^{\mathcal{D}} x^\alpha(\lambda)\varepsilon_\alpha\right\}\varphi(\lambda) \tag{5.1.1}$$

must have a finite number of algebraic solutions. We demonstrated that this is possible when the functions $x^0(\lambda)$ and x^α, $\alpha = 1, \ldots, \mathcal{D}$ belong to $(2\mathcal{D}+3)$-dimensional spaces $R_{2n}\left(\begin{array}{c}\vec{a}\\2\vec{n}\end{array}\right)$ and satisfy some specific conditions which make the system of spectral equations for (5.1.1) compatible and solvable. In this section we show that these conditions admit very natural group-theoretical interpretation allowing us to explain the phenomenon of quasi-exact solvability as a consequence of a certain hidden symmetry present in equation (5.1.1).

In order to reveal this symmetry, let us consider an auxiliary $(\mathcal{D}+1)$-parameter spectral equation of the form

$$\left\{-\frac{\partial^2}{\partial\lambda^2} + x^0(\lambda)\right\}\varphi(\lambda) = \left\{\sum_{\alpha=1}^{\mathcal{D}+1} x^\alpha(\lambda)\varepsilon_\alpha\right\}\varphi(\lambda) \tag{5.1.2}$$

in which $x^0(\lambda) \in R_{2n}\left(\begin{array}{c}\vec{a}\\2\vec{n}\end{array}\right)$ and $x^\alpha(\lambda) \in R_{2n}\left(\begin{array}{c}\vec{a}\\2\vec{n}\end{array}\right)$, $\alpha = 1, \ldots, \mathcal{D}$ are

322

the same functions as in (5.1.1), $x^{\mathcal{D}+1}(\lambda) \in R_{2n}\begin{pmatrix} \vec{a} \\ 2\vec{n} \end{pmatrix}$ is an arbitraily chosen additional function and $\varepsilon_{\mathcal{D}+1}$ is an additional spectral parameter. According to the results of section 3.5, this equation is algebraically solvable and has an infinite number of solutions of the form (3.5.33). In the general case, when $x^0(\lambda)$ and $x^\alpha(\lambda), \alpha = 1, \ldots, \mathcal{D}+1$ are arbitrarily chosen (random) elements of the space $R_{2n}\begin{pmatrix} \vec{a} \\ 2\vec{n} \end{pmatrix}$, the spectral parameter $\varepsilon_{\mathcal{D}+1}$ (as well as the other parameters $\varepsilon_1, \ldots, \varepsilon_{\mathcal{D}}$) takes different values for different solutions of equation (5.1.2). However, when the condition of compatibility imposed on $x^0(\lambda)$ and $x^\alpha(\lambda), \alpha = 1, \ldots, \mathcal{D}$ is satisfied, the situation changes. Indeed, the fact that in this case equation (5.1.1) has several algebraic solutions means that there exist several solutions of equation (5.1.2), for which the parameter $\varepsilon_{\mathcal{D}+1}$ takes one and the same value $\varepsilon_{\mathcal{D}+1} = 0$. But this means that the spectrum of the parameter $\varepsilon_{\mathcal{D}+1}$ is degenerate relative to the spectra of other parameters $\varepsilon_1, \ldots, \varepsilon_{\mathcal{D}}$.

What is the reason for this degeneracy? To answer this question remember that, according to the inverse method of separation of variables, the admissible values of spectral parameters $\varepsilon_\alpha, \alpha = 1, \ldots, \mathcal{D}+1$ can be interpreted as eigenvalues of certain commuting $(\mathcal{D}+1)$-dimensional differential operators $L_\alpha, \alpha = 1, \ldots, \mathcal{D}+1$,

$$L_\alpha \phi = \varepsilon_\alpha \phi \quad , \quad \alpha = 1, \ldots, \mathcal{D}+1. \tag{5.1.3}$$

In this language, the spectrum of operator $L_{\mathcal{D}+1}$ is degenerate relative to the spectra of other operators $L_1, \ldots, L_{\mathcal{D}}$. But this suggests that there exists a certain group of symmetry G, under which the operator $L_{\mathcal{D}+1}$ is invariant:

$$G^{-1}L_{\mathcal{D}+1}G = L_{\mathcal{D}+1}, \tag{5.1.4a}$$

while the other operators $L_1, \ldots, L_{\mathcal{D}}$ are not:

$$G^{-1}L_\alpha G \neq L_\alpha \quad , \quad \alpha = 1, \ldots, \mathcal{D}. \tag{5.1.4b}$$

Moreover, the group G has finite-dimensional representations determining the multiplicities of the degenerate eigenvalues $\varepsilon_{\mathcal{D}+1}$.

Below we demonstrate that such a group really exists. We construct this group for equations (5.1.2) and describe its finite-dimensional representations, determining the multiplicities of the degenerate eigenvalues $\varepsilon_{\mathcal{D}+1}$ and, consequently, the order of the corresponding quasi-exactly solvable models. This programme will be realized in the following five subsections.

1. Spectral degeneration in the generalized MPS equation. First of all, remember that any MPS equation discussed in this book can be viewed as a particular case of a more general equation

$$\frac{\partial^2}{\partial \lambda^2} \varphi(\lambda) = \omega(\lambda) \varphi(\lambda) \tag{5.1.5}$$

for two functions $\varphi(\lambda)$ and $\omega(\lambda)$, provided that $\omega(\lambda)$ belongs to a finite-dimensional functional space and, thus, can be interpreted as a linear combination of a finite number of given basis functions. We identify the coefficients of functions depending on the sort of solution with the spectral parameters.

According to the results of section 3.5, the most general solution of the equation (5.1.5) has the form

$$\omega(\lambda) = F'(\lambda) + F^2(\lambda) + 2 \sum_{i=1}^{M} \frac{F(\lambda) - F(\xi_i)}{\lambda - \xi_i}, \tag{5.1.6}$$

$$\varphi(\lambda) = \prod_{i=1}^{M} (\lambda - \xi_i) \exp\left(\int F(\lambda)\, d\lambda \right), \tag{5.1.7}$$

where $F(\lambda)$ is a certain rational function, M is an arbitrary non-negative integer, and $\xi_1, \ldots, \xi_M$ are the numbers satisfying the system of equations

$$\sum_{k=1}^{M}{}' \frac{1}{\xi_i - \xi_k} + F(\xi_i) = 0, \qquad i = 1, \ldots, M. \tag{5.1.8}$$

For the sake of simplicity we restrict ourselves to discussion of the case when the function $F(\lambda)$ is fixed and satisfies the following condition:

$$\lim_{\lambda \to \infty} \lambda F(\lambda) = F, \tag{5.1.9}$$

where F is a finite number. This condition gives us the possibility of revealing the presence of spectral degenerations in equation (5.1.6).

Indeed, using formulas (5.1.6) and (5.1.8) we can write

$$\omega(\lambda) = \left(F(\lambda) + \sum_{i=1}^{M} \frac{1}{\lambda - \xi_i} \right)' + \left(F(\lambda) + \sum_{i=1}^{M} \frac{1}{\lambda - \xi_i} \right)^2. \tag{5.1.10}$$

Taking into account (5.1.9), we find the relation

$$\lim_{\lambda \to \infty} \lambda^2 \omega(\lambda) = (F + M)(F + M - 1), \tag{5.1.11}$$

which shows that there exists a certain linear combination of spectral parameters

$$\varepsilon = \lim_{\lambda \to \infty} \lambda^2 \omega(\lambda) \qquad (5.1.12)$$

depending only on the number M and not on the numbers $\xi_i, i = 1, \ldots, M$. We can consider this linear combination as a new spectral parameter, choosing an appropiate basis in the functional space.

According to statement 3.11, the number of solutions of system (5.1.8) is $\frac{(M+N-1)!}{M!(N-1)!}$, where N is the dimension of the space to which the rational function $F(\lambda)$ belongs. But this means that for any given M the spectrum of the parameter ε is $\frac{(M+N-1)!}{M!(N-1)!}$-fold degenerate relative to the spectra of other spectral parameters entering into the function $\omega(\lambda)$. This allows us to assert that equation (5.1.5) is an equation with a partially degenerate spectrum and, thus, it can always be transformed to a quasi-exactly solvable model of quantum mechanics. The next step in our procedure is to construct the completely integrable model associated with the generalized MPS equation (5.1.5). We show that this is simply the Gaudin model discussed in section 1.10. We give here a general definition of this model and discuss in detail the properties of its solutions.

2. Generalized Gaudin model.

Let $S^{\pm}(\lambda)$ and $S^0(\lambda)$ be generators of the Gaudin algebra satisfying the commutation relations

$$[S^-(\lambda), S^+(\mu)] = -2 \frac{S^0(\lambda) - S^0(\mu)}{\lambda - \mu}, \qquad (5.1.13a)$$

$$[S^0(\lambda), S^{\pm}(\mu)] = \mp \frac{S^{\pm}(\lambda) - S^{\pm}(\mu)}{\lambda - \mu}. \qquad (5.1.13b)$$

Consider the representation of this algebra, defined by the formulas

$$S^0(\lambda)|0\rangle = F(\lambda)|0\rangle, \qquad (5.1.14a)$$
$$S^-(\lambda)|0\rangle = 0, \qquad (5.1.14b)$$

in which $|0\rangle$ is the lowest-weight vector and $F(\lambda)$ is the corresponding lowest weight. Define the representation space W as the linear span of all vectors of the type

$$|0\rangle, S^+(\xi_1)|0\rangle, S^+(\xi_1)S^+(\xi_2)|0\rangle, \ldots,$$
$$S^+(\xi_1)\ldots S^+(\xi_M)|0\rangle, \ldots, \qquad (5.1.15)$$

where $\xi_1, \xi_2, \ldots$ are arbitrary numbers.

The Gaudin operators are defined by the formula

$$K(\lambda) = S^0(\lambda)S^0(\lambda) - \frac{1}{2}S^+(\lambda)S^-(\lambda) - \frac{1}{2}S^-(\lambda)S^+(\lambda). \qquad (5.1.16)$$

Using relations (5.1.14) it is easy to show that they commute with each other

$$[K(\lambda), K(\mu)] = 0 \qquad (5.1.17)$$

for any λ and μ, and, thus, have a common set of eigenfunctions in the space W.

In spite of the fact that the space W is infinite dimensional, the generalized Gaudin spectral problem in W

$$K(\lambda)\phi = \omega(\lambda)\phi, \qquad \phi \in W \qquad (5.1.18)$$

is exactly solvable.[1] Its Bethe solutions have the form

$$\omega(\lambda) = F'(\lambda) + F^2(\lambda) + 2\sum_{i=1}^{M} \frac{F(\lambda) - F(\xi_i)}{\lambda - \xi_i}, \qquad (5.1.19)$$

$$\phi = \prod_{i=1}^{M} S^+(\xi_i)|0\rangle, \qquad (5.1.20)$$

where M is an arbitrary non-negative integer, and $\xi_1, \ldots, \xi_M$ are the numbers satisfying the system of numerical equations

$$\sum_{k=1}^{M}{}' \frac{1}{\xi_i - \xi_k} + F(\xi_i) = 0 \quad, \quad i = 1, \ldots, M. \qquad (5.1.21)$$

This can be demonstrated as follows. Using the commutation relations (5.1.14), and taking there $\mu = \lambda$, we obtain

$$[S^+(\lambda), S^-(\lambda)] = 2\frac{\partial}{\partial \lambda}S^0(\lambda). \qquad (5.1.22)$$

From conditions (5.1.19) and (5.1.22) and definition (5.1.16) it follows that

$$K(\lambda)|0\rangle = \{F'(\lambda) + F^2(\lambda)\}|0\rangle. \qquad (5.1.23)$$

---

[1] We use the adjective "generalized" here because Gaudin himself and other authors (Gaudin 1983, Sklyanin 1987, Jurčo 1989) considered only the case corresponding to a non-degenerate rational function $F(\lambda)$. A special form of the operators $S^a(\lambda), a = \pm, 0$, was also considered. Here, no restrictions are imposed on the form of the function $F(\lambda)$ and operators $S^a(\lambda)$.

Besides, we have

$$[K(\lambda), S^+(\mu)] = \frac{2}{\mu - \lambda}[S^+(\lambda)S^0(\mu) - S^+(\mu)S^0(\lambda)]. \qquad (5.1.24)$$

Let us now act by the operator $K(\lambda)$ on the vector (5.1.20). Transferring this operator to the right and using formulas (5.1.22)–(5.1.24) we obtain

$$K(\lambda)\phi = [F'(\lambda) + F^2(\lambda)]\phi$$

$$+ \sum_{i=1}^{M} \frac{2}{\lambda - \xi_i} S^+(\xi_1) \ldots S^+(\xi_i)S^0(\lambda)S^+(\xi_{i+1}) \ldots S^+(\xi_M)|0\rangle$$

$$- \sum_{i=1}^{M} \frac{2}{\lambda - \xi_i} S^+(\xi_i) \ldots S^+(\xi_{i-1})S^+(\lambda)S^0(\xi_i)S^+(\xi_{i+1}) \ldots S^+(\xi_M)|0\rangle.$$

$$(5.1.25)$$

The operators $S^0(\lambda)$ and $S^0(\xi_i)$ appearing in the second and third groups of terms can also be transferred to the right by means of the commutation conditions (5.1.13b). Taking into account formula (5.1.14a) we obtain

$$K(\lambda)\phi = \left\{ F'(\lambda) + F^2(\lambda) + 2 \sum_{i=1}^{M} \frac{F(\lambda) - F(\xi_i)}{\lambda - \xi_i} \right.$$

$$\left. + 2 \sum_{i=1}^{M} \frac{1}{\lambda - \xi_i} \left[\sum_{k=1}^{M}{}' \frac{1}{\xi_i - \xi_k} + F(\xi_i) \right] \phi \right\}$$

$$- 2 \sum_{i=1}^{M} \frac{1}{\lambda - \xi_i} \left[\sum_{k=1}^{M}{}' \frac{1}{\xi_i - \xi_k} + F(\xi_i) \right]$$

$$\times S^+(\xi_1) \ldots S^+(\xi_{i-1})S^+(\lambda)S^+(\xi_{i+1}) \ldots S^+(\xi_M)|0\rangle.$$

$$(5.1.26)$$

Equating to zero the coefficients for the terms not proportional to ϕ we arrive at the system (5.1.21). The remaining terms determine the eigenvalues of the operators $K(\lambda)$, which, obviously, have the form (5.1.19). This completes the derivation.

3. Transition from the generalized MPS equation to the generalized Gaudin model. It is not difficult to see that solutions of the generalized Gaudin model have the same functional structure as solutions of the generalized MPS equation. Indeed, we see that expressions (5.1.6) and (5.1.19) for $\omega(\lambda)$ coincide. Equations (5.1.8) and (5.1.21) for the numbers $\xi_i, i = 1, \ldots, M$ also coincide, and the solutions (5.1.7) and (5.1.20) are described by formulas having similar (factorized) form. Such a coincidence cannot be casual. It indicates that there is a deep connection between equations (5.1.5) and (5.1.18). Below we demonstrate that this is really so and show that equation (5.1.18) can actually be obtained from (5.1.5) by means of the inverse method of separation of variables.

First of all, note that any rational function $F(\lambda)$ can be interpreted as a limiting case of a certain non-degenerate rational function of the form

$$F(\lambda) = \sum_{\alpha=1}^{N} \frac{f_\alpha}{\lambda - a_\alpha}. \tag{5.1.27}$$

This makes it possible to restrict ourselves by discussing here only the non-degenerate case which, on the one hand, is rather simple and, on the other hand, does not lead to loss of generality.

Substituting (5.1.27) into expressions (5.1.5)–(5.1.8) we reduce equation (5.1.5) to the form

$$\left\{ -\frac{\partial^2}{\partial \lambda^2} + \left(\sum_{\alpha=1}^{N} \frac{f_\alpha}{\lambda - a_\alpha} \right)' + \left(\sum_{\alpha=1}^{N} \frac{f_\alpha}{\lambda - a_\alpha} \right)^2 \right. $$
$$\left. + \sum_{\alpha=1}^{N} \frac{e_\alpha}{\lambda - a_\alpha} \right\} \varphi(\lambda) = 0 \tag{5.1.28}$$

and obtain the following expressions for its solutions

$$\varphi(\lambda) = \prod_{\alpha=1}^{N} (\lambda - a_\alpha)^{f_\alpha} \prod_{i=1}^{M} (\lambda - \xi_i), \tag{5.1.29}$$

$$e_\alpha = 2 \sum_{i=1}^{M} \frac{f_\alpha}{\xi_i - a_\alpha}, \tag{5.1.30}$$

in which the numbers $\xi_i, i = 1, \ldots, M$ satisfy the constraints

$$\sum_{k=1}^{M} {}' \frac{1}{\xi_i - \xi_k} + \sum_{\alpha=1}^{N} \frac{f_\alpha}{\xi_i - a_\alpha} = 0, \qquad i = 1, \ldots, M. \tag{5.1.31}$$

According to the general prescriptions given in section 3.3, let us reproduce equation (5.1.28) N times, rewriting it as the system

$$\left\{ -\frac{\partial^2}{\partial \lambda_i^2} + \left(\sum_{\alpha=1}^{N} \frac{f_\alpha}{\lambda_i - a_\alpha} \right)' + \left(\sum_{\alpha=1}^{N} \frac{f_\alpha}{\lambda_i - a_\alpha} \right)^2 + \sum_{\alpha=1}^{N} \frac{e_\alpha}{\lambda_i - a_\alpha} \right\} \phi = 0,$$
$$i = 1, \ldots, N \tag{5.1.32}$$

in which $\lambda_i, i = 1, \ldots, N$ are independent variables, and

$$\phi = \prod_{i=1}^{N} \varphi(\lambda_i) = \prod_{i=1}^{N} \left\{ \prod_{\alpha=1}^{N} (\lambda_i - a_\alpha)^{f_\alpha} \prod_{k=1}^{M} (\lambda_i - \xi_k) \right\}. \qquad (5.1.33)$$

Now, we introduce N differential operators $L_\alpha, \alpha = 1, \ldots, N$ determined as solutions of the system of N linear equations

$$\sum_{\alpha=1}^{N} \frac{1}{\lambda_i - a_a} L_\alpha = \frac{\partial^2}{\partial \lambda_i^2} - \left(\sum_{\alpha=1}^{N} \frac{f_\alpha}{\lambda_i - a_\alpha} \right)' - \left(\sum_{\alpha=1}^{N} \frac{f_\alpha}{\lambda_i - a_\alpha} \right)^2 ,$$

$$i = 1, \ldots, N. \qquad (5.1.34)$$

Substituting (5.1.34) into (5.1.32) we obtain

$$\sum_{\alpha=1}^{N} \frac{1}{\lambda_i - a_\alpha} (L_\alpha - e_\alpha)\phi = 0, \qquad i = 1, \ldots, N \qquad (5.1.35)$$

which, due to the non-degeneracy of the matrix $(\lambda_i - a_\alpha)^{-1}$, gives

$$L_\alpha \phi = e_\alpha \phi, \qquad \alpha = 1, \ldots, N. \qquad (5.1.36)$$

Let us now construct a new operator $L(\lambda)$, depending on the parameter λ:

$$L(\lambda) = \left(\sum_{\alpha=1}^{N} \frac{f_\alpha}{\lambda - a_\alpha} \right)' - \left(\sum_{\alpha=1}^{N} \frac{f_\alpha}{\lambda - a_\alpha} \right)^2 + \sum_{\alpha=1}^{N} \frac{L_\alpha}{\lambda - a_\alpha}. \qquad (5.1.37)$$

From formula (5.1.36) it follows that

$$\left\{ L(\lambda) - \left(\sum_{\alpha=1}^{N} \frac{f_\alpha}{\lambda - a_\alpha} \right)' - \left(\sum_{\alpha=1}^{N} \frac{f_\alpha}{\lambda - a_\alpha} \right)^2 - \sum_{\alpha=1}^{N} \frac{e_\alpha}{\lambda - a_\alpha} \right\} \phi = 0$$

$$(5.1.38)$$

or, equivalently,

$$L(\lambda)\phi = \omega(\lambda)\phi. \qquad (5.1.39)$$

Thus, we have shown that the operator $L(\lambda)$ has the same spectral properties as the Gaudin operator $K(\lambda)$.

Now it remains to show that $L(\lambda)$ coincides with $K(\lambda)$ and that the function (5.1.33) admits the representation (5.1.20). For this it is sufficient to find an explicit form of $L(\lambda)$ by solving equation (5.1.34).

Acting on (5.1.34) by the operator

$$\sum_{i=1}^{N} \frac{\prod_{\gamma=1}^{N}(\lambda_i - a_\gamma)}{\prod_{k=1}^{N}{}'(\lambda_i - \lambda_k)(\lambda_i - a_\beta)} \qquad (5.1.40)$$

and using the identity

$$\sum_{i=1}^{N} \frac{\prod_{\gamma=1}^{N}(\lambda_i - a_\gamma)}{\prod_{k=1}^{N}{}'(\lambda_i - \lambda_k)(\lambda_i - a_\beta)(\lambda - a_\alpha)} = \left\{ \begin{array}{ll} 0, & \beta \neq \alpha \\ \gamma t_\alpha^{-1}, & \beta = \alpha \end{array} \right. \qquad (5.1.41)$$

in which

$$t_\alpha = \gamma \prod_{i=1}^{N}(\lambda_i - a_\alpha) \left\{ \prod_{\beta=1}^{N}{}'(a_\beta - a_\alpha) \right\}^{-1}, \qquad \alpha = 1, \ldots, N, \qquad (5.1.42)$$

we obtain explicit expressions for L_α which, in terms of the new variables t_α, have the form

$$L_\alpha = \gamma \left[t_\alpha \frac{\partial^2}{\partial t_\alpha^2} - \frac{f_\alpha(f_\alpha - 1)}{t_\alpha} \right] + 2 \sum_{\beta=1}^{N}{}' \left\{ \left(t_\alpha \frac{\partial}{\partial t_\alpha} \right) \left(t_\beta \frac{\partial}{\partial t_\beta} \right) \right.$$

$$- f_\alpha f_\beta - \frac{1}{2} t_\alpha \left(t_\beta \frac{\partial^2}{\partial t_\beta^2} - \frac{f_\beta(f_\beta - 1)}{t_\beta} \right)$$

$$\left. - \frac{1}{2} \left(t_\alpha \frac{\partial^2}{\partial t_\alpha^2} - \frac{f_\alpha(f_\alpha - 1)}{t_\alpha} \right) t_\beta \right\} \frac{1}{a_\alpha - a_\beta}.$$

$$(5.1.43)$$

The last step is to substitute these expressions into (5.1.37). Making elementary transformations we find that the operator $L(\lambda)$ can be represented in the same form as $K(\lambda)$:

$$L(\lambda) = S^0(\lambda)S^0(\lambda) - \frac{1}{2}S^+(\lambda)S^-(\lambda) - \frac{1}{2}S^-(\lambda)S^+(\lambda) \qquad (5.1.44)$$

where

$$S^+(\lambda) = \sum_{\alpha=1}^{N} \frac{t_\alpha}{\lambda - a_\alpha} - \gamma, \qquad (5.1.45a)$$

$$S^0(\lambda) = \sum_{\alpha=1}^{N} \frac{t_\alpha \frac{\partial}{\partial t_\alpha}}{\lambda - a_\alpha}, \qquad (5.1.45b)$$

$$S^-(\lambda) = \sum_{\alpha=1}^{N} \frac{t_\alpha \frac{\partial^2}{\partial t_\alpha^2} - \frac{f_\alpha(f_\alpha - 1)}{t_\alpha}}{\lambda - a_\alpha}. \qquad (5.1.45c)$$

It is not difficult to verify that the operators (5.1.45) satisfy the commutation relations (5.1.14), i.e., form the Gaudin algebra. Therefore, $L(\lambda) = K(\lambda)$. Rewriting (5.1.45a) in terms of the initial variables λ_i, $i = 1, \dots, N$:

$$S^+(\lambda) = \frac{-\gamma \prod_{i=1}^N (\lambda_i - \lambda)}{\prod_{\beta=1}^N (\alpha_\beta - \lambda)},\qquad (5.1.46)$$

we see that the function (5.1.33) can be rewritten in the form

$$\phi \sim \prod_{i=1}^M S^+(\xi_i) \prod_{\alpha,k=1}^N (\lambda_k - a_\alpha)^{f_\alpha} \sim \prod_{i=1}^M S^+(\xi_i) \prod_{\alpha=1}^N t_\alpha^{f_\alpha}. \qquad (5.1.47)$$

Comparing (5.1.47) with (5.1.20), we see that these formulas coincide if

$$|0\rangle = \prod_{\alpha=1}^N t_\alpha^{f_\alpha}. \qquad (5.1.48)$$

In order to make sure that (5.1.48) is actually the lowest-weight vector, we must verify formulas (5.1.15). Acting on (5.1.48) by the operators (5.1.45b) and (5.1.45c) and using (5.1.27) we obtain:

$$S^0(\lambda)|0\rangle = \left(\sum_{\alpha=1}^N \frac{t_\alpha \frac{\partial}{\partial t_\alpha}}{\lambda - a_\alpha} \right) \prod_{\alpha=1}^N t_\alpha^{f_\alpha} = F(\lambda)|0\rangle, \qquad (5.1.49a)$$

$$S^-(\lambda)|0\rangle = \left(\sum_{\alpha=1}^N \frac{t_\alpha \frac{\partial^2}{\partial t_\alpha^2} - \frac{f_\alpha(f_\alpha - 1)}{t_\alpha}}{\lambda - a_\alpha} \right) \prod_{\alpha=1}^N t_\alpha^{f_\alpha} = 0. \qquad (5.1.49b)$$

This completes the proof of the equivalence of the generalized MPS equation (5.1.5) and the generalized Gaudin equation (5.1.18).

Before continuing our discussion, note that γ is an arbitrary parameter and, without loss of generality, we can take

$$\gamma = 0. \qquad (5.1.50)$$

Note also that from a practical point of view it is more convenient to deal with the homogeneously transformed Gaudin operators, defined by the formula

$$K(\lambda) = U L(\lambda) U^{-1}, \qquad (5.1.51)$$

where U is a λ-independent operator of the form

$$U = \prod_{\alpha=1}^N t_\alpha^{f_\alpha}. \qquad (5.1.52)$$

These operators have the same spectral properties as the untransformed ones, and can be represented in the form (5.1.16) with

$$S^+(\lambda) \;=\; \sum_{\alpha=1}^{N} \frac{t_\alpha}{\lambda - a_\alpha}, \tag{5.1.53a}$$

$$S^o(\lambda) \;=\; \sum_{\alpha=1}^{N} \frac{t_\alpha \frac{\partial}{\partial t_\alpha} + f_\alpha}{\lambda - a_\alpha}, \tag{5.1.53b}$$

$$S^-(\lambda) \;=\; \sum_{\alpha=1}^{N} \frac{t_\alpha \frac{\partial^2}{\partial t_\alpha^2} + 2f_\alpha \frac{\partial}{\partial t_\alpha}}{\lambda - a_\alpha}. \tag{5.1.53c}$$

In this case the vacuum vector takes an especially simple form:

$$|0\rangle = 1, \tag{5.1.54}$$

and the solutions defined by formula (5.1.20) become polynomial in t_α.

4. Gaudin models as magnetic chains. It is not difficult to see that the operators

$$S_\alpha^+ = t_\alpha, \qquad S_\alpha^0 = t_\alpha \frac{\partial}{\partial t_\alpha} + f_\alpha, \qquad S_\alpha^- = t_\alpha \frac{\partial^2}{\partial t_\alpha^2} + 2f_\alpha \frac{\partial}{\partial t_\alpha}, \tag{5.1.55}$$

entering into (5.1.54) satisfy the following commutation relations:

$$\begin{aligned}
[S_\alpha^-, S_\beta^+] &= 2\delta_{\alpha\beta} S_\beta^0, \\
[S_\alpha^0, S_\beta^\pm] &= \pm \delta_{\alpha\beta} S_\beta^\pm,
\end{aligned} \tag{5.1.56}$$

and form the algebra

$$L = sl(2) \oplus \ldots \oplus sl(2) \qquad (N \text{ times}). \tag{5.1.57}$$

Rewriting expressions (5.1.54) as

$$\vec{S}(\lambda) = \sum_{\alpha=1}^{N} \frac{\vec{S}_\alpha}{\lambda - a_\alpha}, \tag{5.1.58}$$

and substituting them into (5.1.16) we obtain for $K(\lambda)$:

$$K(\lambda) = \sum_{\alpha=1}^{N} \frac{(-f_\alpha)(-f_\alpha + 1)}{(\lambda - a_\alpha)^2} + 2\sum_{\alpha=1}^{N} \frac{H_\alpha}{\lambda - a_\alpha} \tag{5.1.59}$$

where

$$H_\alpha = \sum_{\beta=1}^{N}{}' \frac{\vec{S}_\alpha \vec{S}_\beta}{a_\alpha - a_\beta} = \sum_{\beta=1}^{N}{}' \frac{S_\alpha^0 S_\beta^0 - \frac{1}{2} S_\alpha^+ S_\beta^- - \frac{1}{2} S_\alpha^- S_\beta^+}{a_\alpha - a_\beta}$$

(5.1.60)

are the operators acting in the direct product $W = W_1 \otimes \ldots \otimes W_N$ of representation spaces of the algebra $sl(2)$. They obviously commute with each other, $[H_\alpha, H_\beta] = 0$, and, thus, can be interpreted as hamiltonians of a completely integrable non-local spin system on a finite one-dimensional lattice.

As we see, the parameters a_α enter explicitly into the hamiltonian and play the role of coupling constants characterizing the strength of the interaction between spins located at different sites. The parameters f_α do not appear explicitly in the hamiltonian. They are included in the definition of the generators, characterizing the representations in which they act. They are related to the "spins" of infinite-dimensional irreducible representations of the algebra $sl(2)$ which can be realized in the spaces of all analytic functions regular near the origin. In these spaces there exist the vectors of lowest weight $|0\rangle_\alpha = 1$ such that $S_\alpha^-|0\rangle_\alpha = 0$. The eigenvalues of the operator for the z-projection of the spin S_α^0 on $|0\rangle$ is $-f_\alpha$, so that f_α is the spin of the irreducible representation of the algebra $sl(2)$ with opposite sign. This is confirmed by the fact that the eigenvalues of the Casimir operator $\vec{S}_\alpha^2$ on $|0\rangle_\alpha$ are $(-f_\alpha)(-f_\alpha + 1)$. The representations (5.1.55) are infinite dimensional, owing to the absence of a vector of highest weight, i.e., an analytic function, regular near the origin, on which the operator S_α^+ would give zero.

The models of magnetic systems of the type (5.1.60) are not local spin systems. In the hamiltonians of these systems each spin interacts with all other spins, i.e., there is a long-range force and the situation is apparently a typical semi-classical one. This is also confirmed by the fact that the complete integrability of these models is related to the solutions of not the usual quantum Yang–Baxter equation (the triangle equation), but the so-called classical triangle equation arising in the limit $\hbar \to 0$. Remember that the quantum scattering matrix $S_{\alpha\beta}(\lambda)$ is related to the classical matrix $X_{\alpha\beta}(\lambda)$ as

$$S_{\alpha\beta} \approx 1 + \hbar X_{\alpha\beta}(\lambda).$$

(5.1.61)

The classical triangle equation for the matrix $X_{\alpha\beta}(\lambda)$ has the form

$$[X_{\alpha\beta}(\lambda_1), X_{\beta\gamma}(\lambda_2)] + [X_{\beta\gamma}(\lambda_2), X_{\gamma\alpha}(\lambda_3)]$$
$$+ [X_{\gamma\alpha}(\lambda_3), X_{\alpha\beta}(\lambda_1)] = 0, \quad \lambda_1 + \lambda_2 + \lambda_3 = 0.$$

(5.1.62)

It is known that the operators related to the solutions of this equation as

$$h_\alpha = \sum_{\beta=1}^{N}{}' X_{\alpha\beta}(a_\alpha - a_\beta) \tag{5.1.63}$$

commute with each other and can be interpreted as hamiltonians of a certain completely integrable model. It is easily seen that the operators

$$X_{\alpha\beta}(\lambda) = \frac{\vec{S}_\alpha \vec{S}_\beta}{\lambda} \tag{5.1.64}$$

satisfy the equation (5.1.62). The substitution of (5.1.64) into (5.1.63) gives us the hamiltonians (5.1.60) obtained above by other methods.

We have therefore succeeded in relating the non-degenerate generalized MPS equations (5.1.5) to magnetic systems based on algebras of the form $sl(2) \oplus \ldots \oplus sl(2)$. We see that, by solving the spectral problem for these systems, we can obtain an exhaustive amount of information on the spectra of the associated exactly and quasi-exactly solvable systems.

So far we have considered only the non-degenerate case. It can be shown that an analogous correspondence holds also when degeneracy is present, but in this case the magnetic systems are based on other (contracted) Lie algebras. As before, the hamiltonians of degenerate magnetic systems can be obtained from the Gaudin operator $K(\lambda)$ defined by (5.1.16). The operators $S^\pm(\lambda)$ and $S^0(\lambda)$ entering into (5.1.16) satisfy the same commutation relations as in (5.1.14). However, in the degenerate case the form of these operators is different from (5.1.54) and is determined by the form of the function $F(\lambda)$.

We know that any rational function $F(\lambda)$ can be represented as the sum

$$F(\lambda) = \sum_{\alpha=1}^{N} f_\alpha r^\alpha(\lambda) \tag{5.1.65}$$

in which $r^\alpha(\lambda)$ are certain elementary rational functions. According to (5.1.54) the generators of the Gaudin algebra must have the same functional structure as $F(\lambda)$ and therefore can be sought in the form

$$\vec{S}(\lambda) = \sum_{\alpha=1}^{N} \vec{S}_\alpha r^\alpha(\lambda) \tag{5.1.66}$$

where $\vec{S}_\alpha$ are certain unknown operators.

Now note that elementary rational functions satisfy the following conditions

$$\frac{r^\alpha(\lambda) - r^\beta(\mu)}{\lambda - \mu} = \sum_{\beta,\gamma=1}^{N} C^\alpha_{\beta\gamma} r^\beta(\lambda) r^\gamma(\mu) \qquad (5.1.67)$$

where $C^\alpha_{\beta\gamma} = C^\alpha_{\gamma\beta}$ are certain structure constants. Substituting (5.1.66) into the commutation relations (5.1.14) and using (5.1.67) we find the commutation relations immediately for operators $\vec{S}_\alpha$:

$$[S^-_\alpha, S^+_\beta] = -2 \sum_{\gamma=1}^{N} C^\gamma_{\alpha\beta} S^0_\gamma,$$

$$[S^0_\alpha, S^\pm_\beta] = \mp \sum_{\gamma=1}^{N} C^\gamma_{\alpha\beta} S^\pm_\gamma, \qquad (5.1.68)$$

from which it follows that they form a finite-dimensional Lie algebra $\mathbf{L}'$ being a contraction of the algebra $\mathbf{L}$ (5.1.57). Substitution of (5.1.66) into the Gaudin operator $K(\lambda)$ leads to the expression

$$K(\lambda) = \sum_{\alpha,\beta=1}^{N} \{S^0_\alpha S^0_\beta - \frac{1}{2} S^+_\alpha S^-_\beta - \frac{1}{2} S^-_\alpha S^+_\beta\} r^\alpha(\lambda) r^\beta(\lambda) \qquad (5.1.69)$$

which is bilinear in generators $\vec{S}_\alpha$ and can again be interpreted as a generating function for the hamiltonians of a certain magnetic system.

5. Hidden symmetry of Gaudin models. We have already mentioned that the phenomenon of quasi-exact solvability is a consequence of the presence of a degeneracy in the system of spectral parameters. For example, the degeneracy of the spectral parameter ε, defined by (5.1.12), relative to the other spectral parameters entering into the function $\omega(\lambda)$ leads to the existence of quasi-exactly solvable models of order equal to the degree of degeneracy. The parameter ε is the eigenvalue of the operator

$$K = \lim_{\lambda \to \infty} \lambda^2 K(\lambda) \qquad (5.1.70)$$

and, therefore, in order to find the symmetry related to this degeneracy we should consider operators which commute with K, but not with each of $K(\lambda)$ separately. To find such operators, note that K can be represented in the form

$$K = \vec{S} \cdot \vec{S}, \qquad (5.1.71)$$

where

$$\vec{S} = \lim_{\lambda \to \infty} \lambda \vec{S}(\lambda). \tag{5.1.72}$$

Using (5.1.14) it is easy to demonstrate that the operators $\vec{S}$ form the algebra $sl(2)$:

$$[S^-, S^+] = 2S^0,$$
$$[S^0, S^\pm] = \pm S^\pm, \tag{5.1.73}$$

and thus, K in (5.1.71) is the Casimir operator for this algebra. From the same commutation relations (5.1.14) it follows that K commutes with all operators of the form

$$K(\nu, \mu) = \vec{S}(\nu)\vec{S}(\mu) \tag{5.1.74}$$

where ν and μ are arbitrary parameters. At the same time, the operators (5.1.74) do not commute with the Gaudin operator $K(\lambda)$, if $\nu \neq \mu$. Therefore, the operators $K(\nu, \mu)$ can be viewed as generating functions for the generators of the algebra associated with the hidden symmetry responsible for the degeneracy. Attempts to make this algebra closed convince us that it is infinite dimensional. This suggests that we have constructed not the symmetry algebra L_{sym}, but its universal enveloping algebra $U(L_{sym})$. The transition from $U(L_{sym})$ to L_{sym} can easily be realized if we use differential representations of the generators of the Gaudin algebra given for the non-degenerate case by (5.1.54). Substituting them into (5.1.16) we obtain

$$K(\lambda) = \sum_{\alpha,\beta=1}^{N} \frac{t_\alpha t_\beta (\frac{\partial}{\partial t_\alpha} - \frac{\partial}{\partial t_\beta})^2 + 2(f_\alpha t_\beta - f_\beta t_\alpha)(\frac{\partial}{\partial t_\alpha} - \frac{\partial}{\partial t_\beta})}{(\lambda - a_\alpha)(\lambda - a_\beta)}. \tag{5.1.75}$$

Taking

$$t_\alpha = -x_\alpha, \qquad \alpha = 1, \dots, N-1,$$
$$t_N = x_1 + \dots + x_{N-1} + c x_N, \tag{5.1.76}$$

where c is an arbitrary parameter, it is not difficult to see that the operators $K(\lambda)$ rewritten in terms of new variables x_α do not contain derivatives in x_N. Therefore, x_N can be considered as an external parameter which can be excluded from consideration by taking $c = 0$. This does not lead to loss of generality. Using (5.1.76),(5.1.75) and (5.1.70) we find

$$K = \left(\sum_{\mu=1}^{N-1} x_\mu \frac{\partial}{\partial x_\mu}\right)^2 + 2\left(\sum_{\alpha=1}^{N} f_\alpha\right) \sum_{\mu=1}^{N-1} x_\mu \frac{\partial}{\partial x_\mu}, \tag{5.1.77}$$

while the expression for $K(\lambda)$ becomes

$$K(\lambda) = \sum_{\alpha,\beta=1}^{N} \sum_{\mu,\nu,\lambda,\rho=1}^{N-1} \frac{A_{\alpha\beta}^{\mu\nu\lambda\rho}}{(\lambda - a_\alpha)(\lambda - a_\beta)} \left(x_\mu \frac{\partial}{\partial x_\nu} \right) \left(x_\lambda \frac{\partial}{\partial x_\rho} \right).$$

$$(5.1.78)$$

Obviously, the operator K commutes with all operators $x_\mu \frac{\partial}{\partial x_\nu}$ which form the algebra $gl(N-1)$. We know that $gl(N-1)$ can be represented as a direct sum of the algebra $gl(1)$, with the generator

$$J = \sum_{\mu=1}^{N-1} x_\mu \frac{\partial}{\partial x_\mu}, \qquad (5.1.79)$$

and the algebra $sl(N-1)$, with generators

$$\{I_i\} = \left\{ x_\mu \frac{\partial}{\partial x_\nu}, \quad \mu \neq \nu; \quad x_\mu \frac{\partial}{\partial x_\mu} - x_{\mu+1} \frac{\partial}{\partial x_{\mu+1}} \right\}. \qquad (5.1.80)$$

Therefore, we can write

$$K = J^2 + 2 \left(\sum_{\alpha=1}^{N} f_\alpha \right) J, \qquad J \in gl(1) \qquad (5.1.81)$$

and

$$K(\lambda) = \sum_{\alpha,\beta=1}^{N} \frac{1}{(\lambda - a_\alpha)(\lambda - a_\beta)}$$
$$\times \left\{ \sum_{i,k=1}^{N(N-2)} A_{\alpha\beta}^{ik} I_i I_k + \sum_{i=1}^{N(N-2)} A_{\alpha\beta}^i I_i J + A_{\alpha\beta} J^2 \right\},$$
$$J \in gl(1), \quad I_i \in sl(N-1). \qquad (5.1.82)$$

From (5.1.81) and (5.1.82) it follows that the role of the symmetry algebra L_{sym} for the operator K is played by the algebra $sl(N-1)$, the irreducible representations of which are realized in the spaces of homogeneous polynomials in $x_1, \ldots, x_{N-1}$. These representations are characterized by the signatures $(M, 0, \ldots, 0)$ and their dimensions are equal to $\frac{(M+N-2)!}{M!(N-2)!}$, where $M = 0, 1, 2, \ldots$ are the orders of homogenous polynomials. The same obviously remains valid for the degenerate case, since the generators I_i of the algebra $sl(N-1)$ do not depend on the parameters of the system a_α and f_α entering into the function $F(\lambda)$.

This completes the procedure of finding the symmetry algebra L_{sym} responsible for the phenomenon of quasi-exact solvability of models associated with the generalized MPS equation (5.1.5).

6.   The classical electrostatic analogue and the magnetic monopole. In the preceding subsection we discussed group-theoretical properties of quasi-exactly solvable models associated with the degenerate function

$$F(\lambda) = \sum_{\alpha=1}^{N} \frac{f_\alpha}{\lambda - a_\alpha}. \tag{5.1.83}$$

Here we consider the same models from the point of view of their classical interpretation.

Following the general prescriptions given in chapters 1 and 2, we start with the system of numerical equations describing a distribution of wavefunction nodes ξ_j. According to formula (5.1.31), these equations have the form

$$\sum_{k=1}^{M}{}' \frac{1}{\xi_j - \xi_k} + \sum_{\alpha=1}^{N} \frac{f_\alpha}{\xi_j - a_\alpha} = 0, \qquad j = 1, \dots, M. \tag{5.1.84}$$

where a_α and f_α are, in general, complex numbers:

$$a_\alpha = a_{1\alpha} + i a_{2\alpha}, \quad f_\alpha = f_{1\alpha} + i f_{2\alpha}. \tag{5.1.85}$$

Therefore, ξ_j should also be taken to be complex:

$$\xi_j = \xi_{1j} + i \xi_{2j}. \tag{5.1.86}$$

Substitution of (5.1.85) and (5.1.86) into (5.1.84) leads to a system of real equations, which can be written as

$$\sum_{k=1}^{M}{}' \frac{\vec{\xi}_j - \vec{\xi}_k}{|\vec{\xi}_j - \vec{\xi}_k|^2}$$
$$+ \sum_{\alpha=1}^{N} f_{1\alpha} \frac{\vec{\xi}_j - \vec{a}_\alpha}{|\vec{\xi}_j - \vec{a}_\alpha|^2} + \sum_{\alpha=1}^{N} f_{2\alpha} \hat{\varepsilon} \frac{\vec{\xi}_j - \vec{a}_\alpha}{|\vec{\xi}_j - \vec{a}_\alpha|^2} = 0,$$
$$j = 1, \dots, M, \tag{5.1.87}$$

where $\xi_j = (\xi_{1j}, \xi_{2j})$ and $a_\alpha = (a_{1\alpha}, a_{2\alpha})$ are real two-dimensional vectors, and $\hat{\varepsilon}$ is the matrix rotating the vectors by 90° counterclockwise. Equation

(5.1.87) can be interpreted as the condition for an extremum of the function

$$W(\vec{\xi_1}, \ldots, \vec{\xi_M}) = -\sum_{i<k}^{M} q_i q_k \ln |\vec{\xi_i} - \vec{\xi_k}| - \sum_{i=1}^{M} q_i U(\vec{\xi_i}), \qquad (5.1.88)$$

where $q_i = 1$ are unit numbers and

$$U(\vec{\xi}) = -\sum_{\alpha=1}^{N} f_{1\alpha} \ln |\vec{\xi} - \vec{a_\alpha}| - \sum_{\alpha=1}^{N} f_{2\alpha} \varphi(\vec{\xi} - \vec{a_\alpha}). \qquad (5.1.89)$$

Here $\varphi(\vec{\xi})$ is the angular coordinate of the vector $\vec{\xi}$. It is not difficult to see that (5.1.89) is none other than the potential of a two-dimensional (logarithmic) Coulomb system consisting of M particles with coordinates $\vec{\xi_i}$ and unit charges $q_i = 1$ moving in the potential $U(\vec{\xi_i})$. The latter is generated by N stationary particles with coordinates $\vec{a_\alpha}$ and two types of charge: ordinary electric charges $f_{1\alpha}$, and magnetic charges $f_{2\alpha}$ creating a vortex electrostatic field. It can be verified that $f_{1\alpha}$ and $f_{2\alpha}$ do actually correspond to electric and magnetic charges by writing down the potential produced by a single particle located at the origin,

$$U = f_1 \ln |\vec{\xi}| + f_2 \varphi(\vec{\xi}), \qquad (5.1.90)$$

and noting that this potential can be obtained from the equations of (2+1)-dimensional magnetoelectrodynamics

$$\partial^\mu F_{\mu\nu} = j_\nu, \quad \partial^\mu F_\mu^* = g, \qquad (5.1.91)$$

with

$$F_{\mu\nu} = \partial_\mu A_\nu - \partial_\nu A_\mu, \quad F_\mu^* = \frac{1}{2} \epsilon_{\mu\nu\lambda} F^{\nu\lambda} \qquad (5.1.92)$$

in the static limit. In fact, taking

$$g \sim f_2 \delta(|\vec{\xi}|), \quad j_0 \sim f_1 \delta(|\vec{\xi}|), \quad j_{1,2} = 0, \qquad (5.1.93)$$

and finding the static solution of (5.1.91) in the class of functions of the form

$$A_0 = U, \quad A_{1,2} = 0, \qquad (5.1.94)$$

we obtain (5.1.90).

We therefore see that the problem of finding a solution to the system of algebraic equations (5.1.84) is equivalent to the problem of finding the equilibrium positions of a system of Coulomb particles moving in the field of stationary dyons (i.e., particles having both an electric and magnetic charge). In general, this problem is quite complicated. However, in the special case in which the parameters a_α and f_α either are real or are complex conjugate pairs (we recall that this is the condition for the quantum mechanical potential to be real) it simplifies considerably. In fact, the presence of a Z_2-symmetry in the system leads to the existence of a straight line (coinciding in the present case with the real λ-axis) on which all the Coulomb forces (from the stationary dyons) are longitudinal. The problem of the equilibrium of the particles moving on this line therefore becomes one dimensional, which allows us to seek solutions for (5.1.84) in the class of real numbers.

5.2   Partial separation of variables

In this section we formulate a simple Lie-algebraic method for constructing quasi-exactly solvable models. This method, which we shall refer below to as "the method of partial separation of variables", is based on the observation that any completely integrable quantum system satisfying some special symmetry conditions can be reduced to a class of quasi-exactly solvable equations of one- or multi-dimensional quantum mechanics.

In order to understand the essence of this method, it is reasonable to start our discussion with the quasi-exactly solvable models described in chapter 4. These models are distinguished by the fact that they admit total separation of variables and occupy an intermediate position between one-dimensional multi-parameter spectral equations and multi-dimensional one-parameter spectral equations describing some completely integrable system.

Let us denote by $E(n,m)$ a spectral linear differential equation of dimension m involving n spectral parameters. From the results of chapter 3 we know that any multi-parameter spectral equation of the type $E(N,1)$ is related to some completely integrable system of one-parameter spectral equations of the type $E(1,N)$.

This relationship has been discussed in section 5.1, where we constructed the completely integrable model associated with the multi-parameter spectral equation (5.1.28). Remember that this equation being of the type $E(N,1)$ involves N spectral parameters $e_1,\ldots,e_N$, admitting two kinds of degeneracy: 1) the infinite degeneracy of the combination $E_0 = e_1+\ldots+e_N$, which is equal to zero for all solutions of equation (5.1.28), and 2) the finite degeneracy of the other combination $E_1 = a_1e_1 + \ldots + a_Ne_N$, which only depends on M and thus takes the same values for solutions having equal quantum number M. The remaining $N-2$ parameters are non-degenerate and take different values for different solutions.

We demonstrated that the system of completely integrable equations $E(1, N)$ associated with equation (5.1.28) is nothing else than the Gaudin model based on the Lie algebra $sl(2)\oplus\ldots\oplus sl(2)$ (N times) and characterized by N commuting integrals of motion $h_1, \ldots, h_N$ which are the coefficients for the pole terms $(\lambda - a_1)^{-1}, \ldots, (\lambda - a_N)^{-1}$ in the expansion of the generating function $K(\lambda)$. The eigenvalues of these operators coincide with the admissible values of spectral parameters $\varepsilon_1, \ldots, e_N$. Therefore, the spectrum of the operator $H_0 = h_1 + \ldots + h_N$ consists of a single infinitely degenerate zero eigenvalue E_0.[2] The spectrum of the operator $H_1 = a_1 h_1 + \ldots + a_N h_N$ has a finite degree of degeneracy. We explained this degeneracy as the consequence of the hidden $sl(N - 1)$ symmetry presenting in the Gaudin model and realized in the form of finite-dimensional representations.

In this language the quasi-exactly solvable models studied in chapter 4 can be considered as equations of the type $E(3, D - 2)$ with three spectral parameters, one of which is identically zero and can be excluded from the consideration; another has a degenerate spectrum and should be included in the potential, while the last one having a non-degenerate spectrum plays the role of the energy. One can say that quasi-exactly solvable equations occupy an intermediate position between equation (5.1.28) and the corresponding Gaudin model.

Indeed, on the one hand, they can be obtained from (5.1.28) by means of an incomplete inverse procedure of separation of variables which only eliminates $D - 3$ of $D - 1$ superfluous spectral parameters. Such a method for conctructing quasi-exactly solvable models we discussed in detail in chapter 4. On the other hand, we see that quasi-exactly solvable models can be obtained immediately from the Gaudin model by means of an incomplete but direct procedure of separation of variables in it. This procedure eliminates two degrees of freedom in the Gaudin model and leads to the appearance of two additional spectral parameters playing the role of separation constants. It is not difficult to see that the reducibility of the Gaudin model to quasi-exactly solvable models can be explained as a consequence of a global $sl(2)$ symmetry in the Gaudin model which is responsible for the partial separation of variables.

In order to demonstrate this fact, let us consider again the

---

[2] Note that for $\gamma \neq 0$ the operators $h_1, \ldots, h_N$ are linearly independent and their sum is a non-zero operator having zero eigenvalue. If $\gamma = 0$ the operators $h_1, \ldots, h_N$ become linearly dependent and their sum is identically zero. This does not mean that the model ceases to be completely integrable, because for zero value of γ there exists the large-λ limit of expressions $\lambda \vec{S}(\lambda)$, so that the model acquires an additional global $sl(2)$ symmetry with generators $\vec{S} = \lim \lambda S(\lambda)$ (see section 5.1). In this case, the role of the Nth integral of motion can be played by any element $\vec{S}$ of this algebra.

representation of the Gaudin algebra given by the generators

$$\vec{S}(\lambda) = \sum_{\alpha=1}^{N} \frac{\vec{S}_\alpha}{\lambda - a_\alpha} \qquad (5.2.1)$$

and characterized by the lowest-weight function

$$F(\lambda) = \sum_{\alpha=1}^{N} \frac{f_\alpha}{\lambda - a_\alpha}. \qquad (5.2.2)$$

Remember that the operators

$$K(\lambda) = \vec{S}(\lambda)\vec{S}(\lambda) = \sum_{\alpha,\beta=1}^{N} \frac{\vec{S}_\alpha \vec{S}_\beta}{(\lambda - a_\alpha)(\lambda - a_\beta)} \qquad (5.2.3)$$

possess global $sl(2)$ symmetry whose generators are

$$\vec{S} = \sum_{\alpha=1}^{N} \vec{S}_\alpha, \qquad (5.2.4)$$

and the lowest weight of the corresponding representation is

$$F = \sum_{\alpha=1}^{N} f_\alpha. \qquad (5.2.5)$$

Obviously, in the general case, this representation is infinite dimensional.

Using commutation relations (5.1.14) of the Gaudin algebra, it is not difficult to prove that the Bethe solutions (5.1.20) of the Gaudin spectral equation

$$K(\lambda)\phi = \omega(\lambda)\phi, \qquad \phi \in W\{F(\lambda)\} \qquad (5.2.6)$$

have the following remarkable property:

$$S^- \phi_M = 0, \qquad S^0 \phi_M = (F + M)\phi_M, \qquad (5.2.7)$$

which enables us to regard the Bethe solutions as the lowest vectors of representations of the symmetry algebra with lowest weights $F + M$, $M = 0, 1, 2, \ldots$.

Let us now denote by $\Phi_M\{F\}$ the spaces that consist of all vectors $\phi \in W\{F(\lambda)\}$ that satisfy the conditions

$$S^- \phi = 0, \qquad S^0 \phi = (F + M)\phi. \qquad (5.2.8)$$

It follows from the commutativity of the operators $K(\lambda)$ and $\vec{S}$ that if $\phi \in \Phi_M\{F\}$, then also $K(\lambda)\phi \in \Phi_M\{F\}$. Therefore, the spaces $\Phi_M\{F\}$ are invariant with respect to the action of the operators $K(\lambda)$. Accordingly, the spectral problems

$$K(\lambda)\phi = \omega(\lambda)\phi, \qquad \phi \in \Phi_M\{F\}, \tag{5.2.9}$$

are defined for all $M = 0, 1, 2, \ldots$.

It is readily seen that the spaces $\Phi_M\{F\}$ are finite dimensional. To prove this, we consider the auxiliary spaces $\Phi'_M\{F\}$, which are formed from vectors that satisfy only the second equation (5.2.8). It is obvious that the basis in $\Phi_M\{F\}$ is supplied by vectors of the form

$$S^+_{\alpha_1} \ldots S^+_{\alpha_M} |0\rangle, \qquad \alpha_1, \ldots, \alpha_M = 1, \ldots, N. \tag{5.2.10}$$

The number of such vectors determines the dimension of $\Phi'_M\{F\}$, which is

$$\dim \Phi'_M\{F\} = \frac{(M + N - 1)!}{(N - 1)! M!}. \tag{5.2.11}$$

Since

$$\Phi_M\{F\} \subset \Phi'_M\{F\}, \tag{5.2.12}$$

it follows that

$$\dim \Phi_M\{F\} < \dim \Phi'_M\{F\}. \tag{5.2.13}$$

Using the first condition in (5.2.8), we find that

$$\dim \Phi_M\{F\} = \dim \Phi'_M\{F\} - \dim \Phi'_{M-1}\{F\} = \frac{(M + N - 2)!}{(N - 2)! M!}. \tag{5.2.14}$$

It follows from this formula that for each M the spectral equations (5.2.9) have precisely $(M + N - 2)![(N - 2)! M!]^{-1}$ solutions.

In accordance with equations (5.2.7), the Bethe solutions of the Gaudin equation (5.2.6) belong to the spaces $\Phi_M\{F\}$ and are therefore simultaneously solutions of the finite-dimensional spectral problems (5.2.9). To resolve the question of the completeness of these solutions, it is sufficient to count the number of admissible sets of numbers ξ_n that satisfy the equation (5.1.31). It follows from analysis of (5.1.31) by the Coulomb analogy method that for given M the required number is also equal to $(M + N - 2)![(N - 2)! M!]^{-1}$. From this we conclude that for all M all solutions of equations (5.2.9) are completely described by the Bethe formulas (5.1.19) and (5.1.20) or, equivalently, by formulas

$$\phi = \phi_M(\xi) = \prod_{i=1}^{M} \left(\sum_{\alpha=1}^{N} \frac{S^+_\alpha}{\xi_i - a_\alpha} \right) |0\rangle \tag{5.2.15}$$

and

$$\omega(\lambda) = \omega_M(\lambda) =$$

$$(\sum_{\alpha=1}^{N} \frac{f_\alpha}{\lambda - a_\alpha} + \sum_{i=1}^{M} \frac{1}{\lambda - \xi_i})' + (\sum_{\alpha=1}^{N} \frac{f_\alpha}{\lambda - a_\alpha} + \sum_{i=1}^{M} \frac{1}{\lambda - \xi_i})^2, \quad (5.2.16)$$

with ξ_i satisfying the system (5.1.31).

Now our aim is to show that equations (5.2.9) can be reduced to the form of quasi-exactly solvable second-order differential equations. This can be done by means of the projection method described in chapter 1 or, in other words, by means of a partial separation of variables in these equations. The first step which should be made is to rewrite all generators entering into the definition of the operators $K(\lambda)$ and spaces $\Phi_M\{F\}$ in the differential form.

In previous section we have considered a differential realization of the generators of the algebra $sl(2)$ given by formulas (5.1.55). Here we consider another, more convenient differential realization of these generators. It can be obtained from (5.1.55) by means of the Fourier transformation and is given by the formulas

$$S_\alpha^- = \frac{\partial}{\partial t_\alpha}, \qquad S_\alpha^0 = t_\alpha \frac{\partial}{\partial t_\alpha} + f_\alpha, \qquad S_\alpha^+ = t_\alpha^2 \frac{\partial}{\partial t_\alpha} + 2t_\alpha f_\alpha. \quad (5.2.17)$$

It is worth stressing that all generators in (5.2.17) are first-order differential operators, which is especially important for the applicability of the projection method described in section 1.9. As before, generators (5.2.16) realize representations with lowest weights f_α. The unit function plays the part of the lowest vector: $|0\rangle = 1$.

Substituting (5.2.16) into (5.2.3), we see that $K(\lambda)$ take the form of N-dimensional differential operators of second order:

$$K(\lambda) = \sum_{\alpha,\beta=1}^{N} \frac{(t_\alpha - t_\beta)^2 \frac{\partial^2}{\partial t_\alpha \partial t_\beta} + 2(t_\alpha - t_\beta)(f_\alpha \frac{\partial}{\partial t_\beta} - f_\beta \frac{\partial}{\partial t_\alpha})}{(\lambda - a_\alpha)(\lambda - a_\beta)}. \quad (5.2.18)$$

Accordingly, equations (5.2.9) become differential equations. In the considered non-degenerate case, their solutions given by formulas (5.2.15) take the form

$$\phi = \prod_{m=1}^{M} \left(\sum_{\alpha}^{N} \frac{t_\alpha^2(\partial/\partial t_\alpha) + 2f_\alpha t_\alpha}{\xi_m - a_\alpha} \right) 1, \quad (5.2.19)$$

i.e., they are homogeneous polynomials in $t_1, \ldots, t_N$ of degree M.

The simplest way to describe the spaces $\Phi_M\{F\}$ to which these polynomials belong is to solve explicitly equations (5.2.8). Using formulas (5.2.4), (5.2.5) and (5.2.17), we can rewrite equations (5.2.8) in the differential form:

$$\left(\sum_{\alpha=1}^{N}\frac{\partial}{\partial t_\alpha}\right)\phi = 0, \qquad \left(\sum_{\alpha=1}^{N}t_\alpha\frac{\partial}{\partial t_\alpha}\right)\phi = M\phi. \tag{5.2.20}$$

The general solution of this system can be represented in the factorized form

$$\phi = \phi_M\psi, \tag{5.2.21}$$

where

$$\phi_M = (t_{N-1} - t_N)^M \tag{5.2.22}$$

is a particular solution of the system (5.2.20), and

$$\psi = \psi\left(\frac{t_1 - t_N}{t_{N-1} - t_N},\ldots,\frac{t_{N-2} - t_N}{t_{N-1} - t_N}\right) \tag{5.2.23}$$

is the general solution of the homogeneous system. As we see, this solution depends effectively on only $N - 2$ variables:

$$x_\alpha = \frac{t_\alpha - t_N}{t_{N-1} - t_N}, \qquad \alpha = 1,\ldots,N - 2. \tag{5.2.24}$$

The fact that (5.2.21) belongs to $\Phi_M\{F\}$ restricts the arbitrariness in choice of the functions ψ. These functions must be polynomials of degree M in the variables x_α. We denote the space of such polynomials by Ψ_M. Since $\Phi_M\{F\}$ is invariant with respect to the action of the operators $K(\lambda)$, the result of applying $K(\lambda)$ to $\phi_M\psi$ (where $\psi \in \Psi_M$) must again have the form $\phi_M\psi'$ (where $\psi' \in \Psi_M$). Therefore

$$K(\lambda)[\phi_M\psi] = \phi_M K_M(\lambda)\psi, \tag{5.2.25}$$

where $K_M(\lambda)$ is some $(N - 2)$-dimensional differential operator of second order that acts only on the variables x_α and can be viewed as a projection of $K(\lambda)$ onto $\Phi_M\{F\}$. This operator can be represented in the form

$$K_M(\lambda) = \sum_{\alpha,\beta=1}^{N-2} P_{\alpha\beta}(\lambda, M, \vec{x})\frac{\partial^2}{\partial x_\alpha\partial x_\beta}$$

$$+ \sum_{\alpha=1}^{N-2} Q_\alpha(\lambda, M, \vec{x})\frac{\partial}{\partial x_\alpha} + R_{\alpha\beta}(\lambda, M, \vec{x}), \tag{5.2.26}$$

where P, Q and R are polynomials in $x_1, \ldots, x_{N-2}$ of degrees 3, 2, and 1, respectively. Substituting (5.2.21) into (5.2.9) and using (5.2.25), we find that the spectral problems (5.2.26) are equivalent to differential spectral equations

$$K_M(\lambda)\psi = \omega_M(\lambda)\psi, \qquad \psi \in \Psi_M. \tag{5.2.27}$$

Now denote by Ψ the set of all analytic functions of the variables $x_1, \ldots, x_{N-2}$. Then it is obvious that the equations

$$K_M(\lambda)\psi = \omega_M(\lambda)\psi, \qquad \psi \in \Psi \tag{5.2.28}$$

with $K_M(\lambda)$ defined by formula (5.2.26) can be regarded as quasi-exactly solvable second-order differential equations. For each $M = 0, 1, 2, \ldots$, they have $(M + N - 2)![(N - 2)!M!]^{-1}$ exact solutions, which lie in the class of polynomials of degree M and are described by Bethe formulas.

Thus we have essentially formulated a new method for constructing quasi-exactly solvable spectral differential equations. We see that the operators $K_M(\lambda)$ commute with each other,

$$[K_M(\lambda), K_M(\mu)] = 0, \tag{5.2.29}$$

and the number of independent "hamiltonians" associated with $K_M(\lambda)$ is $N - 2$. The coincidence of the number of commuting "hamiltonians" with the number of variables in second-order differential equations (5.2.28) means that each of these equations admits a total separation of variables, and thus, the constructed quasi-exactly solvable equations coincide with those that have already been discussed in detail in chapter 4. In other words, in the case of the Gaudin models associated with algebra $sl(2)$, the method does not give new quasi-exactly solvable equations.

Fortunately, the class of completely integrable Gaudin models is not exhausted by the $sl(2)$ case. It is known that any simple (or semi-simple) Lie algebra $\mathcal{L}$ generates a set of Gaudin models that can be solved exactly within the Bethe *ansatz*. The integrals of motion for these models (the hamiltonians) can be represented as certain multi-dimensional second-order differential operators. Each such hamiltonian admits a global symmetry algebra which coincides with the generating algebra $\mathcal{L}$. The generators of this algebra can be realized as the first-order differential operators acting in the same functional space as the corresponding hamiltonians. Collecting all these facts, we can easily see that they together form the condition of the partial separability of variables in the generalized Gaudin model. This allows one to assert that the reasonings given above for the $sl(2)$ case can be extended to the general one, so that starting with the general Gaudin models and performing in them the procedure of partial separation of variables we can obtain wide classes of new quasi-exactly solvable models of quantum mechanics. The detailed description of this generalized reduction procedure will be given in the following sections.

The discussion is organized as follows. Sections 5.3 and 5.4 are devoted to the discussion of some properties of simple Lie algebras that are needed for the formulation of the approach in the general case. In these sections, much attention is devoted to the algorithms for constructing differential realizations of representations of Lie algebras and the choice of a special basis convenient for exact solution of the generalized Gaudin models. In section 5.5 we give the solutions for the models of Gaudin magnets in the framework of the algebraic Bethe anzatz. In section 5.6 we formulate the method of reduction of the Gaudin problems to multi-dimensional quasi-exactly solvable differential equations, and in section 5.7 propose a simple method for reduction of obtained equations to the Schrödinger form. In the final section 5.8 we discuss the possibility of dealgebraization of the reduction procedure.

5.3 Some properties of simple Lie algebras

1. Cartan–Weyl basis. Let $\mathcal{L}_r$ be a simple Lie algebra of rank r and dimension d_r and $I_a, a \in \Omega_r$, be basis elements of it satisfying the commutation relations

$$[I_a, I_b] = \sum_{c \in \Omega_r} \Gamma^c_{ab} I_c; \quad a, b \in \Omega_r. \tag{5.3.1}$$

From the practical point of view, it is most convenient to take the Cartan–Weyl basis, which is based on the decomposition

$$\mathcal{L}_r = \mathcal{L}_r^- \oplus \mathcal{L}_r^0 \oplus \mathcal{L}_r^+. \tag{5.3.2}$$

We denote the elements of the $((d_r - r)/2)$-dimensional subalgebras $\mathcal{L}_r^{\pm}$ by I_a, $a \in \Delta_r^{\pm}$, where $\Delta_r^{\pm}$ are the sets of positive and negative roots α of the algebra $\mathcal{L}_r$. We denote the elements of the r-dimensional Cartan subalgebra $\mathcal{L}_r^0$ by I_i, $i \in N_r$, where N_r is the set of numbers $1, \dots, r$. Thus, the d_r-element set of indices that label the basis elements of the algebra $\mathcal{L}_r$ can be represented in the form $\Omega_r = \Delta_r^- \cup N_r \cup \Delta_r^+$.

The most important commutation relations in the chosen basis are

$$[I_i, I_a] = (\alpha, \pi_i) I_\alpha, \quad i \in N_r \ \ \alpha \in \Delta_r^{\pm}. \tag{5.3.3}$$

Here, π_i, $i \in N_r$ are simple roots.

In the algebra $\mathcal{L}_r$ we introduce the bilinear form $\langle I_a, I_b \rangle$, which satisfies the subsidiary condition

$$\langle [I_a, I_b], I_c \rangle = \langle I_a, [I_b, I_c] \rangle. \tag{5.3.4}$$

Such a definition fixes the form only up to a factor. We choose this factor in such a way as to satisfy the requirements

$$\langle I_i, I_k \rangle = \gamma_{ik}, \quad \langle I_\alpha, I_\beta \rangle = \varepsilon_{\alpha\beta}, \tag{5.3.5}$$

where $\gamma_{ik} \equiv (\pi_i, \pi_k)$ is the matrix of scalar products of the simple roots (it is non-degenerate by virtue of the linear independence of the set π_i), and $\varepsilon_{\alpha\beta}$ is the matrix whose elements are unity for $\alpha + \beta = 0$ and zero in all the remaining cases. In what follows, the matrix

$$g_{ab} \equiv \langle I_a, I_b \rangle \tag{5.3.6}$$

will play the part of a metric tensor (obviously, non-degenerate), which will be used to raise and lower the indices that label the elements of the algebra $\mathcal{L}_r$. For example,

$$I_a = \sum_{b \in \Omega_r} g_{ab} I^b. \tag{5.3.7}$$

If equation (5.3.7) is written out in terms of its components, it takes the form

$$I_i = \sum_{k \in N_r} \gamma_{ik} I^k, \quad I_{\pm\alpha} = I^{\mp\alpha}. \tag{5.3.8}$$

The metric tensor (5.3.6) (which differs from the Killing–Cartan tensor only by a factor) is convenient in that its components g_{ab} do not depend on whether the generators I_a and I_b are regarded as elements of the algebra $\mathcal{L}_r$ or some subalgebra of it. It is obvious that this does not apply to the inverse tensor and to entities with superscripts. Bearing this in mind, but not wishing to burden the text with redundant notation, we shall retain for the elements conjugate to I_a the same notation I^a irrespective of the method of conjugation that is used – with respect to the algebra or a subalgebra of it. Of course, we shall attempt to make it clear from the context which particular case we have.

Equations (5.3.3)–(5.3.5) make it possible to recover uniquely all the commutation relations in the algebra $\mathcal{L}_r$ not yet given. They have the form

$$[I_i, I_k] = 0, \quad i, k \in N_r \ ; \tag{5.3.9}$$

$$[I_\alpha, I_{-\alpha}] = \sum_{i \in N_r} (\alpha, \pi_i) x^i, \quad \alpha \in \Delta_r^\pm; \tag{5.3.10}$$

$$[I_\alpha, I_\beta] = \Gamma_{\alpha\beta} I_{\alpha+\beta}, \quad \alpha, \beta \in \Delta_r^\pm, \quad \alpha + \beta \in \Delta_r^\pm, \tag{5.3.11}$$

where $\Gamma_{\alpha\beta}$ are calculable structure constants.

We define the quadratic Casimir operator as follows:

$$K_r = \sum_{a \in \Omega_r} I^a I_a, \qquad (5.3.12)$$

or, with allowance for the commutation relations (5.3.10),

$$K_r = \sum_{i \in N_r} I^i \left[I_i + \sum_{\alpha \in \Delta_r^+} (\alpha, \pi_i) \right] + 2 \sum_{\alpha \in \Delta_r^+} I^\alpha I_\alpha. \qquad (5.3.13)$$

Since

$$\sum_{\alpha \in \Delta_r^+} \alpha = \sum_{i \in N_r} \nu^i \pi_i \qquad (5.3.14)$$

(where ν^i is a certain set of non-negative integers that characterize the algebra $\mathcal{L}_r$), we have

$$K_r = \sum_{i \in N_r} (I^i + \nu^i) I_i + 2 \sum_{\alpha \in \Delta_r^+} I^\alpha I_\alpha. \qquad (5.3.15)$$

The representations of the algebras $\mathcal{L}_r$ with highest weight are determined as follows:

$$I_i |0\rangle = \Lambda_{0i} |0\rangle, \quad i \in N_r ; \qquad (5.3.16)$$
$$I_\alpha |0\rangle = 0, \quad \alpha \in \Delta_r^+. \qquad (5.3.17)$$

The set Λ_{0i} is called the highest weight, and $|0\rangle$ is called the highest vector. Let M^i, $i \in N_r$, be a set of non-negative integers. We denote by $|M\rangle$ the linear hull of vectors of the form

$$I_{\alpha_1} \ldots I_{\alpha_K} |0\rangle, \quad \alpha_1, \ldots, \alpha_K \in \Delta_r^-, \qquad (5.3.18)$$

in which $\alpha_1 + \ldots + \alpha_K = - \sum_{i \in N_r} M^i \pi_i$. Obviously, the spaces $|M\rangle$ are eigenspaces with respect to the elements of the Cartan subalgebra:

$$I_i |M\rangle = (\Lambda_{0i} - M_i)|M\rangle. \qquad (5.3.19)$$

The linear hull of all such spaces,

$$W\{\Lambda_0\} = \oplus_{M \geq 0} |M\rangle, \qquad (5.3.20)$$

forms the representation space of the algebra $\mathcal{L}_r$ with highest weight Λ_0.

2. Realization of representations of Lie algebras in the form of differential operators. We have already mentioned that in our scheme a decisive role is played by the possibility of realizing representations of Lie algebras in the form of first-order differential operators. There exists an opinion that the construction of such realizations is a rather complicated matter. This is not all so. The general principles of the construction are rather simple, although the algorithms described in the literature are not always carried through to explicit formulas, especially for the higher algebras. We shall attempt to fill this gap by constructing explicit expressions for differential operators which realize arbitrary representations of arbitrary simple Lie algebras.

Let G_r be the Lie group of the Lie algebra $\mathcal{L}_r$ and let

$$g(x) \in G_r, \quad x \in R_{d_r} \tag{5.3.21}$$

be the elements of the group, parametrized by the vectors $x \equiv \{x^a\}, a \in \Omega_r$, of the d_r-dimensional space R_{d_r}. We choose the parametrization in such a way that the following equations hold:

$$g(0) = 1, \tag{5.3.22}$$

$$\left.\frac{\partial g(x)}{\partial x}\right|_{x=0} = I, \tag{5.3.23}$$

where $I \equiv \{I_a\}, a \in \Omega_r$, are the generators for the algebra $\mathcal{L}_r$ that we introduced earlier. One of the possible methods of parametrization consists of the choice

$$g(x) = \exp(xI), \tag{5.3.24}$$

but this method is not unique and, as we shall see later, not the most convenient. Actually, the general scheme that will be presented in this section does not depend on the particular method of parametrization.

For the elements x, y of the space R_{d_r} we define the binary operation $x \dot{+} y$ in accordance with the formula

$$g(x)g(y) = g(x \dot{+} y). \tag{5.3.25}$$

This operation, for which we have choosen the symbol $\dot{+}$ (not to be confused with the direct sum, for which the symbol $\oplus$ is reserved), possesses the following properties:

(a) for all $x, y, z \in R_{d_r}$

$$x \dot{+} (y \dot{+} z) = (x \dot{+} y) \dot{+} z; \tag{5.3.26}$$

(b) for all $x \in R_{d_r}$ there exists a zero element $0 \in R_{d_r}$ such that

$$x \dot{+} 0 = 0 \dot{+} x = x \qquad (5.3.27)$$

(c) for all $x \in R_{d_r}$ there exists a unique inverse element $(\dot{-}x) \in R_{d_r}$ such that

$$x \dot{+} (\dot{-}x) = (\dot{-}x) \dot{+} x = 0. \qquad (5.3.28)$$

In what follows, in place of $x \dot{+} (\dot{-}y)$ we shall simply write $x \dot{-} y$. It is readily verified that

$$\dot{-}(x \dot{+} y) = \dot{-}y \dot{-} x. \qquad (5.3.29)$$

If the conditions (5.3.22) and (5.3.23) are satisfied, the point 0 corresponds to the usual zero of the space R_{d_r}. In the general case, $-x \neq \dot{-}x$, although for some parametrizations (see, for example, (5.3.24)) the equation $-x = \dot{-}x$ can hold. For commutative groups, we have $x \dot{+} y = x + y$, i.e., in this case the binary operation that we have introduced can be identified with ordinary addition of vectors in R_{d_r}.

Let Φ_r be the space of functions on the group G_r and let

$$\phi(x) \in \Phi_r, \ x \in R_{d_r} \qquad (5.3.30)$$

be the elements of this space. We define the operators $\hat{g}(\varepsilon), \varepsilon \in R_{d_r}$, which are linear on Φ_r, by means of the formulas

$$\hat{g}(\varepsilon)\phi(x) = \phi(x \dot{+} \varepsilon). \qquad (5.3.31)$$

It follows from this definition that

$$\hat{g}(\varepsilon_2)\hat{g}(\varepsilon_1) = \hat{g}(\varepsilon_1 \dot{+} \varepsilon_2), \qquad (5.3.32)$$

i.e., the operators $\hat{g}(\varepsilon), \varepsilon \in R_{d_r}$, form a representation of the group G_r. In accordance with the formulas (5.3.22) and (5.3.23), we have

$$\hat{g}(0) = \hat{1}; \qquad (5.3.33)$$

$$\left. \frac{\partial \hat{g}(x)}{\partial x} \right|_{x=0} = \hat{I}, \qquad (5.3.34)$$

where $\hat{I} = \{\hat{I}_a\}, a \in \Omega_r$, are the generators of the corresponding representations of the algebra $\mathcal{L}_r$. Differentiating (5.3.31) with respect to ε and setting $\varepsilon = 0$, we find

$$\hat{I} = \hat{T}(x)\frac{\partial}{\partial x}, \qquad (5.3.35)$$

where

$$\hat{T}(x) = \frac{\partial}{\partial \varepsilon} \otimes (x \dot{+} \varepsilon)|_{\varepsilon=0}. \tag{5.3.36}$$

For further simplification of equations (5.3.35) and (5.3.36), we note that any vector of the space R_{d_r} can be represented in the form of the expansion

$$x = x^- \dot{+} x^0 \dot{+} x^+, \tag{5.3.37}$$

where $x^\pm \equiv \{x^\alpha\}, \alpha \in \Delta_r^\pm$, and $x^0 \equiv \{x^i\}, i \in N_r$, are the vectors of dimensions $(d_r - r)/2$ and r associated with the generators $I_\pm \equiv \{I_\alpha\}, \alpha \in \Delta_r^\pm$, and $I_0 \equiv \{I_i\}, i \in N_r$, of the algebra $\mathcal{L}_r$ in the Cartan–Weyl basis. Equations (5.3.35) and (5.3.36) can now be rewritten in the form

$$\hat{I} = \hat{T}^-(x)\frac{\partial}{\partial x^-} + \hat{T}^0(x)\frac{\partial}{\partial x^0} + \hat{T}^+(x)\frac{\partial}{\partial x^+}, \tag{5.3.38}$$

where

$$\hat{T}^\pm(x) = \frac{\partial}{\partial \varepsilon} \otimes (x \dot{+} \varepsilon)^\pm|_{\varepsilon=0}; \quad \hat{T}^0(x) = \frac{\partial}{\partial \varepsilon} \otimes (x \dot{+} \varepsilon)^0|_{\varepsilon=0}. \tag{5.3.39}$$

It follows from the obvious equation

$$x \dot{+} \varepsilon = \{x^- \dot{+} [x^0 \dot{+} (x^+ \dot{+} \varepsilon)^-]^-\}$$
$$+ \{x^0 \dot{+} (x^+ \dot{+} \varepsilon)^0\} + \{(x^+ \dot{+} \varepsilon)^+\} \tag{5.3.40}$$

that

$$\begin{aligned} (x \dot{+} \varepsilon)^+ &= (x^+ \dot{+} \varepsilon)^+; \\ (x \dot{+} \varepsilon)^0 &= x^0 \dot{+} (x^+ \dot{+} \varepsilon)^0; \\ (x \dot{+} \varepsilon)^- &= x^- \dot{+} [x^0 \dot{+} (x^+ \dot{+} \varepsilon)^-]^-. \end{aligned} \tag{5.3.41}$$

Therefore

$$\begin{aligned} \hat{T}^+(x) &= \frac{\partial}{\partial \varepsilon} \otimes (x^+ \dot{+} \varepsilon)^+|_{\varepsilon=0}; \\ \hat{T}^0(x) &= \frac{\partial}{\partial \varepsilon} \otimes (x^+ \dot{+} \varepsilon)^0|_{\varepsilon=0}; \\ \hat{T}^-(x) &= \frac{\partial}{\partial \varepsilon} \otimes \{x^- \dot{+} (x^+ \dot{+} \varepsilon)^-]^-\}|_{\varepsilon=0}. \end{aligned} \tag{5.3.42}$$

We see that the matrices $\hat{T}^+(x)$ and $\hat{T}^0(x)$ depend only on the variables x^+. Therefore, considering the action of the operators (5.3.38) on the class of functions of the form

$$\phi(x) = \exp(x^0 \Lambda_0)\psi(x^+), \tag{5.3.43}$$

we obtain

$$\hat{I}\exp(x^0\Lambda_0)\psi(x^+) = \exp(x^0\Lambda_0)\hat{I}(\Lambda_0)\psi(x^+). \tag{5.3.44}$$

The operators

$$\hat{I}(\Lambda_0) = \hat{t}^+(x^+)\frac{\partial}{\partial x^+} + \hat{t}^0(x^+)\Lambda_0, \tag{5.3.45}$$

in which

$$\hat{t}^+(x^+) = \frac{\partial}{\partial\varepsilon}\otimes(x^+\dot{+}\varepsilon)^+|_{\varepsilon=0},$$

$$\hat{t}^0(x^+) = \frac{\partial}{\partial\varepsilon}\otimes(x^+\dot{+}\varepsilon)^0|_{\varepsilon=0} \tag{5.3.46}$$

act on the space of functions of the $(d_r - r)/2$ variables x^+, are inhomogeneous first-order differential operators, and realize a certain representation of the algebra $\mathcal{L}_r$. To identify this representation, we note that by virtue of the obvious formulas

$$\frac{\partial}{\partial\varepsilon^+}\oplus(x^+\dot{+}\varepsilon^+)^0\bigg|_{\varepsilon^+=0} = 0,$$

$$\frac{\partial}{\partial\varepsilon^0}\oplus(x^+\dot{+}\varepsilon^0)^0\bigg|_{\varepsilon^0=0} = 1, \tag{5.3.47}$$

we have

$$\hat{I}_+(\Lambda_0) = \hat{t}^+_+(x^+)\frac{\partial}{\partial x^+}; \tag{5.3.48}$$

$$\hat{I}_0(\Lambda_0) = \hat{t}^+_0(x^+)\frac{\partial}{\partial x^+} + \Lambda_0; \tag{5.3.49}$$

$$\hat{I}_-(\Lambda_0) = \hat{t}^+_-(x^+)\frac{\partial}{\partial x^+} + \hat{t}^0_-(x^+)\Lambda_0. \tag{5.3.50}$$

Therefore, the operators (5.3.48)–(5.3.50) describe a representation of the algebra $\mathcal{L}_r$, with highest weight Λ_0. The unit function $|0\rangle \equiv 1$ here plays the part of the highest vector. It is readily seen that the operators (5.3.48)–(5.3.50) can also be obtained as infinitesimal operators of the representation $G_r(\Lambda_0)$ of the group G_r determined by the formula

$$\hat{g}(\varepsilon,\Lambda_0)\psi(x^+) = \exp[(x\dot{+}\varepsilon)^0\Lambda_0]\psi[(x^+\dot{+}\varepsilon)^+]. \tag{5.3.51}$$

In what follows, we shall frequently need to work with the operators

$$\hat{I}(\Lambda^1_0,\ldots,\Lambda^N_0) \equiv \sum_{A=1}^{N}\hat{I}(\Lambda^A_0)$$

$$= \sum_{A=1}^{N}\left\{\hat{t}^+(x^+_A)\frac{\partial}{\partial x^+_A} + \hat{t}^0(x^+_A)\Lambda^A_0\right\}, \tag{5.3.52}$$

which realize a representation of the algebra $\mathcal{L}_r$ on the class of functions that depend on N vector variables. By analogy with (5.3.51), these operators can be interpreted as the infinitesimal operators of the following group transformations:

$$
\begin{aligned}
\hat{g}(\varepsilon, \Lambda_0^1, &\quad \ldots \quad , \Lambda_0^N)\psi(x_1^+, \ldots, x_N^+) \\
&= \exp\left\{\sum_{A=1}^{N}(x_A^+ + \varepsilon)^0 \Lambda_0^A\right\} \\
&\times \quad \psi[(x_1^+ + \varepsilon)^+, \ldots, (x_N^+ + \varepsilon)^+].
\end{aligned}
\tag{5.3.53}
$$

If we choose as the highest vector the unit function $|0\rangle \equiv 1$, then the highest weight will be $\Lambda_0^1 + \ldots + \Lambda_0^N$. However, in what follows we shall need to consider representations of the algebra of operators (5.3.52) with arbitrary higher weights, which can be written in the form $\Lambda_0^1 + \ldots + \Lambda_0^N - M_0$. In this case, which form can the highest vector $|0\rangle$ have? To answer this question, we must solve the system of equations

$$
\hat{I}_+(\Lambda_0^1, \ldots, \Lambda_0^N)\psi = 0;
$$

$$
\hat{I}_0(\Lambda_0^1, \ldots, \Lambda_0^N)\psi = \left(\sum_{A=1}^{N}\Lambda_0^A - M_0\right)\psi
\tag{5.3.54}
$$

for the functions ψ. Essentially, we must find functions ϕ invariant with respect to transformations of the subgroup $\hat{g}(\varepsilon^+, 0, \ldots, 0)$ and transforming homogeneously (with the addition of the factor $\exp(-\varepsilon^0 M_0)$) under transformation the subgroup $\hat{g}(\varepsilon^0, 0 \ldots, 0)$:

$$
\psi(x_1^+ + \varepsilon^+)^+, \ldots, (x_N^+ + \varepsilon^+)^+] = \psi(x_1^+, \ldots, x_N^+);
\tag{5.3.55}
$$

$$
\psi[(x_1^+ + \varepsilon^0)^+, \ldots, (x_N^+ + \varepsilon^0)^+] = e^{-\varepsilon^0 M_0}\psi(x_1^+, \ldots, x_N^+).
\tag{5.3.56}
$$

To solve (5.3.55), we use the equation

$$
(x_A^+ + \varepsilon^+)^+ = x_1^+ + \varepsilon^+,
\tag{5.3.57}
$$

from which it follows that

$$
(x_A^+ + \varepsilon^+) \dot{-} (x_B^+ + \varepsilon^+) = x_A^+ + \varepsilon^+ \dot{-} \varepsilon^+ \dot{-} x_B^+ = x_A^+ \dot{-} x_B^+.
\tag{5.3.58}
$$

Thus, the functions $x_A^+ \dot{-} x_N^+$ are invariant with respect to the transformation $g(\varepsilon^+, 0, \ldots, 0)$. The following combinations of them are functionally independent:

$$
\zeta_A = x_A^+ \dot{-} x_N^+, \quad A = 1, \ldots, N-1.
\tag{5.3.59}
$$

Therefore, the most general solution of equations (5.3.55) and (5.3.56) has the form

$$\psi = \psi(\zeta_1, \ldots, \zeta_{N-1}). \tag{5.3.60}$$

We now consider how the components of the vectors $\zeta_A = \{\zeta_A^\alpha\}$, $\alpha \in \Delta_r^+$, change under a transformation of the subgroup $g(\varepsilon^0, 0, \ldots, 0)$. For the components $x_A = \{x_A^\alpha\}$, we have

$$(x_A^+ \dotplus \varepsilon^0)^\alpha = \dot{-}\varepsilon^0 \dotplus x_A^\alpha \dotplus \varepsilon^0 = \exp\{-\varepsilon^0(\pi_0, \alpha)\} x_A^\alpha, \tag{5.3.61}$$

where $\pi_0 \equiv \{\pi_i\}$, $i \in N_r$. The components of the vectors $\zeta_A = x_A \dot{-} x_N$ also transform homogeneously:

$$\zeta_A^\alpha \rightarrow \exp\{-\varepsilon^0(\pi_0, \alpha)\} \zeta_A^\alpha. \tag{5.3.62}$$

This means that all quantities of the form $[\zeta_{A_1}^{\alpha_1} \ldots \zeta_{A_K}^{\alpha_K}] \times [\zeta_{B_1}^{\beta_1} \ldots \zeta_{B_L}^{\beta_L}]^{-1}$, where $\alpha_1 + \ldots + \alpha_K = \beta_1 + \ldots \beta_L$, $\alpha_k, \beta_l \in \Delta_r^+$, are invariant with respect to the transformations $g(\varepsilon^0, 0, \ldots, 0)$. As functionally independent variables, we can choose the following $(N-2)(d_r - r)/2$ variables:

$$\eta = \left\{ \frac{\zeta_A^\alpha}{\prod_{i \in N_r} (\zeta_{N-1}^{\pi_i})^{(\alpha, \pi^i)}} \right\} \quad \alpha \in \Delta_r^+, \ A = 1, \ldots, N-2, \tag{5.3.63}$$

and also $(d_r - r)/2$ variables of the form

$$\nu = \left\{ \frac{\zeta_{N-1}^\alpha}{\prod_{i \in N_r} (\zeta_{N-1}^{\pi_i})^{(\alpha, \pi^i)}} \right\} \quad \alpha \in \Delta_r^+, \ \alpha \neq \pi_i, \ i \in N_r. \tag{5.3.64}$$

Introducing the notation

$$\zeta^i \equiv \zeta_{N-1}^{\pi_i}, \ i \in N_r, \tag{5.3.65}$$

we find that the functions

$$\psi = \prod_{i \in N_r} (\zeta^i)^{M^i} \psi(\eta, \nu) \tag{5.3.66}$$

realize the most general solution of the system (5.3.55), (5.3.56). Any of these functions can play the part of the highest vector for representation of the algebra of operators (5.3.52) with highest weight $\Delta_0^1 + \ldots + \Delta_0^N - M_0$.

We now obtain explicit expressions for the matrices $\hat{t}^+(x^+)$ and $\hat{t}^0(x^+)$ under the assumption that the parametrization is gaussian:

$$g(x) = \exp \left\{ \sum_{\alpha \in \Delta_r^-} x^\alpha I_\alpha \right\} \exp \left\{ \sum_{i \in N_r} x^i I_i \right\} \exp \left\{ \sum_{\alpha \in \Delta_r^+} x^\alpha I_\alpha \right\}. \tag{5.3.67}$$

In this case, the equation for the matrices $\hat{t}^+(x^+)$ and $\hat{t}^0(x^+)$ takes the form

$$\exp\left\{\sum_{i\in N_r} x^i I_i\right\} \exp\left\{\sum_{\alpha\in\Delta_r^+} x^\alpha I_\alpha\right\} \exp\{\varepsilon^a I_a\}$$

$$= \exp\left\{\sum_{i\in N_r}(x^i I_i + \varepsilon^a t_a^i(x^+)I_i\right\} \exp\left\{\sum_{\alpha\in\Delta_r^+}(x^\alpha I_\alpha + \varepsilon^a t_a^\alpha(x^+)I_\alpha\right\}.$$

$$(5.3.68)$$

Applying the well known formula

$$\exp(A+B) \;=\; \exp A\Big[T\exp\int_0^1 d\tau\,\exp(-\tau A)B\exp(\tau A)\Big]$$

$$(5.3.69)$$

to the right-hand side of (5.3.68), expanding both sides in series in powers of ε^a, and retaining only the first powers in ε^a, we obtain after trivial manipulations

$$t_a^i(x^+) = \left\langle \exp\left(\mathrm{ad}\sum_{\alpha\in\Delta_r^+} x^\alpha I_\alpha\right) I_a, I^i\right\rangle, \qquad i\in N_r. \qquad (5.3.70)$$

With regard to the matrix $t_a^\alpha(x^+)$, it can be found from the system of algebraic equations

$$t_a^\alpha(x^+)\int_0^1 d\tau\left\langle \exp\left(\mathrm{ad}\sum_{\beta\in\Delta_r^+} x^\beta I_\beta\right) I_\alpha, I^\gamma\right\rangle$$

$$= \left\langle \exp\left(\mathrm{ad}\sum_{\beta\in\Delta_r^+} x^\beta I_\beta\right) I_\alpha, I^\gamma\right\rangle, \quad a\in\Omega_r,\ \gamma\in\Delta_r^+. \qquad (5.3.71)$$

We now note that the matrix

$$\int_0^1 d\tau\left\langle \exp\left(\mathrm{ad}\,\tau\sum_{\beta\in\Delta_r^+} x^\beta I_\beta\right) I_\alpha, I^\gamma\right\rangle, \qquad (5.3.72)$$

which occurs in equation (5.3.71), will have triangular form if the roots α and γ are arranged in non-descending order of their heights. On the principal diagonal of (5.3.72) there will be units. The determinant of such a matrix is equal to unity, and therefore its inverse matrix consists of the minors. Since each minor is a finite polynomial in x^α (this also applies to the right-hand side of (5.3.71)), the matrices $\hat{t}^+(x^+)$ and $\hat{t}^0(x^+)$ will also be polynomial in x^α.

The differential realizations of the representations of the algebra $\mathcal{L}_r$ that we have obtained are not particulary convenient from the practical point of view, since some work is needed to reduce them to explicit form. This is because the gaussian decomposition that we have used (and that is also mainly discussed in the literature) is not completely suitable for this purpose. In the following sections, we shall consider more convenient decompositions, and on their basis we shall derive explicit expressions suitable for any simple Lie algebra.

To conclude this section, we consider how the matrices $\hat{t}^+(x^+)$ and $\hat{t}^0(x^+)$ transform under homogeneous transformations of the components of the parameter x^+:

$$x^\alpha \to \exp\{-\varepsilon^0(\pi_0, \alpha)\}x^\alpha. \tag{5.3.73}$$

Since each component x^α acquires a factor $\exp[-\varepsilon^0(\pi_0, \alpha)]$, the only non-vanishing terms in the decomposition (5.3.72) will be the terms that acquire the factor $\exp\{-\varepsilon^0[\pi_0, (\gamma - \alpha)]\}$. This means that the components of the considered matrices will transform as

$$\left.\begin{array}{l} t_\beta^\alpha(x^+) \to \exp\{-\varepsilon^0(\pi_0, (\alpha - \beta))\}t_\beta^\alpha(x^+); \\ t_i^\alpha(x^+) \to \exp\{-\varepsilon^0(\pi_0, \alpha)\}t_i^\alpha(x^i). \end{array}\right\} \tag{5.3.74}$$

Similary, we can show that

$$t_\alpha^i(x^+) \to \exp\{\varepsilon^0(\pi_0, \alpha)\}t_\alpha^i(x^+). \tag{5.3.75}$$

It follows from this that the differential operators $\hat{I}_\alpha(\Lambda_0), \alpha \in \Delta_r^\pm$, and $\hat{I}_i(\Gamma_0), i \in N_r$, which realize representations of the algebra $\mathcal{L}_r$ with the highest weight Λ_0, transform in accordance with the rules

$$\left.\begin{array}{l} t_\alpha(\Lambda_0) \to \exp\{\varepsilon^0(\pi_0, \alpha)\}\hat{I}_\alpha(\Lambda_0), \quad \alpha \in \Delta_r^\pm, \\ \hat{I}_i(\Lambda_0) \to \hat{I}_i(\Lambda_0), \quad i \in N_r. \end{array}\right\} \tag{5.3.76}$$

5.4   Special decomposition in simple Lie algebras

We shall say that a simple root of the algebra $\mathcal{L}_r$ is singular if it occurs in the decomposition of any root with a coefficient whose modulus does not exceed unity.

In figure 5.1 we give the Dynkin diagrams of the simple Lie algebras and identify with black circles the vertices associated with singular simple roots.

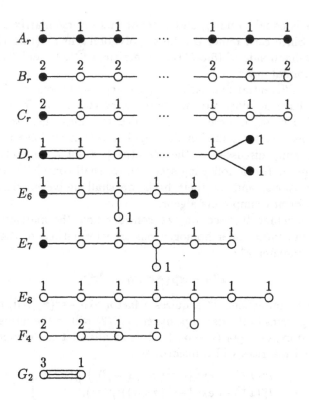

Figure 5.1. The Dynkin diagrams with singular simple roots (black circles).

We see that for the algebras A_r all the simple roots are singular, while for the algebras E_8, F_4 and G_2 there are no such roots. In what follows, we shall consider only algebras that have at least one singular root. These are the algebras $A_r (r \geq 1), B_r (r \geq 2), C_r (r \geq 3), D_r (r \geq 4), E_6$, and E_7. We shall call them the singular algebras.

The removal from the algebra $\mathcal{L}_r$ of the extreme singular root (together with all non-simple roots containing it) transforms the algebra into the simple subalgebra $\mathcal{L}_{r-1}$. For example,

$$\left. \begin{array}{l} A_r \to A_{r-1}, \ B_r \to B_{r-1}, \ C_r \to A_{r-1}, \\[2mm] D_r \begin{array}{c} \nearrow D_{r-1} \\ \searrow A_{r-1} \end{array}, \ E_7 \to E_6, \ E_6 \to D_5 . \end{array} \right\} \tag{5.4.1}$$

In what follows, we shall ascribe the number r to the eliminated singular root. We denote the sets of roots containing in the decomposition a root with coefficients ± 1 by $\Sigma_r^{\pm}$.

Singular roots are remarkable in that the elements I_α associated with the roots $\alpha \in \Sigma_r^{\pm}$ form commutative subalgebras, which we shall denote by $\mathcal{E}_{\pm r}$. The commutativity of $\mathcal{E}_{\pm r}$ follows directly from the definition of a simple root. It is obvious that

$$\dim \mathcal{E}_{\pm r} = \frac{1}{2}(d_r - d_{r-1} - 1). \tag{5.4.2}$$

The algebras $\mathcal{E}_{\pm r}$ are conjugate with respect to the bilinear form (5.3.6). For the elements $E_{\pm r} = \{E_{\pm r,\alpha}\}$, $\alpha \in \Sigma_r^{\pm}$, of these subalgebras the following normalization conditions are satisfied:

$$\langle E_{\pm r} \otimes E_{\mp r} \rangle = \hat{I}_r \tag{5.4.3}$$

where $\hat{I}_r$ is the unit matrix of dimension $\dim \mathcal{E}_r$. Since the dimensions of the Cartan subalgebras $\mathcal{L}_r^0$ and $\mathcal{L}_{r-1}^0$ differ by unity,

$$\mathcal{L}_r^0 = \mathcal{L}_{r-1}^0 \oplus H_r, \tag{5.4.4}$$

where H_r is an element of $\mathcal{L}_r^0$ that does not belong to $\mathcal{L}_{r-1}^0$. For unique determination of this element, we require it to be self-adjoint with respect to the form (5.3.6), and this is equivalent to the conditions

$$\langle H_r, I_i \rangle = 0, \quad i \in N_{r-1}; \tag{5.4.5}$$
$$\langle H_r, H_r \rangle = 1. \tag{5.4.6}$$

It is obvious that the element H_r must commute with all elements of the subalgebra $\mathcal{L}_{r-1}$.

Summarizing what was said above, we can conclude that for all singular Lie algebras the following decomposition holds:

$$\mathcal{L}_r = (\mathcal{E}_{-r} \oplus H_r \oplus \mathcal{E}_{+r}) \oplus \mathcal{L}_{r-1}, \tag{5.4.7}$$

where $\mathcal{E}_{\pm r}$ are commutative conjugate subalgebras of dimensions $(d_r - d_{r-1} - 1)/2$, H_r is a self-adjoint element of the Cartan subalgebra, and $\mathcal{L}_{r-1}$ is a simple Lie algebra.

The non-vanishing commutation relations in (5.4.7) have the form

$$[\mathcal{L}_{r-1}, \mathcal{L}_{r-1}] = \mathcal{L}_{r-1}, \tag{5.4.8}$$
$$[\mathcal{L}_{r-1}, \mathcal{E}_{\pm r}] = \mathcal{E}_{\pm r}, \tag{5.4.9}$$
$$[H_r, \mathcal{E}_{\pm r}] = \mathcal{E}_{\pm r}, \tag{5.4.10}$$
$$[\mathcal{E}_{\pm r}, \mathcal{E}_{\mp r}] = H_r \oplus \mathcal{L}_{r-1}. \tag{5.4.11}$$

We now can write out these relations in more detail. For (5.4.8), we have

$$[I_a, I_b] = \sum_{c \in \Omega_{r-1}} \Gamma_{ab}^c I_c, \quad a, b \in \Omega_{r-1}. \tag{5.4.12}$$

The relations (5.4.9) can be rewritten in the form

$$[I_a, E_{\pm r}] = -\hat{I}_a(\mathcal{E}_{\pm r}) E_{\pm r}, \quad a \in \Omega_{r-1}, \tag{5.4.13}$$

where $\hat{I}_a(\mathcal{E}_{\pm r})$ are matrices that play the part of structure constants. Using the Jacobi identity

$$[[I_a, E_{\pm r}], I_b] + [[E_{\pm r}, I_b], I_a] + [[I_b, I_a], E_{\pm r}] = 0 \tag{5.4.14}$$

and equations (5.4.12) and (5.4.13), we find

$$[\hat{I}_a(\mathcal{E}_{\pm r}), \hat{I}_b(\mathcal{E}_{\pm r})] = \sum_{c \in \Omega_{r-1}} \Gamma_{ab}^c \hat{I}_c(\mathcal{E}_{\pm r}), \quad a, b \in \Omega_{r-1}. \tag{5.4.15}$$

Thus, the matrices $\hat{I}_a(\mathcal{E}_{\pm r})$, $a \in \Omega_{r-1}$, form representations of the algebra $\mathcal{L}_{r-1}$ of dimension dim $\mathcal{E}_{\pm r}$ and are realized in the spaces of the commutative subalgebras $\mathcal{E}_{\pm r}$. It follows from the simplicity of the algebra $\mathcal{L}_{r-1}$ that

$$\mathrm{Sp}\,\hat{I}_a(\mathcal{E}_{\pm r}) = 0. \tag{5.4.16}$$

In addition, it is readily seen that

$$\hat{I}_a(\mathcal{E}_{+r}) + \widetilde{\hat{I}}_a(\mathcal{E}_{-r}) = 0, \tag{5.4.17}$$

where the tilde denotes the transpose.

In what follows, we shall need only representations realized by the matrices $\hat{I}_a(\mathcal{E}_{+r})$ in the subalgebra $\mathcal{E}_{+r}$. For their identification, we consider equation (5.4.13) with the plus sign. Labelling the roots $\alpha \in \Delta_{r-}^+$ in order of non-decrease of their heights, we see that in this case the matrices $\hat{I}_\alpha(\mathcal{E}_{+r})$, $\alpha \in \Delta_{r-}^+$, have an upper triangular form, while the matrices $\hat{I}_i(\mathcal{E}_{+r})$, $i \in N_{r-1}$, are diagonal. Therefore the role of highest vector will be played by a vector of dimension dim $\mathcal{E}_r$ for which only the first component is non-zero. This means that the corresponding highest weight must be determined by the result of commutation of the elements $I_i, i \in N_{r-1}$, with the element $\mathcal{E}_r$ corresponding to the positive root of the lowest possible height. This is the root π_r. From this we conclude that the highest vector of the representation (5.4.15) will be an $(r-1)$-dimensional vector γ_{ir}, $i \in N_{r-1}$.

Further, using equations (5.4.5) and (5.4.6), we find for the commutation relations (5.4.10)

$$[H_r, E_{\pm r}] = \pm Q_r E_{\pm r}, \tag{5.4.18}$$

where

$$Q_r = \langle H_r, I_r \rangle. \tag{5.4.19}$$

With regard to the relations (5.4.11), they can be written in the form

$$[E_{\pm r} \oplus E_{\mp r}] = \pm Q_r H_r \hat{I}_r - \sum_{a \in \Omega_{r-1}} I^a \hat{1}_a(\mathcal{E}_{\pm r}). \tag{5.4.20}$$

Here, I^a is understood as the element that is the adjoint of the element I_a with respect to the bilinear form of the subalgebra $\mathcal{L}_{r-1}$. Applying to (5.4.20) the relation (5.4.16), we obtain

$$[E_{\pm r} E_{\mp r}] = \pm R_r H_r, \tag{5.4.21}$$

where

$$R_r = Q_r \dim \mathcal{E}_r. \tag{5.4.22}$$

We also introduce the important formula

$$Q_r^2 \hat{1}_r^{(2)} + \sum_{a \in \Omega_{r-1}} \hat{I}_a(\mathcal{E}_{\pm r}^{(1)}) \oplus \hat{I}^a(\mathcal{E}_{\pm r}^{(2)})$$

$$= Q_r^2 \hat{1}_r^{(2)} \oplus \hat{1}_r^{(1)} + \sum_{q \in \Omega_{r-1}} \hat{I}_a(\mathcal{E}_{\pm r}^{(2)} \oplus \hat{I}^a(\mathcal{E}_{\pm r}^{(1)}). \tag{5.4.23}$$

Here, the indices (1) and (2) identify the numbers of the spaces in which the operators $\hat{I}_r$ and $\hat{I}_a(\mathcal{E}_{\pm r})$ act. Finally, we write out the form of the Casimir operator in the decomposition (5.4.7):

$$K_r = H_r^2 + R_r H_r + E_{-r} E_{+r} + K_{r-1}. \tag{5.4.24}$$

We now turn to the procedure for eliminating the singular simple roots that separates from the algebra $\mathcal{L}_r$ the simple subalgebra $\mathcal{L}_{r-1}$ in accordance with the scheme (5.4.1). Each resulting subalgebra contains its own singular root, and therefore the elimination procedure can be continued. As a result, we arrive at the possible chains depicted in figure 5.2.

We agree to label the simple roots of the algebra in such a way that in each successive subalgebra along the chain the singular root has the maximal number. As a result, we arrive at the decomposition

$$\mathcal{L}_r = \oplus_{s=1}^r (\mathcal{E}_{-s} \oplus H_s \oplus \mathcal{E}_{+s}). \tag{5.4.25}$$

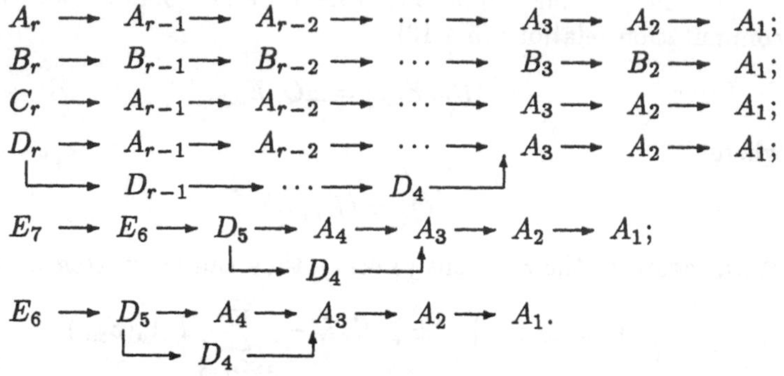

362   *Completely integrable Gaudin models*

$$
\begin{aligned}
A_r &\to A_{r-1} \to A_{r-2} \to \cdots \to A_3 \to A_2 \to A_1; \\
B_r &\to B_{r-1} \to B_{r-2} \to \cdots \to B_3 \to B_2 \to A_1; \\
C_r &\to A_{r-1} \to A_{r-2} \to \cdots \to A_3 \to A_2 \to A_1; \\
D_r &\to A_{r-1} \to A_{r-2} \to \cdots \to A_3 \to A_2 \to A_1; \\
&\hookrightarrow D_{r-1} \to \cdots \to D_4 \\
E_7 &\to E_6 \to D_5 \to A_4 \to A_3 \to A_2 \to A_1; \\
&\hookrightarrow D_4 \\
E_6 &\to D_5 \to A_4 \to A_3 \to A_2 \to A_1. \\
&\hookrightarrow D_4
\end{aligned}
$$

Figure 5.2. The chains of subalgebras corresponding to various schemes of elimination of the singular simple roots.

Introducing the notation

$$\mathcal{E}_0 \equiv \oplus_{s=1}^{r} H_s, \tag{5.4.26}$$

we can write equation (5.4.25) in a different way:

$$\mathcal{L}_r = \oplus_{s=-r}^{r} \mathcal{E}_s. \tag{5.4.27}$$

The commutation relations in $\mathcal{L}_r$ can now be represented in the form

$$[\mathcal{E}_q, \mathcal{E}_p] = \mathcal{E}_p, \quad 0 < |q| < |p|; \tag{5.4.28}$$

$$[\mathcal{E}_q, \mathcal{E}_{-q}] = \oplus_{s=-q}^{q} \mathcal{E}_s, \quad 0 < |q|. \tag{5.4.29}$$

To write out fully these commutation relations, we denote the generators of the subalgebra $\mathcal{E}_{\pm s}$ by $E_{\pm s}$. Then we have

$$\left. \begin{aligned} [E_q, E_p] &= -\hat{E}_q(\mathcal{E}_p) E_p \\ [H_q, E_p] &= -\hat{H}_q(\mathcal{E}_p) E_p \end{aligned} \right\} \, 0 < |q| < |p|; \tag{5.4.30}$$

$$[H_q, E_{\pm q}] = \pm Q_q E_{\pm q}; \tag{5.4.31}$$

$$[E_{\pm p} \otimes E_{\mp p}] = \pm Q_p H_p \hat{1}_p - \sum_{q-1}^{p-1} H_q \hat{H}_q(\mathcal{E}_{\pm p})$$

$$- \sum_{q=1}^{p-1} E_{-q} \hat{E}_{+q}(\mathcal{E}_{\pm p}) - \sum_{q=1}^{p-1} E_{+q} \hat{E}_{-q}(\mathcal{E}_{\pm p}). \tag{5.4.32}$$

The elements $E_{\pm s}$ and H_s satisfy the normalization conditions

$$\langle H_q, H_p \rangle = \delta_{qp}, \tag{5.4.33}$$

$$\langle E_q \oplus E_{-q} \rangle = \hat{1}_q. \tag{5.4.34}$$

The set of matrices $\hat{H}_q(\mathcal{E}_p)$, $\hat{E}_{\pm q}(\mathcal{E}_p)$, $q = 1, \ldots, p-1$, realizes the representation of the algebra $\mathcal{L}_{p-1}$ of dimension $\dim \mathcal{E}_p$ with highest weights $\Lambda_{0s} = \gamma_{sp}$, $s \in N_{p-1}$.

We introduce the matrix S_q^i, which relates H_q to the basis elements I_i:

$$H_q = \sum_{i=1}^{r} S_q^i I_i. \tag{5.4.35}$$

It is readily seen that by virtue of (5.4.5) the matrix S_q^i is triangular:

$$S_q^i = 0, \quad i > q. \tag{5.4.36}$$

Substituting (5.4.35) in (5.4.33), we obtain the helpful formula

$$\sum_{i,k \in N_r} S_p^i S_q^k \gamma_{ik} = \delta_{pq} \tag{5.4.37}$$

which in conjunction with the condition (5.4.36) enables us to determine the matrix S_q^i uniquely (by means of the standard Gram–Schmidt orthogonalization procedure). Many properties of the considered basis can be expressed in terms of the matrix S_q^i. For example, we have

$$\hat{E}_q(\mathcal{E}_p)|0\rangle = 0, \quad q \in N_{p-1}; \tag{5.4.38a}$$

$$\hat{H}_q(\mathcal{E}_p)|0\rangle = S_{qp}|0\rangle, \quad q \in N_{p-1}. \tag{5.4.38b}$$

It is also easy to show that

$$Q_q = S_{qq} \tag{5.4.39}$$

and

$$\sum_{q=1}^{r} S_q^i R_q = \nu^i. \tag{5.4.40}$$

The last formula can be deducted by comparing (5.3.15) with the other expression for the Casimir operator

$$K_r = \sum_{q=1}^{r} H_q^2 + \sum_{q=1}^{r} R_q H_q + 2 \sum_{q=1}^{r} E_{-q} E_{+q} \tag{5.4.41}$$

which follows from (5.4.24).

We now turn to the construction of differential realizations of the singular simple Lie algebras based on the decomposition (5.4.26). In accordance with this decomposition, any element of the group can be parametrized as follows:

$$g(x) = \prod_{q=r-1}^{r} \exp(x_q E_q). \qquad (5.4.42)$$

We shall read the product in (5.4.42) from the left to the right. Note that any partial product $\exp(x_s E_s) \ldots \exp(x_r E_r)$ forms a subgroup. For this reason, equation (5.4.42) can be split into $2r + 1$ formulas of the form

$$\prod_{q=s}^{r} \exp(x_q E_q) \exp(\varepsilon_s E_s) = \sum_{q=s}^{r} \exp[x_q E_q + \varepsilon_s (E_s | x_q) E_q].$$

$$(5.4.43)$$

After simple manipulations, each of these formulas can be rewritten in one of the following forms:

$$\left(\prod_{q=p}^{r} \exp(x_q E_q) \right) E_s \left(\prod_{q=p}^{r} \exp(x_q E_q) \right)^{-1} = (E_s | x_p) E_p$$

$$+ \sum_{n=p+1}^{r} (E_s | x_n) \left(\prod_{q=p}^{n-1} \exp(x_q E_q) \right) E_n \left(\prod_{q=p}^{n-1} \exp(x_q E_q) \right)^{-1}$$

$$+ \sum_{n=s}^{p-1} (E_s | x_n) \left(\prod_{q=n}^{p-1} \exp(x_q E_q) \right)^{-1} E_n \left(\prod_{q=n}^{p-1} \exp(x_q E_q) \right),$$

$$s = -r, \ldots, +r. \quad (5.4.44)$$

Multiplying both sides of (5.4.44) by E_{-p} , we find

$$(E_s | x_p) = \left\langle E_s \oplus \left(\prod_{q=p}^{r} \exp(-\operatorname{ad} x_q E_q) \right) E_{-p} \right\rangle. \qquad (5.4.45)$$

From this it is easy to obtain differential representations for the operators

$$E_s = \sum_{p=1}^{r} \left\langle E_s \oplus \left(\prod_{q=p}^{r} \exp(-\operatorname{ad} x_q E_q) \right) E_{-p} \right\rangle \frac{\partial}{\partial x_p}. \qquad (5.4.46)$$

To simplify these expressions, we project the operators (5.4.46) onto the class of functions

$$\psi = \exp(x_0 h_0)\psi(x_1, \ldots, x_r), \tag{5.4.47}$$

as a result of which they take the form

$$E_{+s} = \frac{\partial}{\partial x_s} - \sum_{a=s+1}^{r} \langle E_s \otimes \operatorname{ad}(x_a E_a) E_{-a}\rangle \frac{\partial}{\partial x_a}, \tag{5.4.48a}$$

$$H_s = h_s - \sum_{a=s}^{r} \langle H_s \otimes \operatorname{ad}(x_a E_a) E_{-a}\rangle \frac{\partial}{\partial x_a}, \tag{5.4.48b}$$

$$E_{-s} = \sum_{a=1}^{s} \left\langle E_{-s} \otimes \prod_{q=a}^{s} \exp(-\operatorname{ad} x_q E_q) H_a \right\rangle h_a$$

$$+ \sum_{a=1}^{s} \left\langle E_{-s} \otimes \prod_{q=a}^{s} \exp(-\operatorname{ad} x_q E_q) E_{-a} \right\rangle \frac{\partial}{\partial x_a}$$

$$+ \sum_{a=s+1}^{r} \left\langle E_{-s} \otimes \exp(-\operatorname{ad} x_a E_a) E_{-a} \right\rangle \frac{\partial}{\partial x_a}. \tag{5.4.48c}$$

These are the required differentials of the representations of the Lie algebras with highest weights Λ_0. Here, h_a are the eigenvalues of the operators H_a associated with the highest weights by means of the matrix $\hat{S}$:

$$h_a = \sum_{i \in N_r} S_a^i \Lambda_{0i}. \tag{5.4.49}$$

5.5   The generalized Gaudin model and its solutions

1. Definition of the Gaudin model in the general case. With the simple Lie algebra $\mathcal{L}_r$ we associate the infinite-dimensional Gaudin algebra $G(\mathcal{L}_r)$, whose generators $I_a(\lambda)$, $a \in \Omega_r$ depend on the complex parameter λ and satisfy the commutation relations

$$[I_a(\lambda), I_b(\mu)] = \sum_{c \in \Omega_r} \Gamma_{ab}^c \frac{I_c(\lambda) - I_c(\mu)}{\lambda - \mu}. \tag{5.5.1}$$

Using (5.5.1), we associate the Cartan–Weyl decomposition (5.3.2) in the algebra $\mathcal{L}_r$ with the analogous decomposition in the algebra $G(\mathcal{L}_r)$:

$$G(\mathcal{L}_r) = G(\mathcal{L}_r^-) \oplus G(\mathcal{L}_r^0) \oplus G(\mathcal{L}_r^+). \tag{5.5.2}$$

We denote the elements of the subalgebras $G(\mathcal{L}_r^{\pm})$ and $G(\mathcal{L}_r^0)$ by $I_a(\lambda)$, $a \in \Delta_r^{\pm}$, and $I_i(\lambda)$, $i \in N_r$, respectively. We define a representation of the Gaudin algebra by means of the formulas

$$\left. \begin{array}{l} I_\alpha(\lambda)|0\rangle = 0, \quad \alpha \in \Delta_r^+; \\ I_i(\lambda)|0\rangle = F_i(\lambda)|0\rangle, \quad i \in N_r. \end{array} \right\} \tag{5.5.3}$$

Here, $|0\rangle$ is the highest vector of the representation, and the functions $F_i(\lambda)$, $i \in N_r$, play the part of the components of the highest weight. Let $M = \{M^i\}$, $i \in N_r$, be sets of non-negative integers. With each such set, we associate the space $|M\rangle$, which is formed from all possible linear combinations of vectors of the form

$$I_{\alpha_1}(\lambda_1)\ldots I_{\alpha_K}(\lambda_K)|0\rangle, \tag{5.5.4}$$

where $\lambda_1,\ldots,\lambda_K$ are arbitrary complex numbers, and $\alpha_1,\ldots,\alpha_K \in \Delta_r^-$ are different sets of negative roots of the algebra $\mathcal{L}_r$ that satisfy the restriction

$$\alpha_1 + \alpha_2 + \ldots + \alpha_K = -\sum_{i \in N_r} M^i \pi_i. \tag{5.5.5}$$

Subject to this restriction, the number of roots in the set can be arbitrary. If the operators $I_\alpha(\lambda)$ in equation (5.5.4) satisfy the conditions (5.5.3), then the space $W_r\{F(\lambda)\}$ of the representation of the Gaudin algebra realized by them is defined as

$$W_r\{F(\lambda)\} = \oplus_{M \geq 0}|M\rangle. \tag{5.5.6}$$

Following Gaudin (1983), we introduce the operators

$$K_r(\lambda) = \sum_{a,b \in \Omega_r} g^{ab} I_a(\lambda) I_b(\lambda), \tag{5.5.7}$$

which depend on λ and in their structure recall the Casimir operators (5.3.12) in the algebra $\mathcal{L}_r$. However, in reality (5.5.7) are not Casimir operators for the Gaudin algebra $G(\mathcal{L}_r)$, since they do not commute with all of its elements. They possess a different remarkable property:

$$[K_r(\lambda), K_r(\mu)] = 0, \tag{5.5.8}$$

i.e., they form a commutative family. For this reason, the spectral problem

$$K_r(\lambda)\phi_r = E_r(\lambda)\phi_r, \quad \phi_r \in W_r\{F(\lambda)\}, \tag{5.5.9}$$

is completely integrable. In the following section, we shall show that it can be solved exactly for all singular simple Lie algebras in the framework of the algebraic Bethe *ansatz*. In the expression, we shall follow the studies given by Ushveridze (1990b, c, 1992). For other methods of solving the Gaudin problem for the simple classical Lie algebras, see, for example, Jurčo (1989).

2. The Gaudin algebra in the special basis. Beginning with this section, we restrict the treatment to only the singular simple Lie algebras, i.e., the algebras $A_r (r \geq 1), B_r (r \geq 2), C_r (r \geq 3), D_r (r \geq 4), E_6$, and E_7. By virtue of the correspondence noted in the previous section between these algebras and the Gaudin algebras, the latter have decompositions analogous to the decompositions (5.4.7):

$$G(\mathcal{L}_r) = \{G(\mathcal{E}_{r-1}) \oplus G(H_r) \oplus (\mathcal{E}_{+r})\} \oplus G(\mathcal{L}_{r-1}). \qquad (5.5.10)$$

We denote the elements of the subalgebras $G(\mathcal{E}_{\pm r}), G(H_r)$, and $G(\mathcal{L}_{r-1})$ by $E_{\pm r}(\lambda), H_r(\lambda)$, and $I_a(\lambda)$, where $a \in \Omega_{r-1}$. Then in accordance with figure 5.1 and the results of the previous section, the following commutation relation holds for them:

$$[I_a(\lambda), I_b(\zeta)] = \sum_{c \in \Omega_{r-1}} \Gamma^c_{ab} \frac{I_c(\lambda) - I_c(\zeta)}{\lambda - \zeta}, \quad a, b \in \Omega_{r-1}; \qquad (5.5.11)$$

$$[I_a(\lambda), E_{\pm r}(\zeta)] = -\hat{I}_a(\mathcal{E}_{\pm r}) \frac{E_{\pm r}(\lambda) - E_{\pm r}(\zeta)}{\lambda - \zeta}, \quad a \in \Omega_{r-1}; \qquad (5.5.12)$$

$$[I_a(\lambda), H_r(\zeta)] = 0, \quad a \in \Omega_{r-1}; \qquad (5.5.13)$$

$$[H_r(\lambda), E_{\pm r}(\zeta)] = \pm Q_r \frac{E_{\pm r}(\lambda) - E_{\pm r}(\zeta)}{\lambda - \zeta}; \qquad (5.5.14)$$

$$[H_r(\lambda), H_r(\zeta)] = 0; \qquad (5.5.15)$$

$$[E_{\pm r}(\lambda), \otimes E_{\pm r}(\zeta)] = 0; \qquad (5.5.16)$$

$$[E_{\pm r}(\lambda), \otimes E_{\mp r}(\zeta)]$$
$$= \pm Q_r \frac{H_r(\lambda) - H_r(\zeta)}{\lambda - \zeta} \hat{1}_r \quad - \sum_{a \in \Omega_{r-1}} \frac{I_a(\lambda) - I_a(\zeta)}{\lambda - \zeta} \hat{I}^a(\mathcal{E}_{\pm r}).$$
$$(5.5.17)$$

The superscript a in (5.5.17) is understood in the sense of conjugation with respect to the bilinear form of the algebra $\mathcal{L}_{r-1}$. We have the formula

$$[E_{\pm r}(\lambda) E_{\mp r}(\lambda)] = \pm R_r H'_r(\lambda), \qquad (5.5.18)$$

which is the Gaudian analogue of (5.4.22). For the operators $K_r(\lambda)$, we have

$$K_r(\lambda) = H_r^2(\lambda) + R_r H_r'(\lambda) + 2E_{-r}(\lambda)E_{+r}(\lambda) + K_{r-1}(\lambda).$$

$$(5.5.19)$$

We now determine the vacuum subspace $W_{r-1}\{F(\lambda)\}$ of the representation space $W_r\{F(\lambda)\}$ as a linear hull of the vectors

$$|0\rangle, \; I_{\alpha_1}(\lambda_1)|0\rangle, \; I_{a_1}(\lambda_1)I_{a_2}(\lambda_2)|0\rangle, \ldots, \qquad (5.5.20)$$

in which the numbers $\lambda_1, \lambda_2, \ldots$ are arbitrary, and $a_1, a_2, \ldots \in \Omega_{r-1}$. It is readily verified that the elements ϕ_{r-1} of the vacuum subspace possess the properties

$$\left. \begin{array}{l} E_{+r}(\lambda)\phi_{r-1} = 0; \\ H_r(\lambda)\phi_{r-1} = h_r(\lambda)\phi_{r-1}. \end{array} \right\} \qquad (5.5.21)$$

Here, $h_r(\lambda)$ are the eigenvalues of the operator $H_r(\lambda)$ on $|0\rangle$, and they are equal to

$$h_r(\lambda) = \sum_{i \in N_r} S_r^i F_i(\lambda), \qquad (5.5.22)$$

and S_r^i is the matrix introduced earlier in (5.4.35).

3. The Bethe solution of the generalized Gaudin problem. We now turn to the construction of exact solutions of the spectral problem (5.5.9). Let M^r be a fixed natural number, and $\zeta_{r,i}$, $i = 1, \ldots, M^r$ be as yet unknown numerical parameters. We shall seek an eigenvector of the Casimir–Gaudin operator $K_r(\lambda)$ in the form

$$\phi_r = E_{-r}(\zeta_{r,1}) \otimes \ldots \otimes E_{-r}(\zeta_{r,M^r})\phi_{r-1}, \qquad (5.5.23)$$

where ϕ_{r-1} is a tensor of rank M^r contracted with the vectors of the generators E_{-r} of the algebra $G(\mathcal{E}_{-r})$ with respect to all indices. We shall regard the components of this tensor as elements of the vacuum subspace $W_{r-1}\{F(\lambda)\}$. Using equation (5.5.21), we can write

$$K_r(\lambda)\phi_{r-1} = [h_r^2(\lambda) + R_r h_r'(\lambda)]\phi_{r-1}$$
$$+ K_{r-1}(\lambda)\phi_{r-1}. \qquad (5.5.24)$$

We now calculate the result of applying the operator $K_r(\lambda)$ to the vector ϕ_r. For this, we require the commutation relations

$$[K_r(\lambda), E_{-r}(\zeta)] = -\frac{Q_r}{(\lambda - \zeta)}\{E_{-r}(\lambda)H_r(\zeta) - E_{-r}(\zeta)H_r(\lambda)\}$$

$$- \frac{1}{\lambda - \zeta} \sum_{a \in \Omega_{r-1}} \hat{I}^a(\mathcal{E}_{-r}) \{ E_{-r}(\lambda) I_a(\zeta)$$

$$- E_{-r}(\zeta) I_a(\lambda) \}; \tag{5.5.25}$$

$$[I_a(\lambda), E_{-r}(\zeta)] = \frac{1}{\lambda - \zeta} \hat{I}_a(\mathcal{E}_{-r}) \{ E_{-r}(\lambda) - E_{-r}(\zeta) \}; \tag{5.5.26}$$

$$[H_r(\lambda), E_{-r}(\zeta)] = -\frac{Q_r}{\lambda - \zeta} \{ E_{-r}(\lambda) - E_{-r}(\zeta) \}, \tag{5.5.27}$$

which can be readily deduced from the relations (5.5.1) and the definition (5.5.7). In the expression

$$K_r(\lambda)\phi_r = K_r(\lambda) E_{-r}(\zeta_{r,1}) \otimes \ldots \otimes E_{-r}(\zeta_{r,M^r})\phi_{r-1} \tag{5.5.28}$$

we move the operator $K_r(\lambda)$ to the right-hand side and, using the formulas (5.5.24) and (5.5.25), we obtain

$$
\begin{aligned}
K_r(\lambda)\phi_r &= [h_r^2(\lambda) + R_r h_r'(\lambda)]\phi_r \\
&+ E_{-r}(\zeta_{r,1}) \otimes \ldots \otimes E_{-r}(\zeta_{r,M^r}) K_{r-1}(\lambda)\phi_{r-1} \\
&- 2Q_r \sum_{m=1}^{M^r} \frac{1}{\lambda - \zeta_{r,m}} E_{-r}(\zeta_{r,1}) \otimes \ldots \otimes \{ E_{-r}(\lambda) H_r(\zeta_{r,m}) \\
&- E_{-r}(\zeta_{r,m}) H_r(\lambda) \} \otimes \ldots \otimes E_{-r}(\zeta_{r,M^r})\phi_{r-1} \\
&+ 2 \sum_{a \in \Omega_{r-1}} \sum_{m=1}^{M^r} \frac{1}{\lambda - \zeta_{r,m}} E_{-r}(\zeta_{r,1}) \oplus \ldots \oplus \{ E_{-r}(\lambda) I_a(\zeta_{r,m}) \\
&- E_{-r}(\zeta_{r,m}) I_a(\lambda) \} \oplus \ldots \oplus E_{-r}(\zeta_{r,M^r}) \hat{I}_a(\mathcal{E}_r)\phi_{r-1}. \tag{5.5.29}
\end{aligned}
$$

Further, shifting the operators $H_r(\lambda)$, $H_r(\zeta_{r,m})$, $I_a(\lambda)$, $I_a(\zeta_{r,m})$, $a \in \Omega_{r-1}$, to the right by means of the commutation relations (5.5.26) and (5.5.27), and using (5.5.21), we find that the expression $K_r(\lambda)\phi_r$ can be represented as a sum of terms of two types:

$$K_r(\lambda)\phi_r = \{ K_r(\lambda)\phi_r \}_1 + \{ K_r(\lambda)\phi_r \}_2. \tag{5.5.30}$$

The terms identified by the index 1 are proportional to the vector ϕ_r and, therefore, determine an eigenvalue $E_r(\lambda)$ of the operator $K_r(\lambda)$. The remaining terms, identified by the index 2, are not proportional to the vector ϕ_r and they must be annihilated. We begin by considering the first group of terms. They have the form

$$
\{K_r(\lambda)\phi_r\}_1 = \left\{ \left[h_r(\lambda) + Q_r \sum_{m=1}^{M^r} \frac{1}{\lambda - \zeta_{r,m}} \right]^2 \right.
$$
$$
+ R_r \left[h_r(\lambda) + Q_r \sum_{m=1}^{M^r} \frac{1}{\lambda - \zeta_{r,m}} \right]' \left. \right\} \phi_r
$$
$$
+ E_{-r}(\zeta_{r,1}) \otimes \ldots \otimes E_{-r}(\zeta_{r,M^r}) \tilde{K}_{r-1}(\lambda)\phi_{r-1},
$$
$$(5.5.31)$$

where

$$
\tilde{K}_{r-1}(\lambda) = \sum_{a,b \in \Omega_{r-1}} g^{ab} \tilde{I}_a(\lambda) \tilde{I}_b(\lambda). \qquad (5.5.32)
$$

It is readily verified that the operators

$$
\tilde{I}_a(\lambda) = I_a(\lambda) - \sum_{m=1}^{M^r} \frac{\hat{I}_a(\mathcal{E}_r)}{\lambda - \zeta_{r,m}}, \quad a \in \Omega_{r-1}, \qquad (5.5.33)
$$

satisfy the same commutation relations as the operators $I_a(\lambda)$. Therefore, $\tilde{K}_{r-1}(\lambda)$, defined by (5.5.32), is a Gaudin operator for the algebra $G(\mathcal{L}_{r-1})$. The set of vectors ϕ_{r-1} can be considered as the space $W_{r-1}\{\tilde{F}(\lambda)\}$ of a representation of the Gaudin algebra characterized by highest weight with components

$$
\tilde{F}_i(\lambda) = F_i(\lambda) - \sum_{m-1}^{M^r} \frac{\gamma_{ri}}{\lambda - \zeta_{r,m}}, \quad i \in N_{r-1}. \qquad (5.5.34)
$$

Therefore, the spectral problem

$$
\tilde{K}_{r-1}(\lambda)\phi_{r-1} = \tilde{E}_{r-1}(\lambda)\phi_{r-1}, \quad \phi_{r-1} \in W_{r-1}\{\tilde{F}(\lambda)\}, \qquad (5.5.35)
$$

is the Gaudin problem for the algebra $G(\mathcal{L}_{r-1})$. If it has solutions, then we can write

$$
K_r(\lambda)\phi_r = \left\{ \left[h_r(\lambda) + Q_r \sum_{m=1}^{M^r} \frac{1}{\lambda - \zeta_{r,m}} \right]^2 \right.
$$
$$
+ R_r \left[h_r(\lambda) + Q_r \sum_{M=1}^{M^r} \frac{1}{\lambda - \zeta_{r,m}} \right]'
$$
$$
+ \tilde{E}_{r-1}(\lambda) \left. \right\} \phi_r
$$
$$(5.5.36)$$

provided, of course, that the terms of the second type are equal to zero. We write out these terms separately. They have the form

$$\{K_r(\lambda)\phi_r\}_2 = 2 \sum_{m>n} \frac{1}{(\lambda - \zeta_{r,n})(\zeta_{r,n} - \zeta_{r,m})}$$

$$E_{-r}(\zeta_{r,1}) \otimes \ldots \otimes E_{-r}(\lambda) \otimes \ldots \otimes E_{-r}(\zeta_{r,n}) \otimes \ldots \otimes E_{-r}(\zeta_{r,M^r})$$

$$\uparrow n \qquad\qquad \uparrow m$$

$$\left\{ \left[Q_r^2 \hat{1}_r^{(n)} \otimes \hat{1}_r^{(m)} + \sum_{a,b\in\Omega_{r-1}} g^{ab} \hat{I}_a(\mathcal{E}_r^{(n)})\hat{I}_b(\mathcal{E}_r^{(m)}) \right] \right.$$

$$\left. - \left[Q_r^2 \hat{1}_r^{(m)} \otimes \hat{1}_r^{(n)} + \sum_{a,b\in\Omega_{r-1}} g^{ab} \hat{I}_a(\mathcal{E}_r^{(m)})\hat{I}_b(\mathcal{E}_r^{(n)}) \right] \right\} \phi_{r-1}$$

$$- \sum_{n=1}^{M^r} \frac{1}{\lambda - \zeta_{r,n}} E_{-r}(\zeta_{r,1}) \otimes \ldots \otimes E_{-r}(\lambda) \otimes \ldots \otimes E_{-r}(\zeta_{r,M^r})$$

$$\uparrow n$$

$$\left\{ 2Q_r h_r(\zeta_{r,n}) + 2Q_r^2 \sum_{m=1}^{M^r}{}' \frac{1}{\zeta_{r,n} - \zeta_{r,m}} + 2 \sum_{a,b\in\Omega_{r-1}} \hat{I}^a(\mathcal{E}_r)I_a(\zeta_{r,n}) \right.$$

$$\left. -2 \sum_{m=1}^{M^r}{}' \frac{1}{\zeta_{r,n} - \zeta_{r,m}} \sum_{a,b\in\Omega_{r-1}} \hat{I}^a(\mathcal{E}_r)\hat{I}_a(\mathcal{E}_r) \right\} \phi_{r-1}.$$

$$(5.5.37)$$

The numbers with the arrows indicate the serial number of the operator in the product. Using the formulas (5.5.23) and (5.5.32), we find that

$$\{K_r(\lambda)\phi_r\}_2 = \sum_{n=1}^{M^r} \frac{1}{\lambda - \zeta_{r,n}} \otimes \ldots \otimes E_{-r}(\lambda) \otimes \ldots \otimes E_{-r}(\zeta_{r,M^r})$$

$$\left\{ 2Q_r h_r(\zeta_{r,n}) + 2Q_r^2 \sum_{m=1}^{M^r}{}' \frac{1}{\zeta_{r,n} - \zeta_{r,m}} - \mathrm{Res}_{\zeta_{r,n}} \tilde{K}_{r-1}(\lambda) \right\} \phi_{r-1}.$$

$$(5.5.38)$$

Taking into account (5.5.35), we finally find that the condition of vanishing of the terms (5.5.38) has the form

$$2Q_r h_r(\zeta_{r,n}) + 2Q_r^2 \sum_{m=1}^{M^r}{}' \frac{1}{\zeta_{r,n} - \zeta_{r,m}}$$

$$= \mathrm{Res}_{\zeta_{r,n}} \tilde{E}_{r-1}(\lambda), \quad n = 1, \ldots, M^r, \qquad (5.5.39)$$

where by $\text{Res}_\zeta \tilde{E}_{r-1}(\lambda)$ we have denoted the residue of the function $\tilde{E}_{r-1}(\lambda)$ at the simple pole situated at the point ζ.

Thus, we see that the Gaudin problem for the algebra $G(\mathcal{E}_{r-1})$ has been reduced to the analogous problem for the algebra $G(\mathcal{L}_{r-1})$. In its structure, the latter completely repeats the former, and therefore it, in its turn, can be reduced to the Gaudin problem for the algebra $G(\mathcal{L}_{r-2})$, etc, and this goes on until we reach the algebra $G(\mathcal{L}_1) = G(A_1)$, for which the solution of this problem is known (see section 4.15). This enables us to write the solution of the Gaudin problem in the explicit form

$$
\begin{aligned}
\phi_r \quad &= \quad \phi_r(M,\zeta) \\
&= \quad \left\{ \otimes_{i=1}^{M^r} E_{-r}(\zeta_{r,i}) \right\} \left\{ \left\{ \otimes_{i=1}^{M^{r-q}} \left[E_{-(r-1)}(\zeta_{r-1,i}) \right.\right.\right. \\
&\qquad \left.\left.\left. - \sum_{m=1}^{M^r} \frac{\hat{E}_{-(r-1)}(\mathcal{E}_r)}{\zeta_{r-1,i} - \zeta_{r,m}} \right] \right\} \right. \\
&\quad \times \ldots \times \left. \left\{ \otimes_{i=1}^{M^1} \left[E_{-1}(\zeta_{1,i}) - \sum_{q=2}^{r} \sum_{m=1}^{M^q} \frac{E_{-1}(\mathcal{E}_q)}{\zeta_{1,i} - \zeta_{q,m}} \right] \right\} |0\rangle, \right.
\end{aligned} \tag{5.5.40}
$$

$$
\begin{aligned}
E_r(\lambda) \quad &= \quad E_r(M,\zeta;\lambda) \\
&= \quad \sum_{s=1}^{r} \{ h_s^2(\lambda) + R_s h_s'(\lambda) \} \\
&\quad + \quad 2 \sum_{p=1}^{r} \sum_{m=1}^{M^p} S_{sp} \frac{h_s(\lambda) - h_s(\zeta_{p,m})}{\lambda - \zeta_{p,m}},
\end{aligned} \tag{5.5.41}
$$

where $\zeta_{p,m}$, $m = 1,\ldots,M^p$, $p = 1,\ldots,r$ are numbers that satisfy the system of equations

$$
\sum_{p=1}^{r} \sum_{m=1}^{M^p}{}' \frac{\gamma_{sp}}{\zeta_{s,n} - \zeta_{p,m}} + \sum_{q=1}^{r} S_{qs} h_q(\zeta_{s,n}) = 0. \tag{5.5.42}
$$

Using the properties of the matrix S_{pq} and the results of section 5.4, we can rewrite the last two expressions in terms of the highest weights $F(\lambda)$ of the representation of the Gaudin algebra:

$$
\begin{aligned}
E_r(M,\zeta;\lambda) \quad &= \quad \sum_{i \in N_r} \left[F^i(\lambda) + \nu^i \frac{\partial}{\partial \lambda} \right] F_i(\lambda) \\
&\quad + \quad 2 \sum_{p=1}^{r} \sum_{i=1}^{M^p} \frac{F_p(\lambda) - F_p(\zeta_{p,i})}{\lambda - \zeta_{p,i}};
\end{aligned} \tag{5.5.43}
$$

$$\sum_{k=1}^{r} \sideset{}{'}\sum_{m=1}^{M^k} \frac{(\pi_i, \pi_k)}{\zeta_{i,n} - \zeta_{k,m}} + F_i(\zeta_{i,n}) = 0,$$

$$n = 1, \ldots, M^i, \quad i = 1, \ldots, r. \tag{5.5.44}$$

5.6 Quasi-exactly solvable equations

1. Symmetry of the Gaudin model. We assume that the functions $F_i(\lambda)$, $i \in N_r$, are such that the limits

$$F_i = -\lim_{\lambda \to \infty} \lambda F_i(\lambda), \quad i \in N_r, \tag{5.6.1}$$

exist. Then by virtue of (5.5.3) and (5.5.1) there must also exist the operator limits

$$I_a = -\lim_{\lambda \to \infty} \lambda I_a(\lambda), \quad a \in \Omega_r. \tag{5.6.2}$$

Using equations (5.5.1), (5.5.7), and (5.6.2), we find that the operators (5.6.2) form the algebra $\mathcal{L}_r$:

$$[I_a, I_b] = \sum_{c \in \Omega_r} \Gamma_{ab}^c I_c, \quad a, b \in \Omega_r, \tag{5.6.3}$$

and commute with the family of operators $K_r(\lambda)$:

$$[K_r(\lambda), I_a] = 0, \quad a \in \Omega_r. \tag{5.6.4}$$

This means that the Gaudin model possesses the global symmetry algebra $\mathcal{L}_r$.

We now consider what representations of the symmetry algebra can be realized in the space $W_r\{F(\lambda)\}$. For this, we apply to the Bethe solutions (5.5.40) the operators I_α, $\alpha \in \Delta_r^+$, and I_i, $i \in N_r$. Shifting them to the right by means of the commutation relations

$$[I_a, I_b(\lambda)] = \sum_{c \in \Omega_r} \Gamma_{ab}^c I_c(\lambda), \quad a, b \in \Omega_r, \tag{5.6.5}$$

and using the restrictions (5.5.42) on the parameters $\zeta_{p,m}$, we obtain

$$I_i \phi_r(M, \zeta) = (F_i - M_i)\phi_r(M, \zeta), \quad i \in N_r; \tag{5.6.6}$$

$$I_\alpha \phi_r(M, \zeta) = 0, \quad \alpha \in \Delta_r^+. \tag{5.6.7}$$

It can be seen from equations (5.6.6) and (5.6.7) that the Bethe solutions play the role of the highest vectors for representations of the symmetry algebra $\mathcal{L}_r$. The numbers $F_i - M_i$, $i \in N_r$, are the components of the corresponding highest weights. The representations described by equations (5.6.5) and (5.6.6) are finite dimensional. Indeed, applying successively to $\phi_r(M, \zeta)$ the operators I_α, $\alpha \in \Delta_r^-$, we will obtain more and more new solutions of the Gaudin model with the same eigenvalues $E_r(M, \zeta; \lambda)$. This means that the Bethe solutions found in the previous section are only some of all the solutions of the Gaudin problem.

We denote by $\Phi_M\{F\}$ the set of all solutions of the system:

$$I_i\phi = (F_i - M_i)\phi, \quad i \in N_r; \tag{5.6.8}$$

$$I_\alpha\phi = 0, \quad \alpha \in \Delta_r^+; \tag{5.6.9}$$

$$\phi \in W_r\{F(\lambda)\}. \tag{5.6.10}$$

By virtue of (5.6.4), the space $\Phi_M\{F\}$ is invariant with respect to the action of the operators $K_r(\lambda)$. Therefore, the spectral problems

$$K_r(\lambda)\phi_r = E_r(\lambda)\phi_r, \quad \phi_r \in \Phi_M\{F\}, \tag{5.6.11}$$

are defined for all $M = \{M^i\}$, $i \in N_r$. It can be shown that the linear hull of the functions (5.5.40) is identical to the space $\Phi_M\{F\}$, and for this reason the solutions of any of the problems (5.6.11) are completely described by the explicit Bethe formulas (5.5.40), (5.5.43), and (5.5.44).

An important property of the spaces $\Phi_M\{M\}$ is that they are all finite dimensional if $F(\lambda)$ is a rational function. This has the consequence that in the rational case equations (5.6.11) have only a finite number of exact solutions and therefore can be used as starting points in the construction of quasi-exactly solvable problems.

2. Differential form for the Gaudin equations and transition to quasi-exactly solvable equations. The finite dimensionality of the invariant spaces $\Phi_M\{F\}$ is most readily proved in the case when the rational functions $F_i(\lambda)$ are non-degenerate, i.e., they are described by the formulas

$$F_i(\lambda) = -\sum_{A=1}^{N} \frac{f_{Ai}}{\lambda - \sigma_A}, \quad i \in N_r. \tag{5.6.12}$$

If (5.5.1) and (5.5.3) are to agree, the generators of the Gaudin algebra must have an analogous form:

$$I_a(\lambda) = -\sum_{A=1}^{N} \frac{I_{Aa}}{\lambda - \sigma_A}, \quad a \in \Omega_r, \tag{5.6.13}$$

where I_{Aa} are certain operators. Substituting (5.6.13) into the commutation relations (5.5.1), we find that they satisfy the relations

$$[I_{Aa}, I_{Ab}] = \sum_{c \in \Omega_r} \Gamma_{ab}^c I_{Ac}, \quad a, b \in \Omega_r;$$

$$[I_{Aa}, I_{Bb}] = 0, \quad a, b \in \Omega_r, \quad A \neq B, \tag{5.6.14}$$

i.e., they form the algebra $\mathcal{L}_r \oplus \ldots \oplus \mathcal{L}_r$ (N times). In this case, the operators $K_r(\lambda)$ take the form

$$K_r(\lambda) = \sum_{A,B=1}^{N} \frac{\sum_{a,b \in \Omega_r} g^{ab} I_{Aa} I_{Bb}}{(\lambda - \sigma_A)(\lambda - \sigma_B)}, \tag{5.6.15}$$

i.e., they are transformed into the hamiltonians of N-site models of a magnet based on the algebra $\mathcal{L}_r \oplus \ldots \oplus \mathcal{L}_r$ (N times). It is obvious that at site A there acts the representation of the algebra $\mathcal{L}_r$ with highest weight $f_a = \{f_{Ai}\}$, $i \in N_r$. For the numbers F_i, we have in this case

$$F_i = \sum_{A=1}^{N} f_{Ai}, \quad i \in N_r. \tag{5.6.16}$$

The expressions for the symmetry operators become extremely simple:

$$I_a = \sum_{A=1}^{N} I_{Aa}, \quad a \in \Omega_r. \tag{5.6.17}$$

We are now ready to use the differential realizations of the representations of the algebra $\mathcal{L}_r$ obtained in section 5.3:

$$I_{Aa} = \hat{t}_a^+(x_A) \frac{\partial}{\partial x_A} + \hat{t}_a^0(x_A) f_A. \tag{5.6.18}$$

Substitution of (5.6.18) into (5.6.15) transforms $K_r(\lambda)$ into a single-parameter family of differential operators of second order:

$$K_r(\lambda)$$

$$= \sum_{A,B=1}^{N} \frac{\sum_{a,b \in \Omega_r} g^{ab} \left[\hat{t}_a^+(x_A) \frac{\partial}{\partial x_A} + \hat{t}_a^0(x_A) f_A \right] \left[\hat{t}_b^+(x_B) \frac{\partial}{\partial x_B} + \hat{t}_b^0(x_B) f_B \right]}{(\lambda - \sigma_A)(\lambda - \sigma_B)}.$$

$$\tag{5.6.19}$$

The invariant space $\Phi_M\{F\}$ on which the operators (5.6.19) act is transformed accordingly, into the space of functions (polynomials) in the variables x_A, $A = 1, \ldots, N$. To describe the structure of this function space, we use the differential realizations of the symmetry operators I_a, $a \in \Delta_r^+$, and I_i, $i \in N_r$, the explicit form of which can be obtained from (5.6.17) and (5.6.18) with allowance for the formulas (5.3.48)–(5.3.50):

$$I_\alpha = \sum_{A=1}^{N} \hat{t}_\alpha^+(x_A)\frac{\partial}{\partial x_A}, \quad \alpha \in \Delta_r^+;$$

$$I_i = \sum_{A=1}^{N} \hat{t}_i^+(x_A)\frac{\partial}{\partial x_A} + F_i, \quad i \in N_r. \tag{5.6.20}$$

Then it follows from the definition (5.6.8)–(5.6.10) of the space $\Phi_M\{F\}$ that its elements are functions that satisfy the system of first-order differential equations

$$\left\{ \sum_{A=1}^{N} \hat{t}_\alpha^+(x_A)\frac{\partial}{\partial x_A} \right\} \phi = 0, \quad \alpha \in \Delta_r^+; \tag{5.6.21}$$

$$\left\{ \sum_{A=1}^{N} \hat{t}_i^+(x_A)\frac{\partial}{\partial x_A} \right\} \phi = -M_i\,\phi, \quad i \in N_r. \tag{5.6.22}$$

The subsidiary restriction (5.6.10) reduces to the requirement that the solutions of the system (5.6.21), (5.6.22) be sought in the class of polynomials in x_A, $A = 1, \ldots, N$.

In accordance with the results of section 5.3, the general solution of the subsystem (5.6.21) has the form of a function that depends on $N - 1$ vector variables:

$$\zeta_A = x_A - x_N, \quad A = 1, \ldots, N - 1. \tag{5.6.23}$$

If this function is to satisfy the second subsystem (5.6.22) and simultaneously be a polynomial, it must have the form of a linear combination of monomials,

$$\prod_{\alpha \in \Delta_r^+} (\zeta_{1\alpha})^{K_{1\alpha}}(\zeta_{2\alpha})^{K_{2\alpha}} \cdots (\zeta_{N-1,\alpha})^{K_{N-1,\alpha}}, \tag{5.6.24}$$

in which the non-negative integers $K_{A\alpha}$, $A = 0, \ldots, N-1$, $\alpha \in \Delta_r^+$, satisfy the system of equations

$$\sum_{\alpha \in \Delta_r^+} \sum_{A=1}^{N-1} \alpha K_{A\alpha} = \sum_{i \in N_r} M^i \pi_i. \tag{5.6.25}$$

The number of solutions of the system (5.6.25) for $K_{A\alpha}$ for given M^i determines the dimension of the space $\Phi_M\{F\}$. We see that in all cases it is finite.

Now that we have at our disposal the explicit form of the operators $K_r(\lambda)$ and the function spaces $\Phi_M\{F\}$, we can construct differential analogues of the spectral equations (5.6.11). For this, it is sufficient to project $K_r(\lambda)$ onto the spaces $\Phi_M\{F\}$. We go over to the new variables in accordance with the formulas

$$\zeta_A = x_A - x_N, \quad A = 1, \dots, N-1; \quad \zeta_N = x_N; \tag{5.6.26}$$

$$\frac{\partial}{\partial x_A} = \frac{\partial(x_A - x_N)}{\partial x_A}\frac{\partial}{\partial \zeta_A}, \quad A = 1, \dots, N-1;$$

$$\frac{\partial}{\partial x_A} = \frac{\partial}{\partial \zeta_N} + \sum_{A=1}^{N-1}\frac{\partial(x_A - x_N)}{\partial x_N}\frac{\partial}{\partial \zeta_A}. \tag{5.6.27}$$

By virtue of the "translational" invariance of the operator $K_r(\lambda)$ and the functions of the space $\Phi_M\{F\}$, they do not depend explicitly on ζ_N. Therefore, in equations (5.6.26) and (5.6.27) we can set $\zeta_N = 0$, simultaneously omitting the derivates with respect to ζ_N. This gives, in place of (5.6.26) and (5.6.27),

$$x_A = z_A, \quad A = 1, \dots, N-1; \quad x_N = 0; \tag{5.6.28}$$

$$\frac{\partial}{\partial x_A} = \frac{\partial}{\partial \zeta_A}, \quad A = 1, \dots, N-1;$$

$$\frac{\partial}{\partial x_N} = -\sum_{A=1}^{N-1} t_+^+(\zeta_A)\frac{\partial}{\partial \zeta_A} \tag{5.6.29}$$

(in obtaining the last formula, we have used the definition of the matrix $\hat{t}(\zeta)$ given in section 5.3). In addition, we have the equations

$$\hat{t}_+^+(0) = \hat{1}, \quad \hat{t}_0^+(0) = 0, \quad \hat{t}_-^+(0) = 0,$$
$$\hat{t}_+^0(0) = 0, \quad \hat{t}_0^0(0) = \hat{0}, \quad \hat{t}_-^+(0) = 0. \tag{5.6.30}$$

Using the relations (5.6.28)–(5.6.30), we obtain for $K_r(\lambda)$:

$$K_r(\lambda) = \sum_{A,B=1}^{N-1}\frac{\sum_{a,b\in\Omega}g^{ab}I_{Aa}I_{Bb}}{(\lambda - \sigma_A)(\lambda - \sigma_B)}$$

$$+2\sum_{A=1}^{N-1}\frac{\sum_{a,b\in\Omega}g^{ab}I_{Aa}J_b}{(\lambda - \sigma_A)(\lambda - \sigma_N)} + \frac{\sum_{a,b\in\Omega}g^{ab}J_aJ_b}{(\lambda - \sigma_N)^2}. \tag{5.6.31}$$

Here

$$I_{Aa} = \sum_{\alpha \in \Delta_r^+} t_a^\alpha(\zeta_A)\frac{\partial}{\partial\zeta_A^\alpha} + \sum_{i \in N_r} t_A^i(\zeta_A)f_{Ai}, \quad a \in \Omega_r; \qquad (5.6.32)$$

$$\begin{aligned}
J_\alpha &= -\sum_{B=1}^{N-1}\sum_{\beta \in \Delta_r^+} t_\alpha^\beta(\zeta_B)\frac{\partial}{\partial\zeta_B^\beta}, \quad \alpha \in \Delta_r^+; \\
J_i &= f_{Ni}, \quad i \in N_r; \\
J_\alpha &= 0, \quad \alpha \in \Delta_r^-.
\end{aligned} \qquad (5.6.33)$$

The obtained operators (5.6.31) act on the space of homogeneous polynomials in ζ_A, $A = 1,\ldots,N-1$, with basis (5.6.24). In accordance with the results of section 5.3, the elements of this space can be represented in the form

$$\phi = \prod_{i \in N_r} (\zeta_{N-1}^{\pi_i})^{M_i}\psi(\eta,\nu), \qquad (5.6.34)$$

where η is the vector of dimension $(N-2)(d_r-r)/2$ with components

$$\eta_A^\alpha = \frac{\zeta_A^\alpha}{\prod_{i \in N_r}(\zeta_{N-1}^{\pi_i})^{(\alpha,\pi_i)}}, \quad A = 1,\ldots,N-2, \quad \alpha \in \Delta_r^+, \qquad (5.6.35)$$

and ν is the $((d_r-r)/2)$-dimensional vector with components

$$\nu^\alpha = \frac{\zeta_{N-1}^\alpha}{\prod_{i \in N_r}(\zeta_{N-1}^{\pi_i})^{(\alpha,\pi_i)}}, \quad \alpha \in \Delta_r^+. \qquad (5.6.36)$$

For this vector, the only trivial components are the $(d_r - 3r)/2$ components with $\alpha \in \Delta_r^+ - \Pi_r^+$, where Π_r^+ is the set of simple roots of the algebra $\mathcal{L}_r$. The remaining r components with $\alpha \in \Pi_r^+$ are equal to unity. The functions $\psi(\eta,\nu)$ in (5.6.34) are polynomials in the $[(N-2)(d_r-r) + (d_r-3r)]/2$ variables η and ν. The form of these polynomials can be determined from the condition that the elements (5.6.34) belong to the spaces $\Psi_M\{F\}$. We denote the set of allowed polynomials $\psi(\eta,\nu)$ by $\Psi_M\{F\}$.

It follows from the invariance of the spaces $\Psi_M\{F\}$ that the result of applying (5.6.31) to functions of the form (5.6.34) must have the same form. This enables us to write

$$K_r(\lambda)\prod_{i \in N_r}(\zeta_{N-1}^{\pi_i})^{M^i}\phi(\eta,\nu)$$

$$= \prod_{i \in N_r}(\zeta_{N-1}^{\pi_i})^{M^i} K_r(M,\lambda;\eta,\nu)\phi(\eta,\nu), \qquad (5.6.37)$$

where $K_r(M, \lambda; \eta, \nu)$ is a differential operator that acts on the space $\Psi_M\{F\}$ and depends on the non-negative integers M^i. Accordingly, equation (5.6.11) can be written in the form

$$K_r(M, \lambda; \eta, \nu)\phi(\eta, \nu) \;=\; E_r(M, \lambda)\phi(\eta, \nu),$$
$$\phi(\eta, \nu) \;\in\; \Psi_M\{F\}. \tag{5.6.38}$$

If we denote by Ψ the set of all analytic functions of the variables η and ν, then the equations

$$K_r(M, \lambda; \eta, \nu)\phi(\eta, \nu) \;=\; E_r(M, \lambda)\phi(\eta, \nu),$$
$$\phi(\eta, \nu) \;\in\; \Psi \tag{5.6.39}$$

will obviously have in Ψ only a finite number of exact solutions, which are determined by the Bethe functions and lie in the class of polynomials in the variables η and ν of the form $\Psi_M\{F\}$.

Thus, we arrive at an infinite series of quasi-exactly solvable equations associated with Gaudin models on the representations of the algebras $\mathcal{L}_r$ with highest weights $F(\lambda)$.

It now remains to find the explicit form of the operators $K_r(M, \lambda; \eta, \nu)$. For this, we must go over in (5.6.31) to new variables in accordance with the formulas

$$\left.\begin{array}{l}
\zeta_{N-1}^{\pi^i} = \zeta^i, \; i \in N_r; \\[4pt]
\zeta_A^\alpha = \prod_{i \in N_r} (\zeta^i)^{(\alpha, \pi^i)} \eta_A^\alpha, \; A = 1, \ldots, N-2; \; \alpha \in \Delta_r^+; \\[4pt]
\zeta_{N-1}^\alpha = \prod_{i \in N_r} (\zeta^i)^{(\alpha, \pi^i)} \nu^\alpha, \; \alpha \in \Delta_r^+ - \Pi_r^+;
\end{array}\right\} \tag{5.6.40}$$

$$\frac{\partial}{\partial \zeta_{N-1}^{\pi^i}} \;=\; \frac{\partial}{\partial \zeta^i} - \sum_{A=1}^{N-2} \sum_{\alpha \in \Delta_r^+} \frac{(\alpha, \pi^i)}{\zeta^i} \eta_A^\alpha \frac{\partial}{\partial \eta_A^\alpha}$$

$$- \sum_{\alpha \in \Delta_r^+ - \Pi_r^+} \frac{(\alpha, \pi^i)}{\zeta^i} \nu^\alpha \frac{\partial}{\partial \nu^\alpha}, \; i \in N_r;$$

$$\frac{\partial}{\partial \zeta_A^\alpha} \;=\; \prod_{i \in N_r} (\zeta^i)^{-(\alpha, \pi^i)} \frac{\partial}{\partial \eta_A^\alpha}, \; A = 1, \ldots, N-2, \; \alpha \in \Delta_r^+;$$

$$\frac{\partial}{\partial \zeta_{N-1}^\alpha} \;=\; \prod_{i \in N_r} (\zeta^i)^{-(\alpha, \pi^i)} \frac{\partial}{\partial \nu^\alpha}, \; \alpha \in \Delta_r^+ - \Pi_r^+. \tag{5.6.41}$$

Since the operator (5.6.31) is "scale invariant", it can contain a dependence on ζ^i and $\partial/\partial\zeta^i$ only in the form of the combinations $\zeta^i(\partial/\partial\zeta^i)$, $i \in N_r$. As a result of the projection of (5.6.31) onto the class of functions of the form

(5.6.34), these combinations are replaced by numbers M^i. In practice, this is done as follows. We use the scaling property of the coefficient matrices $t_a^\alpha(\zeta)$ and $t_a^i(\zeta)$:

$$
\begin{aligned}
t_a^\alpha(\zeta_A) &= \prod_{i\in N_r}(\zeta^i)^{(\alpha-a,\pi^i)}t_a^\alpha(\eta_A); \\
t_a^i(\zeta_A) &= \prod_{i\in N_r}(\zeta^i)^{-(a,\pi^i)}t_a^i(\eta_A); \\
t_a^\alpha(\zeta_{N-1}) &= \prod_{i\in N_r}(\zeta^i)^{(\alpha-a,\pi^i)}t_a^\alpha(\nu); \\
t_a^i(\zeta_{N-1}) &= \prod_{i\in N_r}(\zeta^i)^{-(a,\pi^i)}t_a^i(\nu).
\end{aligned}
\tag{5.6.42}
$$

Here, $A = 1,\ldots,N-2$, $\alpha \in \Delta_r^+$, $i \in N_r$, $\alpha \in \Omega_r$. Substituting the formulas (5.6.40), (5.6.41) in (5.6.31) and (5.6.32)–(5.6.33), and replacing the operators $\zeta^i(\partial/\partial\zeta^i)$ by the numbers M^i, we find

$$
\begin{aligned}
K_r(M,\lambda;\eta,\nu) &= \sum_{A,B=1}^{N-2}\frac{\sum_{a,b\in\Omega_r}g^{ab}I_{Aa}I_{Bb}}{(\lambda-\sigma_A)(\lambda-\sigma_B)} \\
&+ \sum_{A=1}^{N-2}\frac{\sum_{a,b\in\Omega_r}g^{ab}[\widetilde{K}_b I_{Aa}+I_{Ab}K_a]}{(\lambda-\sigma_A)(\lambda-\sigma_{N-1})} \\
&+ \frac{\sum_{a,b\in\Omega_r}g^{ab}\widetilde{K}_a K_b}{(\lambda-\sigma_{N-1})^2}+2\sum_{A=1}^{N-2}\frac{\sum_{a,b\in\Omega_r}g^{ab}I_{Aa}L_b}{(\lambda-\sigma_A)(\lambda-\sigma_N)} \\
&+ 2\frac{\sum_{a,b\in\Omega_r}g^{ab}I_{Aa}L_b}{(\lambda-\sigma_A)(\lambda-\sigma_N)} \\
&+ 2\frac{\sum_{a,b\in\Omega_r}g^{ab}\widetilde{K}_a L_b}{(\lambda-\sigma_N)(\lambda-\sigma_N)}+\frac{\sum_{a,b\in\Omega_r}g^{ab}L_a L_b}{(\lambda-\sigma_N)^2},
\end{aligned}
\tag{5.6.43}
$$

where

$$
I_{Aa} = \sum_{\alpha\in\Delta_r^+}t_a^\alpha(\eta_A)\frac{\partial}{\partial\eta_A^\alpha}+\sum_{i\in N_r}t_a^i(\eta_A)f_{Ai},\quad a\in\Omega_r,
$$
$$
A=1,\ldots,N-2;
\tag{5.6.44}
$$

$$
\begin{aligned}
K_a\sum_{i\in N_r}t_a^{\pi_i}(\nu)&\left[M^i-\sum_{B=1}^{N-2}\sum_{\beta\in\Delta_r^+}(\beta,\pi^i)\eta_B^\beta\frac{\partial}{\partial\eta_B^\beta}\right. \\
&\left.-\sum_{\beta\in\Delta_r^+-\Pi_r^+}(\beta,\pi^i)\nu^\beta\frac{\partial}{\partial\nu^\beta}\right]+\sum_{\alpha\in\Delta_r^+-\Pi_r^+}t_a^\alpha(\nu)\frac{\partial}{\partial\nu^\alpha}
\end{aligned}
$$

$$+ \sum_{i \in N_r} t_a^i(\nu) f_{N-1,i}, \quad a \in \Omega_r; \qquad (5.6.45)$$

$$\tilde{K}_a = K_a - \sum_{i \in N_r} t_a^{\pi_i}(\nu)(\alpha, \pi^i), \quad a \in \Omega_r; \qquad (5.6.46)$$

$$
L_\alpha = \ - \sum_{B=1}^{N-2} \sum_{\beta \in \Delta_r^+} t_\alpha^\beta(\eta_B) \frac{\partial}{\partial \eta_B^\beta}
$$
$$
- \sum_{i \in N_r} t_\alpha^{\pi_i}(\nu) \left(M^i - \sum_{B=1}^{N-2} \sum_{\beta \in \Delta_r^+} (\beta, \pi^i) \eta_B^\beta \frac{\partial}{\partial \eta_B^\beta} \right.
$$
$$
\left. - \sum_{\beta \in \Delta_r^+ - \Pi_r^+} (\beta, \pi^i) \nu^\beta \frac{\partial}{\partial \nu_B^\beta} \right)
$$
$$
- \sum_{\beta \in \Delta_r^+ - \Pi_r^+} t_\alpha^\beta(\nu) \frac{\partial}{\partial \nu_\beta}, \quad \alpha \in \Delta_r^+;
$$

$$L_i = f_{N_i}, \quad i \in N_r; \quad L_\alpha = 0; \quad \alpha \in \Delta_r^-. \qquad (5.6.47)$$

With this, we complete the construction of the multi-dimensional quasi-exactly solvable differential equations of second order associated with the completely integrable Gaudin models in the case when the functions $F_i(\lambda)$, $i \in N_r$, which play the part of highest weights of the representations of the Gaudin algebra, are rational and non-degenerate. The transition to the degenerate rational case can be made as follows.

We note that all rational functions $F_i(\lambda)$, $i \in N_r$, admit representation in the form

$$F_i(\lambda) = \sum_A \tilde{f}_{Ai} \omega^A(\lambda), \qquad (5.6.48)$$

where $\omega^A(\lambda)$ are elementary rational functions of the form $(\lambda - \sigma)^{-n}$, $\sigma \in C$, $n \in N$. The index A labelling them is in fact a multiple index $A = (\sigma, n)$. The sum in (5.6.48) is assumed to be finite. In the decomposition (5.6.48) we have used only decreasing elementary rational functions because the components of the highest weights $F_i(\lambda)$ must, by hypothesis, be regular at infinity.

For the functions $\omega^A(\lambda)$ we have the composition theorems

$$\frac{\omega^A(\lambda) - \omega^A(\mu)}{\lambda - \mu} = \sum_{B,C} C_{BC}^A \omega^B(\lambda) \omega^C(\mu), \tag{5.6.49}$$

$$\omega^B(\lambda) \omega^C(\lambda) = \sum_A D_A^{BC} \omega^A(\lambda), \tag{5.6.50}$$

in which C_{BC}^A and D_A^{BC} are certain structure constants. The sums over A, B, and C in (5.6.49) and (5.6.50) are also assumed to be finite.

In accordance with (5.5.3) and (5.5.1), the generators of the Gaudin algebra should be sought in an analogous form:

$$I_a(\lambda) = \sum_A \tilde{I}_{Aa} \omega^A(\lambda). \tag{5.6.51}$$

Substituting (5.6.51) in the commutation relations (5.5.1) and using (5.6.49), we obtain commutation relations directly for the coefficient operators $\tilde{I}_{Aa}$:

$$[\tilde{I}_{Aa}, \tilde{I}_{Bb}] = \sum_{c \in \Omega_r} \Gamma_{ab}^c \sum_C C_{AB}^C \tilde{I}_{Cc}. \tag{5.6.52}$$

By virtue of the finiteness of the sum over A in (5.6.51), the operators $\tilde{I}_{Aa}$ form a finite-dimensional Lie algebra. Substituting the expansion (5.6.51) into the expression (5.5.7) for the operators $K_r(\lambda)$, we can reduce them to the form

$$K_r(\lambda) = \sum_{a,b \in \Omega_r} \sum_{A,B,C} g^{ab} D_C^{AB} \omega^C(\lambda) \tilde{I}_{Aa} \tilde{I}_{Bb}. \tag{5.6.53}$$

These are the hamiltonians of magnets based on the finite-dimensional Lie algebra (5.6.52), which can be interpreted as a certain contraction of the algebra $\mathcal{L}_r \oplus \ldots \oplus \mathcal{L}_r$ (N times) if the functions $F_i(\lambda)$, $i \in N_r$, are obtained as a result of the degeneracy of functions of the form (5.6.12).

To construct differential realizations of the operators $\tilde{I}_{Aa}$, we first consider the procedure for going over from the non-degenerate functions $F_i(\lambda)$ to degenerate functions:

$$\sum_{A=1}^N \frac{f_{Ai}}{\lambda - \sigma_a} \to \sum_{A=1}^N \tilde{f}_{Ai} \omega^A(\lambda). \tag{5.6.54}$$

To realize this procedure, we must make a suitable linear substitution:

$$f_{Ai} = \sum_{B=1}^N C_A^B(\sigma_1, \ldots, \sigma_N) \tilde{f}_{Bi}, \quad i \in N_r, \tag{5.6.55}$$

and we must then let the parameters $\sigma_1, \ldots, \sigma_N$ tend to their limiting values, merging all or some of the simple poles of the non-degenerate functions $F_i(\lambda)$. The explicit form of the matrix C_A^B is determined by the specific form of the degeneracy, i.e., by the requirement that the result be identical to the right-hand side of equation (5.6.54).

A similar procedure must be carried out for the operators $I_a(\lambda)$:

$$\sum_{A=1}^{N} \frac{I_{Aa}}{\lambda - \sigma_A} \to \sum_{A=1}^{N} \tilde{I}_{Aa} \omega^A(\lambda). \qquad (5.6.56)$$

Here, it is most convenient to proceed from the non-degenerate operators $I_{A\alpha}$, $\alpha \in \Delta_r^+$, whose differential realizations contain in accordance with equation (5.4.48) the operators $\partial/\partial x_A^\alpha$ as terms. Requiring that the degenerate operators contain as terms analogous operators of differentiation, but now with respect to the new variables, $\partial/\partial \tilde{x}_A^\alpha$, we arrive at the need to consider the limiting process

$$\sum_{A=1}^{N} \frac{\partial/\partial x_A^\alpha}{\lambda - \sigma_A} \to \sum_{A=1}^{N} (\partial/\partial \tilde{x}_A^\alpha) \omega^A(\lambda), \qquad (5.6.57)$$

the structure of which is completely analogous to (5.6.54). This enables us to write down the connection between the derivates $\partial/\partial x_A^\alpha$ and $\partial/\partial \tilde{x}_A^\alpha$:

$$\frac{\partial}{\partial x_A^\alpha} = \sum_{B=1}^{N} C_A^B(\sigma_1, \ldots, \sigma_N) \frac{\partial}{\partial \tilde{x}_B^\alpha}, \quad \alpha \in \Delta_r^+, \qquad (5.6.58)$$

and express the old variables x_A^α in terms of the new ones $\tilde{x}_A^\alpha$:

$$x_A^\alpha = \sum_{B=1}^{N} \tilde{C}_A^B(\sigma_1, \ldots, \sigma_N) \tilde{x}_B^\alpha, \quad \alpha \in \Delta_r^+. \qquad (5.6.59)$$

Here, $\tilde{C}_A^B$ is the matrix that is the inverse of C_A^B. Substituting (5.6.55), (5.6.58), and (5.6.59) in the expressions for the remaining operators I_{Aa} and making the necessary transition to the limiting values $\sigma_1, \ldots, \sigma_N$, we arrive at differential forms of the degenerate operators $\tilde{I}_{Aa}$ that realize the representation of the contracted algebra $\mathcal{L}_r \oplus \ldots \oplus \mathcal{L}_r$ (N times).

3. Coulomb analogy. We return to the algebraic equations (5.6.23), from which we can find the spectra of the Gaudin magnets and the quasi-exactly solvable problems associated with them. The roots of these equations, i.e., the numbers $\zeta_{i,q}$, $q = 1, \ldots, M^i$, $i = 1, \ldots, r$, are, in general, complex. Therefore, it is meaningful to introduce the two-dimensional vectors

$$\zeta_{i,q} = (\mathrm{Re}\ \zeta_{i,q}), \quad q = 1, \ldots, M^i, \quad i = 1, \ldots, r. \qquad (5.6.60)$$

If with them we introduce the notation

$$U_i(\zeta) \equiv \mathrm{Re} \int F_i(\zeta)\, d\zeta, \quad i = 1, \ldots, r, \qquad (5.6.61)$$

then the system (5.5.44) can be interpreted as the condition for an extremum of the function

$$
U(\zeta) = \; - \sum_{i,k=1}^{r} \sum_{q=1}^{M^i} \sum_{p=1}^{M^k} (\pi_i, \pi_k) \ln |\zeta_{i,q} - \zeta_{k,p}|
$$
$$
- \sum_{i=1}^{r} \sum_{p=1}^{M^i} U_i(\zeta_{i,p}). \qquad (5.6.62)
$$

We now note that the function (5.6.62) is none other than the potential of a two-dimensional logarithmic many-particle Coulomb system in an external field. There are altogether r species of particles, labelled by the index $i = 1, \ldots, r$. There are M^i of the particles of the species i. The numbers $\zeta_{i,p}$ denote coordinates of these particles, and the simple roots of the Lie algebra, π_i, play the role of their "vector" charges. Particles of the same species have the same vector charges, and therefore repel each other ($(\pi_i, \pi_i) > 0$), while particles of different species attract each other ($(\pi_i, \pi_k) \le 0$, $i \ne k$). In addition, there are r potentials $-U_i(\zeta_i)$, each of which acts only on the particles of a definite species.

The Coulomb analogy is extremely helpful. We have already had the opportunity to demonstrate this in chapters 1 and 2. The analogy makes it possible in a qualitative analysis of the solutions of quasi-exactly solvable equations to use our classical intuition, which is obviously much more developed than the quantum mechanical intuition.

5.7 Reduction of quasi-exactly solvable differential equations to the Schrödinger form

In previous sections we have formulated a rather general method for constructing multi-dimensional second-order differential operators

$$H = \left\{ \sum_{i,k=1}^{d} P_{ik}(x) \frac{\partial^2}{\partial x_i \partial x_k} + \sum_{i=1}^{d} Q_i(x) \frac{\partial}{\partial x_i} \right\} \qquad (5.7.1)$$

having an infinite number of exactly calculable eigenvalues and eigenfunctions. It would be very temping to interpret operators H as hamiltonians of quasi-exactly solvable models of quantum mechanics. However, the non-hermiticity of these operators and the non-normalizability of their solutions makes such an interpretation impossible.

In order to try to improve the situation, we can replace the operators H by the homogeneously transformed operators $H_S = S^{-1}HS$, also having exactly calculable spectra. Requiring the hermiticity of operators H_S

$$H_S^+ = H_S \qquad (5.7.2)$$

we get the equation for S

$$(SS^+)H^+ = H(SS^+), \qquad (5.7.3)$$

which is solvable in the class of some integral operators. The problem, however, is that not all solutions of this equation are admissible. Indeed, we cannot consider transformations changing the order of the differential operator H or reducing it to an integral form. It is quite obvious that for the operator H_S to be again a second-order differential operator, the transformation S must have the form of a multiplication by an ordinary function of x. Choosing this function in the form

$$S = \{P(x)\}^{1/4}\{U(x)\}^{1/2}, \qquad (5.7.4)$$

where $P(x) \equiv \det\|P_{ik}(x)\|$, and $U(x)$ is an unknown function (vanishing on the boundary of a certain domain $\Omega_d \subset R_d$ in which the matrix $P_{ik}(x)$ is positive definite), and substituting (5.7.4) and (5.7.1) into equation (5.7.3), we can reduce it to the form

$$\sum_{k=1}^{d} P_{ik}(x)\frac{\partial U(x)}{U(x)\partial x_k} + \sum_{k=1}^{d} \frac{\partial}{\partial x_k}P_{ik}(x) = Q_i(x), \qquad i = 1,\ldots,d. \quad (5.7.5)$$

It is not difficult to verify that if conditions (5.7.5) are satisfied, then the operator H_S becomes

$$H_S = \sqrt{P(x)} \sum_{i,k=1}^{d} \frac{\partial}{\partial x_i}\left(\frac{P_{ik}(x)}{\sqrt{P(x)}}\frac{\partial}{\partial x_i}\right) + V(x), \qquad (5.7.6)$$

and, thus, can be interpreted as the hamiltonian of a certain quantum system defined in the domain Ω_d of a d-dimensional (in general, curved) manifold with the metric $\|g_{ik}\| = \|P_{ik}\|^{-1}$. The function $V(x)$ (playing the role of the potential) depends explicitly on $U(x)$ and is described by the formula

$$
V(x) = -P^{\frac{1}{4}}(x)U^{\frac{1}{2}}(x)\left\{ \sum_{i,k=1}^{d} P_{ik}(x)\frac{\partial^2}{\partial x_i \partial x_k} \right.
$$

$$
\left. + \sum_{i=1}^{d} Q_i(x)\frac{\partial}{\partial x_i} \right\} P^{-\frac{1}{4}}(x)U^{-\frac{1}{2}}(x). \tag{5.7.7}
$$

The spectral equation for H_S is quasi-exactly solvable by construction. The eigenvalues of the operator H_S coincide, obviously, with the eigenvalues of the initial operator H, while the eigenfunctions of H_S are connected with the eigenfunctions of H by the formula:

$$
\Psi_S(x) = \{P(x)\}^{1/4}\{U(x)\}^{1/2}\Psi(x). \tag{5.7.8}
$$

The condition of the normalizability of functions Ψ_S,

$$
\int_{\Omega_d} \frac{\Psi_S^2(x)}{\sqrt{P(x)}}\, dx < \infty, \tag{5.7.9}
$$

being rewritten in terms of initial eigenfunctions $\Psi(x)$ is

$$
\int_{\Omega_d} U(x)\Psi^2(x)\, dx < \infty. \tag{5.7.10}
$$

Thus, we see that, in order to guarantee the physical sensibility of the transformed operator H_S, it is sufficient to find function $U(x)$ satisfying both the conditions (5.7.5) and (5.7.10) and vanishing on the boundary of the domain Ω_d in which the metric tensor $P_{ik}(x)$ is positive definite.

Unfortunately, this is not always possible. The main difficulty is to solve the system (5.7.5) which, for $d > 1$, is over-determined and the compatibility requirement for which imposes quite stringent constraints on the allowed form of the functions $P_{ik}(x), Q_i(x)$, and $U(x)$.

Another difficulty is that even if the function $U(x)$ satisfying the system (5.7.5) exists, this does not necessarily mean that the boundary conditions for $U(x)$ in Ω_d are automatically satisfied.

Our assertion lies in the fact that both these difficulties[3] can easily be overcome by dropping the requirement that the dimension of the space

---

[3] These difficulties do not appear in the case of quasi-exactly solvable equations associated with the $sl(2)$ Gaudin model (see section 5.2).

in which the resulting Schrödinger problem is formulated is d (Ushveridze 1989c). In fact, consider the equation for the initial operator H

$$\left\{ \sum_{i,k=1}^{d} P_{ik}(x)\frac{\partial^2}{\partial x_i \partial x_k} + \sum_{i=1}^{d} Q_i(x)\frac{\partial}{\partial x_i} \right\} \Psi(x) = E\Psi(x), \qquad (5.7.11)$$

and rewrite it in the $(d+1)$-dimensional form:

$$\left\{ \sum_{i,k=0}^{d} P_{ik}(x,x_0)\frac{\partial^2}{\partial x_i \partial x_k} + \sum_{i=0}^{d} Q_i(x,x_0)\frac{\partial}{\partial x_i} \right\} \Psi(x) = E\Psi(x). \quad (5.7.12)$$

Here $P_{ik}(x,x_0) \equiv P_{ik}(x)$ and $Q_i(x,x_0) \equiv Q_i(x)$ for all i, $k = 1,\ldots,d$, and $P_{i0}(x,x_0)$ and $Q_0(x,x_0)$ are arbitrary functions of $x = (x_1,\ldots,x_d)$ and of the newly introduced, extra variable x_0. Since equation (5.7.12) has the same form as (5.7.11), but is formulated in $(d+1)$-dimensional space, it can be reduced to the $(d+1)$-dimensional Schrödinger equation if a function $U(x,x_0)$ is found for which the $(d+1)$-dimensional analogues of (5.7.5) are satisfied:

$$\sum_{k=0}^{d} P_{ik}(x,x_0)\frac{\partial U(x,x_0)}{U(x,x_0)\partial x_k} + \sum_{k=0}^{d} \frac{\partial}{\partial x_k} P_{ik}(x,x_0) = Q_i(x,x_0),$$

$$i = 0,1,\ldots,N. \qquad (5.7.13)$$

In contrast to equations (5.7.5), the system of equations (5.7.13) can always be solved, since the components $P_{i0}(x,x_0)$ and $Q_0(x,x_0)$ are arbitrary. The solutions depend on two arbitrary functions, for which it is convenient to choose the function $P_{00}(x,x_0)$, and also the function $U(x,x_0)$, which *a priori* ensures the normalizability of the wavefunctions, and zero boundary conditions for them in a given domain Ω_{d+1} of $(d+1)$-dimensional space in which the spectral problem is formulated. In this case the other unknown functions $P_{i0}(x,x_0)$, $i = 1,\ldots,N$, and $Q_0(x,x_0)$ are found explicitly. Rewriting (5.7.13) as a system

$$\sum_{k=1}^{d} P_{0k}(x,x_0)\frac{\partial \ln U(x,x_0)}{\partial x_k} + P_{00}(x,x_0)\frac{\partial \ln U(x,x_0)}{\partial x_0}$$

$$+ \sum_{k=1}^{d} \frac{\partial}{\partial x_k} P_{0k}(x,x_0) + \frac{\partial P_{00}(x,x_0)}{\partial x_0} = Q_0(x,x_0),$$

$$(5.7.14)$$

$$\sum_{k=1}^{d} P_{ik}(x)\frac{\partial \ln U(x,x_0)}{\partial x_k} + P_{i0}(x,x_0)\frac{\partial \ln U(x,x_0)}{\partial x_0}$$

$$+ \sum_{k=1}^{d} \frac{\partial}{\partial x_k} P_{ik}(x) + \frac{\partial P_{i0}(x,x_0)}{\partial x_0} = Q_i(x), \qquad i = 1,\dots,d$$

$$(5.7.15)$$

and fixing $U(x,x_0)$ and $P_{00}(x,x_0)$, we find

$$P_{i0}(x,x_0) \;=\; U^{-1}(x,x_0)\left\{ \sum_{k=1}^{d} P_{ik}(x)\frac{\partial}{\partial x_k} \right.$$

$$\left. + \; \sum_{k=1}^{d} \frac{\partial}{\partial x_k} P_{ik}(x) - Q_i(x) \right\} \int U(x,x_0)\, \mathrm{d}x_0 \quad (5.7.16)$$

and

$$Q_0(x,x_0) = \left\{ \frac{\partial}{\partial x_0} + \frac{\partial \ln U(x,x_0)}{\partial x_0} \right\} P_{00}(x,x_0)$$

$$+ \sum_{i=1}^{d} \left\{ \frac{\partial}{\partial x_i} + \frac{\partial \ln U(x,x_0)}{\partial x_i} \right\} P_{i0}(x,x_0). \qquad (5.7.17)$$

Thus, we see that every equation of the type (5.7.11) can formally be reduced to the covariant Schrödinger form. Formally — because this form does not necessarily guarantee the positive definiteness of the metric tensor $P_{ik}(x,x_0)$ in the chosen domain Ω_{d+1}. Nevertheless, it is easily seen that the problem of constructing Schrödinger equations with positive definite metric can also be completely solved. Indeed, we know that the neccessary and sufficient condition of positive definiteness of the $(d+1)$-dimensional metric tensor $P_{ik}(x,x_0)$ in the domain Ω_{d+1} is the positive definiteness of the initial d-dimensional tensor $P_{ik}(x)$ and the positivity of the $(d + 1)$-dimensional determinant $P(x,x_0) = \det\|P_{ik}(x,x_0)\|$ in Ω_{d+1}. It is quite obvious that it is always possible to satisfy both these conditions by an appropriate choice of the domain Ω_{d+1} and the functions $U(x,x_0)$ and $P_{00}(x,x_0)$ determining the form of the metric tensor in it.

5.8   Conclusions. Dealgebraization of the method and prospects

Thus, we have completed the exposition of our approach to the problem of quasi-exactly solvability in non-relativistic quantum mechanics. The method developed is suitable for constructing both one-dimensional and multi-dimensional quasi-exactly solvable differential equations, which, using the procedure described in section 5.7, can always be reduced to equations of Schrödinger type. The approach is algebraic. The original objects in it are completely integrable Gaudin models based on the various simple Lie algebras, and they are exactly solvable in the framework of the algebraic Bethe *ansatz*. The global symmetry of these models makes it possible to carry out in them (or, rather, in the differential forms of the corresponding integral equations) a partial separation of the variables, after which these equations become quasi-exactly solvable. Using the Bethe *ansatz* equations describing the spectra of the quasi-exactly solvable quantum mechanical models obtained in this manner, one can show that these models are equivalent to classical models of two-dimensional Coulomb systems in an external field. This connection between three completely different, at first glance, physical systems – models of magnets based on Lie algebras, quasi-exactly solvable quantum mechanical models, and the classical many-particle Coulomb problem – was noted earlier, but we restricted ourselves to a discussion of the case of the algebras $sl(2)$. We now see that the connection also holds in the general case.

An interesting feature of the approach discussed here is that by its very essence it contains a possibility of further generalization. This assertion is important, and therefore it is worth dwelling on it in more detail.

We begin with this question: What role in the approach is played by the complete integrability of the model of a Gaudin magnet? At first glance, everything is founded upon it. However, on closer examination it becomes obvious that its role reduces merely to the possibility of representing the result in a closed Bethe form. This circumstance is undoubtedly helpful, since the Bethe form of expression is the most convenient for carrying out various limiting processes, for example, the passage to the infinite-dimensional ($N \to \infty$) or the exactly non-solvable ($M \to \infty$) cases. At the same time, the functional structure of the result remains the same, so that both the pre-limit and the limit models can be interpreted from the point of view of the Coulomb analogy. However, these facts have only a secondary nature. To the main question, that of whether integrability has fundamental significance for quasi-exactly solvability (i.e., for the possibility of algebraization of the spectral problem), one can answer with confidence: no, it does not. To demonstrate this, we consider the model of a magnet based on the algebra $\mathcal{L}_r \oplus \ldots \oplus \mathcal{L}_r$ with hamiltonian

$$H = \sum_{A,B=1}^{N} \sum_{a,b \in \Omega_r} C^{AB} g^{ab} I_{Aa} I_{Bb}, \qquad (5.8.1)$$

in which I_{Aa}, which act at site A, are the generators of the algebra $\mathcal{L}_r$

with highest weights $f_A = \{f_{Ai}\}$, $i \in N_r$, and C^{AB} are arbitrary numerical coefficients. The arbitrariness of C^{AB} means that we do not require integrability of the model (5.8.1). Despite this, it can also be associated with a certain quasi-exactly solvable model by means of the method discussed in the paper, which for this purpose is completely ready (Ushveridze 1990a).

Indeed, the space W on which the operator H acts is the direct product of the spaces W_A of the representations of the algebra $\mathcal{L}_r$. They, in their turn, can be regarded as direct sums of the subspaces $|M_A\rangle$, defined as the sets of vectors of the form $I_{A\alpha_1} \ldots I_{A\alpha_K} |0\rangle$, provided that the roots $\alpha_1, \ldots, \alpha_K$ are negative and their sum is equal to $- \sum_{i \in N_r} M^i \pi_i$. This means that for W we have the decomposition

$$W = \oplus_{M \geq 0} \Phi_M, \tag{5.8.2}$$

where the spaces Φ_M are determined by the formulas

$$\Phi_M = \oplus_{M_1, \ldots, M_N} \{|M_1\rangle \otimes \ldots \otimes |M_N\rangle\}, \tag{5.8.3}$$

subject to the condition that

$$\sum_{A=1}^{N} M_A^i = M^i, \ i \in N_r. \tag{5.8.4}$$

A key property of the spaces Φ_M is that they are all finite dimensional and invariant with respect to the action of the operator H.

Further, the operator H has the global symmetry group $\mathcal{L}_r$ realized by the operators

$$I_a = \sum_{A=1}^{N} I_{Aa}, \ a \in \Omega_r. \tag{5.8.5}$$

It is easy to show that the spaces Φ_M are eigenspaces with respect to the elements of the Cartan subalgebra of the symmetry algebra $\mathcal{L}_r$:

$$I_i \Phi_M = (F_i - M_i) \Phi_M, \tag{5.8.6}$$

where $F_i = \sum_{A=1}^{N} f_{Ai}$. This means that the sets of vectors satisfying the conditions

$$\begin{aligned} I_\alpha \phi &= 0, \ \alpha \in \Delta_r^+; \\ I_i \phi &= (F_i - M_i)\phi, \ i \in N_r; \ \phi \in W, \end{aligned} \tag{5.8.7}$$

certainly belong to Φ_M and are therefore finite dimensional. These sets, which we denote by Ψ_M, are also invariant with respect to H, and we therefore arrive at the infinite series of equations

$$H\phi = E\phi, \quad \phi \in \Psi_M, \quad M \geq 0, \tag{5.8.8}$$

each of which has only a finite number of solutions. The transition from the algebraic form of these equations to the differential form can be realized by the same method as in the integrable case. This transition can be made in two stages. In the first, allowance is made for the translational invariance of the operator (5.8.1), by virtue of which it is reduced to the form

$$H = \sum_{A,B=1}^{N-1} \sum_{\alpha,\beta \in \Delta_r^+} P_{AB}^{\alpha+\beta}(\zeta) \frac{\partial^2}{\partial \zeta_A^\alpha \partial \zeta_B^\beta}$$

$$+ \sum_{a=1}^{N} \sum_{\alpha \in \Delta_r^+} Q_A^\alpha(\zeta) \frac{\partial}{\partial \zeta_A^\alpha}. \tag{5.8.9}$$

Here, $P_{AB}^{\alpha+\beta}$ and Q_A^α are homogeneous polynomials in the variables ζ_A^α consisting of monomials of the form

$$P^{\alpha+\beta} = \{\zeta_{A_1}^{\alpha_1} \ldots \zeta_{A_K}^{\alpha_K}\}, \quad \alpha_1 + \ldots + \alpha_K = \alpha + \beta;$$
$$Q^\alpha = \{\zeta_{A_1}^{\alpha_1} \ldots \zeta_{A_K}^{\alpha_K}\}, \quad \alpha_1 + \ldots + \alpha_K = \alpha. \tag{5.8.10}$$

The spaces Φ_M on which the operator H acts are linear combination of monomials:

$$\Phi_M = \{\zeta_{A_1}^{\alpha_1} \ldots \zeta_{A_K}^{\alpha_K}\}, \quad \alpha_1 + \ldots + \alpha_K = \sum_{i \in N_r} M^i \pi_i. \tag{5.8.11}$$

In the second stage, we take into account the "scale" invariance (homogeneity) of the operator (5.8.1), which makes possible partial separation of the variables ζ in the spectral equation for H. At the same time, we use the *ansatz* (5.6.34), which transforms the spectral equation for (5.8.9) into a differential equation in a smaller number of variables η and ν. It depends explicitly on the non-negative integers M^i and is quasi-exactly solvable by construction.

It follows from the above derivation that the requirement of integrability of the original model (5.8.1) is indeed redundant. However, at the same time we are forced to recognize that this is also true of the entire algebraic structure of the model, i.e., actually the model itself. Indeed, we could with success start with the operator (5.8.9) acting on the space

(5.8.11), taking as $P^{\alpha+\beta}$ and Q^α the most general polynomials of the form (5.8.10). After partial separation of the variables, we would again obtain a quasi-exactly solvable model. It could be objected here that the algebraic nature is implicitly present in equation (5.8.8), since in determining the spaces (5.8.11) and the coefficient functions in (5.8.9) we used the properties of the root system of the algebra $\mathcal{L}_r$. However, it can be shown that this last thread connecting equations of the type (5.8.8) to Lie algebras can also be readily broken. Indeed, let Δ_r^+ be a finite system of vectors of an r-dimensional space, including a basis of r vectors, Π_r^+, such that all the remaining vectors in Δ_r^+ (if there are any) can be decomposed with respect to Π_r^+ with non-negative integer coefficients. The system Δ_r^+ in general is not a root system. However, if in equation (5.8.10) and (5.8.11) the vectors α_i are assumed to be elements of a root system, then all the arguments that reduce equations (5.8.10) to quasi-exactly solvable form remain valid.

Thus, we arrive at a conclusion which at first glance appears paradoxical: In the formulation of the theory of quasi-exact solvability one can get by perfectly well without a concept such as a Lie algebra. The basic principles of this phenomenon can be understood without going beyond the framework of the analytic approach. The abandonment of the language of symmetries not only simplifies the problem, but, as we have seen above, permits its formulation in a much more general form.

Here, however, there may arise a natural question concerning the status of the method of partial algebraization (Shifman and Turbiner 1989), in the formulation of which a Lie algebra, or, rather, finite-dimensional representations of it, play a decisive role. To answer this question, we consider a typical Shifman–Turbiner hamiltonian:

$$H = \sum_{a,b} P_{ab} S^a S^b + \sum_a Q_a S^a, \qquad (5.8.12)$$

in which S^a are the generators of a finite-dimensional representation of some Lie algebra, and P_{ab} and Q_a are arbitrary numerical coefficients. Remember that the spectral equation for (5.8.12) is quasi-exactly solvable because a finite-dimensional representation space in which the operators S^a act is an invariant subspace for the hamiltonian H. If generators S^a are realized as first-order differential operators acting in the space of polynomials, then the hamiltonian H takes the form of a second-order differential operator and we can speak of quasi-exactly solvable differential equations.

If our hamiltonian (5.8.7) is to take the form (5.8.12), it must be possible to represent the operators S^a in the form

$$S^a = \sum_{\alpha,A} P_A^{a\alpha}(\zeta)\frac{\partial}{\partial \zeta_A^\alpha} + Q^a(\zeta), \qquad (5.8.13)$$

where $P^{a\alpha}(\zeta)$ are homogeneous polynomials in ζ formed from monomials of the form (5.8.10). Only such operators close to make a finite-dimensional Lie algebra without taking us outside the space (5.8.11), i.e., realize on it finite-dimensional representations. However, it is readily seen that such a reduction of hamiltonians of the type (5.8.9) to the rotator hamiltonians (5.8.12) is by no means always possible. This could be prevented by the presence of terms of the form $\zeta_C^{\alpha+\beta}\partial^2/(\partial\zeta_A^\alpha\partial\zeta_B^\beta)$, which are not factorizable, i.e., cannot be represented as a product of two operators of the form (5.4.12). This means that the Shifman–Turbiner algebraic approach is not the most general, i.e., it does not exhaust all possible quasi-exactly solvable models.

Of course, it is as yet clearly premature to claim that the dealgebraized version of our approach discussed here lays claim to the greatest generality. Although we do have arguments for such a claim, this question can only be settled at the theorem level of rigour. Thus, there may be "surprises." In turn, this means that in the theory of quasi-exact solvability it is still early to put the final full stop.

Appendix A

The inverse Schrödinger problem and its solution for several given states

A.1 The one-dimensional case. Three states

We discuss in this section three simple analytic methods of constructing one-dimensional Schrödinger equations

$$\left[-\frac{\partial^2}{\partial x^2} + V(x)\right]\psi(x) = E\psi(x) \qquad (A.1.1)$$

having one, two or three exact solutions with *a priori* specified numbering.

Each of these methods naturally splits into two stages. In the first stage a definite algorithm is stated which allows the construction of the formal solution of the problem. In the second stage attempts are made to take into account boundary conditions and also the standard requirements of normalizability and smoothness of wavefunctions.

1. First stage. Formal consideration. As noted in section 1.1, the number of quasi-exactly solvable equations of first order is functionally large. Now we repeat these reasonings and demonstrate that the same is true for the quasi-exactly solvable equations of second and third order.

1. *One explicit solution.* Introducing the logarithmic derivative of the wavefunction, $y(x) = \psi'(x)/\psi(x)$, we rewrite equation (A.1.1) in the Riccati form:

$$y'(x) + y^2(x) + E = V(x). \qquad (A.1.2)$$

It follows from (A.1.2) that the set of first-order quasi-exactly solvable equations of the type (A.1.1) can be parameterized by the pairs $(E, y(x))$.

Indeed, fixing the number E and function $y(x)$, we can use the equality (A.1.2) to restore the potential $V(x)$, for which equation (A.1.1) has one explicit solution E and $\psi(x) = \exp\left\{\int y(x)\,dx\right\}$.

2. *Two explicit solutions.* Let us consider two Riccati equations

$$y_i'(x) + y_i^2(x) + E_i = V(x), \quad i = 1, 2. \tag{A.1.3}$$

Subtracting the first equation from the second one and introducing new functions

$$z(x) = y_2(x) - y_1(x), \quad g(x) = y_2(x) + y_1(x), \tag{A.1.4}$$

we obtain the relation

$$g(x) = -\frac{z'(x) + E_2 - E_1}{z(x)}, \tag{A.1.5}$$

from which it follows that the set of second-order quasi-exactly solvable equations of type (A.1.1) can be parameterized by the triples $(E_1, E_2, z(x))$. Indeed, fixing arbitraily the numbers E_1 and E_2 and the function $z(x)$ one can construct the functions $g(x)$, and then, using formulas (A.1.4) and (A.1.3), restore $y_i(x)$, $i = 1, 2$ and $V(x)$. Thus, we obtain the equation (A.1.1), having, evidently, two explicit solutions E_i and $\psi_i(x) = \exp\left\{\int y_i(x)\,dx\right\}$, $i = 1, 2$.

3. *Three explicit solutions.* Now let us consider three Riccati equations

$$y_i'(x) + y_i^2(x) + E_i = V(x), \quad i = 1, 2, 3. \tag{A.1.6}$$

As in the previous case, we subtract the first equation from the other two

$$[y_i(x) - y_1(x)]' + [y_i(x) - y_1(x)][y_i(x) + y_1(x)] + E_i - E_1 = 0,$$
$$i = 2, 3,$$
$$\tag{A.1.7}$$

and introduce the auxiliary functions

$$z_i(x) = y_i(x) - y_1(x), \quad g_i(x) = y_i(x) + y_1(x), \quad i = 2, 3. \tag{A.1.8}$$

Substituting (A.1.8) into (A.1.7) we find two relations

$$g_i(x) = -\frac{z_i'(x) + E_i - E_1}{z_i(x)}, \quad i = 2, 3, \tag{A.1.9}$$

from which, using (A.1.8), we obtain

$$y_i(x) = \frac{1}{2}\left[z_i(x) - \frac{z_i'(x) + E_i - E_1}{z_i(x)}\right], \quad i = 2,3,$$

(A.1.10a)

$$y_1(x) = \frac{1}{2}\left[z_2(x) - \frac{z_2'(x) + E_2 - E_1}{z_2(x)}\right]$$

$$= -\frac{1}{2}\left[z_3(x) - \frac{z_3'(x) + E_3 - E_1}{z_3(x)}\right].$$

(A.1.10b)

Introducing the function

$$t(x) = z_3(x)/z_2(x),$$

(A.1.11)

we rewrite (A.1.10b) as

$$[t(x) - 1]z_2^2(x) + \frac{t'(x)}{t(x)}z_2(x) + \frac{E_3 - E_1}{t(x)} - (E_2 - E_1) = 0,$$

(A.1.12)

from which we find

$$z_2(x) = \frac{\frac{t'(x)}{t(x)} \pm \sqrt{\left[\frac{t'(x)}{t(x)}\right]^2 - 4[t(x) - 1]\left[\frac{(E_3 - E_1)}{t(x)} - (E_2 - E_1)\right]}}{2[t(x) - 1]}.$$

(A.1.13)

From (A.1.13) it follows that the set of third-order quasi-exactly solvable equations of type (A.1.1) can be parameterized by the quadruples $(E_1, E_2, E_3, t(x))$. Indeed, fixing arbitrarily three numbers E_1, E_2 and E_3 and the function $t(x)$, we can reconstruct $z_2(x)$. Then, after finding the functions $y_i(x)$, $i = 1, 2, 3$ from (A.1.11) and (A.1.10) we can construct three explicit solutions E_i and $\psi_i(x) = \exp\left\{\int y_i(x)\,dx\right\}$, $i = 1, 2, 3$ of equation (A.1.1) with the potential $V(x)$, reconstructed using (A.1.6) (Ushveridze 1988o, 1989c).

Thus, we have obtained an infinitely (functionally) large set of Schrödinger-type equations with one, two or three explicit solutions. Curiously, to write down the explicit forms of these equations it is sufficient to specify only one arbitrary fuunction, namely $y(x)$, $z(x)$, or $t(x)$. Below we shall refer to such functions as generating functions. Of course, when choosing generating functions one must take care that potentials and wavefunctions satisfying equation (A.1.1) are physically sensible. The problem of constructing such functions will be solved below.

2. Second stage. Physical consideration. Let us assume for definiteness that the potential $V(x)$ is regular in a certain finite interval $[a, b]$ and increases near its ends as

$$V(x) \approx \frac{A(A-1)}{(x-a)^2}, \quad x \to a + 0; \quad V(x) \approx \frac{B(B-1)}{(x-b)^2}, \quad x \to b - 0.$$

$$(A.1.14)$$

In this case, the wavefunctions $\psi(x)$ satisfying (A.1.1) also must be regular in interval $[a, b]$ and must vanish at its end points as

$$\psi(x) \sim (x-a)^A, \quad x \to a + 0; \quad \psi(x) \sim (x-b)^B, \quad x \to b - 0.$$

$$(A.1.15)$$

Note also that any wavefunction $\psi(x)$ corresponding to the nth excited energy level must have n nodes (simple zeros) within the interval $[a, b]$ (the oscillator theorem).

In order to guarantee such (physically sensible) behaviour of wavefunctions $\psi(x)$, it is necessary to impose special constraints on the classes of generating functions $y(x)$, $z(x)$ and $t(x)$. A simple analysis of formulas obtained in the first stage of our consideration shows that these constraints have a local character. They imply the existence of certain "critical" points, in whose neighbourhoods the behaviour of generating functions cannot be arbitrary. Below we shall distinguish the "external" and "internal" critical points. In the former a fulfillment of the boundary conditions for $\psi(x)$ is guaranteed. They coincide with the ends of interval $[a, b]$. The latter lie within the interval $[a, b]$ and determine a nodal structure of functions $\psi(x)$. The number and location of these points cannot be completely arbitrary. Our aim is to describe and classify all the possible forms of their dispositions.

First of all, note that each generating function allows only a very limited number of types of internal critical point. They are for $y(x)$, the negative simple poles, for $z(x)$, the negative and positive simple poles, and for $t(x)$, the negative and positive simple poles, the negative and positive simple zeros, and also the four special types of critical point which will be defined below[1]. Such small diversity of these types gives us the possibility to solve the classification problem without difficulty. This can be done

---

[1] Under the sign of critical point we mean here the sign of inclination of the generating function near this point.

by means of a simple graphical method which is based on the following prescriptions.

The class of admissible generating functions allowing N different types $V_1, V_2, \ldots, V_N$ of critical point must be identified with certain oriented graph having N different vertices of the types $V_1, V_2, \ldots, V_N$. If this class contains the functions allowing at least two neighbouring critical points of types V_i (left point) and V_k (right point), then the corresponding vertices V_i and V_k must be connected by the line directed from V_i to V_k. When the types of neighbouring critical point coincide ($i = k$) we obtain a closed loop. Each graph constructed by means of these prescriptions has the beginning vertex V_1 and the end vertex V_n which correspond to points a and b of interval $[a, b]$ in which the Schrödinger problem is considered. Any admissible way from V_1 to V_N along the directed line can be described as monotone variation of coordinate x from a to b. The distribution of the vertices along such a way determines a possible distribution of critical points within the interval $[a, b]$. Thus, the classification problem for the generation functions is reduced to the problem of classification of all allowed ways in the corresponding oriented graphs.

The analysis of formulas (A.1.2)–(A.1.13) obtained above allows us to construct all needed graphs describing the properties of generating functions $y(x)$, $z(x)$, and $t(x)$. Below, in order to distinguish between the vertices of these graphs, we shall use different italic letters labelled by the signs of corresponding critical points. Now let us consider concrete examples.

1. *Quasi-exactly solvable models of first order. Properties of generating function $y(x)$.*

(a) In the vicinity of the points a and b the functions $y(x)$ must behave as $y(x) \approx \frac{A}{x-a}$, $x \to a + 0$ and $y(x) \approx \frac{B}{x-a}$, $x \to b - 0$. (Critical points of types A_- and B_-, respectively.)

(b) Within the interval $[a, b]$ the function $y(x)$ may have simple poles with positive residues. (Critical points of type P_-.)

(c) At all other points the function $y(x)$ must be regular and can be chosen arbitrarily.

All admissible types of generating function $y(x)$ can be described by the oriented graph depicted in figure A.1. A total number of passages through the vertex P_- determines the order of level E. This result immediately follows from the oscillator theorem. For example, the way

$$A_- B_-$$

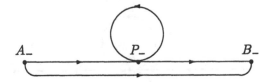

Figure A.1. The oriented graph describing the generating function $y(x)$.

corresponds to the ground state, and the way

$$A_- P_- P_- P_- P_- P_- B_-$$

describes the fifth excited state.

2. *Quasi-exactly solvable models of second order. Properties of generating function $z(x)$.* We assume for definiteness that the nodes of wavefunctions $\psi_1(x)$ and $\psi_2(x)$ do not coincide and $E_1 < E_2$. In this case:

(a) In the vicinity of the points a and b the functions $z(x)$ must behave as $z(x) \approx \frac{E_1 - E_2}{2A-1}(x - a)$, $x \to a + 0$ and $z(x) \approx \frac{E_1 - E_2}{2B-1}(x - b)$, $x \to b - 0$. (Critical points of types A_- and B_-, respectively.)

(b) Within the interval $[a, b]$ function $z(x)$ may have simple poles with residues $+1$ or -1. (Critical points of types P_- and P_+, respectively.)

(c) The function $z(x)$ may have simple zeros in internal points of interval $[a, b]$ if the values of $z'(x)$ in these points are negative and equal to $E_1 - E_2$.

(d) In other points of interval $[a, b]$ the function $z(x)$ must be regular and can be chosen arbitrarily.

All admissible types of functions $z(x)$ can be described by the oriented graph depicted in figure A.2. A total number of passages through the vertex P_+ (or P_-) determines the order of level E_1 (or E_2). As in the previous case this result follows from the oscillator theorem. From this figure it is clear that, for any admissible way from A_- to B_- along the directed lines, the point P_- is passed more often than the point P_+. This fact is in full accordance with the requirement that $E_1 < E_2$. Consider two examples. The way

$$A_- B_-$$

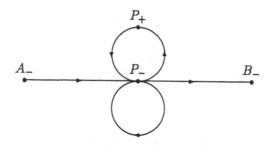

Figure A.2. The oriented graph describing the generating function $z(x)$.

corresponds to the ground and first excited states, and the way

$$A_- P_- P_- P_+ P_- P_- P_- P_+ P_- B_-$$

describes the second and sixth energy levels.

3.   *Quasi-exactly solvable models of third order.   Properties of generating function $t(x)$.* As in the previous case, let us assume that the nodes of wavefunctions $\psi_1(x)$ and $\psi_2(x)$ and $\psi_3(x)$ do not coincide and $E_1 < E_2 < E_3$. In this case:

(a) In the vicinity of the point a the function $t(x)$ must behave as

$$t(x) \approx \frac{E_3 - E_1}{E_2 - E_1} \left[1 + \frac{E_3 - E_2}{(2A+1)(2A+3)}(x-a)^2 \right], \quad x \to a+0$$

provided that the sign of the root in (A.1.13) is negative (critical point of type A_+). In the vicinity of the point b the function $t(x)$ must behave as

$$t(x) \approx \frac{E_3 - E_1}{E_2 - E_1} \left[1 + \frac{E_3 - E_2}{(2B+1)(2B+3)}(x-b)^2 \right], \quad x \to b-0$$

provided that the sign of the root in (A.1.13) is positive (critical point of type B_-).

(b) Within the interval $[a, b]$ the function $t(x)$ may have both simple poles with arbitrary residues and simple zeros with arbitrary inclinations. (Critical points of types $P_\pm$ and $N_\pm$, respectively.) In all such points the root in (A.1.13) changes its sign.

(c) In order to guarantee the positive definiteness of the subradical expression in (A.1.13), the module of the function $t'(x)$ must exceed certain critical values in both the domains $-\infty < t(x) < 0$ and $1 < t(x) < \frac{E_3 - E_1}{E_2 - E_1}$. The points at which these values are arrived at are the critical points. We denote them by $Q_\pm$ and $R_\pm$ for the first and second domains, respectively. In these points the root in (A.1.13) also changes its sign.

(d) At the points where the function $t(x)$ takes the value $(E_3 - E_1)/(E_2 - E_1)$, the sign of its derivative $t'(x)$ must be opposite to the sign of the square root in (A.1.13).

(e) At all other points the function $t(x)$ must be regular and can be chosen arbitrarily.

The admissible types of generating functions $t(x)$ can be described by the oriented graph depicted in figure A.3. Note that any passage through

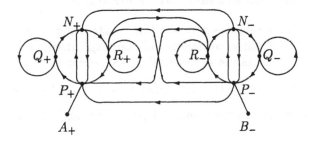

Figure A.3. The oriented graph describing the generating function $t(x)$.

any internal vertex of this graph changes the sign of the root in formula (A.1.13). Note also that the sign of this root must be negative on the lines N_+P_+, R_+P_+, R_-P_+, A_+P_+ and positive on the lines P_-N_-, P_-R_-, P_-R_+, P_-B_-. Therefore, not every way from A_+ to B_- along the directed lines is admissible.

Using the oscillator theorem it is not difficult to obtain the following simple assertions. The total number of vertices P_+ and P_- (belonging to a certain admissible way) in which the root changes its sign from $(-)$ to $(+)$ determines the order of level E_3. The total number of vertices N_+ and N_- (belonging to the same way) in which the root changes its sign from $(+)$ to $(-)$ determines the order of level E_2. Finally, the total number of passages through the lines N_+P_+ and N_+R_+ (with negative sign of the root) and through the lines P_-N_- and R_-N_- (with positive sign of the root) determines the order of level E_1. Consider two examples. The way

$$A_+P_+Q_+N_+N_-Q_-Q_-P_-B_-$$

is admissible and corresponds to the zeroth, first and second energy levels. The more complex way

$$A_+P_+Q_+Q_+N_+P_+N_+R_+R_+P_+Q_+N_+N_-P_-R_-N_-Q_-P_-R_+P_+$$
$$N_+N_-P_-P_+N_+N_-Q_-P_-B_-$$

is also admissible and describes the second, fifth and eighth excited states.

Summarizing the results of this section we can conclude that there exists an infinitely (functionally) large number of one-dimensional Schrödinger equations having one, two or three explicit and physically sensible solutions with *a priori* specified numbering (Ushveridze 1988o).

A.2   The one-dimensional case. Four states

In this section we discuss a method of constructing wide classes of one-dimensional Schrödinger equations with four explicit solutions. The method is based on the use of the generalized Riccati equation

$$y'(\lambda) + a_4(\lambda)y^2(\lambda) + a_3(\lambda)y(\lambda) + a_2(\lambda) + Ea_1(\lambda) = 0, \qquad (A.2.1)$$

in which E is an unknown spectral parameter and $y(\lambda)$ is an unknown function.

1. *First stage.* Instead of solving this equation with respect to E and $y(\lambda)$ for given $a_i(\lambda)$, $i = 1, \ldots, 4$, we state an inverse problem to find such functions $a_i(\lambda)$, $i = 1, \ldots, 4$, for which equation (A.2.1) has several *a priori* given solutions E and $y(\lambda)$. It is not difficult to verify that the needed form of this equation is easily restored when the number of solutions does not exceed four. In fact, substituting four pairs $(E_i, y_i(\lambda))$, $i = 1, \ldots, 4$ into (A.2.1) we obtain the system of four linear algebraic equations

$$\begin{bmatrix} E_1 & 1 & y_1(\lambda) & y_1^2(\lambda) \\ E_2 & 1 & y_2(\lambda) & y_2^2(\lambda) \\ E_3 & 1 & y_3(\lambda) & y_3^2(\lambda) \\ E_4 & 1 & y_4(\lambda) & y_4^2(\lambda) \end{bmatrix} \begin{bmatrix} a_1(\lambda) \\ a_2(\lambda) \\ a_3(\lambda) \\ a_4(\lambda) \end{bmatrix} = - \begin{bmatrix} y_1'(\lambda) \\ y_2'(\lambda) \\ y_3'(\lambda) \\ y_4'(\lambda) \end{bmatrix}, \qquad (A.2.2)$$

which can be solved exactly with respect to functions $a_i(\lambda)$, $i = 1, \ldots, 4$. This gives us the explicit form of the Riccati equation (A.2.1) having four explicit solutions E_i and $y_i(\lambda)$, $i = 1, \ldots, 4$.

2. *Second stage.* Remember that any Riccati equation can easily be reduced to linear form. Indeed, taking

$$y(\lambda) = \frac{1}{a_4(\lambda)} \left\{ \frac{\varphi'(\lambda)}{\varphi(\lambda)} + \frac{1}{2} \left[\frac{a_4'(\lambda)}{a_4(\lambda)} - a_3(\lambda) \right] \right\} \tag{A.2.3}$$

and substituting (A.2.3) into (A.2.1) we obtain the second-order linear differential equation with respect to $\varphi(\lambda)$:

$$-\varphi''(\lambda) + \left\{ \frac{1}{2} \left[a_3(\lambda) - \frac{a_4'(\lambda)}{a_4(\lambda)} \right]' + \frac{1}{4} \left[a_3(\lambda) - \frac{a_4'(\lambda)}{a_4(\lambda)} \right]^2 \right.$$

$$\left. -a_2(\lambda)a_4(\lambda) \right\} \varphi(\lambda) = E a_1(\lambda)a_4(\lambda)\varphi(\lambda). \tag{A.2.4}$$

Going over to the new variable

$$x = \int \sqrt{a_1(\lambda)a_4(\lambda)} \, d\lambda \tag{A.2.5}$$

and introducing the new function

$$\psi(x) = [a_1(\lambda)a_4(\lambda)]^{\frac{1}{4}} \varphi(\lambda), \tag{A.2.6}$$

we transform this equation to the Schrödinger form:

$$\left[-\frac{\partial^2}{\partial x^2} + V(x) \right] \varphi(x) = E\psi(x). \tag{A.2.7}$$

The obtained Schrödinger-type equation is characterized by the potential

$$V(x) = \frac{1}{a_1(\lambda)a_4(\lambda)} \left\{ \frac{1}{2} \left[a_3(\lambda) - \frac{a_4'(\lambda)}{a_4(\lambda)} \right]' + \frac{1}{4} \left[a_3(\lambda) - \frac{a_4'(\lambda)}{a_4(\lambda)} \right]^2 \right.$$

$$\left. -a_2(\lambda)a_4(\lambda) + \frac{1}{4} \left[\frac{(a_2(\lambda)a_4(\lambda))'}{(a_1(\lambda)a_4(\lambda))} \right] - \frac{1}{4} \left[\frac{(a_2(\lambda)a_4(\lambda))'}{(a_1(\lambda)a_4(\lambda))} \right]^2 \right\}$$

$$\tag{A.2.8}$$

and has, evidently, four explicit solutions E_i, $i = 1, \ldots, 4$ and

$$\psi_i(\lambda) = \left[\frac{a_1(\lambda)}{a_4(\lambda)} \right]^{\frac{1}{4}} \exp\left\{ \int a_4(\lambda)y_i(\lambda) \, d\lambda + \frac{1}{2} \int a_3(\lambda) \, d\lambda \right\},$$

$$i = 1, \ldots, 4. \tag{A.2.9}$$

The functions $a_i(\lambda)$, $i = 1, \ldots, 4$ presented in formulas (A.2.8) and (A.2.9) must be restored from the system (A.2.2).

Thus, the class of fourth-order quasi-exactly solvable Schrödinger equations is constructed. We see that it is parametrized by four numbers E_i, $i = 1, \ldots, 4$ and four generating functions $y_i(\lambda)$, $i = 1, \ldots, 4$. Note, however, that the presence of fourth-order determinants in the final explicit expressions complicates the analysis of the obtained solutions from the point of view of their physical sensibility. Nevertheless, it is not difficult to understand that, as in the previous case, the necessary constraints on the generating functions have local character and, therefore, the set of physically sensible fourth-order quasi-exactly solvable models turns out to be functionally large.

A.3   The multi-dimensional case

There are two possible ways of constructing functionally large classes of multi-dimensional quasi-exactly solvable equations. The first way is an evident generalization of that described in section A.2. This is based on the use of the multi-parameter generalized Riccati equation

$$y'(\lambda) + a_4(\lambda)y^2(\lambda) + a_3(\lambda)y(\lambda) + a_2(\lambda) + \sum_{n=1}^{D} E_n a_{1n}(\lambda) = 0, \quad (A.3.1)$$

in which $E_1, \ldots, E_D$ are unknown spectral parameters and $y(\lambda)$ is an unknown function. As before, instead of solving this equation with respect to $E_1, \ldots, E_D$ and $y(\lambda)$, we state an inverse problem of finding all such functions $a_{11}(\lambda), \ldots, a_{1D}(\lambda), a_2(\lambda), a_3(\lambda), a_4(\lambda)$, for which it has several (namely $D + 3$) a priori given solutions $E_{1i}, \ldots, E_{Di}$ and $y_i(\lambda)$, $i = 1, \ldots, D + 3$. Substituting these solutions into equation (A.3.1), we obtain a system of $D + 3$ linear algebraic equations for $D + 3$ unknown functions $a_{11}(\lambda), \ldots, a_{1D}(\lambda), a_2(\lambda), a_3(\lambda), a_4(\lambda)$, which can be solved without dificulty. The obtained Riccati equation can be linearized, after which we arrive at a D-parameter second-order linear differential equation having $D + 3$ explicit solutions. From the results of chapter 3 we know that any such equation can easily be reduced to a D-dimensional Schrödinger equation on a certain curved manifold describing a quasi-exactly solvable model of order $D + 3$. As in the one-dimensional case, the set of obtained models is functionally large.

Another way leading to multi-dimensional models of higher order is based on the following reasoning. Let us consider a D-dimensional second-order linear differential equation

$$\left\{ \sum_{\alpha,\beta}^{D} P_{\alpha\beta}(\vec{x}) \frac{\partial^2}{\partial x_\alpha \partial x_\beta} + \sum_{\alpha}^{D} Q_\alpha(\vec{x}) \frac{\partial}{\partial x_\alpha} + R(\vec{x}) \right\} \phi(\vec{x}) = E\phi(\vec{x}). \quad (A.3.2)$$

The total number of independent coefficient functions in it is $1 + D + \frac{1}{2}D(D+1) = \frac{1}{2}(D+1)(D+2)$. Let us now fix $\frac{1}{2}(D+1)(D+2)$ pairs $(E_i, \phi_i(\vec{x}))$, $i = 1, \ldots, \frac{1}{2}(D+1)(D+2)$, treating them as solutions of equation (A.3.2). Substituting these pairs into (A.3.2) we obtain a system of $\frac{1}{2}(D+1)(D+2)$ linear algebraic equations from which all $\frac{1}{2}(D+1)(D+2)$ coefficient functions can easily be found. As a result, we arrive at a D-dimensional equation of the form (A.3.2) having $\frac{1}{2}(D+1)(D+2)$ *a priori* given solutions. Applying to it the procedure described in section 5.7, we can always reduce this equation to a class of quasi-exactly solvable models in a $(D+1)$-dimensional, in general, curved space. The order of these models will be $\frac{1}{2}(D+1)(D+2)$ and their number will be functionally large.

Appendix B

The generalized quantum tops and exact solvability

B.1 The method

Exactly solvable second-order spectral differential equations play an important role in many branches of mathematical physics. Equations reducible to the Schrödinger form and describing various exactly solvable models of one- and multi-dimensional quantum mechanics on flat and curved manifolds have an especially wide range of applicability. Following Morozov *et al* (1990) we describe here a simple method for constructing such equations. This method is based on the use of the so-called "generalized quantum tops" constructed from the generators of compact Lie groups.

Let G be a compact finite-dimensional Lie group of dimension D and let $S_a, a = 1, \ldots, D$ be generators of this group. Consider the operator (the hamiltonian of the generalized quantum top)

$$H = \sum_{a,b=1}^{D} C^{ab} S_a S_b, \qquad (B.1.1)$$

in which $C^{ab} = C^{ba}$ are some real constants. Assume that the generators S_a, $a = 1, \ldots, D$ are realized as vector fields on a d-dimensional homogeneous space

$$\mathcal{M} = G/G_0, \qquad (B.1.2)$$

where G_0 is a certain stationary subgroup of G. Introducing the coordinates $\xi^\mu, \mu = 1, \ldots, d$ on $\mathcal{M}$, we can write

$$S_a = \sum_{\mu=1}^{d} T_a^\mu(\vec{\xi}) \frac{\partial}{\partial \xi^\mu}. \qquad (B.1.3)$$

The algorithms for constructing the vector fields on $\mathcal{M}$ for compact groups were worked out in detail by Kirillov (1972), Kostant (1977, 1979) and Hurt (1983).

Now note that the compactness of G results in the compactness of G_0. This enables one to introduce on $\mathcal{M}$ a G-invariant metric $\{g_0^{\mu\nu}(\vec{\xi})\}$ with elements given by the following simple formula:

$$g_0^{\mu\nu}(\vec{\xi}) = \sum_{a=1}^{D} C_0^{ab} T_a^{\mu}(\vec{\xi}) T_b^{\nu}(\vec{\xi}), \qquad (B.1.4)$$

in which C_0^{ab} is the Killing–Cartan tensor. It is known that the operators (B.1.3) are anti-hermitian with respect to the metric (B.1.4) when they act on scalars on $\mathcal{M}$:

$$S_a = -S_a^{+}. \qquad (B.1.5)$$

From the definition of the scalar product

$$\langle \phi_1, \phi_2 \rangle = \int \phi_1(\vec{\xi}) \phi_2(\vec{\xi}) \sqrt{g_0(\vec{\xi})} \, d^d\xi, \qquad (B.1.6)$$

in which

$$g_0(\vec{\xi}) = \frac{1}{\det\{g_0^{\mu\nu}(\vec{\xi})\}}, \qquad (B.1.7)$$

it follows that

$$S_a^{+} = -\sum_{\mu=1}^{d} \left[T_a^{\mu}(\vec{\xi}) \frac{\partial}{\partial \xi^{\mu}} - \frac{\partial}{\partial \xi^{\mu}} T_a^{\mu}(\vec{\xi}) - T_a^{\mu}(\vec{\xi}) \left(\frac{\partial}{\partial \xi^{\mu}} \ln \sqrt{g_0(\vec{\xi})} \right) \right]. \quad (B.1.8)$$

Comparing formulas (B.1.3), (B.1.5) and (B.1.8) we see that

$$\sum_{\mu=1}^{d} \frac{\partial}{\partial \xi^{\mu}} T_a^{\mu}(\vec{\xi}) = -\sum_{\mu=1}^{d} T_a^{\mu}(\vec{\xi}) \left(\frac{\partial}{\partial \xi^{\mu}} \ln \sqrt{g_0(\vec{\xi})} \right). \qquad (B.1.9)$$

Using this formula, it is not difficult to show that

$$H = e^{\Phi}(\Delta_g - V)e^{-\Phi} \qquad (B.1.10)$$

where

$$\Delta_g = \sum_{\mu,\nu=1}^{d} \frac{1}{\sqrt{g(\vec{\xi})}} \frac{\partial}{\partial \xi^{\mu}} \left[\sqrt{g(\vec{\xi})} g^{\mu\nu}(\vec{\xi}) \frac{\partial}{\partial \xi^{\nu}} \right] \qquad (B.1.11)$$

is the covariant Laplace operator with the deformed metric

$$g^{\mu\nu}(\vec{\xi}) = \sum_{a,b=1}^{D} C^{ab} T_a^\mu(\vec{\xi}) T_b^\nu(\vec{\xi}) \tag{B.1.12}$$

and

$$g(\vec{\xi}) = \frac{1}{\det\{g^{\mu\nu}(\vec{\xi})\}}. \tag{B.1.13}$$

The "imaginary phase" in (B.1.10) is expressed in terms of the ratio $g(\vec{\xi})/g_0(\vec{\xi})$ as

$$\Phi = \frac{1}{4} \ln \left(\frac{g(\vec{\xi})}{g_0(\vec{\xi})} \right), \tag{B.1.14}$$

and the potential V is given by the formula

$$V = \frac{1}{4} \sum_{\mu\nu=1}^{d} g^{\mu\nu}(\vec{\xi}) \frac{\partial \Phi(\vec{\xi})}{\partial \xi^\mu} \frac{\partial \Phi(\vec{\xi})}{\partial \xi^\nu} - \frac{1}{4} \Delta_g \Phi(\vec{\xi}). \tag{B.1.15}$$

Thus, we have demonstrated that the operator H introduced by formula (B.1.1) is equivalent to the hamiltonian of a certain quantum mechanical model defined on a curved space with the metric (B.1.12).

Let us now show that the quantum model obtained is exactly solvable. This means that the whole spectrum of its hamiltonian can be obtained algebraically.

In order to demonstrate this fact, let us consider the quadratic Casimir operator for the group G:

$$H_0 = \sum_{a,b=1}^{D} C_0^{ab} S_a S_b. \tag{B.1.16}$$

This operator can be interpreted as the Laplace operator on $\mathcal{M}$ and is given by the formula

$$H_0 = \Delta_{g_0} = \sum_{\mu,\nu=1}^{d} \frac{1}{\sqrt{g_0(\vec{\xi})}} \frac{\partial}{\partial \xi^\mu} \left[\sqrt{g_0(\vec{\xi})} g_0^{\mu\nu}(\vec{\xi}) \frac{\partial}{\partial \xi^\nu} \right]. \tag{B.1.17}$$

It is not difficult to see that H_0 is hermitian, non-degenerate and commutes with the hamiltonian H. From this it follows that the Hilbert space in which the hamiltonian H acts splits into a sum of eigensubspaces of the operator H_0. Since H_0 commutes with generators S_a, every such subspace forms a finite-dimensional representation of the group G. The basis functions in these representation spaces are the so-called "generalized spherical harmonics" whose concrete form is known if the coordinates $\vec{\xi}$ are fixed. In the basis of the generalized spherical harmonics the hamiltonian H takes the block diagonal form. Each block has a finite dimension and is completely disconnected from all others. Therefore, the spectral problem for this hamiltonian breaks up into an infinite number of finite-dimensional spectral problems, each of which can be solved algebraically. This completes the procedure of constructing exactly solvable models associated with a given compact Lie group G and its stationary subgroup G_0.

B.2 An example

In order to demonstrate the constructivity of this scheme, we consider a simplest example. Let $G = SO(3)$ and $G_0 = SO(2)$. Then the orbit is the sphere, $\mathcal{M} = S^2$. It is convenient to parametrize this sphere by two coordinates η and ζ in which the generators S_a, $a = 1, 2, 3$ of the group $SO(3)$ take the form:

$$
\begin{aligned}
S_1 &= \eta \frac{\partial}{\partial \zeta} - \zeta \frac{\partial}{\partial \eta}, \\
S_2 &= (1 + \eta^2) \frac{\partial}{\partial \eta} + \eta\zeta \frac{\partial}{\partial \zeta}, \\
S_3 &= -(1 + \zeta^2) \frac{\partial}{\partial \zeta} - \eta\zeta \frac{\partial}{\partial \eta}.
\end{aligned}
\tag{B.2.1}
$$

In this case, the invariant metric tensor on the sphere is given by the formulas

$$
\begin{aligned}
g_0^{11}(\eta, \zeta) &= (\zeta^2 + 1)^2 + \zeta^2\eta^2 + \eta^2, \\
g_0^{22}(\eta, \zeta) &= (\eta^2 + 1)^2 + \zeta^2\eta^2 + \zeta^2, \\
g_0^{12}(\eta, \zeta) = g_0^{21}(\eta, \zeta) &= \eta\zeta(\zeta^2 + \eta^2 + 1).
\end{aligned}
\tag{B.2.2}
$$

Note also that

$$
g_0(\eta, \zeta) = (1 + \eta^2 + \zeta^2)^{-3}.
\tag{B.2.3}
$$

It is not difficult to verify that the generators (B.2.2) are anti-hermitian with respect to the metric (B.2.3).

In these coordinates the elements of the (reducible) representations are

$$\{\zeta^j, \zeta^{j-1}\eta, \ldots, \zeta\eta^{j-1}, \eta^j; \zeta^{j-1}, \zeta^{j-2}\eta, \ldots, \zeta\eta^{j-2}, \eta^{j-1}; \ldots; \zeta, \eta, 1\}$$
$$\times \exp[(-j/2)\ln(1 + \zeta^2 + \eta^2)]. \tag{B.2.4}$$

The dimension of this representation is $(1 + j)(1 + j/2)$. It can easily be decomposed into a sum of irreducible representations of dimensions $1, 3, 5, 7, \ldots$ in which the spherical Laplace operator has the eigenvalues $0, 2, 6, 12, \ldots$.

Let us now consider several special choices of coefficients C^{ab} in (B.1.1) and construct the corresponding two-dimensional exactly solvable models.

(a) $C^{22} = C^{33} = 1$, other coefficients C^{ab} zero. Then

$$
\begin{aligned}
g^{11}(\zeta, \eta) &= \zeta^2\eta^2 + (1 + \zeta^2)^2, \\
g^{22}(\zeta, \eta) &= \zeta^2\eta^2 + (1 + \eta^2)^2, \\
g^{12}(\zeta, \eta) = g^{21}(\zeta, \eta) &= \zeta\eta(2 + \zeta^2 + \eta^2).
\end{aligned}
\tag{B.2.5}
$$

The scalar curvature corresponding to this metric is

$$R = -4(\zeta^2 + \eta^2 + 1). \tag{B.2.6}$$

The potential V turns out to be

$$V = -\frac{3}{8}(\zeta^2 + \eta^2). \tag{B.2.7}$$

It is interesting that in this model the variables can be separated in the polar coordinates, $\zeta = r\cos\theta$, $\eta = r\sin\theta$. In these coordinates the metric takes the form $\{g^{\mu\nu}(\zeta, \eta)\} = \text{diag}\{(r^2 + 1)^2, r^{-2}\}$. In spite of the fact that the potential V in (B.2.7) is negative the eigenfunctions are normalizable. The reason is that the curvature R is non-vanishing and, moreover, tends to infinity at large r.

(b) $C_{22} = C_{33} = C_{23} = 1$, other coefficients zero. In this case

$$
\begin{aligned}
g^{11}(\zeta, \eta) &= (\zeta^2 + 1 - \zeta\eta)^2 + \zeta\eta(1 + \zeta^2), \\
g^{22}(\zeta, \eta) &= (\zeta^2 + \eta^2 - \zeta\eta)^2 + 3\eta(1 + \eta^2), \\
g^{12}(\zeta, \eta) &= g^{21}(\zeta, \eta) = [(\zeta^2 + \eta^2 - 3\eta)(2\zeta\eta - 1) + 3\zeta\eta - 1)]/2.
\end{aligned}
\tag{B.2.8}
$$

The curvature corresponding to this metric is

$$R = -4(\zeta^2 + \eta^2 + 1 - \zeta\eta), \tag{B.2.9}$$

while the potential is

$$V = -\frac{3}{8}(\zeta^2 + \eta^2 - \zeta\eta). \tag{B.2.10}$$

Comparing the results for R and V we see that there is no obvious substitution leading to separation of the variables.

(c) $2C_{12} = 1$, other coefficients zero. We have

$$
\begin{aligned}
g^{11}(\zeta,\eta) &= \zeta\eta^2, \\
g^{22}(\zeta,\eta) &= -\zeta(\eta^2 + 1), \\
g^{12}(\zeta,\eta) = g^{21}(\zeta,\eta) &= \eta(\eta^2 + 1 - \zeta^2)/2.
\end{aligned}
\tag{B.2.11}
$$

This example is made interesting by the fact that both the curvature

$$R = 4\zeta/\eta^2 \tag{B.2.12}$$

and the potential

$$V = -\frac{3\zeta}{8\eta^2} \tag{B.2.13}$$

are singular. Nevertheless, all the wavefunctions in this model are normalizable (Shifman 1989a, Shifman and Turbiner 1989).

Other examples of exactly solvable models associated with the group $SO(3)$ can be found in the paper of Shifman and Turbiner (1989).

Appendix C

The method of raising and lowering operators

C.1 Introduction

It is, unfortunately, impossible to formulate a universal method for constructing and classifying all exactly solvable spectral differential equations. The reason is the absence of an appropriate mathematical definition of the exact solvability: we call a spectral differential equation exactly solvable if all its solutions can be expressed in terms of sufficiently simple functions. However, we do not know how to formalize the notion of the simple function and this prevents us from stating the problem at the mathematical level of rigour[1].

One of the possibilities for avoiding this difficulty is to look for equations solvable in some classes of *a priori* given functions, for example, in the classes of polynomials. This leads to loss of generality but allows one to formulate the problem correctly and solve it by means of more or less usual mathematical methods.

In the one-dimensional case the most general form of second-order spectral differential equations having an infinite number of polynomial solutions is known. This is the hypergeometric equation $H\psi = E\psi$ with

$$H = P_2(x)\frac{\partial^2}{\partial x^2} + P_1(x)\frac{\partial}{\partial x}. \qquad (C.1.1)$$

Here $P_1(x)$ and $P_2(x)$ are arbitrarily fixed first- and second-order polynomials, respectively. The exact solvability of this equation can be interpreted as an algebraic solvability. Indeed, as follows from (C.1.1), the spaces of polynomials of a given order are the invariant subspaces for H,

---

[1] The same relates to quasi-exactly solvable equations.

412

and, therefore, the construction of eigenvalues and eigenfunctions of the operator (C.1.1) is a purely algebraic problem.

In the multi-dimensional case the situation is more complicated. There are several known examples of multi-dimensional generalizations of the hypergeometric equation (Erdélyi and Bateman 1953), but the general methods for constructing and classifying all second-order equations that are exactly solvable in classes of polynomials are not known.

One of the first attempts to formulate a general approach to the problem was based on the use of finite-dimensional representations of Lie algebras (Shifman and Turbiner 1989). It is known that any such representation (characterized by the lowest-weight $\vec{\nu}$) can always be realized in the space of polynomials. In this case the generators $L_a(\vec{\nu})$ of Lie algebras take the form of first-order differential operators (for more details see sections 5.3 and 5.4). It is quite obvious that any bilinear combination of such generators

$$H(\vec{\nu}) = \sum_{a,b} C^{ab} L_a(\vec{\nu}) L_b(\vec{\nu}) + \sum_a C^a L_a(\vec{\nu}) \qquad (C.1.2)$$

has the form of a certain second-order differential operator. As follows from (C.1.2), the representation spaces $\Phi(\vec{\nu})$, in which the generators $L_a(\vec{\nu})$ act, are invariant subspaces for $H(\vec{\nu})$. Due to the finite dimensionality of these spaces, the spectral problem for $H(\vec{\nu})$ in them can be solved algebraically. This gives us a class of second-order differential equations parametrized by the vectors $\vec{\nu}$ and having $\dim \Phi(\vec{\nu})$ exact (polynomial) solutions for any given $\vec{\nu}$.

The form of these equations and the number of their solutions depend explicitly on $\vec{\nu}$. Changing $\vec{\nu}$ we can obtain equations having arbitrarily large (but finite) number of solutions. We call such equations quasi-exactly solvable. The transition to the exactly solvable case can be performed by choosing the coefficients C^{ab} and C^a in such a way as to guarantee the $\vec{\nu}$-independence of operators $H(\vec{\nu})$: $H(\vec{\nu}) = H$. In cases when this is possible, all the spaces $\Phi(\vec{\nu})$ become invariant subspaces for operators H, which leads us to a class of second-order differential equations having a non-limited number of exact (polynomial) solutions (Shifman and Turbiner 1989).

Despite all the ideological advantages of this method (which is known as the method of partial algebraization), we must admit with regret that it is not free from technical difficulties. The first difficulty lies in the fact that the procedure of constructing differential realizations of generators of Lie algebras is, as a rule, rather complicated and leads to very cumbersome explicit expressions, especially for the algebras of higher ranks. The second difficulty is that the procedure of choosing the coefficients C^{ab} and C^a

ensuring the $\vec{\nu}$-independence of operators $H(\vec{\nu})$ is not algorithmized in the general case. The absence of convenient explicit expressions for the generators makes the problem of constructing such an algorithm very difficult.

Below we demonstrate that the origin of these difficulties is that the method of partial algebraization (if it is used for constructing exactly solvable equations) is too overloaded by various unnecessary constraints.

One such constraint is the requirement of a finite dimensionality of the representations of Lie algebras which are used for constructing operators H. In section C.2 we show that the method takes a much more simple and natural form being reformulated in terms of infinite-dimensional representations of Lie algebras. It turns out that such a reformulation automatically solves the problem of choosing the coefficients C^{ab} and C^a entering into the expression for H (Ushveridze 1989f, Doebner and Ushveridze 1992).

The second constraint has more fundamental character. This is that the operators L_a are considered as generators of some Lie algebras. However, we can see that the formulation of the method of partial algebraization does not require the knowledge of any commutation relations between L_a. This suggests that the method may allow another formulation, free from any Lie-algebraic structure. In sections C.3 and C.4 we show that such a dealgebraized version of the method actually exists and is much simpler than the initial (algebraic) one. In particular, it gives the possibility of writing down explicit, compact and practically convenient expressions for the resulting exactly solvable differential equations and, besides, allows one to solve the classification problem for them (Ushveridze 1989f, Doebner and Ushveridze 1992). In section C.5 we consider some examples demonstrating the simplicity and constructivity of our dealgebraized scheme. The last section C.6 discusses the transition to the case of quasi-exactly solvable equations[2].

C.2　The Lie-algebraic approach

Let L be a certain simple Lie algebra of rank r. Consider its arbitrary (infinite-dimensional) representation with lowest-weight $\vec{\nu} = (\nu_1, \ldots, \nu_r)$. The generators of this representation in the Cartan basis we denote by $L_{\vec{\alpha}}^{\pm}$, $\vec{\alpha} \in \aleph$ and L_i^0, $i = 1, \ldots, r$, where $\aleph$ is the set of d positive roots

---

[2] The procedure of reduction of these equations to the Schrödinger form is not discussed here, since it has been well algorithmized in section 5.7. According to this algorithm, any D-dimensional exactly solvable equation of second order can be reduced in infinitely many ways to equations of Schrödinger type on $(D + 1)$-dimensional manifolds, which are, in general, curved.

$\vec{\alpha}_1, \ldots, \vec{\alpha}_d$ of algebra L.

Denoting the lowest-weight vector by $\phi_0(\vec{\nu})$, we can specify this representation by the formulas

$$L_i^0(\vec{\nu})\phi_0(\vec{\nu}) = \nu_i\phi_0(\vec{\nu}), \qquad i = 1, \ldots, r \qquad \text{(C.2.1)}$$

and

$$L_{\vec{\alpha}}^-(\vec{\nu})\phi_0(\vec{\nu}) = 0, \qquad \vec{\alpha} \in \aleph. \qquad \text{(C.2.2)}$$

The corresponding representation space $\Phi(\vec{\nu})$ is the linear span of the vectors $[L_{\vec{\alpha}_1}^+(\vec{\nu})]^{n_1} \ldots [L_{\vec{\alpha}_d}^+(\vec{\nu})]^{n_d}\phi_0(\vec{\nu})$ with $\vec{\alpha}_1, \ldots, \vec{\alpha}_d \in \aleph$ and arbitrary non-negative integers $n_1, \ldots, n_d$. In the general case this space is infinite dimensional.

Let us now consider the class of operators belonging to the universal enveloping algebra of the algebra L and having the following specific form:

$$H(\vec{\nu}) = \sum_{\vec{\alpha} \leq \vec{\beta}} C_{\vec{\alpha}\vec{\beta}}^{(+-)} L_{\vec{\alpha}}^+(\vec{\nu})L_{\vec{\beta}}^-(\vec{\nu}) + \sum_{\vec{\alpha},\vec{\beta}} C_{\vec{\alpha}\vec{\beta}}^{(--)} L_{\vec{\alpha}}^-(\vec{\nu})L_{\vec{\beta}}^-(\vec{\nu})$$

$$+ \sum_{\vec{\alpha}}\sum_i C_{\vec{\alpha}i}^{(0-)} L_i^0(\vec{\nu})L_{\vec{\alpha}i}^-(\vec{\nu}) + \sum_{i,k} C_{ik}^{(00)} L_i^0(\vec{\nu})L_k^0(\vec{\nu})$$

$$+ \sum_{\vec{\alpha}} C_{\vec{\alpha}}^{(-)} L_{\vec{\alpha}}^-(\vec{\nu}) + \sum_i C_i^{(0)} L_i^0(\vec{\nu}). \qquad \text{(C.2.3)}$$

It is not difficult to show that spectral equations for these operators

$$H(\vec{\nu})\phi = E(\vec{\nu})\phi, \qquad \phi \in \Phi(\vec{\nu}) \qquad \text{(C.2.4)}$$

(which, obviously, are infinite dimensional) are exactly (algebraically) solvable.

Indeed, consider the subspaces $\Phi_{N_1,\ldots,N_r}(\vec{\nu})$ of the space $\Phi(\vec{\nu})$ characterized by r non-negative integers $N_1, \ldots, N_r$ and defined as

$$\Phi_{N_1,\ldots,N_r}(\vec{\nu}) = \bigoplus_{n_1,\ldots,n_d} [L_{\vec{\alpha}_1}^+(\vec{\nu})]^{n_1} \ldots [L_{\vec{\alpha}_d}^+(\vec{\nu})]^{n_d}\phi_0(\vec{\nu}) \qquad \text{(C.2.5)}$$

where $\vec{\pi}_i$ are simple roots of algebra L and n_k are non-negative integers satisfying the system of inequalities

$$\sum_k n_k\vec{\alpha}_k \leq \sum_{i=1}^r N_i\vec{\pi}_i. \qquad \text{(C.2.6)}$$

For any given $N_1, \ldots, N_r$ the space $\Phi_{N_1,\ldots,N_r}(\vec{\nu})$ is finite dimensional. Its dimension does not exceed the total number of solutions of the system of inequalities (C.2.5). At the same time, it is easily seen that any space $\Phi_{N_1,\ldots,N_r}(\vec{\nu})$ is an invariant subspace for the operator $H(\vec{\nu})$ defined by formula (C.2.3). This means that the initial infinite-dimensional spectral problem (C.2.4) breaks up into an infinite number of finite-dimensional spectral problems

$$H(\vec{\nu})\phi = E(\vec{\nu})\phi, \qquad \phi \in \Phi_{N_1,\ldots,N_r}(\vec{\nu}), \tag{C.2.7}$$

each of which can be solved algebraically.

Let us now recall that the generators of any representations of Lie algebra L can be realized as first-order differential operators in d variables acting in the space of polynomials. Using such a realization we can rewrite the operator $H(\vec{\nu})$ in a differential form. Due to the bilinearity of $H(\vec{\nu})$ with respect to generators of algebra L, the equation (C.2.4) becomes a second-order differential equation being exactly solvable by construction.

Thus, we have formulated an analogue of the partial algebraization method which is based on the use of infinite-dimensional representations of Lie algebras. We see that this method is technically simpler than that described in section C.1. Its main advantage is that it does not require any special choice of coefficients C in expression (C.2.3). For any values of these coefficients the spectral equation for $H(\vec{\nu})$ is exactly solvable and has infinite number of polynomial solutions with algebraicaly calculable coefficients. The lowest-weight $\vec{\nu}$ plays in this case the role of an additional vector parameter determining some quantitative properties of the spectrum.

Of course, the scheme described above is not free from difficulties associated with the construction of differential realizations of generators of algebra L. However, these difficulties also can easily be overcome and this will be done in the following sections.

C.3   Dealgebraization

Let us associate with any positive root $\vec{\alpha}_i \in \aleph$ of algebra L a certain variable x_i. The number of such variables, coinciding with the number of positive roots, is d. Then the generators of algebra L can be represented in the following form:

$$L_i^0(\vec{\nu}) = \sum_k (\vec{\pi}_i, \vec{\alpha}_k) x_k \frac{\partial}{\partial x_k} + \nu_i \tag{C.3.1}$$

and

$$L_{\vec{\alpha}}^{\pm}(\vec{\nu}) \;=\; \sum_k \sum_{n_1,\ldots,n_d} [C_{\vec{\alpha}}^{\pm}(\vec{\nu})]_k^{n_1\cdots n_d} x_1^{n_1} \ldots x_d^{n_d} \frac{\partial}{\partial x_k}$$

$$+ \sum_{n_1,\ldots,n_d} [C_{\vec{\alpha}}^{\pm}(\vec{\nu})]^{n_1\cdots n_d} x_1^{n_1} \ldots x_d^{n_d} \qquad (C.3.2)$$

where $[C_{\vec{\alpha}}^{\pm}(\vec{\nu})]_k^{n_1\cdots n_d}$ and $[C_{\vec{\alpha}}^{\pm}(\vec{\nu})]^{n_1\cdots n_d}$ are certain constants vanishing if

$$\sum_l n_l \vec{\alpha}_l = \pm \vec{\alpha} + \vec{\alpha}_k \qquad (C.3.3)$$

and

$$\sum_l n_l \vec{\alpha}_l = \pm \vec{\alpha}, \qquad (C.3.4)$$

respectively. It is not difficult to see that for any values of constants C entering into (C.3.2) we have correct commutation relations between the generators of subalgebras L^0 and $L^{\pm}$:

$$[L_i^0, L_{\vec{\alpha}}^{\pm}] = \pm(\vec{\pi}_i, \vec{\alpha}) L_{\vec{\alpha}}^{\pm}. \qquad (C.3.5)$$

In order to establish concrete values of constants C we can use the remaining commutation relations in algebra L.

Surprising enough, the knowledge of these constants is not necessary! The result (that H is an algebraically diagonalizable operator) turns out to be true for any values of parameters C!

In order to make sure that this is really so, let us construct the functional extensions of the spaces $\Phi_{N_1,\ldots,N_r}(\vec{\nu})$. First of all, note that the latter can be represented in the form

$$\Phi_{N_1,\ldots,N_r}(\vec{\nu}) = \bigoplus_{N_1',\ldots,N_r'} \phi_{N_1',\ldots,N_r'}(\vec{\nu}), \qquad N_i' \leq N_i, \quad i = 1,\ldots,r \quad (C.3.6)$$

where the spaces

$$\phi_{N_1',\ldots,N_r'}(\vec{\nu}) = \bigoplus_{n_1,\ldots,n_d} [L_{\vec{\alpha}_1}^+(\vec{\nu})]^{n_1} \ldots [L_{\vec{\alpha}_d}^+(\vec{\nu})]^{n_d} \phi_0(\vec{\nu}), \qquad (C.3.7)$$

whose structure is completely determined by the sets of non-negative integers satisfying the system of Diophantine equations

$$\sum_k n_k \vec{\alpha}_k = \sum_{i=1}^r N_i' \vec{\pi}_i, \qquad (C.3.8)$$

are the eigenspaces of operators $L_i^0(\vec{\nu})$:

$$L_i^0(\vec{\nu}) \phi_{N_1',\ldots,N_r'}(\vec{\nu}) = \lambda_{i,N_1',\ldots,N_r'}(\vec{\nu}) \phi_{N_1',\ldots,N_r'}(\vec{\nu}). \qquad (C.3.9)$$

The corresponding eigenvalues $\lambda_{i,N_1',\ldots,N_r'}(\vec{\nu})$ can easily be found from equations (C.2.1) and (C.2.2) and commutation relations in algebra L. They are

$$\lambda_{i,N_1',\ldots,N_r'}(\vec{\nu}) = \nu_i + \sum_{k=1}^{r}(\pi_i\pi_k)N_k'. \qquad (\text{C.3.10})$$

Denoting by q_{ik} the non-negative integer coefficients of the expansion of the root $\vec{\alpha}_k$ in simple roots π_i and substituting (C.3.1) and (C.3.10) into (C.3.9), we obtain a differential analogue of equation (C.3.9):

$$\sum_{k=1}^{d} q_{ik}x_k \frac{\partial}{\partial x_k}\phi_{N_1',\ldots,N_r'}(\vec{\nu}) = N_i'\phi_{N_1',\ldots,N_r'}(\vec{\nu}). \qquad (\text{C.3.11})$$

From (C.3.11) it follows that

$$\phi_{N_1',\ldots,N_r'}(\vec{\nu}) \subset \phi_{N_1',\ldots,N_r'}, \qquad (\text{C.3.12})$$

where

$$\phi_{N_1',\ldots,N_r'} = \bigoplus_{n_1,\ldots,n_d} \prod_{i=1}^{d}(x_i)^{n_i}, \qquad \sum_{k=1}^{d} q_{ik}m_k = N_i', \qquad i=1,\ldots,r.$$
$$(\text{C.3.13})$$

This enables us to write

$$\Phi_{N_1,\ldots,N_r}(\vec{\nu}) \subset \Phi_{N_1,\ldots,N_r} \qquad (\text{C.3.14})$$

where

$$\Phi_{N_1,\ldots,N_r} = \bigoplus_{n_1,\ldots,n_d} \prod_{i=1}^{d}(x_i)^{n_i}, \qquad \sum_{k=1}^{d} q_{ik}m_k \leq N_i, \qquad i=1,\ldots,r.$$
$$(\text{C.3.15})$$

The spaces $\Phi_{N_1,\ldots,N_r}$ defined by the formulas (C.3.15) we call functional extensions of the spaces $\Phi_{N_1,\ldots,N_r}(\vec{\nu})$. One of the reasons for which these extensions are interesting is that they again are the invariant subspaces for operators (C.2.3). This fact can be verified immediately by substituting differential realizations (C.3.1) and (C.3.2) of generators of algebra L into the expression (C.2.3) for $H(\vec{\nu})$ and acting by the obtained differential operator $H(\vec{\nu})$ on $\Phi_{N_1,\ldots,N_r}$. Performing these manipulations, it is not difficult to reveal another (no less remarkable) feature of the spaces $\Phi_{N_1,\ldots,N_r}$. This is that their invariance properties do not depend on a concrete choice of non-zero values of constants C entering into expressions (C.3.2)!

This important remark automatically avoids the problem of constructing differential realizations of generators of Lie algebras: we do not need the knowledge of such realizations any longer! We see that the requirement that the operators (C.3.1) and (C.3.2) must neccessarily have the form of generators of some Lie algebras is superfluous. In turn, this suggests that the method for constructing multi-dimensional spectral equations exactly solvable in classes of polynomials can be formulated without introducing the notion of Lie algebras, in a purely analytic way. Such a formulation will be given in the following section.

C.4   The analytic approach

Any measurable physical quantity has some dimension. In any physical theory there are several quantities (measurable in a most simple way) whose dimensions $d_1, \ldots, d_r$ we call elementary. Any other quantity in such a theory has a composite dimension, which can be represented in the form $d_1^{\alpha_1} \ldots d_r^{\alpha_r}$, where $\alpha_1, \ldots, \alpha_r$ are certain real numbers. This means that if the set of basis dimensions $d_1, \ldots, d_r$ is fixed, any other dimension $d_1^{\alpha_1} \ldots d_r^{\alpha_r}$ can be completely characterized by the vectors $\vec{\alpha} = (\alpha_1, \ldots, \alpha_r)$. The basic dimensions correspond in this case to basis vectors with the components $\pi_1 = (1, \ldots, 0), \ldots, \pi_r = (0, \ldots, 1)$. For example, in classical mechanics we have three basic dimensions: $[length] \rightarrow (1, 0, 0)$, $[time] \rightarrow (0, 1, 0)$, $[mass] \rightarrow (0, 0, 1)$ and many composite dimensions: $[velocity] \rightarrow (1, -1, 0)$, $[momentum] \rightarrow (1, -1, 1)$, $[force] \rightarrow (1, -2, 1)$, $[energy] \rightarrow (2, -2, 1)$ and so on.

If A is a measurable quantity and $\vec{\alpha}$ is its (vector) dimension, we shall write

$$\vec{vd}\, A = \vec{\alpha}. \tag{C.4.1}$$

The operation $\vec{vd}$ has the following properties:

1. $\vec{vd}\, A^{-1} = -\vec{vd}\, A$,
2. $\vec{vd}\, AB = \vec{vd}\, A + \vec{vd}\, B$
3. $\vec{vd}\, (A + B) = \vec{vd}\, A = \vec{vd}\, B$, if and only if $\vec{vd}\, A = \vec{vd}\, B$.

Note also that not all quantities can be treated as measurable. For example, if A and B are measurable quantities having different dimensions $\vec{vd}\, A \neq \vec{vd}\, B$, then their sum $A + B$ will be non-measurable.

We call the dimension of a certain measurable quantity A positive ($\vec{vd}\, A > 0$), negative ($\vec{vd}\, A < 0$) or zero ($\vec{vd}\, A = 0$), if all the components of the corresponding vector $\vec{\alpha}$ are non-negative, non-positive, or zero, respectively. All other dimensions we shall call sign-indefinite. It is possible to introduce the order relations between different measurable quantities (in the case when their ratio has a sign-definite dimension). So, we can write $\vec{\mathrm{vd}}\, A > \vec{\mathrm{vd}}\, B$ if $\vec{\mathrm{vd}}\, AB^{-1} > 0$.

Let $\aleph$ be a system of d vectors $\vec{\alpha}_1, \ldots, \vec{\alpha}_d$ belonging to a r-dimensional vector space V_r and satisfying the following condition: there exists a basis $\vec{\pi}_1, \ldots, \vec{\pi}_r$ in the space V_r in which all vectors $\vec{\alpha}_i \in \aleph$, $i = 1, \ldots, d$ have non-negative integer coordinates:

$$\vec{\alpha}_i = \sum_{k=1}^{r} q_{ik}\vec{\pi}_k, \qquad q_{ik} \geq 0, \qquad \vec{\alpha}_i \in \aleph, \qquad i = 1, \ldots, d. \qquad \text{(C.4.2)}$$

The number d of vectors $\vec{\alpha}_i$ forming the system $\aleph$ is fixed arbitrarily, so these vectors are not necessarily linearly independent.

Let us now associate with each vector $\vec{\alpha}_i$ a certain variable x_i:

$$\vec{\alpha}_i \to x_i, \qquad \vec{\alpha}_i \in \aleph, \qquad i = 1, \ldots, d \qquad \text{(C.4.3)}$$

and interpret $\vec{\alpha}_i$ as the vector dimension of the variable x_i:

$$\vec{\mathrm{vd}}\, x_i = \vec{\alpha}_i, \qquad \vec{\alpha}_i \in \aleph, \qquad i = 1, \ldots, d. \qquad \text{(C.4.4)}$$

From the property 1 it follows that

$$\vec{\mathrm{vd}}\, \frac{\partial}{\partial x_i} = -\vec{\alpha}_i, \qquad \vec{\alpha}_i \in \aleph, \qquad i = 1, \ldots, d. \qquad \text{(C.4.5)}$$

Formulas (C.4.4) and (C.4.5) enable us to construct the homogeneous (measurable) functions of variables x_i and homogeneous (measurable) differential operators acting on these functions. Under the homogeneous (measurable) functions and operators we mean those whose dimensions are defined and can be computed by means of rules 1, 2 and 3.

Denote by $\phi_{\vec{N}}$ the space of homogeneous polynomials of a given positive vector dimension $\vec{N}$. This space consists of the monomials

$$p_{\,n_1,\ldots,n_d} = x_1^{n_1} \ldots x_d^{n_d} \qquad \text{(C.4.6)}$$

with $n_1, \ldots, n_d \geq 0$ satisfying the system of Diophantine equations:

$$\sum_{i=1}^{d} n_i \vec{\alpha}_i = \vec{N}. \qquad \text{(C.4.7)}$$

The number of solutions of these equations determines the dimension of the space $\phi_{\vec{N}}$. Due to the positivity of $\vec{N}$ and non-negativity of $n_1, \ldots, n_d$, this dimension is finite.

Denote also by $\hat{D}_{\vec{M}}$ the homogeneous differential operators constructed from variables x_i and partial derivatives $\frac{\partial}{\partial x_i}$ and having a given vector dimension $\vec{M}$. (The only requirement for $\vec{M}$ is that it must be sign definite.) These operators can be represented as linear combinations of the elementary homogeneous operators

$$\hat{d} \, {}^{m'_1,\ldots,m'_d}_{m_1,\ldots,m_d} = x_1^{m_1} \ldots x_d^{m_d} \left(\frac{\partial}{\partial x_1} \right)^{m'_1} \ldots \left(\frac{\partial}{\partial x_d} \right)^{m'_d} \qquad (C.4.8)$$

with the non-negative integers $m_1,\ldots,m_d$ and $m'_1,\ldots,m'_d$ satisfying the system of Diophantine equations

$$\sum_{i=1}^{d} (m_i - m'_i)\vec{\alpha}_i = \vec{M}. \qquad (C.4.9)$$

Because of the absence of any limitation on the admissible values of non-negative integers n_i and m_i, this system has, in general, an infinite number of solutions.

The operators $\hat{D}_{\vec{M}}$ have the remarkable property:

$$\hat{D}_{\vec{M}} \phi_{\vec{N}} \subseteq \phi_{\vec{N}+\vec{M}}. \qquad (C.4.10)$$

This property plays a central role in our scheme, since it enables one to introduce the notion of the raising, lowering and neutral operators $\hat{D}_{\vec{M}}$. Indeed, we see that the operators $\hat{D}_{\vec{M}}$ having positive (negative) vector dimension $\vec{M}$ increase (decrease) the dimension of the space $\phi_{\vec{N}}$, so that they can be treated as raising and lowering operators, respectively. At the same time, the operators $\hat{D}_{\vec{0}}$ do not change the dimension of the space $\phi_{\vec{N}}$, and therefore, they play the role of neutral operators.

This observation enables one to construct a wide class of non-trivial differential operators acting in an infinite-dimensional space Φ of all functions of $x_1,\ldots,x_d$ and being completely diagonalizable by means of purely algebraic methods.

Indeed, denote by $\Phi_{\vec{N}}$ the linear span of all the spaces $\phi_{\vec{N}'}$ with $\vec{0} \le \vec{N}' \le \vec{N}$:

$$\Phi_{\vec{N}} = \bigoplus_{0 \le N' \le N} \phi_{\vec{N}'}. \qquad (C.4.11)$$

Due to the finite dimensionality of the spaces $\phi_{\vec{N}'}$, the space $\Phi_{\vec{N}}$ is a finite-dimensional subspace of the space Φ for any $\vec{N} > \vec{0}$:

$$\Phi_{\vec{N}} \subset \Phi, \qquad \dim \Phi_{\vec{N}} < \infty, \qquad \dim \Phi = \infty. \qquad (C.4.12)$$

Denote also by $\hat{D}$ an arbitrary linear combination of various neutral and lowering operators $\hat{D}_{\vec{M}}$, i.e., operators having non-positive vector dimensions $\vec{M}$:

$$\hat{D} = \bigoplus_{\vec{M} \leq \vec{0}} \hat{D}_{\vec{M}}. \tag{C.4.13}$$

Using formulas (C.4.10) and (C.4.11) it is easily seen that for any given $\vec{N}$ the space $\Phi_{\vec{N}}$ is an invariant subspace for the operator $\hat{D}$. This means that the spectral problem

$$\hat{D}\phi = E\phi, \qquad \phi \in \Phi \tag{C.4.14}$$

(which, obviously, is infinite dimensional) breaks up into an infinite number of finite-dimensional spectral problems

$$\hat{D}\phi = E\phi, \qquad \phi \in \Phi_{\vec{N}}, \tag{C.4.15}$$

each of which can be solved algebraically. This allows one to assert that the problem (C.4.14) is exactly solvable and all its solutions have a polynomial form.

Let us now discuss the problem of constructing the eigenvalues of operators $\hat{D}$ having a given quantum number $\vec{N}$. As follows from the definitions (C.4.13), these operators can be interpreted as infinite-dimensional block-triangular matrices, and, therefore, the problem of finding their eigenvalues is reduced to a more simple problem of diagonalization of finite-dimensional blocks standing on the principal diagonal. In order to perform this reduction, we must replace the operator $\hat{D}$ defined by formula (C.4.13) by its block-diagonal part $\hat{D}_{\vec{0}}$, replacing the space $\Phi_{\vec{N}}$ by its highest space component $\phi_{\vec{N}}$. Then the equation for the eigenvalue E can be rewritten in the following simplified form:

$$\hat{D}_{\vec{0}}\phi = E\phi, \qquad \phi \in \phi_{\vec{N}}. \tag{C.4.16}$$

We see that the eigenvalues E with quantum number $\vec{N}$ form a multiplet whose multiplicity is equal to a dimension of the space $\phi_{\vec{N}}$ (which, in turn, is equal to a total number of solutions of the system (C.4.7)). These eigenvalues are analytic functions of the parameters entering into the explicit expression for operator $\hat{D}_{\vec{0}}$, and are, generally, plaited, forming a common Riemann surface with a finite (dim $\phi_{\vec{N}}$) number of sheets. Since the spectral problems (C.4.16) corresponding to different values of $\vec{N}$ are completely independent, the eigenvalues corresponding to different quantum numbers $\vec{N}$ are not plaited and cannot be obtained from each other by means of any analytic continuation.

We know that in many branches of mathematical physics an especially important role is played by the second-order differential operators. The most general form of operators $\hat{D}$ whose order does not exceed two is

$$H = \sum_{i,k=1}^{d} \sum_{n_1,\ldots,n_d \in A_{ik}} C_{n_1,\ldots,n_d}^{ik} x_1^{n_1} \ldots x_d^{n_d} \frac{\partial}{\partial x_i} \frac{\partial}{\partial x_k}$$

$$+ \sum_{i=1}^{D} \sum_{n_1,\ldots,n_d \in A_i} C_{n_1,\ldots,n_d}^{i} x_1^{n_1} \ldots x_d^{n_d} \frac{\partial}{\partial x_i}. \qquad (C.4.17)$$

Here $C_{n_1,..,n_d}^{ik}$ and $C_{n_1,..,n_d}^{i}$ are arbitrary coefficients, and A_{ik} and A_i are the sets of non-negative integers $n_1,\ldots,n_d$, satisfying the constraints:

$$\sum_{l=1}^{d} n_l \vec{\alpha}_l \leq \vec{\alpha}_i + \vec{\alpha}_k \qquad (C.4.18)$$

and

$$\sum_{l=1}^{d} n_l \vec{\alpha}_l \leq \vec{\alpha}_i, \qquad (C.4.19)$$

respectively. The spectral problems for all such operators are exactly solvable in a class of polynomials by construction.

Now let us establish the connection between the analytic approach described in this section and the algebraic one discussed in section C.2.

Consider an infinite lattice generated by the set $\aleph$ and consisting of the vectors $\vec{\gamma} = \sum_{l=1}^{d} n_i \vec{\alpha}_i$ with $\vec{\alpha}_1, \ldots, \vec{\alpha}_d \in \aleph$ and arbitrary integer $n_1, \ldots, n_d$. Denote by $\Lambda_{\vec{\gamma}}(\aleph)$ the classes of all first-order homogeneous operators of dimensions $\vec{\gamma}$. It is not difficult to see that commutation relations between these classes have the form

$$[\Lambda_{\vec{\gamma}}(\aleph), \Lambda_{\vec{\delta}}(\aleph)] \subset \Lambda_{\vec{\gamma}+\vec{\delta}}(\aleph). \qquad (C.4.20)$$

The linear span of all such classes

$$\Lambda(\aleph) = \bigoplus_{\vec{\gamma}} \Lambda_{\vec{\gamma}}(\aleph) \qquad (C.4.21)$$

is closed under the procedure of commutation, and therefore, $\Lambda(\aleph)$ is a certain (infinite-dimensional) Lie algebra. It is quite obvious that the class of operators H can always be represented in the form:

$$H = \bigoplus_{\vec{\gamma}+\vec{\delta}\leq\vec{0}} \Lambda_{\vec{\gamma}}(\aleph)\Lambda_{\vec{\delta}}(\aleph) + \bigoplus_{\vec{\gamma}\leq\vec{0}} \Lambda_{\vec{\gamma}}(\aleph) \qquad (C.4.22)$$

Now note that any simple Lie algebra characterized by the root system $\aleph$ can be interpreted as a finite-dimensional subalgebra of algebra $\Lambda(\aleph)$. For the generators of algebra L we can write:

$$
\begin{aligned}
L_i &\in \Lambda_{\vec{0}}(\aleph); \\
L_{\vec{\gamma}}^{\pm} &\in \Lambda_{\pm\vec{\gamma}}(\aleph), \qquad \vec{\gamma} \in \aleph; \\
L_{\vec{\gamma}} &= 0, \qquad \vec{\gamma} \notin \aleph \qquad \text{and} \qquad \vec{\gamma} \neq \vec{0}.
\end{aligned}
\tag{C.4.23}
$$

Conserving in (C.4.22) the summation over only such elements as have the form (C.4.23) we get instead of (C.4.22) the expression (C.2.3). Thus, we see that the algebraic method discussed in section C.2 is a very particular case of the analytic method described in the present section.

C.5   Examples

Let us now consider some examples demonstrating the constructivity and simplicity of the analytic method described in the previous section.

Example 1.   The set $\aleph$ consists of an unique one-dimensional vector of unit length: $\vec{\alpha} = 1$. Then we have only one variable x associated with this vector and having vector dimension 1. The operator $\frac{\partial}{\partial x}$ has in this case the dimension -1. It is easily seen that for the homogeneous differential operators $x^n(\frac{\partial}{\partial x})^m$ to be of first and second order and to have non-positive dimensions, the numbers n and m must satisfy the constraint $n \leq m \leq 2$. Then the most general form of the operator H is

$$
H = (c_1 x^2 + c_2 x + c_3)\frac{\partial^2}{\partial x^2} + (c_4 x + c_5)\frac{\partial}{\partial x}.
\tag{C.5.1}
$$

This is the operator for the hypergeometric equation which is exactly solvable in the class of polynomials for any values of parameters c_i.

In order to find the spectrum of this operator (in the framework of the method described in the previous section) we must replace H by its "zero-dimensional" part

$$
H_0 = c_1 x^2 \frac{\partial^2}{\partial x^2} + c_4 x \frac{\partial}{\partial x}
\tag{C.5.2}
$$

and then consider the eigenvalue problem for H_0 in the space ϕ_N, which, obviously, is one dimensional and consists of the functions proportional to x^N. This gives us the equation

$$
(c_1 x^2 \frac{\partial^2}{\partial x^2} + c_4 x \frac{\partial}{\partial x})x^N = E x^N
\tag{C.5.3}
$$

whose solutions are

$$E = c_1 N(N-1) + c_4 N, \qquad N = 0, 1, \dots . \qquad (C.5.4)$$

We see that the spectrum of the operator H consists of an infinite number of single (displaited) eigenvalues.

Example 2. The set $\aleph$ consists of two one-dimensional unit vectors: $\vec{\alpha}_1 = \vec{\alpha}_2 = 1$. In this case we have two variables x and y having equal vector dimensions 1 and two derivatives $\frac{\partial}{\partial x}$ and $\frac{\partial}{\partial y}$ having the opposite vector dimension -1. For the homogeneous operators $x^n y^m (\frac{\partial}{\partial x})^k (\frac{\partial}{\partial y})^l$ to be of first or second order and to have non-positive dimensions, the numbers n, m, l, k must satisfy the system of inequalities: $k + l \leq 2$ and $n + m - k - l \leq 0$. This leads us to the following most general expression for the operator H:

$$H = (c_1 x^2 + c_2 xy + c_3 y^2 + c_4 x + c_5 y + c_6) \frac{\partial^2}{\partial x^2}$$

$$+ (c_7 x^2 + c_8 xy + c_9 y^2 + c_{10} x + c_{11} y + c_{12}) \frac{\partial^2}{\partial y^2}$$

$$+ (c_{13} x^2 + c_{14} xy + c_{15} y^2 + c_{16} x + c_{17} y + c_{18}) \frac{\partial^2}{\partial x \partial y}$$

$$+ (c_{19} x + c_{20} y + c_{21}) \frac{\partial}{\partial x} + (c_{22} x + c_{23} y + c_{24}) \frac{\partial}{\partial y}. \qquad (C.5.5)$$

The "zero-dimensional" part of this operator is

$$H_0 = (c_1 x^2 + c_2 xy + c_3 y^2) \frac{\partial^2}{\partial x^2}$$

$$+ (c_7 x^2 + c_8 xy + c_9 y^2) \frac{\partial^2}{\partial y^2}$$

$$+ (c_{13} x^2 + c_{14} xy + c_{15} y^2) \frac{\partial^2}{\partial x \partial y}$$

$$+ (c_{19} x + c_{20} y) \frac{\partial}{\partial x} + (c_{22} x + c_{23} y) \frac{\partial}{\partial y} \qquad (C.5.6)$$

and the homogeneous spaces ϕ_N consisting of the monomials $x^n y^m$ with $n + m = N$ are $(N + 1)$ dimensional. In order to find the eigenvalues of the operator H, we must solve the auxiliary (simplified) spectral problem for the operator H_0 in the corresponding homogeneous spaces ϕ_N. In the general case this problem is equivalent to the problem of the diagonalization of a certain $N+1$ by $N+1$ matrix parametrized by the numbers c_i entering into the expression for H_0. Consider three particular cases:

1. $N = 0$. The elements of ϕ_0 are constants $\phi = a$ and therefore $H_0\phi = E\phi$ gives $E = 0$.

2. $N = 1$. The elements of ϕ_1 have the form $\phi = ax + by$ and therefore the equation $H_0\phi = E\phi$ is equivalent to the system of two algebraic equations

$$c_{19}a + c_{22}b = Ea, \qquad c_{20}a + c_{23}b = Eb, \qquad \text{(C.5.7)}$$

having two solutions for E. These solutions form a two-sheeted Riemann surface.

3. $N = 2$. The elements of ϕ_2 have the form $\phi = ax^2 + by^2 + 2cxy$ and therefore the equation $H_0\phi = E\phi$ is equivalent to the system of three algebraic equations

$$(c_1 + c_{19})a + c_7b + (c_{13} + c_{22})c = Ea,$$
$$c_3a + (c_9 + c_{23})b + (c_{15} + c_{20})c = Eb,$$
$$(c_2 + c_{20})a + (c_8 + c_{22})b + (c_{14} + c_{19} + c_{23})c = Ec \qquad \text{(C.5.8)}$$

having three solutions for E. These solutions form a three-sheeted Riemann surface.

The cases with higher values of N can be considered analogously. We see that the spectrum of the operator H consists of the infinite number of displaited multiplets consisting of 1, 2, 3, ... plaited eigenvalues.

Example 3. The set $\aleph$ consists of two non-equal one-dimensional vectors: $\vec{\alpha}_1 = 2$, $\vec{\alpha}_2 = 3$. In this case we must introduce two different variables x and y having different vector dimensions 2 and 3, and two differential operators $\frac{\partial}{\partial x}$ and $\frac{\partial}{\partial y}$ having opposite vector dimensions -2 and -3. Now the numbers n, m, l, k entering into the homogeneous operators $x^n y^m (\frac{\partial}{\partial x})^k (\frac{\partial}{\partial y})^l$ must satisfy the constraints: $k + l \le 2$ and $2n + 3m - 2l - 3k \le 0$. This gives us the following general form for the operator H:

$$H = (c_1x^2 + c_2y + c_3x + c_4)\frac{\partial^2}{\partial x^2}$$
$$+(c_5x^3 + c_6y^2 + c_7xy + c_8x^2 + c_9y + c_{10}x + c_{11})\frac{\partial^2}{\partial y^2}$$
$$+(c_{12}xy + c_{13}x^2 + c_{14}y + c_{15}x + c_{16})\frac{\partial^2}{\partial x\partial y}$$
$$+(c_{17}x + c_{18})\frac{\partial}{\partial x} + (c_{19}y + c_{20}x + c_{21})\frac{\partial}{\partial y}. \qquad \text{(C.5.9)}$$

The "zero-dimensional" part of this operator is

$$H_0 = c_1x^2\frac{\partial^2}{\partial x^2} + (c_5x^3 + c_6y^2)\frac{\partial^2}{\partial y^2} + c_{12}xy\frac{\partial^2}{\partial x\partial y}$$

$$+c_{17}x\frac{\partial}{\partial x}+c_{19}y\frac{\partial}{\partial y} \qquad \text{(C.5.10)}$$

and the homogeneous spaces ϕ_N consisting of the monomials $x^n y^m$ with $2n + 3m = N$ are $([N/6] + 1)$ dimensional. Repeating previous reasonings we find that the spectrum of the operator H depends of the parameters c_i entering into the expression for H_0 and has the form of an infinite number of displaited multiplets consisting of $[N/6] + 1$ plaited eigenvalues.

Example 4. The set $\aleph$ consists of two non-collinear two-dimensional vectors $\vec{\alpha}_1 = (1, 2)$, $\vec{\alpha}_2 = (2, 5)$. We have again two variables x and y of vector dimensions $(1, 2)$ and $(2, 5)$, respectively, and two derivatives $\frac{\partial}{\partial x}$ and $\frac{\partial}{\partial y}$ of opposite vector dimensions $(-1, -2)$ and $(-2, -5)$. Now the constraints for the numbers n, m, l, k entering into the homogeneous operators $x^n y^m (\frac{\partial}{\partial x})^k (\frac{\partial}{\partial y})^l$ take the form: $k + l \leq 2$, $n + 2m - l - 2k \leq 0$ and $2n + 5m - 2l - 5k \leq 0$. Solving this system we find the most general form of the operator H:

$$H = (c_1 x^2 + c_2 x + c_3)\frac{\partial^2}{\partial x^2}$$

$$+(c_4 x^4 + c_5 x^3 + c_6 x^2 y + c_7 x^2 + c_8 xy + c_9 x + c_{10}y^2 + c_{11}y + c_{12})\frac{\partial^2}{\partial y^2}$$

$$+(c_{13}x^3 + c_{14}x^2 + c_{15}xy + c_{16}x + c_{17}y + c_{18})\frac{\partial^2}{\partial x \partial y}$$

$$+(c_{19}x + c_{20})\frac{\partial}{\partial x} + (c_{21}x^2 + c_{22}x + c_{23}y + c_{24})\frac{\partial}{\partial y}.$$

$$\text{(C.5.11)}$$

The "zero-dimensional" part of the operator H is

$$H_0 = c_1 x^2 \frac{\partial^2}{\partial x^2} + c_{10}y^2 \frac{\partial^2}{\partial y^2} + c_{15}xy\frac{\partial^2}{\partial x \partial y}$$

$$+c_{19}x\frac{\partial}{\partial x} + c_{23}y\frac{\partial}{\partial y} \qquad \text{(C.5.12)}$$

and the homogeneous spaces $\phi_{\vec{N}}$ with $\vec{N} = (N_1, N_2)$ consisting of the monomials $x^{5N_1 - 2N_2} y^{N_2 - 2N_1}$ are one dimensional for any N_1 and N_2 such that $5N_1 - 2N_2 \geq 0$ and $N_2 - 2N_1 \geq 0$. This means that the spectrum of the operator H consists of an infinite number of single (displaited) eigenvalues. Their explicit form, which can easily be obtained from (C.5.12), is

$$E = c_1(5N_1 - 2N_2)(5N_1 - 2N_2 - 1)$$

$$+c_{10}(N_2 - 2N_1)(N_2 - 2N_1 - 1)$$

$$+c_{15}(5N_1 - 2N_2)(N_2 - 2N_1)$$
$$+c_{19}(5N_1 - 2N_2) + c_{23}(N_2 - 2N_1) \tag{C.5.13}$$

with $5N_1 - 2N_2 \geq 0$ and $N_2 - 2N_1 \geq 0$.

The cases with other more complex sets $\aleph$ can be considered analogously. We see that any such set generates a certain class of second-order differential operators with exactly (algebraically) calculable spectra.

C.6   Construction of quasi-exactly solvable equations

In this section we demonstrate that the class of neutral operators $D_{\vec{0}}$ introduced in section C.4 can be used as a starting point by constructing differential operators for quasi-exactly solvable models of one- and multi-dimensional quantum mechanics.

In order to show this, let us first remember the result of section C.4, according to which any differential operator $H \in D_{\vec{0}}$ having zero vector dimension possesses a finite-dimensional invariant subspace $\phi_{\vec{N}}$ formed by all homogeneous polynomials of a given physical dimension $\vec{N}$. Therefore, the spectral problem for H in $\phi_{\vec{N}}$

$$H\phi = E\phi, \quad \phi \in \Phi_{\vec{N}} \tag{C.6.1}$$

is formulated correctly and has a finite number of algebraic solutions.

Let us now consider an r-component vector operator defined by the formula

$$\vec{J} = \sum_{l=1}^{d} \vec{\alpha}_l x_l \frac{\partial}{\partial x_l} \tag{C.6.2}$$

and having zero vector dimension, $\mathrm{vd}\vec{J} = \vec{0}$. It is not difficult to see that all the components of this operator commute with each other and with H,

$$[H, \vec{J}] = 0, \tag{C.6.3}$$

and form therefore a commutative symmetry algebra of equation (C.6.1).

Define $\Psi_{\vec{N}}$ as the set of all solutions of the system

$$\vec{J}\phi \equiv \left\{ \sum_{l=1}^{d} \vec{\alpha}_l x_l \frac{\partial}{\partial x_l} \right\} \phi = \vec{N}\phi, \tag{C.6.4}$$

and note that

$$\phi_{\vec{N}} \subset \Psi_{\vec{N}}. \tag{C.6.5}$$

Obviously, the set $\Psi_{\vec{N}}$ can be viewed as a linear space. The general form of its elements is given by the formula

$$\phi = \varphi_{\vec{N}} f(z_1, \ldots, z_{d-r}), \tag{C.6.6}$$

where $\varphi_{\vec{N}}$ is an arbitrarily fixed solution of the inhomogeneous system (C.6.4) and $f(z_1, \ldots, z_{d-r})$ is an arbitrary function of $d - r$ independent solutions of its homogeneous analogue. In order to obtain explicit expressions for these functions, it is convenient to introduce the numbers λ_l, $l = 1, \ldots, d$ and μ_l^k, $l = 1, \ldots, d$, $k = 1, \ldots, d - r$ satisfying the system of algebraic equations

$$\sum_{l=1}^{d} \vec{\alpha}_l \lambda_l = \vec{N} \tag{C.6.7}$$

and

$$\sum_{l=1}^{d} \vec{\alpha}_l \mu_l^k = 0, \quad k = 1, \ldots, d - r. \tag{C.6.8}$$

In terms of these numbers the needed functions take the form

$$\varphi_{\vec{N}} = \prod_{l=1}^{d} x_l^{\lambda_l} \tag{C.6.9}$$

and

$$z_k = \prod_{l=1}^{d} x_l^{\mu_l^k}, \quad k = 1, \ldots, d - r. \tag{C.6.10}$$

Denoting the set of functions $f(z_1, \ldots, z_{d-r})$ by F, and taking into account the condition

$$\dim \Psi_{\vec{N}} = \dim F, \tag{C.6.11}$$

we can conclude that, due to the arbitrariness of functions $f(z_1, \ldots, z_{d-r})$, the space $\Psi_{\vec{N}}$ is infinite dimensional.

Now note that, according to formulas (C.6.3) and (C.6.4), the space $\Psi_{\vec{N}}$ is invariant under the action of operator H, and that this enables us to consider the following spectral problem:

$$H\phi = E\phi, \quad \phi \in \Psi_{\vec{N}}. \tag{C.6.12}$$

We know from (C.6.1) and (C.6.5) that this problem has a finite number $(= \dim \phi_{\vec{N}})$ of exact (polynomial) solutions belonging to the space $\phi_{\vec{N}} \subset \Psi_{\vec{N}}$. At the same time, other solutions of equation (C.6.12) (lying outside the space $\phi_{\vec{N}}$ and therefore having a non-polynomial structure) remain unknown to us. This means that we deal with a typical quasi-exactly solvable equation of order $\dim \phi_{\vec{N}}$.

In order to reduce the obtained equation to a more appropriate form, we use the projection method described in sections 1.9, 5.2 and 5.6.

Following this method, consider the set of all such functions $f \in F$ for which (C.6.6) is an element of the space $\phi_{\vec{N}}$. Denote this set by $F_{\vec{N}}$. Evidently,

$$\dim F_{\vec{N}} = \dim \phi_{\vec{N}} \leq \infty. \tag{C.6.13}$$

Using formula (C.6.6), we can rewrite (C.6.12) as:

$$H_{\vec{N}} f = E f, \quad f \in F, \tag{C.6.14}$$

with

$$H_{\vec{N}} = \varphi_{\vec{N}}^{-1} H \varphi_{\vec{N}}. \tag{C.6.15}$$

The fact that the elements of the space F are functions of only the $d - r$ variables $z_1, \ldots, z_{d-r}$ allows one to replace the operator $H_{\vec{N}}$ by its projection on F. After this, $H_{\vec{N}}$ takes the form of a $(d - r)$-dimensional differential operator in variables $z_1, \ldots, z_{d-r}$ with the coefficient functions depending on the same variables and also on the vector $\vec{N}$ determining the number $(\dim F_{\vec{N}})$ of exactly calculable eigenvalues and eigenfunctions of $H_{\vec{N}}$.

Let us restrict ourselves to the case of second-order differential operators H. In this case the most general form of the operator H is

$$H = \sum_{i,k=1}^{D} \sum_{n_1,\ldots,n_D \in B_{ik}} C_{n_1,\ldots,n_D}^{ik} x_1^{n_1} \ldots x_D^{n_D} \frac{\partial}{\partial x_i} \frac{\partial}{\partial x_k}$$

$$+ \sum_{i=1}^{D} \sum_{n_1,\ldots,n_D \in B_i} C_{n_1,\ldots,n_D}^{i} x_1^{n_1} \ldots x_D^{n_D} \frac{\partial}{\partial x_i} \tag{C.6.16}$$

where $C_{n_1,\ldots,n_D}^{ik}$ and $C_{n_1,\ldots,n_D}^{i}$ are arbitrary real coefficients, and B_{ik} and B_i are the sets of non-negative integers $n_1, \ldots, n_D$ satisfying the constraints

$$\sum_{l=1}^{d} \vec{\alpha}_l n_l = \vec{\alpha}_i - \vec{\alpha}_k \tag{C.6.17}$$

and

$$\sum_{l=1}^{d} \vec{\alpha}_l n_l = \vec{\alpha}_i. \tag{C.6.18}$$

Applying to this operator the reduction procedure we obtain a second-order differential operator

$$H_{\vec{N}} = \sum_{i,k=1}^{d-r} A_{\vec{N}}^{ik}(\vec{z}) \frac{\partial^2}{\partial z_i \partial z_k} + \sum_{i=1}^{d-r} A_{\vec{N}}^{i}(\vec{z}) \frac{\partial}{\partial z_i} + A_{\vec{N}}(\vec{z}) \tag{C.6.19}$$

with easily computable coefficient functions, describing a certain quasi-exactly solvable spectral equation. This equation can easily be reduced to the Schrödinger form. For this it is sufficient to apply to (C.6.19) the procedure described in section 5.7. The order of the quasi-exactly solvable models obtained is given by the number of solutions of the system of Diophantine equations (C.4.7).

Appendix D

Lie-algebraic hamiltonians and quasi-exact solvability

A González-López[1], N Kamran[2] and P J Olver[3]

D.1   Introduction

In this appendix, we shall give a brief summary of recent results of the three authors into the mathematical theory and classification of quasi-exactly solvable Schrödinger operators.   We will emphasize new results in the multi-dimensional case, even though additional results, including the complete solution to the "normalizability problem" for quasi-exactly solvable operators on the line, have also followed from our research.   We refer the interested reader to a more complete survey paper by González-López *et al* (1993d), to appear in a volume of *Contemporary Mathematics* devoted to a conference on quasi-exactly solvable problems and related issues held in Springfield, MO, USA, in March, 1992.

Lie-algebraic and Lie group-theoretical methods have played a significant role in the development of quantum mechanics since its inception.   In the classical applications, the Lie group appears as a symmetry group of the hamiltonian, and the associated representation theory provides an algebraic means for computing the spectrum.   Of particular importance are the exactly solvable problems, such as the harmonic oscillator or the hydrogen atom, whose point spectrum can be completely determined using purely algebraic methods. The fundamental concept of a "spectrum generating algebra" was introduced by Arima

[1] Universidad Complutense, Madrid, Spain.
[2] McGill University, Montreal, Canada.
[3] University of Minnesota, USA.

and Iachello (1976, 1978), to study nuclear physics, and subsequently, by Iachello, Alhassid, Gürsey, Levine, Wu and their collaborators, was also successfully applied to molecular dynamics and spectroscopy (Iachello and Levine 1982, Levine 1988), and scattering theory (Alhassid *et al* 1983, 1984, 1986). The Schrödinger operators amenable to the algebraic approach assume a "Lie-algebraic form", meaning that they belong to the universal enveloping algebra of the spectrum generating algebra. Lie-algebraic operators reappeared in the discovery of Turbiner, Ushveridze, Shifman and their collaborators (Turbiner 1988a, Ushveridze 1988c,d, Shifman and Turbiner 1989) of a new class of physically significant spectral problems, which they named "quasi-exactly solvable", having the property that a (finite) part of the point spectrum can be determined using purely algebraic methods. This is an immediate consequence of the additional requirement that the hidden symmetry algebra preserve a finite-dimensional representation space consisting of smooth wavefunctions. In this case, the hamiltonian restricts to a linear transformation on the representation space, and hence the associated eigenvalues can be computed by purely algebraic methods, meaning matrix eigenvalue calculations. Finally, one must decide the "normalizability" problem of whether the resulting "algebraic" eigenfunctions are square integrable and therefore represent true bound states of the system. Connections with conformal field theory (Morozov *et al* 1990, Gorsky 1991, Shifman 1992), and the theory of orthogonal polynomials (Turbiner 1992a, b), lend additional impetus for the study of such problems.

In higher dimensions, much less is known than in the one-dimensional case. Only a few special examples of quasi-exactly solvable problems in two dimensions have appeared in the literature to date (Shifman and Turbiner 1989), all of which are constructed using semi-simple Lie algebras. Complete lists of finite-dimensional Lie algebras of differential operators are known in two (complex) dimensions; there are essentially 24 different classes, some depending on parameters. The quasi-exactly solvable condition imposes a remarkable quantization constraint on the cohomology parameters classifying these Lie algebras. This phenomenon of the "quantization of cohomology" has recently been given an algebro-geometric interpretation (González-López *et al* 1993a). Any of the resulting quasi-exactly solvable Lie algebras of differential operators can be used to construct new examples of two-dimensional quasi-exactly solvable spectral problems. An additional complication is that, in higher dimensions, not every elliptic second-order differential operator is equivalent to a Schrödinger operator (i.e., minus laplacian plus potential), so not every Lie-algebraic operator can be assigned an immediate physical meaning. The resulting "closure conditions" are quite complicated to solve, and so the problem of completely classifying quasi-exactly solvable Schrödinger operators in two dimensions appears to be too difficult to solve in full generality. A variety of interesting examples are given in the paper of González-López *et al* (1992b), and we present a few particular cases of interest here.

D.2   Quasi-exactly solvable Schrödinger operators

Let M denote an open subset of Euclidean space $\mathbf{R}^n$ with coordinates $x = (x^1, \ldots, x^n)$. The time-independent Schrödinger equation for a differential operator $\mathcal{H}$ is the eigenvalue problem

$$\mathcal{H}[\psi] = \lambda \psi. \tag{D.2.1}$$

In the quantum mechanical interpretation, a (self-adjoint) differential operator $\mathcal{H}$ plays the role of the quantum "hamiltonian" of the system. A non-zero wavefunction $\psi(x)$ is called *normalizable* if it is square integrable, i.e., lies in the Hilbert space $L^2(\mathbf{R}^n)$, and so represents a physical bound state of the quantum mechanical system, the corresponding eigenvalue determining the associated energy level. While it is of great interest to know the bound states and energy levels of a given operator, complete explicit lists of eigenvalues and eigenfunctions are known for only a handful of classical "exactly solvable" operators, such as the harmonic oscillator. For the vast majority of quantum mechanical problems, the spectrum can, at best, only be approximated by numerical computation. The quasi-exactly solvable systems occupy an important intermediate station, in that a finite part of the spectrum can be computed by purely algebraic means.

   To describe the general form of a quasi-exactly solvable problem, we begin with a finite-dimensional Lie algebra $\mathbf{g}$ spanned by r linearly independent first-order differential operators

$$T^a = \sum_{i=1}^{n} \xi^{ai}(x)\, \frac{\partial}{\partial x^i} + \eta^a(x), \quad a = 1, \ldots, r, \tag{D.2.2}$$

whose coefficients ξ^{ai}, η^a are smooth functions of x. The Lie algebra assumption requires that the commutator between two such operators can be written as a linear combination of the operators: $[T^a, T^b] = T^a T^b - T^b T^a = \sum_c C^c_{ab} T^c$, where the C^c_{ab} are the structure constants of the Lie algebra $\mathbf{g}$. Note that each differential operator is a sum, $T^a = \mathbf{v}^a + \eta^a$, of a *vector field* $\mathbf{v}^a = \sum \xi^{ai} \partial/\partial x^i$ (which may be zero) and a *multiplication operator* η^a.

   A differential operator is said to be *Lie algebraic* if it lies in the universal enveloping algebra $U(\mathbf{g})$ of the Lie algebra $\mathbf{g}$, meaning that it can be expressed as a polynomial in the operators T^a. In particular, a second-order differential operator is Lie algebraic if it can be written as a quadratic combination

$$-\mathcal{H} = \sum_{a,b} c_{ab} T^a T^b + \sum_a c_a T^a + c_0, \tag{D.2.3}$$

for certain constants c_{ab}, c_a, c_0. (The minus sign in front of the hamiltonian is taken for later convenience.) If some of the operators T^a generating the Lie algebra are pure multiplication operators, then one could allow higher-degree combinations in (D.2.3); however, it is not hard to show that such Lie-algebraic operators can always be re-expressed in a quadratic form, (D.2.3), for some possibly larger Lie algebra g, and so we are not losing any generality with the form (D.2.3). Note that the commutator $[T^c, \mathcal{H}]$ of the hamiltonian with any generator of g, while still of the same Lie-algebraic form, is not in general a multiple of the hamiltonian $\mathcal{H}$ (unless, for example, $\mathcal{H}$ happens to be a Casimir operator for g). Therefore, the "hidden symmetry algebra" g is not a symmetry algebra in the traditional sense. Lie-algebraic operators appeared in the early work of Iachello, Levine, Alhassid, Gürsey and collaborators in the algebraic approach to scattering theory (Alhassid *et al* 1983, 1984, 1986, Levine 1988).

The condition of quasi-exact solvability imposes an additional constraint on the Lie algebra and hence on the types of operator which are allowed. A Lie algebra of first-order differential operators g will be called *quasi-exactly solvable* if it possesses a finite-dimensional representation space (or module) $\mathcal{N} \subset C^\infty$ consisting of smooth functions; this means that if $\psi \in \mathcal{N}$ and $T^a \in$ g, then $T^a(\psi) \in \mathcal{N}$. A differential operator $\mathcal{H}$ is called *quasi-exactly solvable* if it lies in the universal enveloping algebra of a quasi-exactly solvable Lie algebra of differential operators. Clearly, the module $\mathcal{N}$ is an invariant space for the hamiltonian $\mathcal{H}$, i.e., $\mathcal{H}(\mathcal{N}) \subset \mathcal{N}$, and hence $\mathcal{H}$ restricts to a linear matrix operator on $\mathcal{N}$. We will call the eigenvalues and corresponding eigenfunctions for the restriction $\mathcal{H}|\mathcal{N}$ *algebraic* since they can be computed by algebraic methods for matrix eigenvalue problems. (This does not mean that these functions are necessarily "algebraic" in the traditional pure mathematical sense.) Note that the number of such "algebraic" eigenvalues and eigenfunctions equals the dimension of $\mathcal{N}$. So far we have not imposed any normalizability conditions on the algebraic eigenfunctions, but, if they are normalizable, then the corresponding algebraic eigenvalues give part of the point spectrum of the differential operator.

It is of great interest to know when a given differential operator is in Lie-algebraic or quasi-exactly solvable form. There is not, as far as we know, any direct test on the operator in question that will answer this in general. Consequently, the best approach to this problem is to effect a complete classification of such operators under an appropriate notion of equivalence. In order to classify Lie algebras of differential operators, and hence Lie-algebraic and quasi-exactly solvable Schrödinger operators, we need to precisely specify the allowable changes of variables.

Definition 1. Two differential operators are *equivalent* if they can be mapped into each other by a combination of change of independent variable,

$$\bar{x} = \varphi(x), \tag{D.2.4}$$

and "gauge transformation"

$$\overline{\mathcal{H}} = e^{\sigma(x)}\mathcal{H}e^{-\sigma(x)}. \tag{D.2.5}$$

The transformations (D.2.4), (D.2.5), have two key properties. First, they respect the commutator between differential operators, and therefore preserve their Lie algebra structure. Second, they preserve the spectral problem (D.2.1) associated with the differential operator $\mathcal{H}$, so that if $\psi(x)$ is an eigenfunction of $\mathcal{H}$ with eigenvalue λ, then the transformed (or "gauged") function

$$\overline{\psi}(\bar{x}) = e^{\sigma(x)}\psi(x), \qquad \text{where} \qquad \bar{x} = \varphi(x), \tag{D.2.6}$$

is the corresponding eigenfunction of $\overline{\mathcal{H}}$ having the same eigenvalue. Therefore this notion of equivalence is completely adapted to the problem of classifying quasi-exactly solvable Schrödinger operators. The gauge factor $\mu(x) = e^{\sigma(x)}$ in (D.2.5) is *not* necessarily unimodular, i.e., $\sigma(x)$ is not restricted to be purely imaginary, and hence does not necessarily preserve the normalizability properties of the associated eigenfunctions. Therefore, the problem of normalizability of the resulting algebraic wavefunctions must be addressed.

Definition 2. A quasi-exactly solvable Schrödinger operator is called *normalizable* if every algebraic eigenfunction is normalizable. It is called *partially normalizable* if some of the algebraic eigenfunctions are normalizable.

Let us summarize the basic steps that are required in order to obtain a complete classification of quasi-exactly solvable operators and their algebraic physical states.

1. Classify finite-dimensional Lie algebras of differential operators.
2. Determine which Lie algebras are quasi-exactly solvable.
3. Solve the equivalence problem for differential operators.

4. Determine normalizability conditions.

5. Solve the associated matrix eigenvalue problem.

In the remainder of this survey, we will concentrate on the multi-dimensional case and refer the reader to the paper of González-López *et al* (1993b) for the solution to the normalizability problem for one-dimensional quasi-exactly solvable Schrödinger operators.

D.3 The equivalence of differential operators

Consider a second-order linear differential operator

$$-\mathcal{H} = \sum_{i,j=1}^{n} g^{ij}(x) \frac{\partial^2}{\partial x^i \partial x^j} + \sum_{i=1}^{n} h^i(x) \frac{\partial}{\partial x^i} + k(x), \qquad \text{(D.3.1)}$$

defined on an open subset $M \subset \mathbf{R}^n$. We are interested in studying the problem of when two such operators are equivalent under the combination of change of variables and gauge transformations (D.2.4), (D.2.5). Of particular importance is the question of when $\mathcal{H}$ is equivalent to a Schrödinger operator, which we take to mean an operator $\mathcal{S} = -\Delta + V(x)$, where Δ denotes either the flat space laplacian or, more generally, the Laplace–Beltrami operator over a curved manifold. (Operators on Riemannian manifolds with non-zero curvature can be viewed as constrained quantum mechanical systems, e.g., a particle moving on a sphere (Balazs and Voros 1986).) This definition of Schrödinger operator excludes the introduction of a magnetic field, which, however, can also be handled by these methods. There is an essential difference between one-dimensional and higher-dimensional spaces in the solution to the equivalence problem for second-order differential operators because in higher space dimensions, it is no longer true that every second-order differential operator is locally equivalent to a Schrödinger operator of the form $-\Delta + V(x)$, where Δ is the flat space laplacian. Explicit equivalence conditions were first found by Cotton (1900). Since the symbol of a linear differential operator is invariant under coordinate transformations, we begin by assuming that the operator is elliptic, meaning that the symmetric matrix $\widehat{\mathbf{g}}(x) = \left(g^{ij}(x)\right)$ determined by the leading coefficients of $-\mathcal{H}$ is positive definite. Owing to the induced transformation rules under the change of variables (D.3.1), we interpret the inverse matrix $\mathbf{g}(x) = \widehat{\mathbf{g}}(x)^{-1} = \left(g_{ij}(x)\right)$ as defining a Riemannian metric

$$ds^2 = \sum_{i,j=1}^{n} g_{ij}(x) \, \mathrm{d}x^i \, \mathrm{d}x^j, \qquad \text{(D.3.2)}$$

on the subset $M \subset \mathbf{R}^n$. We will follow the usual tensor convention of raising and lowering indices with respect to the Riemannian metric (D.3.2). We rewrite the differential operator (D.3.1) in a more natural coordinate-independent form

$$\mathcal{H} = -\sum_{i,j=1}^{n} g^{ij}(\nabla_i - A_i)(\nabla_j - A_j) + V, \qquad \text{(D.3.3)}$$

where ∇_i denotes covariant differentiation using the associated Levi–Civita connection. Physically, $A(x) = (A_1(x), \ldots, A_n(x))$ can be thought of as a (generalized) magnetic vector potential; in view of its transformation properties, we define the associated *magnetic one-form*

$$\omega = \sum_{i=1}^{n} A_i(x) \, dx^i. \qquad \text{(D.3.4)}$$

(Actually, to qualify as a physical vector potential, A must be purely imaginary and satisfy the stationary Maxwell equations, but we need not impose this additional physical constraint in our definition of the mathematical magnetic one-form (D.3.4).) The explicit formulas relating the covariant form (D.3.3) to the standard form (D.3.1) of the differential operator are

$$A^i = \sum_{j=1}^{n} g^{ij} A_j = -\frac{h^i}{2} + \frac{1}{2\sqrt{g}} \sum_{j=1}^{n} \frac{\partial(\sqrt{g}\, g^{ij})}{\partial x^j},$$

$$V = -k + \sum_{i=1}^{n} \left[A_i A^i - \frac{1}{\sqrt{g}} \frac{\partial}{\partial x^i} (\sqrt{g}\, A^i) \right], \qquad \text{(D.3.5)}$$

where $g(x) = \det(g_{ij}(x)) > 0$. Each second-order elliptic operator then is uniquely specified by a metric, a magnetic one-form and potential function $V(x)$. In particular, if the magnetic form vanishes, so $A = 0$, then $\mathcal{H}$ has the form of a Schrödinger operator $\mathcal{H} = -\Delta + V$, where Δ is the Laplace–Beltrami operator associated with the metric (D.3.2).

The application of a gauge transformation (D.2.5) does not affect the metric or the potential; however the magnetic one-form is modified by an exact one-form: $\omega \mapsto \omega + d\sigma$. Consequently, the "magnetic two-form" $\Omega = d\omega$, whose coefficients represent the associated magnetic field, *is* unaffected by gauge transformations.

Theorem D.1. *Two elliptic second-order differential operators $\mathcal{H}$ and $\overline{\mathcal{H}}$ are (locally) equivalent under a change of variables $\bar{x} = \varphi(x)$ and gauge* *transformation (D.2.5) if and only if their metrics, their magnetic two-forms, and their potentials are mapped to each other*

$$\varphi^*(\mathrm{d}\bar{s}^2) = \mathrm{d}s^2, \qquad \varphi^*(\overline{\Omega}) = \Omega, \qquad \varphi^*(\overline{V}) = V. \tag{D.3.6}$$

(Here φ^* denotes the standard pull-back action of φ on differential forms; in particular, $\varphi^*(\overline{V}) = \overline{V} \circ \varphi$.)

   In particular, an elliptic second-order differential operator is equivalent to a Schrödinger operator $-\Delta + V$ if and only if its magnetic one-form is closed: $\mathrm{d}\omega = \Omega = 0$. Moreover, since the curvature tensor associated with the metric is invariant, the Laplace–Beltrami operator Δ will be equivalent to the flat space laplacian if and only if the metric $\mathrm{d}s^2$ is flat, i.e., has vanishing Riemannian curvature tensor.

D.4   The Lie algebras of differential operators

In this section, we summarize what is known about the classification problem for Lie algebras of first-order differential operators. Any finite-dimensional Lie algebra **g** of first order differential operators has a basis of the form

$$T^1 = \mathbf{v}^1 + \eta^1(x), \dots, T^r = \mathbf{v}^r + \eta^r(x),$$
$$T^{r+1} = \zeta^1(x), \dots, T^{r+s} = \zeta^s(x), \tag{D.4.1}$$

cf. (D.2.2). Here $\mathbf{v}^1, \dots, \mathbf{v}^r$ are linearly independent vector fields spanning an r-dimensional Lie algebra **h**. The functions $\zeta^1(x), \dots, \zeta^s(x)$ define multiplication operators, and span an abelian subalgebra $\mathcal{M}$ of the full Lie algebra **g**. Since the commutator $[\mathbf{v}^i, \zeta^k] = \mathbf{v}^i(\zeta^j)$ is a multiplication operator, which must belong to **g**, we conclude that **h** acts on $\mathcal{M}$, which is a finite-dimensional **h**-module (representation space) of smooth functions, that is $\mathbf{v}^i(\zeta^j) = \sum_k b_k^{ij} \zeta^k$ for constants b_k^{ij}. The functions $\eta^a(x)$ must satisfy additional constraints in order that the operators (D.4.1) span a Lie algebra; we find

$$[\mathbf{v}^i + \eta^i, \mathbf{v}^j + \eta^j] = [\mathbf{v}^i, \mathbf{v}^j] + \mathbf{v}^i(\eta^j) - \mathbf{v}^j(\eta^i). \tag{D.4.2}$$

Now, since **h** is a Lie algebra, $[\mathbf{v}^i, \mathbf{v}^j] = c_{ij}^k \mathbf{v}^k$, where c_{ij}^k are the structure constants. Thus the above commutator will belong to **g** if and only if the right-hand side is equal to $c_{ij}^k(\mathbf{v}^k + \eta^k)$ up to a linear combination of the functions ζ^l.

   These conditions can be conveniently re-expressed using the basic theory of Lie algebra cohomology (Jacobson 1962). Define the "one-cochain" on the vector field Lie algebra **h** by the linear map $F: \mathbf{h} \to C^\infty$

which satisfies $\langle F ; \mathbf{v}^a \rangle = \eta^a$. Since we can add in any constant coefficient linear combination of the ζ^bs to the η^as without changing the Lie algebra $\mathbf{g}$, we should interpret the η^as as lying in the quotient space $C^\infty / \mathcal{M}$, and hence regard F as a $(C^\infty / \mathcal{M})$-valued cochain. In view of (D.4.2), the collection of differential operators (D.4.1) spans a Lie algebra if and only if the cochain F satisfies

$$\mathbf{v}\langle F ; \mathbf{w} \rangle - \mathbf{w}\langle F ; \mathbf{v} \rangle - \langle F ; [\mathbf{v}, \mathbf{w}] \rangle \in \mathcal{M} \quad \text{for all} \quad \mathbf{v}, \mathbf{w} \in \mathbf{h}. \quad (D.4.3)$$

The left-hand side of (D.4.3) is just the evaluation $\langle \delta_1 F ; \mathbf{v}, \mathbf{w} \rangle$ of the *coboundary* of the 1-cochain F, hence (D.4.3) expresses the fact that the cochain F must be a $(C^\infty / \mathcal{M})$-valued *cocycle*. A 1-cocycle is itself a coboundary, written $F = \delta_0 \sigma$ for some $\sigma(x) \in C^\infty$ if and only if $\langle F ; \mathbf{v} \rangle = \mathbf{v}(\sigma)$ for all $\mathbf{v} \in \mathbf{h}$, where $\mathbf{v}(\sigma)$ is considered as an element of $C^\infty / \mathcal{M}$. It can be shown that two cocycles will differ by a coboundary $\delta_0 \sigma$ if and only if the corresponding Lie algebras are equivalent under the gauge transformation (D.2.5). Therefore two cocycles lying in the same *cohomology class* in the cohomology space $H^1(\mathbf{h}, C^\infty / \mathcal{M}) = \text{Ker}\, \delta_1 / \text{Im}\, \delta_0$, will give rise to equivalent Lie algebras of differential operators. In summary, then, we have the following fundamental characterization of Lie algebras of first-order differential operators.

Theorem D.2. *There is a one-to-one correspondence between equivalence classes of finite-dimensional Lie algebras $\mathbf{g}$ of first-order differential operators on M and equivalence classes of triples $[\mathbf{h}, \mathcal{M}, [F]]$, where*

1. *$\mathbf{h}$ is a finite-dimensional Lie algebra of vector fields,*
2. *$\mathcal{M} \subset C^\infty$ is a finite-dimensional $\mathbf{h}$-module of functions,*
3. *$[F]$ is a cohomology class in $H^1(\mathbf{h}, C^\infty / \mathcal{M})$.*

Based on theorem D.2, there are three basic steps required to classify finite-dimensional Lie algebras of first-order differential operators. First, one needs to classify the finite-dimensional Lie algebras of vector fields $\mathbf{h}$ up to changes of variables; this was done by Lie in one and two dimensions under the assumption that the Lie algebra has no singularities — not every vector field in the Lie algebra vanishes at a common point. (Lie further claimed to have completed the classification in three dimensions (Lie 1893), but the complete results were never published.) Secondly, for each of these Lie algebras, one needs to classify all possible finite-dimensional $\mathbf{h}$-modules $\mathcal{M}$ of C^∞ functions. Trivial modules, valid for any Lie algebra of vector fields, are the zero module $\mathcal{M} = 0$, which consists of the zero function alone, and that containing just the constant functions, which we write $\mathcal{M} = \{1\}$. Finally, for each of the modules $\mathcal{M}$, one needs to determine the first cohomology space $H^1(\mathbf{h}, C^\infty/\mathcal{M})$. As the tables at the end of this appendix indicate, the cohomology classes are parametrized by one or more continuous parameters or, in a few cases, smooth functions.

It is then a fairly straightforward matter to determine when a given Lie algebra satisfies the quasi-exactly solvable condition that it admit a non-zero finite-dimensional module $\mathcal{N} \subset C^\infty$. A simple lemma says that we can always, without loss of generality, take the Lie algebra $\mathbf{g}$ to be represented by a triple $[\mathbf{h}, \{1\}, [F]]$ with $\mathcal{M} = \{1\}$. (Indeed, $\mathbf{g}$ admits a finite-dimensional module if and only if $\mathcal{M} = \{1\}$ or $\mathcal{M} = 0$, and, in the latter case, $\mathbf{g}$ can always be enlarged to include constant functions without destroying its quasi-exact solvability.) Remarkably, *in all known cases*, the cohomology parameters are "quantized", the quasi-exact solvability requirement forcing them to assume at most a discrete set of distinct values. This intriguing phenomenon of "quantization of cohomology" has been geometrically explained in terms of line bundles on complex surfaces González-López *et al* (1993a).

In one dimension, there is essentially only one Lie algebra of first-order differential operators, represented by $\partial_x, x\partial_x + c, x^2\partial_x + 2cx$, where $c \in \mathbf{R}$ is a parameter representing the cohomology (Kamran and Olver 1990). This Lie algebra satisfies the requirement of quasi-exact solvability if and only if the cohomology parameter is quantized, $c = -\frac{1}{2}n$, where $n \in \mathbf{N}$ is a non-negative integer (which plays the physical role of spin). The corresponding module is $\mathcal{N} = \mathcal{P}^{(n)}$, the space of polynomials of degree less than or equal to n.

Theorem D.3. *Every (non-singular) finite-dimensional quasi-exactly solvable Lie algebra of first-order differential operators in one (real or complex) variable is, locally, equivalent to a subalgebra of one of the Lie algebras*

$$\widehat{\mathbf{g}}_n = \mathrm{Span}\{ \partial_x, \quad x\partial_x, \quad x^2\partial_x - nx, \quad 1 \},\tag{D.4.4}$$

where $n \in \mathbf{N}$. For $\widehat{\mathbf{g}}_n$, the associated module $\mathcal{N} = \mathcal{P}^{(n)}$ consists of the polynomials of degree at most n.

Turning to the two-dimensional classification, a number of additional complications present themselves. First, as originally shown by Lie, there are many more equivalence classes of finite-dimensional Lie algebras of vector fields. Moreover, the classification results in $\mathbf{R}^2$ and $\mathbf{C}^2$ are no longer the same — here we just present the complex case. (The real classification has recently been completed, by González-López *et al* (1993c).) Another complication is that the modules $\mathcal{M}$ for the vector field Lie algebras are no longer necessarily spanned by monomials, a fact that makes the determination of the cohomology considerably more difficult. Tables 1–3 at the end of the appendix summarize our classification results for finite-dimensional Lie algebras of differential operators in two complex variables (González-López *et al* 1991a, 1992a). Lie's classification of non-singular finite-dimensional Lie algebras of vector fields on $\mathbf{C}^2$ is summarized in table D.1; see the book by Lie (1924) for the original version. The first column exhibits a basis of the algebra, and the second indicates its structure as an abstract Lie algebra. Table D.2 describes the different finite-dimensional modules for each of these Lie algebras. The first column tells whether the module is necessarily spanned by monomials, i.e., single terms of the indicated form. (In cases 5 and 20, we have monomials unless $\alpha \in \mathbf{Q}^+$ or $r < \alpha \in \mathbf{Q}^+$ are positive rational numbers, respectively.) The second column indicates a typical term in a basis element for the module — the non-monomial basis elements will be linear combinations of terms of the indicated type; i, j always denote non-negative integers. The third column either indicates ranges of indices which must be included, or, in the case of an arrow, indicates other indices which must be included if the given one is. For instance, in case 19, if the monomial $x^i y^j \mathrm{e}^{\mu x}$ belongs to the module, so must the monomials $x^{i-1} y^j \mathrm{e}^{\mu x}$ and $x^{i+r} y^{j+1} \mathrm{e}^{(\mu+\lambda)x}$ (provided $i > 0$ and/or $j > 0$) for each exponent λ appearing in the Lie algebra. Finally, $R_k^{m,n}(z)$ denotes the polynomial

$$R_k^{m,n}(z) = \frac{\mathrm{d}^k}{\mathrm{d}z^k}(z-1)^{m+n}(z+1)^m, \qquad (\text{D.4.5})$$

which, for $n = 0$, is a multiple of the ultraspherical Gegenbauer polynomial $C_{2m-k}^{k-m+(1/2)}(z)$ (Erdélyi and Bateman 1953). Table D.3 describes the cohomology spaces $H^1(\mathbf{h}, C^\infty/\mathcal{M})$ for each of the Lie algebras and corresponding modules. The first column indicates the dimension of the cohomology space, and the second column gives a representative cocycle of each non-trivial cohomology class. Only the vector fields $\mathbf{v}^a$ which are actually modified are indicated, i.e., those for which $\eta^a = \langle F ; \mathbf{v}^a \rangle \not\equiv 0$, cf. (D.4.1). In case 4, $\mathrm{Div}\,\mathcal{M} = \{f_x + g_y \,|\, f, g \in \mathcal{M}\}$. Finally, table D.4 describes the quantization condition resulting from the quasi-exactly solvability assumption that, assuming $\mathcal{M} = \{1\}$, the Lie algebra admit a finite-dimensional module $\mathcal{N}$. If the cohomology is trivial, so $\mathbf{g}$ is spanned by vector fields and the constant functions, then it automatically satisfies the quasi-exactly solvable condition, with the associated finite-dimensional modules being explicitly described in table D.2. The maximal algebras, namely Case 11, $\mathbf{sl}(2) \oplus \mathbf{sl}(2)$, Case 15, $\mathbf{sl}(3)$, and Case 24, $\mathbf{gl}(2) + \mathbf{R}^r$, play an important role in Turbiner's theory of differential equations in two dimensions with orthogonal polynomial solutions.

D.5   Two-dimensional problems

There are a number of additional difficulties in the two-dimensional problem which do not appear in the scalar case. First, there are several different classes of quasi-exactly solvable Lie algebras available. Even more important is the fact that, according to theorem D.3, there are non-trivial "closure conditions" which must be satisfied in order that the magnetic one-form associated with a given hamiltonian be closed and hence the operator be equivalent, under a gauge transformation (D.2.5), to a Schrödinger operator. Unfortunately, in all but trivial cases, the closure conditions associated with a quasi-exactly solvable hamiltonian (D.2.3) corresponding to the generators of one of the quasi-exactly solvable Lie algebras on our list are *non-linear algebraic equations* in the coefficients c_{ab}, c_a, c_0, and it appears to be impossible to determine their general solution. Nevertheless, there are useful simplifications of the general closure conditions which can be effectively used to generate large classes of planar quasi-exactly solvable and exactly solvable Schrödinger operators, both for flat space as well as curved metrics.

Suppose that the Lie algebra g is spanned by linearly independent first-order differential operators as in (D.2.2). Substituting these into the general Lie-algebraic form (D.2.3), we find that the operator assumes the form (D.3.1) with

$$g^{ij} = \sum_{a,b=1}^{r} c_{ab}\xi^{ai}\xi^{bj},$$

$$h^i = \sum_{a,b=1}^{r} \left[c_{ab}\left(\xi^{aj}\frac{\partial \xi^{bi}}{\partial x^j} + 2\eta^a\xi^{bi} \right) + c_a\xi^{ai} \right],$$

$$k = \sum_{a,b=1}^{r} \left[c_{ab}\left(\xi^{aj}\frac{\partial \eta^b}{\partial x^j} + \eta^a\eta^b \right) + c_a\eta^a \right]. \tag{D.5.1}$$

The magnetic form ω, (D.3.4), and potential V for the covariant form (D.3.3) of the hamiltonian are then computed using formulas (D.3.5). The *closure conditions* $d\omega = 0$ are equivalent to the solvability of the system of partial differential equations

$$\sum_{a,b=1}^{r} c_{ab}\xi^{ai} \sum_{j=1}^{n}\left(\xi^{bj}\frac{\partial \tau}{\partial x^j} + \frac{\partial \xi^{bj}}{\partial x^j} \right) = \sum_{a=1}^{r}\xi^{ai}\left[2\sum_{b=1}^{r}c_{ab}\eta^b + c_a \right],$$

$$i = 1,\ldots,n, \tag{D.5.2}$$

for a scalar function $\tau(x)$, given by $\tau = 2\sigma + \frac{1}{2}\log\det g$ in terms of the gauge factor e^σ required to place the operator in Schrödinger form. The closure conditions (D.5.2) are extremely complicated to solve in full generality, but a useful subclass of solutions can be obtained from the *simplified closure conditions*

$$\sum_{i=1}^{n}\left(\xi^{ai}\frac{\partial\tau}{\partial x^i}+\frac{\partial\xi^{ai}}{\partial x^i}\right)-2\eta^a=k^a,\qquad a=1,\ldots,r \qquad (D.5.3)$$

where $k^1,\ldots,k^r$ are constants. Any solution $\tau(x)$ of equations (D.5.3) will generate an infinity of solutions to the full closure conditions (D.5.2), with c_{ab} arbitrary, and $c_a=\sum_b c_{ab}k^b$. The case $k^a=0$ and **g** semi-simple was investigated by Morozov *et al* (1990). Although the simplified closure conditions can be explicitly solved for such Lie algebras, with the exception of **so**(3), their solutions are found to generate quasi-exactly solvable Schrödinger operators that are *not* normalizable, and hence are of limited use. Note that even when the simplified closure conditions do not have any acceptable solutions, the *full* closure conditions (D.5.2) may be compatible and may give rise to normalizable operators.

Consider, in the first place, the Lie algebra $\mathbf{g}\simeq\mathbf{sl}(2)\oplus\mathbf{sl}(2)$ of type 11 spanned by the first-order differential operators

$$T^1=\partial_x,\quad T^2=\partial_y,\quad T^3=x\partial_x,$$
$$T^4=y\partial_y,\quad T^5=x^2\partial_x-nx,\quad T^6=y^2\partial_y-my, \qquad (D.5.4)$$

where $n,m\in\mathbf{N}$. The particular choice

$$(c_{ab})=\begin{pmatrix}2&1&0&0&0&1\\1&2&0&0&1&0\\0&0&3&0&0&0\\0&0&0&3&0&0\\0&1&0&0&1&1\\1&0&0&0&1&1\end{pmatrix}, \qquad (D.5.5)$$

$$(c_a)=\Big(0,0,-(1+4n),-(1+4m),0,0\Big), \qquad (D.5.6)$$

$$c_0=\tfrac{3}{4}+m^2+n^2, \qquad (D.5.7)$$

of Lie-algebraic coefficients lead to a quasi-exactly solvable hamiltonian with Riemannian metric

$$g^{11}=(1+x^2)(2+x^2),\quad g^{12}=(1+x^2)(1+y^2),$$
$$g^{22}=(1+y^2)(2+y^2), \qquad (D.5.8)$$

which has complicated curvature, and potential

$$4V=-y^2-\frac{(1+2n)(3+2n)}{1+x^2}-\frac{(1+2m)(3+2m)}{1+y^2}$$

$$-\frac{17 + 12y^2 - y^4 + 2xy(6 + 5y^2)}{3 + x^2 + y^2}$$

$$+\frac{5(3 + 2xy)(1 + y^2)(2 + y^2)}{(3 + x^2 + y^2)^2}. \tag{D.5.9}$$

A second interesting solution is

$$(c_{ab}) = \begin{pmatrix} 1 & 1 & 0 & 0 & 0 & 1 \\ 1 & 4 & 0 & 0 & 1 & 0 \\ 0 & 0 & 2 & 0 & 0 & 0 \\ 0 & 0 & 0 & 8 & 0 & 0 \\ 0 & 1 & 0 & 0 & 1 & 1 \\ 1 & 0 & 0 & 0 & 1 & 4 \end{pmatrix}, \tag{D.5.10}$$

$$(c_a) = \left(0, 0, -2n, -8m, 0, 0\right), \tag{D.5.11}$$

$$c_0 = -n - 4m. \tag{D.5.12}$$

The Riemannian metric is

$$g^{11} = (1 + x^2)^2, \qquad g^{12} = (1 + x^2)(1 + y^2),$$
$$g^{22} = 4(1 + y^2)^2, \tag{D.5.13}$$

which has zero curvature. The potential $V = 0$ also vanishes. The coordinate transformation

$$\bar{x} = \frac{1}{2\sqrt{3}}(4\arctan x - \arctan y), \qquad \bar{y} = \frac{1}{2}\arctan y, \tag{D.5.14}$$

maps the whole plane onto a bounded rectangle, so this example describes the physical situation of a free particle confined to a bounded rectangle. The algebraic eigenfunctions have the form

$$\psi(\bar{x}, \bar{y}) = \frac{P(x, y)}{(1 + x^2)^{\frac{n}{2}}(1 + y^2)^{\frac{m}{2}}}, \tag{D.5.15}$$

where P is a polynomial of degree less than or equal to n in x and less than or equal to m in y, and we re-express x, y using (D.5.14).

The Lie algebra of type 15 provides a realization of $\mathbf{sl}(3, \mathbf{R})$ in terms of first-order planar differential operators, given by

$$T^1 = \partial_x, \quad T^2 = \partial_y, \quad T^3 = x\partial_x,$$
$$T^4 = y\partial_x, \quad T^5 = x\partial_y, \quad T^6 = y\partial_y,$$
$$T^7 = x^2\partial_x + xy\partial_y - nx,$$
$$T^8 = xy\partial_x + y^2\partial_y - ny, \tag{D.5.16}$$

with $n \in \mathbf{N}$ admits the finite-dimensional module consisting of polynomials of total degree in x and y less than or equal to n. The quasi-exactly solvable hamiltonian

$$
\begin{aligned}
\mathcal{H} \;=\; & \left(T^1\right)^2 + \left(T^2\right)^2 + 2\left(T^7\right)^2 + 2\left(T^8\right)^2 \\
& + T^1 T^7 + T^7 T^1 + T^2 T^8 + T^8 T^2 \\
& + (3 + 2n)\left(T^3 + T^6\right)
\end{aligned}
\tag{D.5.17}
$$

has contravariant metric coefficients

$$
g^{11} = x^2(1 + \rho) + 1, \qquad g^{12} = xy(1 + \rho),
$$
$$
g^{22} = y^2(1 + \rho) + 1,
\tag{D.5.18}
$$

where $\rho = 1 + 2(x^2 + y^2)$, whose Gaussian curvature $\kappa = -2\rho$ is negative everywhere. The potential is given by

$$
4V = -3\rho - (7 + 16n + 8n^2) + \frac{14 + 24n + 8n^2 + (22 + 24n + 8n^2)\rho}{\rho^2 + 1}.
\tag{D.5.19}
$$

If we look for solutions of the Schrödinger equation depending only on the "radial" coordinate ρ, we end up with an effectively one-dimensional Schrödinger operator

$$
\begin{aligned}
-\widehat{\mathcal{H}} \;=\; & 4(\rho - 1)(\rho^2 + 1)\frac{\mathrm{d}^2}{\mathrm{d}\rho^2} + \left[(6 - 4n)\rho^2 + (8n + 4)\rho - 4n - 2\right]\frac{\mathrm{d}}{\mathrm{d}\rho} \\
& + \left[(n^2 - n)\rho - n^2 - n\right]
\end{aligned}
\tag{D.5.20}
$$

which does appear among the list of purely one-dimensional quasi-exactly solvable hamiltonians, albeit with a different cohomology parameter. The question of whether the class of one-dimensional quasi-exactly solvable Schrödinger operators can be significantly enlarged via looking at reductions of two-dimensional quasi-exactly solvable hamiltonians remains unanswered.

Next let $\mathbf{g}$ be the non-compact Lie algebra of type 24 for $r = 1$, spanned by

$$
T^1 = \partial_x, \quad T^2 = \partial_y, \quad T^3 = x\partial_x, \quad T^4 = x\partial_y, \quad T^5 = y\partial_y, \tag{D.5.21}
$$

and

$$
T^6 = x^2\partial_x + xy\partial_y - nx, \qquad \text{where} \qquad n \in \mathbf{N}. \tag{D.5.22}
$$

This Lie algebra admits the finite-dimensional module $\mathcal{N}$ spanned by the monomials $x^i y^j$ with $i + j \le n$. The Lie-algebraic coefficients

$$(c_{ab}) \;=\; \begin{pmatrix} 1 & 0 & 0 & 0 & 0 & 0 \\ 0 & 1 & 0 & 0 & 0 & 0 \\ 0 & 0 & 2 & 0 & 1 & 0 \\ 0 & 0 & 0 & 1 & 0 & 0 \\ 0 & 0 & 1 & 0 & 1 & 0 \\ 0 & 0 & 0 & 0 & 0 & 1 \end{pmatrix}, \tag{D.5.23}$$

$$(c_a) \;=\; \Big(0, 0, -2n, 0, -1 - 2n, 0\Big), \tag{D.5.24}$$

$$c_0 \;=\; n^2 + n + 1, \tag{D.5.25}$$

give a quasi-exactly solvable operator with flat Riemannian metric

$$g^{11} = (1 + x^2)^2, \qquad g^{12} = xy(1 + x^2),$$
$$g^{22} = (1 + x^2)(1 + y^2). \tag{D.5.26}$$

Flat coordinates are given by

$$\bar{x} = \arctan x, \qquad \bar{y} = \operatorname{arcsinh} \frac{y}{\sqrt{1 + x^2}}, \tag{D.5.27}$$

which maps the plane to an open infinite strip $(-\pi/2, \pi/2) \times \mathbf{R}$. The potential in these coordinates is

$$V(\bar{x}, \bar{y}) = -\frac{(n+1)(n+2)}{2} \operatorname{sech}^2 \bar{y}, \tag{D.5.28}$$

which is simply a Pöschl–Teller potential in $\bar{y}$. The algebraic eigenfunctions take the form

$$\psi(\bar{x}, \bar{y}) = \frac{\cos^n \bar{x}}{\cosh^{n+1} \bar{y}} \, P(\tan \bar{x}, \sec \bar{x} \sinh \bar{y}), \tag{D.5.29}$$

where $P(x, y)$ is a polynomial of total degree at most n in (x, y). Notice that this potential, the preceding zero potential, and, indeed, all other flat potentials that we have found satisfy a conjecture of Turbiner: any quasi-exactly solvable hamiltonian on a flat manifold in more than one dimension is necessarily separable. However, we do not know whether this conjecture holds in general.

Finally, let **g** be a general Lie algebra of type 24, spanned by the first-order differential operators (D.5.21),

$$T^6 = x^2 \partial_x + rxy \partial_y - nx, \qquad T^{6+i} = x^{i+1} \partial_y, \qquad i = 1, \dots, r - 1. \tag{D.5.30}$$

The module $\mathcal{N}$ is spanned by the monomials $x^i y^j$ with $i + rj \leq n$. For $m \in \mathbf{N}$, $A, B > 0$, the Schrödinger operator with metric

$$g^{11} = Ax^2 + B, \qquad g^{12} = (1 + m)Axy,$$
$$g^{22} = (Ax^2 + B)^m + A(1 + m)^2 y^2, \qquad (\text{D.5.31})$$

and potential

$$V = -\frac{\lambda AB(1 + m)^2 (Ax^2 + B)^m}{(Ax^2 + B)^{1+m} + AB(1 + m)^2 y^2}, \qquad m \leq r \neq 2(m + 1),$$

$$(\text{D.5.32})$$

is normalizable and quasi-exactly solvable with respect to $\mathbf{g}$, provided that the parameter λ is large enough. The metric in this case has constant negative Gaussian curvature $\kappa = -A$. Furthermore, since the potential V does not depend on the cohomology parameter n, the above hamiltonian is *exactly solvable*. Moreover, the potential is also independent of r; hence we have constructed a single exactly solvable hamiltonian which is associated with an infinite number of inequivalent Lie algebras of arbitrarily large dimension.

Table D.1. Finite-dimensional Lie algebras of vector fields in $\mathbf{C}^2$.

	Generators	Structure
1.	$\{\partial_x\}$	$\mathbf{C}$
2.	$\{\partial_x, x\partial_x\}$	$\mathbf{h}_2$
3.	$\{\partial_x, x\partial_x, x^2\partial_x\}$	$\mathrm{sl}(2)$
4.	$\{\partial_x, \partial_y\}$	$\mathbf{C}^2$
5.	$\{\partial_x, \partial_y, x\partial_x + \alpha y\partial_y\}$ $\alpha \neq 0$	$\mathbf{C} + \mathbf{C}^2$
6.	$\{\partial_x, \partial_y, x\partial_x, y\partial_y\}$	$\mathbf{h}_2 \oplus \mathbf{h}_2$
7.	$\{\partial_x, \partial_y, x\partial_x - y\partial_y, y\partial_x, x\partial_y\}$	$\mathrm{sl}(2) + \mathbf{C}^2$
8.	$\{\partial_x, \partial_y, x\partial_x, y\partial_x, x\partial_y, y\partial_y\}$	$\mathrm{gl}(2) + \mathbf{C}^2$
9.	$\{\partial_x, \partial_y, x\partial_x, x^2\partial_x\}$	$\mathrm{gl}(2)$
10.	$\{\partial_x, \partial_y, x\partial_x, y\partial_y, x^2\partial_x\}$	$\mathrm{sl}(2) \oplus \mathbf{h}_2$
11.	$\{\partial_x, \partial_y, x\partial_x, y\partial_y, x^2\partial_x, y^2\partial_y\}$	$\mathrm{sl}(2) \oplus \mathrm{sl}(2)$
12.	$\{\partial_x + \partial_y, x\partial_x + y\partial_y, x^2\partial_x + y^2\partial_y\}$	$\mathrm{sl}(2)$
13.	$\{\partial_x, 2x\partial_x - y\partial_y, x^2\partial_x - xy\partial_y\}$	$\mathrm{sl}(2)$
14.	$\{\partial_x, x\partial_x, y\partial_y, x^2\partial_x - xy\partial_y\}$	$\mathrm{gl}(2)$
15.	$\{\partial_x, \partial_y, x\partial_x, y\partial_x, x\partial_y, y\partial_y,$ $x^2\partial_x + xy\partial_y, xy\partial_x + y^2\partial_y\}$	$\mathrm{sl}(3)$
16.	$\{\xi_1(x)\partial_y, \ldots, \xi_r(x)\partial_y\}$[a]	$\mathbf{C}^r$
17.	$\{\xi_1(x)\partial_y, \ldots, \xi_r(x)\partial_y, y\partial_y\}$[a]	$\mathbf{C} + \mathbf{C}^r$
18.	$\{\partial_x, x^i e^{\lambda x}\partial_y \mid 0 \leq i \leq r_\lambda\}$[b]	$\mathbf{C} + \mathbf{C}^r$
19.	$\{\partial_x, y\partial_y, x^i e^{\lambda x}\partial_y \mid 0 \leq i \leq r_\lambda\}$[b]	$\mathbf{C}^2 + \mathbf{C}^r$
20.	$\{\partial_x, \partial_y, x\partial_x + \alpha y\partial_y, x\partial_y, \ldots, x^r\partial_y\}$[c]	$\mathbf{h}_2 + \mathbf{C}^{r+1}$
21.	$\{\partial_x, \partial_y, x\partial_y, \ldots, x^{r-1}\partial_y, x\partial_x + (ry + x^r)\partial_y\}$[c]	$\mathbf{C} + (\mathbf{C} + \mathbf{C}^r)$
22.	$\{\partial_x, \partial_y, x\partial_x, x\partial_y, y\partial_y, x^2\partial_y, \ldots, x^r\partial_y\}$[c]	$(\mathbf{h}_2 \oplus \mathbf{C}) + \mathbf{C}^{r+1}$
23.	$\{\partial_x, \partial_y, 2x\partial_x + ry\partial_y, x\partial_y,$ $x^2\partial_x + rxy\partial_y, x^2\partial_y, \ldots, x^r\partial_y\}$[c]	$\mathrm{sl}(2) + \mathbf{C}^{r+1}$
24.	$\{\partial_x, \partial_y, x\partial_x, x\partial_y, y\partial_y,$ $x^2\partial_x + rxy\partial_y, x^2\partial_y, \ldots, x^r\partial_y\}$[c]	$\mathrm{gl}(2) + \mathbf{C}^{r+1}$

[a] $r > 1$.

[b] $r = \sum r_\lambda$.

[c] $r \geq 1$.

Table D.2. Finite-dimensional modules for Lie algebras of vector fields.

	Monomials?	Generators	Rules
1.	No	$x^i e^{\lambda x} g(y)$	$(i, \lambda, g) \longrightarrow (i-1, \lambda, g)$,
2.	Yes	$x^i g(y)$	$(i, g) \longrightarrow (i-1, g)$
3.	Yes	$g(y)$	
4.	No	$x^i y^j e^{\lambda x + \mu y}$	$(i, j, \lambda, \mu) \longrightarrow$
			$(i-1, j, \lambda, \mu), (i, j-1, \lambda, \mu)$
5.	Yes[a]	$x^i y^j$	$(i, j) \longrightarrow (i-1, j), (i, j-1)$
6.	Yes	$x^i y^j$	$(i, j) \longrightarrow (i-1, j), (i, j-1)$
7.	Yes	$x^i y^j$	$0 \leq i + j \leq n$
8.	Yes	$x^i y^j$	$0 \leq i + j \leq n$
9.	Yes	$y^j e^{\mu y}$	$0 \leq j \leq n_\mu$
10.	Yes	y^j	$0 \leq j \leq n$
11.	Yes	1	
12.	No	$f_{mk}(x, y)^c$	$0 \leq k \leq 2m$
13.	Yes	$x^i y^n$	$0 \leq i \leq n,\ n \in S$
14.	Yes	$x^i y^n$	$0 \leq i \leq n,\ n \in S$
15.	Yes	1	
16.	No	$y^j g(x)$	$(j, g) \longrightarrow (j-1, g \cdot \xi_k)$
17.	Yes	$y^j g(x)$	$(j, g) \longrightarrow (j-1, g \cdot \xi_k)$
18.	No	$x^i y^j e^{\mu x}$	$(i, j, \mu) \longrightarrow$
			$(i-1, j, \mu), (i + r_\lambda, j-1, \mu + \lambda)$
19.	Yes	$x^i y^j e^{\mu x}$	$(i, j, \mu) \longrightarrow$
			$(i-1, j, \mu), (i + r_\lambda, j-1, \mu + \lambda)$
20.	Yes[b]	$x^i y^j$	$(i, j) \longrightarrow (i-1, j), (i + r, j-1)$
21.	Yes	$x^i y^j$	$(i, j) \longrightarrow (i-1, j), (i + r, j-1)$
22.	Yes	$x^i y^j$	$(i, j) \longrightarrow (i-1, j), (i + r, j-1)$
23.	Yes	1	
24.	Yes	1	

[a] Unless $\alpha \in \mathbf{Q}^+$

[b] Unless $r < \alpha \in \mathbf{Q}^+$

[c] Here $f_{mk}(x, y) = (x - y)^{m-k} R_k^{m,0} \left(\frac{x+y}{x-y} \right)$.

Table D.3. Cohomologies for Lie algebras of vector fields.

	Dimension	Representatives
1.	0	
2.	∞	$x\partial_x + h(y), h \notin \mathcal{M}$
3.	∞	$x\partial_x + h(y), \quad x^2\partial_x + 2xh(y)$
4.	$\dim(\mathcal{M}/\operatorname{Div}\mathcal{M}) < \infty$	$\partial_y + h(x,y), \; h_x \in \mathcal{M}, h \neq \psi_y$ with $\psi_x \in \mathcal{M}$
5.	$0 \, (\alpha \notin \mathbf{Q}^-), 1$ or $0 \, (\alpha \in \mathbf{Q}^-)$	$x\partial_x + \alpha y\partial_y + c_1 x^i y^j \quad$ or $\quad \partial_y + c_1 x^i y^j$
6.	$0 \, (\mathcal{M} \neq 0), 2 \, (\mathcal{M} = 0)$	$x\partial_x + c_1, \quad y\partial_y + c_2$
7.	$0 \, (\mathcal{M} \neq \{1\}), 1 \, (\mathcal{M} = \{1\})$	$\partial_y + 2c_1 x, \quad y\partial_x + c_1 y^2, \quad x\partial_y + c_1 x^2$
8.	$0 \, (\mathcal{M} \neq 0), 1 \, (\mathcal{M} = 0)$	$x\partial_x + c_1, \quad y\partial_y + c_1$
9.	1	$x\partial_x + c_1, \quad x^2\partial_x + 2c_1 x$
10.	$1 \, (\mathcal{M} \neq 0), 2 \, (\mathcal{M} = 0)$	$x\partial_x + c_1, \quad x^2\partial_x + 2c_1 x, \quad y\partial_y + c_2$
11.	2	$x\partial_x + c_1, x^2\partial_x + 2c_1 x, y\partial_y + c_2,$ $y^2\partial_y + 2c_2 y$
12.	1	$x^2\partial_x + y^2\partial_y + c_1(x - y)$
13.	1	$x^2\partial_x - xy\partial_y + c_1 y^{-2}$
14.	$0 \, (1 \in \mathcal{M}), 1 \, (1 \notin \mathcal{M})$	$x\partial_x + c_1, \quad y\partial_y + 2c_1$
15.	1	$x\partial_x + c_1, \quad x^2\partial_x + xy\partial_y + 3c_1 x,$ $y\partial_y + c_1, \quad xy\partial_x + y^2\partial_y + 3c_1 y$
16.	$\infty^r + k, \quad k < \infty$	$\xi_i(x)\partial_y + f_i(x)y^j$
17.	∞	$y\partial_y + f(x)$
18.	$< \infty$	$x^k e^{\lambda x}\partial_y + c_{i,j}^{\lambda,k} x^{i+k} y^j e^{\lambda x}$
19.	1	$y\partial_y + c_1 x^n$
20.	$0 \, (\alpha \notin \mathbf{Q}), 1$ or $0 \, (\alpha \in \mathbf{Q})$	$x\partial_x + \alpha y\partial_y + c_1 x^i y^j,$ or $\xi_k\partial_y + c_k x^{i+k} y^j, \; k \geq l \geq 0$
21.	$0 \, (\mathcal{M} \neq 0), 1 \, (\mathcal{M} = 0)$	$x\partial_x + (ry + x^r y)\partial_y + c_1$
22.	$0 \, (\mathcal{M} \neq 0), 2 \, (\mathcal{M} = 0)$	$x\partial_x + c_1, \quad y\partial_y + c_2$
23.	$1 \, (r > 2)$	$2x\partial_x + ry\partial_y + c_1, \quad x^2\partial_x + rxy\partial_y + c_1 x$
	$2 \, (r = 2)$	$x\partial_x + y\partial_y + c_1, \quad x\partial_y + c_2,$ $x^2\partial_x + 2xy\partial_y + 2c_1 x + 2c_2 y,$ $x^2\partial_y + 2c_2 x$
24.	$1 \, (\mathcal{M} \neq 0), 2 \, (\mathcal{M} = 0)$	$x\partial_x + c_1, y\partial_y + c_2,$ $x^2\partial_x + rxy\partial_y + (2c_1 + rc_2)x$

Table D.4. Quasi-exactly solvable Lie algebras of differential operators.

	Quantization condition	Module
1.	0	
2.	0	
3.	$h = -\frac{n}{2}, \quad n \geq 0,$	$\{x^i g(y) \mid i \leq n, g \in S\}$
4.	0	
5.	0	
6.	0	
7.	0	
8.	0	
9.	$c_1 = -\frac{n}{2}, \quad n \geq 0,$	$\{x^i y^j e^{\mu y} \mid i \leq n, j \leq m_\mu\}$
10.	$c_1 = -\frac{n}{2}, \quad n \geq 0,$	$\{x^i y^j \mid i \leq n, j \leq m\}$
11.	$c_1 = -\frac{n}{2}, c_2 = -\frac{m}{2}, \quad n, m \geq 0,$	$\{x^i y^j \mid i \leq n, j \leq m\}$
12.	$c_1 = \frac{n}{2},$	$\{f_{mnk}(x, y) \mid 0 \leq k \leq 2m + n\}$[a]
13.	0	
14.	0	
15.	$c_1 = -\frac{n}{3}, \quad n \geq 0,$	$\{x^i y^j \mid i + j \leq n\}$
16.	0	
17.	0	
18.	0	
19.	0	
20.	0	
21.	0	
22.	0	
23.	$c_1 = -n, \quad n \geq 0,$	$\{x^i y^j \mid i + rj \leq n, j \leq l\}$
24.	$c_1 = -\frac{n}{2}, c_2 = 0, \quad n \geq 0,$	$\{x^i y^j \mid i + rj \leq n, j \leq l$

[a] Here $f_{mnk}(x, y) = (x - y)^{m + \frac{n}{2} - k} R_k^{m,n} \left(\frac{x+y}{x-y}\right)$, and $m \in S = \{m \mid m \geq \max(0, -n)\}$, where S is a finite set of integers. There is no positivity restriction on n.

References

Abramowitz M and Stegun I A (ed) 1965 *Handbook of Mathematical Functions,* (New York: Dover)

Alhassid Y, Engel J and Wu J 1984 *Phys. Rev. Lett.* **53** 17–20

Alhassid Y, Gürsey F and Iachello F 1983 *Ann. Phys., NY* **148** 346–80

Alhassid Y, Gürsey F and Iachello F 1986 *Ann. Phys., NY* **167** 181–200

Andrianov A A, Borisov N Y and Ioffe M V 1984 *Teor. Mat. Fiz.* **61** 183–99

Arima A and Iachello F 1976 *Ann. Phys., NY* **99** 253–317

Arima A and Iachello F 1978 *Ann. Phys., NY* **111** 201–38

Avron J and Simon B 1978 *Ann. Phys., NY* **110** 85–102

Bagrov V G and Vshivtsev A S 1986 Preprint 31, Siberian Division, USSR Academy of Sciences, Tomsk

Balazs N L and Voros A 1986 *Phys. Rep.* **143** 109–240

Bargmann V 1949 *Rev. Mod. Phys.* **21** 488–93

Bargmann V and Moshinsky M 1961 *Nucl. Phys.* **23** 177–91

Baxter R 1972 *Ann. Phys., NY* **70** 323–37

Bazhanov V V 1985 *Phys. Lett.* **159B** 321–4

Bender C, Happ J J and Svetitsky B 1974 *Phys. Rev.* D **9** 2324–30

Bender C M and Wu T T 1969 *Phys. Rev.* **184** 1231–60

Bethe H A 1931 *Z. Phys.* B **71** 205–26

Blanch G and Glemm D S 1969 *Math. Comp.* **23** 97–108

Blecher M H and Leach P G L 1987 *J. Phys.* A **20** 5923–7

Bogomolny E B, Fateev V A and Lipatov L N 1980 *Sov. Sci. Rev.* **18** 247–324

Bonner J C and Fischer M E 1964 *Phys. Rev.* A **135** 640–58

Bulayevsky L M 1975 *Usp. Fiz. Nauk* **115** 263–300

Cotton É 1900 *Ann. École Norm.* **17** 211–44

Doebner H D and Ushveridze A G 1992 (unpublished)

Dyson F 1952 *Phys. Rev.* **85** 631–2

Erdélyi A and Bateman H 1953 *Higher Transcendental Functions* (New York: McGraw-Hill)

Flessas G P 1981 *Phys. Lett.* **83A** 121–8

Flessas G P 1982 *J. Phys.* A **15** L97–9

Flügge S 1971 *Practical Quantum Mechanics* (Berlin: Springer)

Gallas J A C 1988 *J. Phys.* A **21** 3393–9

Gaudin M 1976 *J. Physique* **37** 1087–98

Gaudin M 1983 *La Fonction d'Onde de Bethe* (Paris: Masson)

Gaudin M, McCoy B M and Wu T T 1981 *Phys. Rev.* D **23** 417–9

Gelfand I M and Levitan B M 1951 *Dokl. AN SSSR* **77** 557–65

Gendenshtein L E 1983 *JETP Lett.* **38** 356–60

Gendenshtein L E and Krive I V 1985 *Usp. Fiz. Nauk* **146** 553–90

Gershenson M E and Turbiner A V 1982 *Sov. J. Nucl. Phys.* **35** 839–46

González-López A, Hurtubise J, Kamran N and Olver P J 1993a *Comptes Rendus Acad. Sci.* **316** 1307–12

González-López A, Kamran N and Olver P J 1991a *Differential Geometry, Global Analysis and Topology. Canadian Math. Soc. Conference Proceedings* vol.12 (Providence, RI: American Mathematical Society) pp 51–84

González-López A, Kamran N and Olver P J 1991b *J. Phys.* A **24** 3995–4008

González-López A, Kamran N and Olver P J 1992a *American J. Math.* **114** 1163–85

González-López A, Kamran N and Olver P J 1992b Preprint, University of Maryland

González-López A, Kamran N and Olver P J 1993b *Commun. Math. Phys.* **153** 117–46

González-López A, Kamran N and Olver P J 1993c *Commun. Math. Phys.* (to appear)

González-López A, Kamran N and Olver P J 1993d *Contemp. Math.* (to appear)

González-López A, Kamran N and Olver P J 1993e Preprint, University of Minnesota

Gorsky A 1991 *JETP Lett.* **54** 289–92

Gorsky A and Selivanov 1992 *Mod. Phys. Lett.* A **7** 2601–9

Graffi S and Grecchi V 1978 *J. Math. Phys.* **19** 1002–6

Graffi S, Grecchi V and Simon B 1970 *Phys. Lett.* **32B** 631–4

Hioe F T, McMillan D and Montroll E W 1976 *J. Math. Phys.* **17** 1520–603

Hioe F T, McMillan D and Montroll E W 1978 *Phys. Rep.* C **43** 305–35

Hislop D, Wolfaardt M F and Leach P L G 1990 *J. Phys.* A **23** L1109–12

Hunter C and Guerrieri B 1981 *Studies Appl. Math.* **64** 113–26

Hunter C and Guerrieri B 1982 *Studies Appl. Math.* **66** 217–24

Hurt N 1983 *Geometrical Quantization in Action* (London: Riedel)

Iachello F and Levine R 1982 *J. Chem. Phys.* **77** 3046–55

Infeld L and Hull T E 1951 *Rev. Mod. Phys.* **23** 21–68

Jacobson N 1962 *Lie Algebras* (New York: Interscience)

Japaridze G I, Nersesyan A A and Wiegmann P B 1984 *Nucl. Phys.* B **230** 511–47

Jerome D and Schulz H J 1982 *Adv. Phys.* **31** 299–490

Jurčo V E 1989 *J. Math. Phys.* **30** 1289–91

Kamran N and Olver P J 1990 *J. Math. Anal. Appl.* **145** 342–56

Kazakov D I and Shirkov D V 1980 *Fortschr. Phys.* B **28** 465–99

Kirillov A A 1972 *Elements of the Theory of Representations* (Moscow: Nauka)

Korepin V E 1982 *Commun. Math. Phys.* **86** 391–418

Korepin V E 1984 *Commun. Math. Phys.* **94** 93–113

Kostant B 1977 *Quantization and Representation Theory of Lie Groups* (Oxford: Pergamon)

Kostant B 1979 *London Math. Lect. Notes* **34** 216–54

Koudinov A V and Smondyrev M A 1983 Preprint JINR E2-83-412, Dubna

Kulish P P, Reshetikhin N Y and Sklyanin E K 1981 *Lett. Math. Phys.* **5** 393–403

Lakhtakia A 1989 *J. Phys.* A **22** 1701–4

Lanczos G 1950 *J. Res. Nat. Bur. Stand.* **45** 255–72

Landau L D and Lifshitz E M 1977 *Quantum Mechanics: Nonrelativistic Theory* (Oxford: Pergamon)

Leach P G L 1984 *J. Math. Phys.* **25** 974–83

Leach P G L 1985 *Physica* D **17** 331–3

Leach P G L, Flessas G P and Gorringe V M 1989 *J. Math. Phys.* **30** 406–11

Levine R D 1988 *Mathematical Frontiers in Computational Chemical Physics. Volumes in Mathematics and its Applications* vol 15 (New York: Springer) pp 245–61

Levitan B M 1987 *Matem. Sbornik* **132** 73–98

Leznov A N 1984 *Lett. Math. Phys.* **8** 5–8

Leznov A N and Savelycv M V 1985 *Group Methods of Integration of Nonlinear Dynamical Systems* (Moscow: Nauka)

Lie S 1893 *Theorie der Transformationsgruppen* vol 3 (Leipzig: Teubner)

Lie S 1924 *Gruppenregister. Gesammelte Abhandlungen* vol 5 (Leipzig: Teubner)

Lipatov L N 1976 *Zh. Eksp. Teor. Fiz.* **71** 2010–26

Luther A 1976 *Phys. Rev.* B **14** 2153–62

Luther A 1977 *Phys. Rev.* B **15** 403–12

Maglaperidze T I and Ushveridze A G 1988 *Sov. Phys.–Lebedev Inst. Rep.* **12** 44–7

Maglaperidze T I and Ushveridze A G 1989a *Sov. Phys.–Lebedev Inst. Rep.* **10** 56 9

Maglaperidze T I and Ushveridze A G 1989b Preprint FIAN 162, Lebedev Physical Indtitute, Moscow

Maglaperidze T I and Ushveridze A G 1989c Preprint FIAN 163, Lebedev Physical Institute, Moscow

Maglaperidze T I and Ushveridze A G 1990 *Mod. Phys. Lett.* A **5** 1883–9

Magyari E 1981 *Phys. Lett.* **81A** 116–9

McKean H and Trubowitz E 1981 *Commun. Math. Phys.* **82** 471–88

Morozov A Y, Perelomov A M, Rosly A A, Shifman M A and Turbiner A V 1990 *Int. J. Mod. Phys.* A **5** 803–32

Moshinsky M 1962 *J. Math. Phys.* **4** 1128–39

Natanson G A 1971 *Vestn. Leningr. Univ.* **10** 22–31

Natanzon G A 1978 *Teor. Mat. Fiz.* **38** 219–30

Ogievetski E, Reshetikhin N Y and Wiegmann P B 1987 *Nucl. Phys.* **280** 45–96

Olshanetsky M A and Perelomov A M 1983 *Phys. Rep.* **94** 313–404

Plekhanov E B, Suzko A S and Zakhariev B N 1982 *Ann. Phys., Lpz* **39** 313–96

Popov V S, Eletsky V L and Turbiner A V 1977 *JETP Lett.* **26** 193–5

Popov V S, Eletsky V L and Turbiner A V 1978 *Zh. Eksp. Teor. Fiz.* **74** 445–66

Pöschl J and Trubowitz E 1987 *Pure and Applied Mathematics* vol 30 (Boston: Academic)

Price P J 1954 *Proc. Phys. Soc.* **67** 383–5

Rampal A and Datta K 1984 *J. Math. Phys.* **24** 860–9

Razavy M A 1981 *Phys. Lett.* **82A** 7–9

Reshetikhin N Y 1985 *Teor. Mat. Fiz.* **63** 347–66

Rudyak B V and Zakhariev B N 1987 *Inverse Problems* **3** 125–33

Shanley P E 1986 *Phys. Lett.* **117A** 161–5

Shaverdyan B S and Ushveridze A G 1983 *Phys. Lett.* **123B** 316–9

Shifman M A 1989a *Int. J. Mod. Phys.* A **4** 2897–952

Shifman M A 1989b *Int. J. Mod. Phys.* A **4** 3305–10

Shifman M A 1989c *Int. J. Mod. Phys.* A **4** 3311–6

Shifman M A 1992 Preprint TPI-MINN-92/16-T, University of Minnesota, Minneapolis

Shifman M A and Turbiner A V 1989 *Commun. Math. Phys.* **126** 347–65

Simon B 1970 *Ann. Phys.* **58** 76–137

Singh V, Biswas S N and Datta K 1978 *Phys. Rev.* D **18** 1901–9

Sklyanin E K 1987 *Zap. Nauchn. Semin. LOMI* **164** 151–70

Solovyev E A 1981 *Sov. Phys.–JETP* **54** 893–901

Solyom J 1979 *Adv. Phys.* **28** 201–303

Sutherland E 1970 *J. Math. Phys.* **11** 3183–6

Takhtajan L A and Faddeev L D 1979 *Usp. Mat. Nauk* **34** 13–63

Takhtajan L A and Faddeev L D 1981 *Zap. Nauchn. Semin. LOMI* **109** 134–78

Taylor D R and Leach P G L 1989 *J. Math. Phys.* **30** 1525–9

Thacker H B 1981 *Rev. Mod. Phys.* **53** 253–85

Toombs G A 1978 *Phys. Rep.* **40** 181–240

Tsvelick A M and Wiegmann P B 1983 *Adv. Phys.* **32** 453–713

Turbiner A V 1984 *Sov. Phys.–Usp.* **27** 668–708

Turbiner A V 1988a *Commun. Math. Phys.* **118** 467–74

Turbiner A V 1988b *Sov. Phys.–JETP* **67** 230–41

Turbiner A V 1992a *J. Phys.* A **25** L1087–93

Turbiner A V 1992b *J. Math. Phys.* **33** 3989–93

Turbiner A V and Ushveridze A G 1986 Preprint ITEP-169, Moscow

Turbiner A V and Ushveridze A G 1987 *Phys. Lett.* **126A** 181–3

Turbiner A V and Ushveridze A G 1988a *J. Math. Phys.* **29** 2053–63

Turbiner A V and Ushveridze A G 1988b *Sov. Phys.–Lebedev Inst. Rep.* **12** 48–51

Turchetti G 1971 *Nuovo Cimento* B **4** 313–41

Ushveridze A G 1987a *J. Phys.* A **20** 5145–55

Ushveridze A G 1987b (unpublished)

Ushveridze A G 1988a *J. Phys.* A **21** 955–70

Ushveridze A G 1988b *J. Phys.* A **21** 1601–5

Ushveridze A G 1988c *Sov. Phys.–Lebedev Inst. Rep.* **2** 50–3

Ushveridze A G 1988d *Sov. Phys.–Lebedev Inst. Rep.* **2** 54–8

Ushveridze A G 1988e *Sov. Phys.–Lebedev Inst. Rep.* **3** 20–2

Ushveridze A G 1988f *Sov. Phys.–Lebedev Inst. Rep.* **7** 14–8

Ushveridze A G 1988g *Sov. Phys.–Lebedev Inst. Rep.* **8** 34–8
Ushveridze A G 1988h *Sov. Phys.–Lebedev Inst. Rep.* **9** 26–9
Ushveridze A G 1988i Preprint FIAN 33, Lebedev Physical Institute, Moscow
Ushveridze A G 1988k Preprint FIAN 96, Lebedev Physical Institute, Moscow
Ushveridze A G 1988l Preprint FIAN 134, Lebedev Physical Institute, Moscow
Ushveridze A G 1988m Preprint FIAN 158, Lebedev Physical Institute, Moscow
Ushveridze A G 1988n Preprint FIAN 190, Lebedev Physical Institute, Moscow
Ushveridze A G 1988o Preprint FTT-11, Institute of Physics, Tbilisi
Ushveridze A G 1988p Preprint FTT-12, Institute of Physics, Tbilisi
Ushveridze A G 1989a *Phys. Scr.* **39** 30–3
Ushveridze A G 1989b *Bulg. J. Phys.* **16** 137–50
Ushveridze A G 1989c *Sov. J. Part. Nucl.* **20** 504–28
Ushveridze A G 1989d *Sov. Phys.–Lebedev Inst. Rep.* **3** 47–9
Ushveridze A G 1989e *Sov. Phys.–Lebedev Inst. Rep.* **4** 39–41
Ushveridze A G 1989f *Sov. Phys.–Lebedev Inst. Rep.* **6** 35–8
Ushveridze A G 1989g *Sov. Phys.–Lebedev Inst. Rep.* **6** 39–42
Ushveridze A G 1989h *Sov. Phys.–Lebedev Inst. Rep.* **10** 52–5
Ushveridze A G 1989i *Sov. Phys.–Lebedev Inst. Rep.* **10** 56–9
Ushveridze A G 1990a *Mod. Phys. Lett.* A **5** 1891–9
Ushveridze A G 1990b *Sov. Phys.–Lebedev Inst. Rep.* **3** 53–5
Ushveridze A G 1990c Preprint FTT-16, Institute of Physics, Tbilisi
Ushveridze A G 1991a *Mod. Phys. Lett.* A **6** 739–42
Ushveridze A G 1991b *Mod. Phys. Lett.* A **6** 977–9
Ushveridze A G 1992 *Sov. J. Part. Nucl.* **23** 25–51
Vainshtein A I 1964 Preprint INP, Siberian Division, USSR Academy of Sciences, Novosibirsk
Vanden Berghe G and De Meyer H E 1989 *J. Phys.* A **22** 1705–12
Wiegmann P B 1981 *J. Phys.* C **14** 1463–78
Wilkinson J H 1965 *The Algebraic Eigenvalue Problem* (Oxford: Clarendon)
Witten E 1982 *Nucl. Phys.* B **202** 253–317
Zakhariev B N and Suzko A S 1985 *Potentials and Quantum Scattering. The Direct and Inverse Problems* (Moscow: Energoatomizdat)
Zamolodchikov A B 1987 (unpublished)
Zamolodchikov A B and Zamolodchikov Al B 1978 *Nucl. Phys.* B **133** 525–35
Zamolodchikov A B and Zamolodchikov Al B 1979 *Ann. Phys., NY* **120** 253–91
Zaslavsky O V and Ulyanov V V 1984 *Sov. Phys.–JETP* **60** 991–1002
Zeldovich Y B 1956 *Zh. Eksp. Teor. Fiz.* **31** 1101–3
Zhelobenko D P 1965 *Lectures on the Theory of Lie Groups* (Dubna: JINR)
Znojil M 1982 *J. Phys.* A **15** 2111–7
Znojil M 1983 *J. Phys.* A **16** 279–83
Znojil N 1984 *J. Phys.* A **17** 3441–9
Znojil M 1990 *Selected Topics in QFT and Mathematical Physics* (Singapore: World Scientific) p 376
Znojil M and Leach P G L 1992 *J. Math. Phys.* **33** 2785–94

Index

9 780367 402167